WOOD FORMATION IN TREES

This book is dedicated to all who have ever gazed at a tree, and wondered.

WOOD FORMATION IN TREES
Cell and Molecular Biology Techniques

Edited by

Nigel Chaffey

IACR-Long Ashton Research Station
University of Bristol, UK

LONDON AND NEW YORK

First published 2002 by Taylor & Francis

2 Park Square, Milton Park, Abingdon, Oxfordshire OX14 4RN
52 Vanderbilt Avenue, New York, NY 10017

Routledge is an imprint of the Taylor & Francis Group, an informa business

First issued in paperback 2019

This edition published in the Taylor & Francis e-Library, 2004.

British Library Cataloguing in Publication Data
A catalogue record for this book is available from the British Library

Library of Congress Cataloging in Publication Data
A catalog record has been requested.

ISBN 978-0-415-27215-5 (hbk)
ISBN 978-0-367-39640-4 (pbk)

CONTENTS

FOREWORD

Tree material is reported to be difficult to study for a number of reasons. In a tree, living and dead cells, thin-walled and thick-walled cells coexist in an intricate pattern whose every element has physical properties different from its neighbours. This fundamental heterogeneity and the presence of very long, fragile, living cells renders the manipulation of such a heterogeneous material extremely delicate. Furthermore, trees, except in equatorial countries, are subjected to a seasonal cycle. Some structural and physiological changes can only be observed for a few weeks so that one may have to wait for an entire year before being able to repeat an experiment.

The transfer of techniques from single cells, or from organs, of herbaceous species to trees has often been fraught with unforeseen difficulties. As early as 1919, Bailey advised 'the development of special techniques' for cambial studies. It is especially disheartening when working for the first time with woody plants to spend days (if not weeks) trying to adjust well-proven, standard techniques to tree material. I remember too well my early days when all the sections I cut from tree branches in spring perversely turned into some kind of pulp, even before reaching the staining stage, and the only comment I received was 'They shouldn't'. It can be easily understood why young (and not so young) scientists decline to explore such an uncharted territory. Difficulties can be circumvented by working on less complex systems, which are more amenable to the exigencies of modern research. But cambial cells lose all their specific characters when they are cultured in batches; and *Zinnia* mesophyll cells, although a truly marvellous system to study lignification, are not suitable for investigating the early events leading to the commitment of cambial derivatives into xylem cells.

I was then thrilled when I heard of Dr Nigel Chaffey's proposal to edit a laboratory manual for the study of the developmental biology of trees. At last, the tree student will know where to find both practical help and moral comfort: practical help because this book will provide him with detailed technical 'recipes' and precise references, moral comfort because he will no longer feel alone in a hostile field. Each of the 16 techniques chapters is written by one (or several) specialist(s) in the field knowing what it is to be faced every day with the problems of handling tree material. Following a general introductory chapter, the second chapter provides precisely the necessary background to understand the problems encountered in working with woody material. Eleven reasons against 'studying the tree' are listed, explained and convincingly answered. The beginner will know that however arduous the task, he will not feel 'intimidated' any more. The general technical survey is completed with chapters 3 and 4 which deal with standard techniques for the study of wood anatomy and ultrastructure, a thorough knowledge of both being required for devising experiments and understanding their results, whatever the degree of sophistication of the methodological approaches.

Each of the following chapters highlights one of these approaches so that most of the techniques currently available for biochemical analysis, *in situ* localisation, and molecular biology are covered. Details are given on recent refinements in sampling methods allowing one to obtain well defined and tissue-specific samples for biochemical analysis and quantitative histochemistry. The interest of different microscopy techniques such as epifluorescence, deep etching, confocal laser scanning microscopy and secondary ion mass spectrometry microscopy is clearly underlined. Emphasis is put on *in situ* localisation techniques (immunolocalisation, *in situ* hybridisation, reporter gene histochemistry) used in parallel with purely biochemical and molecular approaches. Experience acquired on tree material by authors working at the bench allows them to guide the reader step by step towards his goal. The authors really pass on to the reader the tricks of the trade that they have devised and gradually perfected. The 'recipes' are thus often highly personalised, which explains why some technical variants exist between the authors. The reader will have to choose according to his own material and specific aims.

All colleagues working on trees, when told about the project, were keenly interested, thinking how much time could have been gained if such a book had been published before. I am then sure that 'Wood formation in Trees: Cell and Molecular Biology Techniques' will be extremely well received and will become a much sought after 'bench-side companion' for many scientists. I hope, and I wish, that it will also help to encourage many young scientists to study tree biology, a fascinating domain which needs to be thoroughly explored.

Anne-Marie Catesson

PREFACE

Very often when you have an idea, you have trouble remembering where it came from, or when it came. Not so with this book. The inspiration came to me at a meeting on forest tree physiology in the north of Finland on the afternoon of 6 September, 1997. Fortunately (or, perhaps, disappointingly), in the time it has taken to turn the idea into reality, the climate that originally spawned the idea still exists: trees are recognised as being important to our future well-being, but they are little understood. It is the sincere hope of all the contributors to this book that the techniques included here, which show what cell and molecular biology can be done even with an organism as demanding as a tree, will encourage a new age of inquiry into the biology of these fascinating plants.

Although all involved with the project have come to talk of the book as 'the wood cook book', it would be wrong to view it merely as a collection of recipes. Rather, it is a distillation of many years of experience of working with trees and of perfecting techniques so that they can work as well in these plants as they do in other less demanding organisms. Each contribution also gives a flavour of the insights into tree biology that have been achieved with the technique, and some thoughts on future trends and applications. Still, to take the cooking analogy further, 'the proof of the pudding is in the eating', and if the contributions in this book help to convert others to study the cell and molecular biology of trees, we can count it a success.

In co-ordinating a project such as this it is impossible to do everything oneself, and there are several people to whom I am most grateful.

I thank my fellow contributors for the dedication they have shown over the many months of the writing. Convincing them of the worth of the book was the easy part, but I hope they will forgive me the numerous Emails with which they were cajoled in my efforts to keep the project, and them, on schedule. The readiness with which they all agreed to participate helped to convince me that the idea was worthwhile, and I thank them all for that encouragement. In similar vein, I thank all those scientists who supported the project in the form of personal endorsements: it made the task of persuading a publisher so much easier! A special thank you is also due to Anne-Marie Catesson and Jozef Samaj who were particularly industrious in soliciting such endorsements. It is also a pleasure to have persuaded Anne-Marie to write the foreword to this book. Her research into cambial cell biology in the latter half of the 20th century is without equal and her influence is evident in many of the contributions in this volume.

I had very little idea quite what I was letting myself in for when I decided to embark on the task of editing such a diverse book (although colleagues did try to warn me of the magnitude of the task!). However, along the way I have learnt a lot about books, computers, and people, and thank all my contacts at the publisher's for helping me to negotiate the maze that is book production. I would also like to record my thanks to Dr Robin Smith of Springer for his encouragement during the early stages of finding a publisher for the book.

Helen in the library at Long Ashton Research Station is thanked for her tenacity in tracking down the most obscure interlibrary loan requests, which were, of course, always required 'immediately'.

I thank Peter Barlow and John Barnett for, unwittingly I'm sure, starting the whole thing off by giving me the chance to get back into botanical research, and allowing me more licence than I could have hoped for to follow my nose in matters ligneous.

A big thank you is also due to the creators of the internet and, particularly, Email, for making the task of co-ordinating such a diverse group of people much easier than their geographical dispersion would otherwise have dictated.

Finally, since one should keep the really important things to the end, I apologise to my wife, Chris, and my children, Nicholas and Kate, for all the time I spent away from them over the months it has taken to see this project through to its conclusion. I know I cannot make up the lost time, but I thank them for their understanding, and hope that one day they will think that it has been worthwhile?

CONTRIBUTORS

Tannis Beardmore
NRCan Canadian Forest Service – Atlantic Region
1350 Regent St. S.
Fredericton
New Brunswick E3B 5P7
Canada
Email: tbeardmore@nrcan.gc.ca

Marianne Bordenave
Unité de Physiologie Cellulaire et Moléculaire des
Plantes
Université Pierre et Marie Curie
4 place Jussieu
F-75252 Paris Cedex 05
France
Email: bordenave@ijm.jussieu.fr

Alain Michel Boudet
UMR 5546 CNRS – UPS
Pôle de Biotechnologie Végétales
24 chemin de Borde Rouge
BP17 Auzeville
F-31326 Castanet Tolosan
France
Email: amboudet@cict.fr

Anne-Marie Catesson
Laboratoire de Physiologie Végétale Appliquée
Université Pierre et Marie Curie
4 place Jussieu
F-75252 Paris Cedex 05
France
Email: Anne-Marie.Catesson@ens.fr

Nigel Chaffey
IACR-Long Ashton Research Station
Department of Agricultural Sciences
University of Bristol
Long Ashton
Bristol BS41 9AF
UK
Email: nigel.chaffey@bbsrc.ac.uk

Luigi De Filippis
Department of Environmental Sciences
University of Technology
Sydney
PO Box 123
Broadway NSW 2007
Australia
Email: Lou.DeFilippis@uts.edu.au

Eric Duverger
Antenne Scientifique Universitaire de Chartres
Université d'Orléans
21 rue de Loigny la Bataille
F-28000 Chartres
France
eric.duverger@univ-orleans.fr

Fabienne E. Ermel
Unité de Physiologie Cellulaire et Moléculaire des
Plantes
Université Pierre et Marie Curie
4 place Jussieu
F-75252 Paris Cedex 05
France
Tel: +33-(0)1-44-27-76-40

Marie-Laure Follet-Gueye
SCUEOR
Faculté des Sciences
Université de Rouen
F-76821 Mont-Saint-Aignan Cedex
France
marie-laure.follet-gueye@univ-rouen.fr

Ryo Funada
Department of Forest Science
Faculty of Agriculture
Hokkaido University
Sapporo 060-8589
Japan
Email: funada@for.agr.hokudai.ac.jp

Renée Goldberg
Unité de Physiologie Cellulaire et Moléculaire
des Plantes
Université Pierre et Marie Curie
4 place Jussieu
F-75252 Paris Cedex 05
France
Email: goldberg@ccr.jussieu.fr

Jacqueline Grima-Pettenati
UMR 5546 CNRS – UPS
Pôle de Biotechnologie Végétales
24 chemin de Borde Rouge
BP 17 Auzeville
F-31326 Castanet Tolosan
France
Email: grima@cict.fr

Siegfried Hauch
Physiological Ecology of Plants
University of Tübingen
Auf der Morgenstelle 1
D-72076 Tübingen
Germany
Tel: + 49-7071-2976941

Simon Hawkins
Antenne Scientifique Universitaire de Chartres
Université d'Orléans
21, rue de Loigny la Bataille
F-28000 Chartres
France
Email: simon.hawkins@univ-orleans.fr

Takao Itoh
Wood Research Institute
Kyoto University
Uji
Kyoto 611-0011
Japan
Email: titoh@kuwri.kyoto-u.ac.jp

Elisabeth Magel
Physiological Ecology of Plants
University of Tübingen
Auf der Morgenstelle 1
D-72076 Tübingen
Germany
Email: elisabeth.magel@uni-tuebingen.de

Fabienne Micheli
Institut Jacques Monod
Laboratoire d'Enzymologie en Milieu Structuré
Université Pierre et Marie Curie
4 place Jussieu
F-75252 Paris Cedex 05
France
Email: micheli@toulouse.inra.fr

Gilles Pilate
Amélioration, Génétique et Physiologie Forestiére
INRA
Avenue de la Pomme de Pin BP20619
F-45160 Ardon
France
Email: pilate@orleans.inra.fr

Sharon Regan
Department of Biology
Carleton University
1125 Colonel By Drive
Ottawa
Ontario K1S 5B6
Canada
SharonRegan@pigeon.carleton.ca

Kim Rensing
Department of Botany
University of British Columbia
Vancouver
British Columbia V6T 1Z4
Canada
Krensing@interchange.ubc.ca

Luc Richard
Unité de Physiologie Cellulaire et Moléculaire des
Plantes
Université Pierre et Marie Curie
4 place Jussieu
F-75252 Paris Cedex 05
France
Email: richard@ijm.jussieu.fr

Jozef Samaj
Institute of Plant Genetics and Biotechnology
Slovak Academy of Sciences
Akademická 2, PO Box 39A
SK-950 07 Nitra
Slovak Republic
Email: nrgrsama@savba.sk

Björn Sundberg
Department of Forest Genetics & Plant Physiology
Swedish University of Agricultural Sciences
SE-901 83 Umeå
Sweden
Email: bjorn.sundberg@genfys.slu.se

Keiji Takabe
Graduate School of Agriculture
Kyoto University
Kyoto 606-8502
Japan
Email: kjtakabe@kais.kyoto-u.ac.jp

Claes Uggla
BioAgri AB
Box 914
SE-751 09 Uppsala
Sweden
Email: claes.uggla@bioagri.slu.se

Suzanne Wetzel
NRCan Canadian Forest Service – Ontario Region
1219 Queen St. E.
Sault Ste. Marie
Ontario P6A 5M7
Canada
Email: swetzel@nrcan.gc.ca

Carrie-Ann Whittle
Department of Biology
University of Dalhousie
Halifax
Nova Scotia B3H 3J5
Canada
Email: cwhittle@is2.dal.ca

ABBREVIATIONS USED IN THIS BOOK

A	adenine (nucleotide)
A	amp (absorbance, when followed by wavelength)
AAS	atomic absorption spectrometry
Ab	antibody
AMPS	ammonium persulphate
AP	alkaline phosphatase
approx.	approximately
ASA	American Standards Association (film speed)
ATP	adenosine triphosphate
BCA	bicinchoninic acid
BCIP	5-bromo-4-chloro-3-indolylphosphate
BF	bright-field
BIS	bis-acrylamide
BMM	butyl-methylmethacrylate
bp	base pair
BSA	bovine serum albumen
4CL	4-coumarate co-enzyme A ligase
C	cytosine (nucleotide)
C	carbon
C3H	coumarate 3-hydroxylase
C4H	cinnamate-4-hydroxylase
°C	degrees Celsius (centigrade)
CAD	cinnamyl alcohol dehydrogenase
Ca	calcium
CCC	chambered crystalliferous cell
CCD	charge-coupled device
CCoAOMT	caffeoyl-co-enzyme A O-methyltransferase
CCR	cinnamoyl co-enzyme A reductase
CCRC	complex carbohydrate research center
cDNA	complementary deoxyribonucleic acid
CDTA	diaminocyclohexane tetraacetic acid
cER	cortical endoplasmic reticulum
cf.	compare
CFS	cryo-fixation/substitution
CHS	chalcone synthase
CL	cellulosic lamella
CLP	chlorallactophenol
CLSM	confocal laser scanning microscopy
CMF(s)	cortical microfilament(s)
CMT(s)	cortical microtubule(s)
CoA	co-enzyme A
C-OMT	caffeate/5-hydroxyferulate O-methyltransferase
CTAB	cetyltrimethyl-ammonium-bromide
CTP	cytosine triphosphate
CW	cell wall
CY	cyanine

CZ	cambial zone
d	day(s)
1-D	1-dimensional
2-D	2-dimensional
3-D	3-dimensional
DAPI	4, 6′-diamidino-2-phenylindole
ddH$_2$0	double-distilled water
DDRT	differential display reverse transcriptase
DE	degree of esterification
DEPC	diethylpyrocarbonate
DER	diglycidyl ether of propylene glycol
DF	dark-field
dH$_2$O	distilled water
DIC	differential interference contrast ('Nomarski')
DIG	digoxigenin
diX indigo	5,5′-dibromo-4,4′-dichloro-indigo
DMAE	dimethylaminoethanol
DMF	dimethyl formamide
DMP	2,2-dimethoxypropane
DMSO	dimethyl sulphoxide
DNA	deoxyribonucleic acid
DNase	deoxyribonuclease
dNTP	deoxyribonucleotide triphosphate
DPX	distyrene, dibutyl phthalate, xylene
DS	dextran sulphate
DTT	dithiothreitol
DW	dry weight
EC	Enzyme Commission
EDTA	ethylenediaminetetraacetic acid
e.g.	for example
EGTA	ethylene glycol-bis-(-aminoethyl ether) N,N,N′,N′-tetraacetic acid
EM	electron microscopy
EMF(s)	endoplasmic microfilament(s)
EMMEG	ethylene glycol monomethyl ether
EMT(s)	endoplasmic microtubule(s)
ER	endoplasmic reticulum
ERL	vinyl cyclohexene dioxide
etc.	etcetera
EtOH	ethyl alcohol (ethanol, alcohol)
Eu	eucalyptus (re genes, e.g. EuCAD)
F5H	ferulate 5-hydroxylase
FAA	formalin:acetic acid:(ethyl) alcohol
F-actin	filamentous actin
FE-SEM	field emission scanning electron microscopy
FG	fish skin gelatin
FITC	fluorescein *iso*thiocyanate
FRET	fluorescence resonance energy transfer
FTIR	Fourier transform infra-red
FW	fresh weight
g	gravity
G	guanine (nucleotide)
G	guaiacyl

G6P-DH	glucose-6-phosphate dehydrogenase
GA	Golgi apparatus (dictyosome)
G-actin	globular actin
GC	gas chromatography
GFP	green fluorescent protein
GRM	GUS-reaction medium
GTP	guanidine triphosphate
GUS	β-glucuronidase
h	hour(s)
^{3}H	tritium
H^+	proton
HdGS	'homology-dependent gene silencing'
HEPES	N-2-hydroxyethylpiperazine-N'-2-ethanesulfonic acid
Hg	mercury
HG	homogalacturonan
HPLC	high performance/pressure liquid chromatography
IEF	iso-electric focusing
i.e.	that is
Ig	immunoglobulin class
IIF	indirect immunofluorescence
IR	infra-red
ISH	*in situ* hybridisation
IUFRO	International Union of Forestry Research Organisations
JIM	John Innes monoclonal antibody
K	potassium
Ka	concentration of an activator at which half Vmax occurs
Kb	kilobase
kDa	kilodaltons
keV	thousand-electron volts
Ki	concentration of inhibitor that achieves half the maximum inhibition
Km	Michaeli's constant
Kpi	potassium phosphate buffer
l	litre
lb/sq.in	pounds per square inch
LM	light microscopy;
LM	Leeds monoclonal antibody (when followed by a number)
LN_2	liquid nitrogen
LR	London Resin
μg	microgram
μm	micrometre
μM	micromolar
mA	milliamp
M	molar
MAb	monoclonal antibody
MAP	microtubule-associated protein
MBS	m-maleimidobenzoyl N-hydroxysuccinimide ester
MES	4-morpholinoethanesulfonic acid
Mg	magnesium
min	minutes(s)
mm	millimetre
mM	millimolar
MO	magnetic-optical

mRNA	messenger ribonucleic acid
MS	mass spectrometry
MTOC	microtubule-organising centre
MTSB	microtubule-stabilising buffer
MUG	4-methyl-umbelliferyl-D-glucuronide
MW	molecular weight
Na	sodium
NA	numerical aperture
NAD	nicotinamide adenine dinucleotide
NADH	nicotinamide adenine dinucleotide (reduced)
NADP	nicotinamide adenine dinucleotide phosphate
NADPH	nicotinamide adenine dinucleotide phosphate (reduced)
NB	note well
NBT	nitroblue tetrazolium
NEPHGE	nonequilibrium pH gradient electrophoresis
ng	nanogram
NGS	normal goat serum
nm	nanometre
nmol	nanomoles
No.	number (usually catalogue number)
NSA	nonenyl succinic anhydride
NTE	NaCl-tris-EDTA
NTP	nucleotide triphosphate
OD	optical density
OPB	Operon Technology random primer kit B
6PG-DH	6-phosphogluconate dehydrogenase
p.	page
P	phospho-
P	primary cell wall
Pa	pascal
PAb	polycolonal antibody
PAF	p-formaldehyde
PAGE	polyacrylamide gel electrophoresis
PAL	phenylalanine ammonia-lyase
PAM	monoclonal phage antibody
PA-S	periodic acid-Schiff's reaction
PATAg	periodic acid-thiocarbohydrazide-silver proteinate
PBS	phosphate-buffered saline
PBSA	phosphate-buffered saline plus bovine serum albumen, etc [Chaffey]
PBSB	phosphate-buffered saline plus bovine serum albumen, etc [Funada]
PCR	polymerase chain-reaction
PCR	partially-coated reticulum
PD	phloem-oriented derivative
PEG	polyethylene glycol
PF	protoplasmic fracture
pg	picogram
pH	measure of the acidity/alkalinity of a solution
pI	isoelectric point
PIPES	piperazine-N,N'-bis-[2-ethylsulphonic acid]
PL	plasmalemma (plasma membrane)
PME	pectin methylesterase
pmol	picomole

PMSF	phenylmethylsulphonyl fluoride
PO	propylene oxide
pp.	pages
ppb	parts per billion
PPB	pre-prophase band
ppm	parts per million
Pt	Populus tremuloides (re genes, e.g. Pt4CL1)
PTA	phosphotungstic acid
PVC	polyvinyl chloride
PVP	polyvinylpyrrolidone
PVPP	polyvinylpolypyrrolidone
RAPD	random amplification of polymorphic DNA
RFDE	rapid-freezing-deep-etching
RGI	rhamnogalacturonan I
RGII	rhamnogalacturonan II
RH	relative humidity
RLS	radial longitudinal section
RNA	ribonucleic acid
RNase	ribonuclease
rpm	revolutions per minute
RT	room temperature (20–25°C)
RT-PCR	reverse transcription-polymerase chain reaction
S	syringyl
S	secondary cell wall
S_1, S_2, S_3	layers within the secondary cell wall
SDS	sodium dodecyl sulphate
sec	second(s)
SEM	scanning electron microscopy
SIMS	secondary ion mass spectrometry
sp.	species (singular)
spp.	species (plural)
SSC	sodium chloride/sodium citrate solution
ssp.	sub-species
STEM	scanning/transmission electron microscopy
SuSy	sucrose synthase
SVS	secondary vascular system (assemblage of secondary phloem, vascular cambium and secondary xylem)
T	thymine (nucleotide)
TAE	tris-acetate-EDTA
TBA	tertiary butyl alcohol (2-methoxy-2-propanol)
TBE	tris-borate-EDTA
TBS	tris-buffered saline
TBS-B	tris-buffered saline plus 0.1% BSA
TBS-T	tris-buffered saline plus Tween
TCH	thiocarbohydrazide
TE	tris-EDTA
TEM	transmission electron microscopy
TEMED	N,N,N′,N′-tetramethylethylenediamine
tER	tubular endoplasmic reticulum
TGN	*trans*-Golgi network
TIFF	tagged image file
TLC	thin-layer chromatography

TLS   tangential longitudinal section
Tris   tris-[hydroxymethyl]aminomethane
TRITC   tetramethyl rhodamine isothiocyanate
tRNA   transfer ribonucleic acid
TS   transverse (cross) section
TTBS   tris-buffered saline with Tween 20
TTP   thymidine triphosphate
U   units (of enzyme activity)
UBC   University of British Columbia random primer kit
UDP   uridine 5' diphosphate
UV   ultra violet
V   volt
VIS   visible light
vol.   volume
VSP(s)   vegetative storage protein(s)
v/v   volume/volume basis
w/v   weight/volume basis
w/w   weight/weight basis
X-gluc   5-bromo-4-chloro-3-indolyl-β-D-glucuronide
XD   xylem-oriented derivative
ZIO   zinc-iodide-osmium-tetroxide

Introduction

Wood Formation in Trees ed. N. Chaffey
© 2002 Taylor & Francis
Taylor & Francis is an imprint of the Taylor & Francis Group
Printed in Singapore.

Nigel Chaffey

IACR – Long Ashton Research Station, Department of Agricultural Sciences, University of Bristol, Long Ashton, Bristol BS41 9AF, UK

WHY STUDY TREES?

The importance of trees

Trees are the longest-lived members of the natural world (Thomas, 2000). They are the largest of living things (Thomas, 2000). They provide fuel for light, warmth, and cooking (Burley and Plumptre, 1985). They are a valuable source of raw materials (e.g. Cannell and Jackson, 1985; Carruthers *et al.*, 1994). They give us food (e.g. Vaughan and Geissler, 1998). They provide drugs (e.g. aspirin, quinine, strychnine, Taxol®) to relieve our misery. They moderate the environment and contribute to climatic stability (Kohlmaier *et al.*, 1998).

However, of all the good things we get from trees, 'Wood is, perhaps, nature's most wonderful gift to humanity' (Kumar, 1994). Weight-for-weight, wood is as strong as steel (Vincent, 1996), and has been used by Man in hundreds of different ways for hundreds of years. Whether as timber for house- or ship-building, wood-pulp for paper manufacture, a long-term sink for atmospheric carbon (Berna, 1998; Kohlmaier *et al.*, 1998; Phillips *et al.*, 1998; Ceulemans *et al.*, 1999), or as an environmentally cost-effective, renewable source of energy (Carruthers, 1994b; El Bassam, 1996; Haygreen and Bowyer, 1996), it continues to have a unique role in the present, and future, well-being of the planet and all those who inhabit it.

The utility of trees is undeniable. It is therefore surprising that we know so little about their biology (Savidge, 1985; Whetten and Sederoff, 1991).

Old arguments in favour of the study of trees

Throughout the last century, concerns were being raised about the need to study trees. In 1913 Gerry recognised that, 'In order to undertake investiga-tion of the relations existing between structure and properties and uses [of wood], a thorough knowledge of plant anatomy is essential'. It was predicted that from that anatomical insight would come the wisdom which would help to ensure that, 'The reduction of waste and increase in efficiency due to intelligent utilization will help towards the realization of that much-desired object, the proper conservation of resources' (Gerry, 1913).

Forty years later, Bailey (1952) made another strong case for study of trees, by asserting that, 'the growth of forests on nonagricultural lands is so significant to the future welfare of man as potential sources of diversified organic substances and of stored solar energy that a sustained and comprehensive effort should now be made towards a better understanding of developmental processes in the formation of wood'. This plea was subsequently reiterated by Berlyn (1979), who proclaimed that, 'knowledge of the mechanisms that control the pattern of differentiation of xylem cells is of central importance as man attempts to maximize the productivity from forest ecosystems'. And more recently Haigler (1994) stated quite simply that, 'for both basic and applied reasons, understanding how tracheary element differentiation is regulated is of paramount importance in plant biology'.

New arguments in favour of the study of trees

Demand for trees is increasing

Despite all these wise words, it is still the case that too little biological work is carried out on trees. Why? Is it perhaps the very majesty of trees, particularly their great size and antiquity, which has fooled us into believing that they will always be there? Unfortunately, this is unlikely to be the case, since the demand for wood is forecast to increase quite substantially in future (Elliott, 1985; Carruthers, 1994a; Haygreen and Bowyer, 1996; Whiteman, 1996). This increase in demand is likely to be particularly dramatic in the context of biomass energy. As non-renewable (at least within realistic time-scales!) fossil fuel reserves become depleted, there is a serious revival of interest in the use of wood as an alternative energy source (e.g. Goldstein, 1980; Klass, 1998).

Can't we just grow more trees?

Notwithstanding problems due to loss of forests and of land available for growing trees (e.g. Bues, 1990), there is considerable pressure to grow more trees. In response to this, trees have either been 'improved', or non-indigenous species planted, with the result that nowadays trees grow a lot faster than hitherto (e.g. Bues, 1990), and the required, harvestable volume of wood is produced more quickly than before. However, this accelerated growth has not resulted in a corresponding shortening of the normal life cycle of the tree, which for the first 5–25 years undergoes a juvenile period of growth (Haygreen and Bowyer, 1996). The wood – juvenile wood – produced during this period is different to that of adult trees (Haygreen and Bowyer, 1996). Hence, the wood produced by these fast-growing trees is principally juvenile (Zobel and Sprague, 1998). Additionally, trees generally are being harvested earlier and earlier, with the consequence that the proportion of juvenile wood in the harvested product is increased (Zobel and Sprague, 1998).

Whether rightly or wrongly, it is generally still the perception that, 'By most measures, juvenile wood is lower in quality than mature wood …' (Haygreen and Bowyer, 1996). However, as the trend to use such wood seems set to continue, and since that perception persists, there is an urgent need to understand the cell (and molecular!) biology of juvenile wood formation (e.g. Brazier, 1985; Zobel and Sprague, 1998). Furthermore, associated with high growth rate is the high incidence of gelatinous fibres (e.g. White and Robards, 1965; Isebrands and Bensend, 1972), whose presence cause variability in wood quality with consequent processing problems (e.g. Peszlen, 1996). If we are to overcome this problem, we clearly need to know more about the cell biology of gelatinous fibres.

Or use the available trees more economically?

It has been appreciated for a long time that more or less waste can result from the way in which a tree trunk is cut up (Haygreen and Bowyer, 1996), and even today, new methods are being devised which can reduce the amount of waste (e.g. star-cutting – Tickell, 1998).

We are trying to be more environmentally responsible, as indicated by the development of processes that use the 'waste' lignin, as a fuel in paper-making processes (McCarthy and Islam, 2000), in the manufacture of polyurethane (Thring et al., 1997), or as a partial replacement for epoxy resin in manufacture of printed circuit boards (Anon., 1997). It is even possible to mix the lignin with cellulose fibres to produce mouldable 'wood', 'Arboform' (Mackenzie, 1999). However, as imaginative as these ideas are, they do not provide long-term solutions, and do not contribute to solving the problems created by the development of *new* uses of timber, which often require different types of trees or kinds of wood.

Molecular biology to the rescue!

What will persuade people to carry out research on trees? Several driving forces have been recognised (Chaffey, 2001). However, probably the most important one relates to the wood-pulp industry.

One of the major preoccupations of those in the wood-pulp/paper industry is lignin. Lignin is a ubiquitous component of wood cell walls, but its presence compromises the production of high quality, non-discolouring, white paper (e.g. see the chapter by Hawkins et al. in this volume). Although lignin can be removed, this is achieved at substantial cost, and at the risk of causing serious environmental damage with the harmful chemicals that are used (e.g. O'Connell et al., 1998). Consequently, development of trees where the lignin is easier to remove, or which contain less lignin, has been the subject of intense interest from molecular-geneticists in recent years (e.g. O'Connell et al., 1998; Halpin et al., 2000; Lapierre et al., 2000).

Thus, the needs of modern industrial processes would be better served by wood that is more technologically/environmentally friendly. It is also the case that, 'Forestry operations would be most effective if trees were tailored to match specific end uses' (Dinus and Welt, 1997).

So, we need more wood, but we need the right kind of wood. However, as Whetten and Sederoff (1991) have pointed out, 'Our knowledge of wood development is inadequate to exploit the potential of genetic engineering for wood modification.' How can we satisfy these different interests? Understanding the cell biology of wood formation would be a very good starting point!

Although, ultimately, molecular biology may hold the key, a more fundamental understanding of the cell biology of the process will help to show us the lock. In reality, of course, both approaches will be needed, and will complement one another. That the scope for more molecular studies can come from an increase in the knowledge of the cell biology of xylogenesis is indicated by recent work on the cytoskeleton and its potential for creating 'designer wood' (Chaffey, 2000).

HOW DO WE STUDY WOOD FORMATION IN TREES?

Model systems

Xylogenesis (wood formation) has long been recognised as an ideal model system to study plant-cell differentiation (e.g. Torrey et al., 1971); Roberts, 1976). Unfortunately, the obvious candidate for study, the Tree, has largely been rejected in favour of 'systems' that are more amenable to study. Hence, to date much of the work carried out in the name of xylogenesis has been performed with in vitro systems, such as Zinnia mesophyll cells (e.g. McCann et al., 2000). Whilst I acknowledge the benefits of such an approach (Chaffey, 1999, 2001), such systems have many shortcomings and are not necessarily good models for the process that takes place in the tree (Chaffey, 1999).

It is appropriate here to remind ourselves how complicated the secondary vascular system is. This complexity arises from the combination of the following:

- the presence of six widely recognised stages to xylogenesis (Bailey, 1952), from cell division in the cambium to heartwood formation,
- the fact that wood is an intimate mixture of dead and long-lived cells,
- the existence of many different wood and cambial cell types (e.g. Figure 1), some of which can inter-convert, and with many subtle variations within different species, and embracing such characteristics as storied/non-storied cambium; softwoods/hardwoods; tropical/temperate trees; and, diffuse-porous/ring-porous wood,
- there is a seasonal component to cambial activity and wood-cell production, at least in temperate species,
- it comprises two differently oriented sub-systems – axial and radial,
- all the above, to a greater or lesser extent, also applies to the phloem tissue.

How should we study wood formation in trees?

In view of this complexity, it is perhaps small wonder that study of trees has been largely avoided for so long in favour of more amenable plants and tissue systems. However, because of this complexity, it is incumbent upon all those who desire to understand the processes that occur within trees to study trees. Barriers that might have prevented more widespread study of trees in the past have largely disappeared (see Chapter 2, 'An introduction to the problems of working with trees', in this volume). Thus, although models will still have an important role to play, one can realistically predict a future where the shift is away from use of model systems toward use of model species (e.g. Chaffey, 2001).

In keeping with adoption of poplar as the model hardwood tree (see Chapter 2, 'An introduction to the problems of working with trees', in this volume), several contributions in this volume use trees of the genus Populus in their work (e.g. the chapters by Chaffey, Hawkins et al., Micheli et al., Regan and Sundberg, and Samaj and Boudet). However, mindful of the danger of devoting all one's attention to just one type of tree, other tree types are also well represented, e.g. gymnosperms (chapters by Funada, Rensing, Takabe, and Uggla and Sundberg), Robinia (chapters by Magel, and Magel et al.), Eucalyptus (chapters by Itoh, and Magel et al.), and Fagus (chapters by Follet-Gueye, and Takabe).

Current status of tree cell biology

Hierarchical approach

The many and varied uses of wood are a reflection of the properties of the individual cells that make up the wood (e.g. de Zeeuw, 1965; Berlyn, 1979).

This biodiversity is a result of the diverse pathways of differentiation of the derivatives of just two types of precursor cells, the fusiform and ray cambial initials (Figure 1).

In considering the ways in which the process of secondary vascular differentiation (e.g. Table 1) could be monitored, Northcote (1993, 1995) identified changes in cytology, cell wall composition, enzyme activities, and gene expression as markers of the process. Or, to express this in another way: anatomical, biochemical, and molecular-genetical, the hierarchy of levels of enquiry used in this book.

Any scheme is artificial, since we need to study all aspects of the whole process to achieve proper understanding. However, this hierarchical approach emphasises an important point about any developmental study, the constraints of the 'techno-temporal' dimension. Each level of enquiry gets us progressively closer to the genetical level of study. Such progression gives us ever finer detail. However, to achieve this detail, we are reliant upon

advances in techniques and equipment. Thus, each level is of necessity a reflection of the question(s) to be answered; which in turn is a function of the techniques available at the time. Thus, the 'maturity' of any developmental study can be gauged by the level of enquiry at which it is currently being studied. In that respect, wood cell biology has had a very long childhood, but is now, at last, on the brink of adulthood!

Anatomical

Wood anatomy has been much studied for many hundreds of years and is well documented (see Iqbal, 1995; and the Wood microscopical techniques chapter, and the transmission electron microscopy (TEM) chapters by Chaffey, and Rensing in this volume). Similarly, cambial structure and ultrastructure have been much studied and often reviewed (e.g. Larson, 1994; references in Chaffey, 2001; see also the TEM chapter by Chaffey in this volume). And even phloem has been the

SECONDARY PHLOEM   VASCULAR CAMBIUM   SECONDARY XYLEM

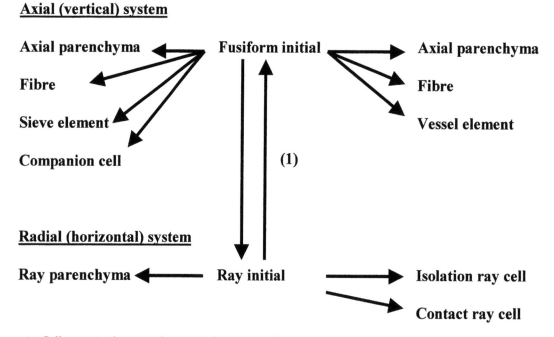

Figure 1 Cell types in the secondary vascular system of the angiosperm tree *Aesculus hippocastanum* L.
Arrows indicate pathways of differentiation within the secondary vascular tissues. (1) is an inter-conversion of cambial initials which is possible (per Philipson *et al.*, 1971) but has not yet been observed in *Aesculus*.

Table 1 Some of the changes that take place as cambial derivatives undergo secondary vascular differentiation in *Aesculus hippocastanum*

Feature	Change effected	Cell type affected
Shape/size	Height/width ratio altered	All
	Flattened at ends	Some sieve and vessel elements, axial parenchyma
	Widened	All
	Shortened	Axial parenchyma
Cell walls	Thickened	All
	More-ordered	All
	Multi-layered	($S_{1, 2, 3}$ layers, xylem cells)
	Additional components (e.g. lignified)	Xylem cells and phloem fibres
	± elaborated (e.g. pits, tertiary thickenings, sieve and perforation plates)	All
	± loss of plasmodesmata (e.g. inter-vessel pits, vessel-ray contact pits)	Contact ray cells and vessel elements
Cytoplasm	Nuclei repositioned	All
	Cytoskeleton rearranged	All
	Total loss of contents	Fibres, vessel elements
	Selective loss of contents	Sieve elements
	No loss of contents	Xylem and phloem parenchyma

subject of much study (e.g. Behnke and Sjölund, 1990; Iqbal, 1995).

But, despite the long history of such studies, there is still a great deal to be done, and new discoveries are being made at this level (such as the essential similarity of root and shoot cambia – see the TEM chapter by Chaffey in this volume). Further insights into wood formation are also being obtained by advances in electron microscopy (see the TEM chapters by Chaffey, and Rensing, and the chapter by Itoh in this volume), and use of other imaging techniques, such as secondary ion-mass spectrometry (see the chapter by Follet-Gueye in this volume).

Anatomical/Biochemical

The relatively recent application of immunological procedures to anatomical study demonstrates the phenomenon whereby the coming together of different techniques gives rise to new ways of looking at processes. In this case 'structural cell biology' is the result, and the study of the role of the cytoskeleton one of its major achievements to date (see the cytoskeletal chapters by Chaffey, and Funada in this volume). Several other chapters in this volume also occupy this niche at the interface between traditional anatomy and biochemistry (Beardmore *et al.* – storage proteins; Micheli *et al.* –

cell wall chemistry and biochemistry; Samaj and Boudet – enzymes of lignification; Takabe – cell wall synthesis and development).

Biochemical

Biochemistry proper has played a big part in unravelling the process of wood formation in trees, yet much still remains to be done. Fortunately, as techniques are developed and/or refined, more is possible. A good example of the benefit to biochemical study of trees is the development of a cryo-sectioning technique (see the chapter by Uggla and Sundberg). This is extremely useful for the preparation of anatomically well-defined samples, and to date has been used to good effect in investigating differences in sugar levels (see the chapter by Magel in this volume), hormone content (see the chapter by Uggla and Sundberg in this volume), enzyme isoform distributions (see the chapter by Micheli *et al.* in this volume), and identifying homeobox genes (Hertzberg and Olsson, 1998) within the secondary vascular tissues of trees.

Molecular-genetical

In some respects this is the newest type of investigation, but its relevance and importance to study of xylogenesis is evident in a recent book dealing with cell and molecular biology of wood formation (Savidge *et al.*, 2000). However, in other respects, e.g. regarding the ability to genetically transform forest trees (e.g. Parsons *et al.*, 1986; Filatti *et al.*, 1987), it is older than techniques such as immuno-localisation of cytoskeletal proteins in trees. Indeed transformation of trees is considered to be so well developed, and widely practised (see e.g. chapters in Klopfenstein *et al.* 1997), that no chapter is devoted to the technique in this volume. However, in terms of analysis of genetic events within trees, the subject is very much in its infancy. Accordingly, in this book it is currently represented by only three chapters, reporter gene histochemistry (Hawkins *et al.*), PCR-RAPD (Magel *et al.*), and *in situ* hybridis-ation (Regan and Sundberg). Although, generally, present knowledge at this level is rather meagre, it is the one where understanding is set to advance most rapidly in the immediate future (Chaffey, 2001).

Bridging the gap

There is an obvious paradox in our current study of wood formation – great expertise in genetic-transformation of trees yet a lack of real under-standing of the cell biology of wood formation. Laudable though such progress in molecular tech-niques is, this near headlong rush to embrace new technology has meant that the basic, fundamental, underpinning cell biology has been seriously neglected. The resultant knowledge gap has previ-ously been alluded to by Whetten and Sederoff (1991). Recognition that this disparity exists is one thing; doing something about it is another. If we are to make serious progress in understanding the developmental biology of wood formation, we need to close that gap. Such basic work is essential if we are to understand the molecular side of things, e.g. as the cellular environment within which the genetic events take place, or consider experimenting with the process.

The whole is greater than the sum of the parts

As anatomists, or biochemists, or molecular bio-logists, there is the danger that we treat our own particular level of enquiry as an end in itself. However, as Northcote (1993) reminds us, 'The changes that occur [during differentiation] depend

on an ordered and sequential read out of informa-tion in the genome. Thus any marker of differentia-tion monitors either directly or indirectly the control of gene transcription and subsequent trans-lation.' Whilst this does not make us all molecular biologists, it does emphasise the fact that, as biolo-gists, we must bear in mind the need to study developmental processes at all levels to achieve genuine understanding.

This more holistic approach was anticipated many years ago by Bailey (1952) when he stressed that, 'To be successful such an effort [study of the biological processes in wood formation] must involve a much closer and broader integration of research in different scientific disciplines'. Although that is beginning to happen today, as evident from such studies as that of Catesson *et al.* (1995), it isn't happening enough.

Few greater challenges exist in plant biology than to unravel the mysteries of the secondary vas-cular system. It is hoped that more widespread use of the diverse techniques included in this volume will encourage a more complementary approach to the study of the cell and molecular biology of wood formation.

REFERENCES

Anon. (1997) Paper solution. *New Scientist*, **156**, 15.

Bailey, I.W. (1952) Biological processes in the formation of wood. *Science*, **115**, 255–259.

Berlyn, G.P. (1979) Physiological control of differentia-tion of xylem elements. *Wood and Fiber*, **11**, 109–126.

Behnke, H.-D. and Sjölund, R.D. (1990) *Sieve Elements: Comparative Structure, Induction and Development.* Berlin: Springer.

Berna, G. (1998) *Integrated Biomass System.* Luxembourg: European Commission, Directorate-General Science, Research and Development.

Brazier, J.D. (1985) Juvenile wood. In *Xylorama: Trends in Wood Research*, edited by L.J. Kucera, pp. 25–32. Basel, Boston, Stuttgart: Birkhäuser Verlag.

Bues, C.T. (1990) Wood quality of fast and normal growing trees. In *Fast Growing and Nitrogen Fixing Trees*, edited by D. Werner and P. Müller, pp. 340–353. Stuttgart, New York: Gustav Fischer Verlag.

Burley, J. and Plumptre, R.A. (1985) Trees as producers of fuel. In *Attributes of Trees as Crop Plants*, edited by M.G.R. Cannell and J.E. Jackson, pp. 271–280. Huntingdon, UK: Institute of Terrestrial Ecology, NERC.

Cannell, M.G.R. and Jackson, J.E. (eds) (1985) *Attributes of Trees as Crop Plants.* Huntingdon, UK: Institute of Terrestrial Ecology, NERC.

Carruthers, S.P. (1994a) Fibres. In *Crops for Industry and Energy, CAS Report 15*, edited by S.P. Carruthers,

F.A. Miller and C.M.A. Vaughan, pp. 92–108. Reading: Centre for Agricultural Study, University of Reading.

Carruthers, S.P. (1994b) Solid biofuels: energy coppice. In *Crops for Industry and Energy, CAS Report 15*, edited by S.P. Carruthers, F.A. Miller and C.M.A. Vaughan, pp. 150–167. Reading: Centre for Agricultural Study, University of Reading.

Carruthers, S.P., Miller, F.A. and Vaughan, C.M.A. (eds) (1994) *Crops for Industry and Energy, CAS Report 15*. Reading: Centre for Agricultural Study, University of Reading.

Catesson, A.-M., Bonnemain, J.L., Eschrich, W. and Magel, E. (1995) The cambium and its derivative tissues: biochemical changes in relation to cell differentiation and seasonal activity. In *EUROSILVA: Contribution to Forest Tree Physiology*, edited by H. Sandermann, Jr and M Bonnet-Masimbert, pp. 57–77. Paris: INRA.

Ceulemans, R., Janssens, I.A. and Jach, M.E. (1999) Effects of CO_2 enrichment on trees and forests: lessons to be learned in view of future ecosystem studies. *Annals of Botany*, **84**, 577–590.

Chaffey, N.J. (1999) Cambium: old challenges – new opportunities. *Trees*, **13**, 138–151.

Chaffey, N.J. (2000) Cytoskeleton, cell walls and cambium: new insights into secondary xylem differentiation. In *Cell and Molecular Biology of Wood Formation*, edited by R. Savidge, J. Barnett and R. Napier, pp. 31–42. Oxford: Bios Scientific Publishers.

Chaffey, N.J. (2001) Cambial cell biology comes of age. In *Trends in European Forest Tree Physiology Research*, edited by S. Huttunen, J. Bucher, P. Jarvis, R. Matyssek and B. Sundberg. Kluwer Academic Press (in press).

de Zeeuw, C. (1965) Variability in wood. In *Cellular Ultrastructure of Woody Plants*, edited by W.A. Côté, Jr, pp. 457–471. Syracuse: Syracuse University Press.

Dinus, R.J. and Welt, T. (1997) Tailoring fiber properties to paper manufacture: Recent developments. *TAPPI Journal*, **80**, 127–139.

El Bassam, N. (1996) *Renewable Energy: Potential Energy Crops for Europe and the Mediterranean Region*. Rome: Food and Agriculture Organization of the United Nations.

Elliott, G.K. (1985) Wood properties, and future requirements for wood products. In *Attributes of Trees as Crop Plants*, edited by M.G.R. Cannell and J.E. Jackson, pp. 545–552. Huntingdon, UK: Institute of Terrestrial Ecology, NERC.

Filatti, J.J., Sellmer, J., McCown, B., Haissig, B. and Comai, L. (1987) *Agrobacterium* mediated transformation and regeneration of *Populus*. *Molecular and General Genetics*, **206**, 192–199.

Gerry, E. (1913) Microscopic properties of woods in relation to properties and uses. *Proceedings of the Society of American Foresters*, **8**, 159–174.

Goldstein, I.S. (1980) New technology for new uses of wood. *TAPPI*, **63**, 105–108.

Haigler, C. (1994) From signal transduction to biophysics: tracheary element differentiation as a model system. *International Journal of Plant Sciences*, **155**, 248–250.

Halpin, C., Abbott, J. and Barakate, A. (2000) Investigating lignin biosynthesis using transgenic mutant plants. In *Cell and Molecular Biology of Wood Formation*, edited by R. Savidge, J. Barnett and R. Napier, pp. 425–436. Oxford: Bios Scientific Publishers.

Haygreen, J.G. and Bowyer, J.L. (1996) *Forest Products and Wood Science*, 3rd edn. Ames, Iowa: Iowa State University Press.

Hertzberg, M. and Olsson, O. (1998) Molecular characterisation of a novel plant homeobox gene expressed in the maturing xylem zone of *Populus tremula* x *tremuloides Plant Journal*, **16**, 285–295.

Iqbal, M. (ed) (1995) *The Cambial Derivatives*. Berlin: Gebrüder Borntraeger.

Isebrands, J.G. and Bensend, D.W. (1972) Incidence and structure of gelatinous fibers within rapid-growing eastern cottonwood. *Wood and Fiber*, **4**, 61–71.

Klass, D.L. (1998) *Biomass for Renewable Energy, Fuels and Chemicals*. San Diego, London, New York: Academic Press.

Klopfenstein, N.B., Chun, Y.W., Kim, M.-S. and Ahuja, M.R. (eds), Dillon, M.C. Carman, R.C. and Eskew, L.G. (tech. eds) (1997) *Micropropagation, Genetic Engineering, and Molecular Biology of Populus*. General Technical Report RM-GTR-297. Fort Collins, CO: US Dept of Agriculture, Forest Service, Rocky Mountain Forest and Range Experimental Station.

Kohlmaier, G.H., Weber, M. and Houghton, R.A. (1998) *Carbon Dioxide Mitigation in Forestry and Wood Industry*. Berlin, Heidelberg, New York: Springer.

Kumar, S. (1994) Chemical modification of wood. *Wood and Fiber Science*, **26**, 270–280.

Lapierre, C., Pollet, B., Petit-Conil, M., Leplé, C., Boerjan, W. and Jouanin, L. (2000) Genetic engineering of poplar lignins: Impact of lignin alteration on kraft pulping performances. In *Lignins: Historical, Biological, and Materials Perspectives*, edited by W.G. Glasser, R.A. Northey and T.P. Schultz, pp. 145–160. Washington, DC: American Chemical Society.

Larson, P. (1994) *The Vascular Cambium: Development and Structure*. Berlin: Springer.

Mackenzie, D. (1999) Woodwork. *New Scientist*, **161**, 20.

McCann, M.C., Domingo, C., Stacey, N.J., Milioni, D. and Roberts, K. (2000) Tracheary element formation in an *in vitro* system. In *Cell and Molecular Biology of Wood Formation*, edited by J. Savidge, J. Barnett and R. Napier, pp.457–470. Oxford: Bios Scientific Publishers.

McCarthy, J.L. and Islam, A. (2000) Lignin chemistry, technology, and utilization: A brief history. In *Lignins: Historical, Biological, and Materials perspectives*, edited by W.G. Glasser, R.A. Northey and T.P. Schultz, pp. 2–99. Washington, DC: American Chemical Society.

Northcote, D.H. (1993) Measurable changes in plant cells during differentiation. *Journal of Plant Research, Special Issue*, **3**, 109–115.

Northcote, D.H. (1995) Aspects of vascular tissue differentiation in plants: parameters that may be used to monitor the process. *International Journal of Plant Sciences*, **156**, 245–256.

O'Connell, A., Bolwell, P. and Schuch, W. (1998) Impact of forest tree biotechnology on the pulp and paper-making processes in the 21st century. In *Transgenic Plant Research*, edited by K. Lindsey, pp. 175–186. Australia, Canada: Harwood Academic Publishers.

Parsons, T.J., Sinkar, V.P., Stettler, R.F., Nester, E.W. and Gordon, M.P. (1986) Transformation of poplar by *Agrobacterium tumefaciens. Bio/Technology*, **4**, 533–536.

Peszlen, I. (1996) Gelatinous fibres in *Populus* x *euramericana* clones. In *Recent Advances in Wood Anatomy*, edited by L.A. Donaldson, B.G. Butterfield, P.A. Singh and L.J. Whitehouse, pp. 327–333. Rotorua, NZ: New Zealand Forest Research Institute.

Phillips, O.L., Malhi, V., Higuchi, N., Laurance, W.F., Núñez, P.V., Vásquez, R.M., Laurance, S.G., Ferreira, L.V., Stern, M., Brown, S. and Grace, J. (1998) Changes in the carbon balance of tropical forests: evidence from long-term plots. *Science*, **282**, 439–442.

Philipson, W.R., Ward, J.M. and Butterfield, B.G. (1971) *The Vascular Cambium: its Development and Activity*. London: Chapman and Hall Ltd.

Roberts, L.W. (1976) *Cytodifferentiation in Plants: Xylogenesis as a Model System*. Cambridge: Cambridge University Press.

Savidge, R.A. (1985) Prospects for manipulating vascular-cambium productivity and xylem-cell differentia-tion. In *Attributes of Trees as Crop Plants*, edited by M.G.R. Cannell and J.E. Jackson, pp. 208–227. Huntingdon, UK: Institute of Terrestrial Ecology, NERC.

Savidge, R., Barnett, J. and Napier, R. (eds) (2000) *Cell and Molecular Biology of Wood Formation*. Oxford: Bios Scientific Publishers.

Thomas, P. (2000) *Trees: their Natural History*. Cambridge: Cambridge University Press.

Thring, R.W., Vanderlaan, M.N. and Griffin, S.L. (1997) Polyurethanes from Alcell® lignin. *Biomass and Bioenergy*, **13**, 125–132.

Tickell, O. (1998) A cut above. *New Scientist*, **159**, 10.

Torrey, J.G., Fosket, D.E. and Hepler, P.K. (1971) Xylem formation: a paradigm of cytodifferentiation in higher plants. *American Scientist*, **59**, 338–352.

Vaughan, J.G. and Geissler, C. (1998) *The New Oxford Book of Food Plants*. Oxford: Oxford University Press.

Vincent, J. (1996) Tricks of nature. *New Scientist*, **151**, 38–40.

Whetten, R. and Sederoff, R. (1991) Genetic engineering of wood. *Forest Ecology and Management*, **43**, 301–316.

White, D.J.B. and Robards, A.W. (1965) Gelatinous fibres in ash (*Fraxinus excelsior* L.). *Nature*, **205**, 818.

Whiteman, A. (1996) *Revised Forecasts of the Supply and Demand for Wood in the United Kingdom*. Edinburgh: Forestry Commission.

Zobel, B.J. and Sprague, J.R. (1998) *Juvenile Wood in Forest Trees*. Berlin, Heidelberg, New York: Springer.

An Introduction to the Problems of Working with Trees

Wood Formation in Trees ed. N. Chaffey
© 2002 Taylor & Francis
Taylor & Francis is an imprint of the
Taylor & Francis Group
Printed in Singapore.

Nigel Chaffey

IACR – Long Ashton Research Station, Department of Agricultural Sciences, University of Bristol, Long Ashton, Bristol BS41 9AF, UK

ABSTRACT

There are many reasons why trees are considered difficult organisms to study. This chapter briefly considers 11 of them: choice of a model tree species; size, and longevity of trees; presence of interfering, and fluorescing, compounds; diversity of cell types in, accessibility, and 3-dimensionality of, the secondary vascular system; bark slippage; low yield of biochemical components from the secondary vascular system; and the lack of published protocols for cell biological study of trees. Most of these 'problems' have been overcome by the detailed techniques contained in the other chapters of this book.

Key words: bark slippage, cambium, model species, poplar, *Populus*, secondary vascular system (SVS)

INTRODUCTION

The chapters that follow give detailed protocols for many cell and molecular biological techniques for the study of wood formation in trees. This chapter provides some background to the problems encountered in working with trees in general, and the secondary vascular system (SVS) in particular, so that those detailed procedures will be better appreciated. It also tries to dispel some of the myths that have prevented more widespread research on trees.

Table 1 lists 11 'objections' that can be raised against studying the tree. Those points are considered in more detail below.

Table 1 Some of the factors that conspire to make study of trees, particularly the cell and molecular biology of the secondary vascular system (SVS), difficult

1 Lack of a model species
2 Trees are big
3 Trees take too long to grow
4 Trees contain interfering compounds
5 Yields of cell components are low
6 The SVS is difficult to access
7 The active cambium is easily torn
8 The SVS contains many types of cells
9 The SVS is 3-dimensional
10 The SVS contains many naturally fluorescent compounds
11 Lack of detailed, published techniques

Lack of a model species

Although one must never lose sight of the tremendous biodiversity that exists, to attempt to study every organism in detail sufficient to understand its biology is probably impossible. For that reason, one has to be selective and study a handful of species (and hope that they are representative of the full range of life forms). Hence 'model species' are chosen so that research efforts throughout the world can be focussed and critical mass achieved ensuring rapid, and co-ordinated, progress in a given area of biology. Thus the ephemeral crucifer *Arabidopsis thaliana* has come to carry a heavy burden of responsibility as the model plant (e.g. Somerville, 2000). However, such a plant is unlikely to be representative of a woody perennial, such as a tree, or, indeed, of a tremendous range of other plant types. Nevertheless, and particularly as research funds are becoming scarcer, the use of models seems set to continue for many years to come.

To that end, it is important that poplar, and other trees and hybrids of the genus *Populus*, have recently been adopted as the model angiosperm (hardwood) tree (e.g. Stettler *et al.*, 1990, 1996; Klopfenstein *et al.*, 1997). There are many good reasons for this choice, amongst which are the following (see Chaffey, 1999a, 2001 for detailed considerations of these points):

1. *Populus* has a relatively small genome (approx. 5×10^8 base pairs, cf. *Arabidopsis*, approx. 2×10^8 base pairs).
2. A large number of expressed sequence tags (ESTs) have now been produced for hybrid aspen (Sterky *et al.*, 1998), which will greatly increase the molecular-genetical work that is already being carried out using poplar (e.g. identification of homeobox genes in wood formation – Hertzberg and Olsson, 1998; GUS-reporter gene studies – see the chapter by Hawkins *et al.* in this volume; *in situ* hybridisation – see the chapter by Regan and Sundberg in this volume).
3. Poplar is transformable. Poplar was the first forest tree to be genetically transformed (Parsons *et al.*, 1986; Filatti *et al.*, 1987), and the technique is now almost a commonplace event (see chapters in Klopfenstein *et al.*, 1997), and such trees are used in a wide variety of studies of gene function (e.g. Sederoff, 1999).
4. The anatomy of the SVS of *Populus* is relatively straightforward and well known.
5. Immunological approaches are one of the most promising of the 'new techniques' of cell biology (Chaffey, 1999a) and have been successfully applied to the study of a wide variety of aspects of the SVS (e.g. enzymes of lignification – see the chapter by Samaj and Boudet in this volume; cytoskeleton – see the cytoskeletal and wood microscopical techniques chapters by Chaffey in this volume; enzymes of polysaccharide modification – see the chapter by Micheli *et al.* in this volume; polysaccharide analysis – see the chapter by Micheli *et al.* in this volume; and storage proteins – Sauter and van Cleve, 1990).

6. Hormones within poplar are both identifiable and quantifiable (see the chapter by Uggla and Sundberg in this volume)
7. The high growth rate of poplar (e.g. Isebrands and Bensend, 1972), particularly hybrid aspen (Einspahr, 1984), means that processes related to operation of the SVS, e.g. lignification, or formation of tension wood, can be studied in time-scales comparable to the generation time of *Arabidopsis*.
8. Considerations of biomass and renewable source of energy ('green energy') are currently topical. And such issues are likely to be of increasing importance as fossil fuel reserves become depleted, and more pressure is put upon governments by the environmental lobby to explore alternative energy sources. Poplar, and the closely related willow species, are of great interest from this point of view (e.g. Douglas, 1989; Smith *et al.*, 1997). Plus, there exists the potential for improving biomass wood quality, e.g. calorific value, as a result of greater understanding of the biology of the SVS (Chaffey, 2000).

It is acknowledged that poplar does not embrace the full range of tree types (Chaffey, 1999a, 2001), and there appears to be a lack of consensus on a model gymnosperm (Chaffey, 1999b). However, concentrating on one type should allow better focus of tree cell biological research in general, and, perhaps more importantly, help to get such work funded.

Trees are (too) big

Trees are big; they can be very big. And mature trees do take up a lot of space, which is not conducive to their being grown in the laboratory or under controlled conditions. Clearly, field trials must be performed outside, but a great deal of work can be carried out on saplings, which can be raised and maintained quite conveniently in greenhouses (where they are no more difficult to manage than, for example, sunflowers, maize or peas). And, as with most other biological studies, it is rare to use the whole organism; usually, a sample is taken. A trip to the field/forest to take tree samples is not only a welcome escape from the laboratory, it also serves as a valuable reminder that the tree lives in the outside world. Trees do not live in a Petri dish, or a glass jar, or a growth cabinet. To paraphrase Woolhouse (1982), if you want to study trees, there is no substitute but to work with trees!

Trees take a long time to grow

It depends very much on what aspect of tree biology you want to study. Certainly, it is generally the case that sexual maturity is not reached for many years (although early-flowering birch clones have been developed, which flower within a year of rooting – Lemmetyinen et al., 1997). However, the main focus of interest, wood formation, begins very early on in the life of the tree, and can be studied in the taproot of the angiosperm tree *Aesculus hippocastanum* within 6 weeks of sowing the seed.

Related to the long life cycle of trees is the fact that most of the current work is performed on young trees (in the case of transformed trees, they are often less than three years old), hence on juvenile wood. It is widely recognised that such wood differs significantly from the later-formed, mature wood (e.g. Krahmer, 1986), and can give rise to lower quality wood products (Jackson and Megraw, 1986). It cannot be assumed that what holds good for juvenile wood will also be the case with mature wood. There is thus a pressing need to extend studies from saplings to more mature trees. Conversely, there is also a need to understand more fully the biology of juvenile wood as demand for wood results in the harvesting of younger trees (see also the Introduction to this volume).

Presence of interfering compounds

Apart from naturally fluorescent compounds (considered below), other materials exist in the cells of the SVS that either hamper biochemical investigations, or interfere with fixation for more anatomically oriented studies. For example, non-specific binding of probes by cell walls (see the chapter by Regan and Sundberg in this volume), high phenolic content (see the chapters by Beardmore et al., Magel, Micheli et al., and Takabe in this volume), chlorophyll-masking of the blue colour for β-glucuronidase localisation (see the chapter by Hawkins et al. in this volume) all contribute to making study of the cell and molecular biology of trees troublesome. However, neither the substances involved nor the problems

encountered are unique to trees, but they are often more problematic here than in more amenable tissues. The protocols in the chapters that follow have been developed with these problems in mind.

Trees give low yields of cell constituents

It is undeniable that the low ratio of living to dead cells in tissues such as the secondary xylem (e.g. Wang et al., 2000) places constraints on the amounts of biochemicals that can be harvested. However, that is in the nature of the material, and, with the advent of ever more sophisticated sampling and analytical methods, not a major problem. For example, amounts of hormones, such as auxin, sufficient for quantification can be harvested from single sections of developing wood (30 μm thick $\times$ 2 mm $\times$ 12 mm) (Uggla et al., 1996; see the chapter by Uggla and Sundberg in this volume). Such sections can also provide sufficient material for analysis of isoforms of pectin methylesterase (Ermel et al., 2000).

The SVS is often quite deeply embedded within the organ

The location of the SVS is a major hindrance to its *in situ/in vivo* study or ready removal, a problem which is exacerbated for those who work on roots! This not only makes it physically difficult to remove the delicate tissues of interest so that their integrity is maintained, but the time taken to do so may permit degradative changes to occur which compromise ultrastructural and biochemical studies. However, as long as one works reasonably quickly – with hacksaw and razor blade for saplings (see e.g. the transmission electron microscopy (TEM) chapter by Chaffey in this volume), or with hammer, chisel and razor blade for more mature specimens (see the chapter by Uggla and Sundberg in this volume), samples can be successfully removed. This physical barrier to fast excision of material may also seem to present an insurmountable obstacle to successful freeze-fixation techniques for the SVS. That this is not the case is well shown in the TEM chapter by Rensing in this volume. The major problem associated with such techniques is the same as with any other plant material – the depth within the excised block to which usable freezing can be created.

The active cambium tears easily during sampling

Tearing of the 'cambium' is a long-established marker of the spring reactivation of this tissue in temperate trees – when cambial cell walls become thin again after the period of winter dormancy, and cell division is resumed – and is known as 'bark slippage'. It occurs both in stems and roots and is a major problem in sampling the SVS. However, although it is widely believed that the tearing takes place in the cambium, this appears rarely to be the case. Rather, the rupture usually occurs within the zone of developing vascular *derivatives*, where the cells have relatively thin walls and are undergoing radial expansion. Thus, depending on whether phloem or xylem is made first upon resumption of cambial cell division, the cambium will either remain attached to the xylem, or be removed with the bark, respectively. Although generally bothersome, this feature can be useful in obtaining samples enriched in the relevant fraction of the SVS, and has been employed in the preparation of DNA libraries of this material (e.g. Label *et al.*, 2000). Although some comfort may be had in the knowledge that bark slippage is not a feature of the dormant period (late autumn to spring for temperate species), ultimately, one has to accept that it is a fact of life. Therefore, I can only stress the need to be as careful as possible in excising and processing the samples (and to take excess samples to ensure that you retain enough intact for your particular study).

The SVS contains many types of cells

Although the cambium is a meristematic tissue, it differs from other, primary meristems (such as those of the shoot and root apices) in having two distinct types of cells (ray and fusiform initials), both of which are highly vacuolate (e.g. Catesson, 1990). These characteristics pose problems from the point of view of fixation and sampling for microscopical studies, and are particularly exacerbated in the case of the axially elongated fusiform cells. Furthermore, cells of the cambium, and their derivatives at the earliest stages of differentiation, are thin-walled, highly vacuolate, and sandwiched between thick-walled xylem and phloem cells. Accordingly, '... actively dividing and differentiating cells in the cambial zone are virtually impossible to fix so that all cell types are well preserved ... This means that, in general, one must be satisfied

to obtain reasonable preservation of the cambial cells at the expense of the youngest differentiating xylem cells (the most difficult to fix), or *vice versa*' (Robards and Kidwai, 1969).

The above problems are further compounded by the variety of cell types – at least 12 in a typical angiosperm tree such as *Aesculus* (see Figure 1 in the Introduction to this volume) – that are found in the component tissues of the SVS. Not only is each cell type unique in terms of size, shape and cytoplasmic components, but each individual cell is also at a different physiological state/stage of differentiation relative to its neighbour. Although these facts have been recognised for several decades, even today, it is still the case that it is, 'not possible to obtain simultaneous perfect fixation in all cells extending over such a wide zone [cell files from cambium deep into the xylem]' (Robards and Kidwai, 1972). However, to appreciate fully the developmental biology of the tree, we need to understand the cell and molecular biology of each cell type.

So much for the bad news. The good news is that quite stunning images *can* be obtained, even with such demanding techniques as TEM (e.g. Farrar and Evert, 1997). Although it may be necessary to experiment with a range of fixation procedures to find out what suits your material, it is usually the case that relatively standard mixtures of formaldehyde and glutaraldehyde in the 'usual' range of buffers all seem suitable (see Table 2 in the TEM chapter by Chaffey in this volume). The various fixation regimes detailed in this book represent a reasonably good compromise, as judged by the overall quality of the micrographs, and should serve as useful starting points for your own particular study. Furthermore, it may sometimes be necessary to use different fixation protocols to permit better preservation of one cell type over another, or to enhance one cell feature over another. But, this is no more than one would need to do if trying to apply fixation protocols designed e.g. for unicellular algae to herbaceous angiosperms. And, on the plus side, the very robustness of wood cells means that processing them for scanning electron microscopy can be as straightforward as air-drying and metal-coating (Exley *et al.*, 1974; Chaffey *et al.*, 1996).

Broadly speaking, the SVS consists of two cell types: axially elongated (cell long axis oriented parallel to axis of organ) fusiform initials and their derivatives, and cuboid ray initials and their derivatives, which are elongated at right angles to the organ axis, i.e. radially. Furthermore, the distribution of individual cell types is often unpredictable so that any tissue block/sample contains a mixture of axial and radial. Consequently, from a sampling point of view, the processing requirements of the long axial cells take precedence, and the key to successful preparation is to think big! For the most part, for TEM, one is taught to fix cubes of tissue with sides no greater than 1 mm. That strategy is inappropriate to the SVS where individual cells can be up to 3 mm long (e.g. in gymnosperms). Figures 1 in the TEM and cytoskeletal chapters by Chaffey in this volume illustrate an appropriate sampling technique for saplings. The essential point, which is worth emphasising here, is to excise and fix *large* slivers of material, but with a thickness of approx. 1–2 mm. That way the smallest dimension is in the range of 'traditional TEM' – permitting ready access of fixatives to cells, etc., whilst the larger dimensions ensure that entire cells are being fixed. After initial fixation, terminal portions of the slivers are removed (to eliminate damaged regions), before trimming to final block size, a more manageable 3 mm × 3 mm × 1–2 mm. Even if whole cells are not present in these smaller blocks, they will have been present in the initial excised sliver, ensuring that they were fixed and their structure faithfully preserved (see also comments and Figure 3 in the chapter by Hawkins *et al.* in this volume).

The highly vacuolate nature of cambial cells and their derivatives makes mobility of ions a major problem in any study which seeks to understand the relationship of these important cell constituents to developmental processes. However, even for this study, techniques have recently been developed which permit their immobilisation so that their *in vivo* location and concentration can be studied (see the chapter by Follet-Gueye in this volume).

The SVS is 3-dimensional

Of course, all tissue systems are 3-dimensional; however, the point is stressed to emphasise the fact that the SVS is an intimate association of axial cells derived from fusiform cambial cells, such as vessel elements, fibres, and sieve elements, and radial elements derived from ray cambial cells. Therefore, to appreciate fully the interrelations between the component cells of the SVS, it is essential to study the system in all three of the main planes, necessitating the cutting of transverse (or cross) (TS), radial longitudinal (RLS) and tangential longitudinal (TLS) sections. Consequently, considerable care is required in correct orientation of blocks when embedding. The major drawback of LS's relative to TS's is that the slightest 'deficiencies' in fixation always seem much more obvious. However, with care – and luck/practice! – excellent LS's *can* be obtained (e.g. Farrar and Evert, 1997). And LS's are essential for interpretation of e.g., the cytoskeletal arrangements in the SVS (see the cytoskeletal chapters by Funada, and Chaffey in this volume), since a TS only reveals the thin parietal layer of cytoplasm in each cell.

A useful tip for achieving good RLS's is to ensure that slivers are excised from the diameter of the sampled organ. Although this usually means that only two or three slivers can be removed from any portion of stem/root, this practice will save many hours of disappointment in trying to interpret 'oblique RLS's'. However, it will soon become apparent that cells rarely lie perfectly horizontally or vertically in the organ. Thus, cells will appear to weave in and out of the plane of section, because they do! This is particularly dramatic in the case of vessels (e.g. Zimmermann, 1968; Burggraaf, 1972).

Another source of confusing images is that individual cells of the SVS are usually different lengths and frequently arranged in a non-storied arrangement. Non-storied refers to the cambium wherein the component cells overlap laterally so that they terminate at different levels in the transverse plane. Since there appears to be more cytoplasm at the ends of the cells (same thickness of cytoplasm, but smaller cell diameter), they can easily be mistaken for a different cell type, by reference to their more-vacuolate neighbours.

Although the appearance of unexpected cell types in a given section can be irritating to the 'anatomist', for the biochemist it can be particularly problematic. For example, attempts to relate enzyme isoforms (see the chapter by Micheli *et al.* in this volume), or sugars levels (see the biochemistry chapter by Magel in this volume), or hormone levels (see the chapter by Uggla and Sundberg in this volume) to particular tissue fractions are dependent on the ability to know the precise tissue composition of a sample. The cryosectioning technique recently developed by Uggla (see the chapter by Uggla and Sundberg in this volume), and the allied method of estimating tissue composition, which takes account of this problem, is thus a particularly useful – and timely – development.

The SVS contains many naturally fluorescent compounds

Chlorophyll and lignin are well-known fluorescent materials and can be major irritants in (immuno) fluorescence studies of the SVS. Depending on the fluorochromes you are using, native fluorescence may be more or less of a problem, and fluorochromes will need to be chosen with this in mind. Chlorophyll fluorescence is not a problem in roots, a much neglected area of study in the context of the SVS (e.g. Chaffey et al., 1996). However, most people concentrate on the chlorenchymatous aerial portions. Fortunately, the organic solvents used during tissue-dehydration remove most of this offending pigment; but note that some processing procedures do not employ dehydration (see the chapter by Funada in this volume).

Tannins and lignin are also particular problems in the tree SVS, the former tending to be sequestered in vacuoles, e.g. of phloem ray cells, the latter is an integral component of cell walls of most xylem cell types, and phloem fibres. Several strategies are available for dealing with this type of fluorescence: remove it, ignore it, reduce it, or use it. Most of these strategies are dealt with in the chapters that deal with fluorescence techniques in this volume.

Autofluorescence can be reduced – quenched – to more tolerable levels by lightly staining the sections with toluidine blue (see the cytoskeletal chapter by Chaffey in this volume), or by use of sodium borohydride or Schiff's reagent (Carnegie et al., 1980; Christensen, 1998). However, it is worth mentioning that the red fluorescence of lignin under the fluorescein isothiocyanate (FITC) filter combination is a pleasant contrast to the green fluorescence of FITC, and can be useful in indicating where lignification of cells is beginning.

Furthermore, in fluorescence work with the confocal laser scanning microscope (CLSM), dual-channel imaging using the 488 nm laser line for FITC, and the 568 nm laser line for autofluorescence, can give images which are particularly instructive (see Figures 1E, F, H in the Wood microscopical techniques chapter in this volume).

Use of cell wall polysaccharide-digesting enzymes (see the cytoskeletal chapter by Chaffey in this volume) may contribute to partial removal of wall-located fluorescent material (and, possibly, improve access of antibodies to antigenic sites). Other attempts to boost the 'signal-to-noise ratio' include the use of detergent (see the chapter by Funada in this volume) to puncture the plasmalemma and facilitate better penetration of antibodies, blocking of non-specific antibody binding (see the chapters by Beardmore et al., Chaffey, Funada, Micheli et al., and Samaj and Boudet in this volume), and use of the CLSM, particularly with appropriate use of the scan-averaging facility (see the chapters by Funada, and Micheli et al. in this volume). Alternatively, the red autofluorescence can be filtered out for viewing or image-recording purposes.

Although it does not naturally occur in cells of the SVS, use of glutaraldehyde as a fixative is another major source of added fluorescence to the tissues which serves to interfere with the fluorescent signal (e.g. Ruzin, 1999). In view of this, it is recommended that the minimum concentration of glutaraldehyde be used sufficient to achieve satisfactory structural preservation. In immunolocalisation of proteins, it is generally the case that only small amounts of glutaraldehyde can be used, if the antigenicity of the target protein is not to be compromised. Glutaraldehyde at 0.2% is routinely used for immunofluorescent work on cytoskeletal proteins in angiosperms (see the chapter by Chaffey in this volume) and gymnosperms (see the chapter by Funada in this volume). At this concentration, both structural integrity and antigenicity appear to be satisfactorily preserved, and added fluorescence is not a major problem. However, it is worth noting that antibodies are quite often raised against denatured antigen, so that a relatively high level of protein-protein cross-linking by glutaraldehyde can be tolerated by the protein of interest. The fact that glutaraldehyde-induced fluorescence is predominantly found in protein-containing (primarily, living) cells has been used to good effect by Singh et al. (1997). Using the CLSM, these workers imaged fungi invading dead wood by the glutaraldehyde fluorescence of the living fungal hyphae.

Allied to the problem of removing unwanted fluorescence is that of retaining the immuno-fluorescence. Both laboratory-made p-phenylenedi-amine in glycerol, and commercially available anti-fade mountants, such as Vectashield® (see the Wood microscopical techniques chapter in this volume) have been successfully used in tree SVS work.

However, one way to avoid the problem of autofluorescence altogether – and that of fluorescence fading – at the light microscope level is to use immunogold localisation, either with or without silver-enhancement (see the chapters by Beardmore et al., Micheli et al., and Samaj and Boudet in this volume).

Lack of detailed, published techniques

Although the Materials and Methods sections of published papers are supposed to give information sufficient for others to be able to repeat the work, that is not always the case. In particular, it is difficult to get protocols designed for herbaceous plants to work in trees, without substantial modification (e.g. Hawkins et al., 1997), hence this book!

CONCLUDING COMMENTS

The problems of working with trees considered above may appear intimidating. Indeed, the problems are genuine, can be formidable, and should not be underestimated. However, in the majority of cases they have been overcome, and protocols designed which surmount these apparent barriers to study of the tree SVS. Many of those protocols are included in the following chapters.

REFERENCES

Burggraaf, P.D. (1972) Some observations on the course of the vessels in the wood of *Fraxinus excelsior* L. *Acta Botanica Neerlandica*, **21**, 32–47.

Carnegie, J.A., McCully, M.E. and Robertson, H.A. (1980) Embedment in glycol methacrylate at low temperature allows immunofluorescent localization of a labile tissue protein. *Journal of Histochemistry & Cytochemistry*, **28**, 308–310.

Catesson, A.-M. (1990) Cambial cytology and biochemistry. In *The Vascular Cambium*, edited by M. Iqbal, pp. 63–112. Taunton, UK: Research Studies Press Ltd.

Chaffey, N.J. (1999a) Cambium: old challenges – new opportunities. *Trees*, **13**, 138–151.

Chaffey, N.J. (1999b) Wood formation in forest trees: from *Arabidopsis* to *Zinnia*. *Trends in Plant Science*, **4**, 203–204.

Chaffey, N.J. (2000) Cytoskeleton, cell walls and cambium: new insights into secondary xylem differentiation. In *Cell and Molecular Biology of Wood*

Formation, edited by R. Savidge, J. Barnett and R. Napier, pp. 31–42. Oxford: Bios Scientific Publishers.

Chaffey, N.J. (2001) Cambial cell biology comes of age. In *Trends in European Forest Tree Physiology Research*, edited by S. Huttunen, J. Bucher, P. Jarvis, R. Matyssek and B. Sundberg. Kluwer Academic Publishers (in press).

Chaffey, N.J., Barlow, P.W. and Barnett, J.R. (1996) Microtubular cytoskeleton of vascular cambium and its derivatives in roots of *Aesculus hippocastanum* L. (Hippocastanaceae). In *Recent Advances in Wood Anatomy*, edited by L.A. Donaldson, B.G. Butterfield, P.A. Singh and L.J. Whitehouse, pp. 171–183. Rotorua, NZ: New Zealand Forest Research Institute.

Christensen, A.K. (1998) Using borohydride to quench autofluorescence of glutaraldehyde. *Microscopy Today*, **98–1**, 21.

Douglas, G.C. (1989) Poplar (*Populus* spp.). In *Biotechnology in Agriculture and Forestry 5, Trees II*, edited by Y.P.S. Baja, pp. 300–323. Berlin, Heidelberg, New York: Springer.

Einspahr, D.W. (1984) Production and utilization of triploid hybrid aspen. *Iowa State Journal of Research*, **58**, 401–409.

Ermel, F.E. Follet-Gueye, M.-L., Cibert, C., Vian, B., Morvan, C., Catesson, A.-M. and Goldberg, R. (2000) Differential localization of arabinan and galactan side chains of rhamnogalacturonan 1 in cambial derivatives. *Planta*, **210**, 732–740.

Exley, R.R., Butterfield, B.G. and Meylan, B.A. (1974) Preparation of wood specimens for the scanning electron microscope. *Journal of Microscopy*, **101**, 21–30.

Farrar, J.J. and Evert, R.F. (1997) Seasonal changes in the ultrastructure of the vascular cambium of *Robinia pseudoacacia*. *Trees*, **11**, 191–200.

Filatti, J.J., Sellmer, J., McCown, B., Haissig, B. and Comai, L. (1987) *Agrobacterium* mediated transformation and regeneration of *Populus*. *Molecular and General Genetics*, **206**, 192–199.

Hawkins, S., Samaj, J., Lauvergeat, V., Boudet, A. and Grima-Pettenati, J. (1997) Cinnamyl alcohol dehydrogenase (CAD): identification of important new sites of promoter activity in transgenic poplar. *Plant Physiology*, **113**, 321–325.

Hertzberg, M. and Olsson, O. (1998) Molecular characterisation of a novel plant homeobox gene expressed in the maturing xylem zone of *Populus tremula* × *tremuloides*. *Plant Journal*, **16**, 285–295.

Isebrands, J.G. and Bensend, D.W. (1972) Incidence and structure of gelatinous fibers within rapid-growing eastern cottonwood. *Wood and Fiber*, **4**, 61–71.

Jackson, M. and Megraw, R.A. (1986) Impact of juvenile wood on pulp and paper products. In *A Technical Workshop: Juvenile Wood – What Does it Mean to Forest Management and Forest Products?*, pp. 75–81. Madison, WI: Forest Products Research Society.

Klopfenstein, N.B., Chun, Y.W., Kim, M.-S. and Ahuja, M.R. (eds), Dillon, M.C., Carman, R.C. and Eskew, L.G. (tech. eds) (1997) *Micropropagation, Genetic Engineering, and Molecular Biology of Populus*. General

Technical Report RM-GTR-297. Fort Collins, CO: US Dept of Agriculture, Forest Service, Rocky Mountain Forest and Range Experimental Station.

Krahmer, R.L. (1986) Fundamental anatomy of juvenile and mature wood. In *A Technical Workshop: Juvenile Wood – What Does it Mean to Forest Management and Forest Products?*, pp. 12–16. Madison, WI: Forest Products Research Society.

Label, P., Beritognolo, I., Burtin, P., Dehon, L., Couée, I., Breton, C., Charpentier, J.-P., and Jay-Allemand, C. (2000) Cambial activity and xylem differentiation in walnut (*Juglans sp.*). In *Cell and Molecular Biology of Wood Formation*, edited by R. Savidge, J. Barnett and R. Napier, pp. 209–221. Oxford: Bios Scientific Publishers.

Lemmetyinen, J., Keinonen-Mettälä, K., Lännenpää, M., von Weissenberg, K. and Sopanen, T. (1997) Transgenic early-flowering birches. *Abstract from XVIII Congress of SPPS* meeting held in Uppsala, Sweden, 12–17 June, 1997, p. 198.

Parsons, T.J., Sinkar, V.P., Stettler, R.F., Nester, E.W. and Gordon, M.P. (1986) Transformation of poplar by *Agrobacterium tumefaciens*. *Bio/Technology*, **4**, 533–536.

Robards, A.W. and Kidwai, P. (1969) A comparative study of the ultrastructure of resting and active cambium of *Salix fragilis* L. *Planta*, **84**, 239–249.

Robards, A.W. and Kidwai, P. (1972) Microtubules and microfibrils in xylem fibres during secondary cell wall formation. *Cytobiologie*, **6**, 1–21.

Ruzin, S.E. (1999) *Plant Microtechnique and Microscopy*. Oxford, New York: Oxford University Press.

Sauter, J.J. and van Cleve, B. (1990) Biochemical, immunochemical, and ultrastructural studies of protein storage in poplar (*Populus × canadensis* 'robusta') wood. *Planta*, **183**, 92–100.

Sederoff, R. (1999) Building better trees with antisense. *Nature Biotechnology*, **17**, 750–751.

Singh, A., Xiao, Y. and Wakeling, R. (1997) Glutaraldehyde fluorescence useful in confocal studies of fungi. *Microscopy Today*, **97–8**, 16.

Smith, N.O., Maclean, I., Miller, F.A. and Carruthers, S.P. (1997) *Crops for Industry and Energy in Europe*. Luxembourg: European Commission Directorate General XII E-2 Agro-industrial Research Unit.

Somerville, C. (2000) The twentieth century trajectory of plant biology. *Cell*, **100**, 13–25.

Sterky, F., Regan, S., Karlsson, J., Rohde, A., Holmberg, A., Amini, B., Bhalerao, R., Larsson, M., Villarroel, R., Van Montagu, M., Sandberg, G., Olsson, O., Teeri, T.T., Boerjan, W., Gustafsson, P., Uhlén, M., Sundberg, B. and Lundeberg, J. (1998) Gene discovery in the wood-forming tissues of poplar: Analysis of 5,692 expressed sequence tags. *Proceedings of the National Academy of Sciences of the United States of America*, **95**, 13330–13335.

Stettler, R.F., Bradshaw, H.D., Jr, Heilman, P.E. and Hinckley, T.M. (eds) (1996) *Biology of Populus and its Implications for Management and Conservation*. Ottawa: NRC Research Press, National Research Council of Canada.

Stettler, R.F., Heilmann, P.E. and Hinckley, T.M. (1990) Genetics/physiology of short rotation biomass production with *Populus*. In *Fast Growing Trees and Nitrogen Fixing Trees*, edited by D. Werner and P. Müller, pp. 194–200. Stuttgart, New York: Gustav Fischer Verlag.

Uggla, C., Moritz, T., Sandberg, G. and Sundberg, B. (1996) Auxin as a positional signal in pattern formation in plants. *Proceedings of the National Academy of Sciences of the United States of America*, **93**, 9282–9286.

Wang, S.X., Hunter, W. and Plant, A. (2000) Isolation and purification of functional total RNA from woody branches and needles of sitka spruce and white spruce. *BioTechniques*, **28**, 292, 294–296.

Woolhouse, H.W. (1982) Leaf senescence. In *The Molecular Biology of Plant Development*, edited by H. Smith and D. Grierson, pp. 256–281. Oxford: Blackwell Scientific Publications.

Zimmermann, M.H. (1968) Physiological aspects of wood anatomy. *IAWA Bulletin*, **1968/2**, 11–14.

3 Wood Microscopical Techniques

Wood Formation in Trees ed. N. Chaffey
© 2002 Taylor & Francis
Taylor & Francis is an imprint of the Taylor & Francis Group
Printed in Singapore.

Nigel Chaffey

IACR – Long Ashton Research Station, Department of Agricultural Sciences, University of Bristol, Long Ashton, Bristol BS41 9AF, UK

ABSTRACT

This chapter provides detailed protocols for a wide range of light and fluorescence microscopy techniques for examination of the anatomy and cytochemistry of cells of the secondary vascular system of woody species. Techniques covered include staining of unfixed hand-sections, and of fixed and embedded sections, maceration, and epifluorescence and confocal laser scanning microscopy.

Key words: anatomical techniques, confocal laser scanning microscopy, epifluorescence microscopy, light microscopy, maceration, wood

INTRODUCTION

For almost as long as microscopy has been pursued, the anatomy of wood has been a source of fascination and inspiration (e.g. Grew, 1682). The chapters that follow each deal with specialised aspects of the cell and molecular biology of the secondary vascular system (SVS) of trees. Often, however, it is necessary to apply more general techniques to woody tissues, either to gain a wider appreciation of the anatomy of the subject material, or to provide an adjunct to other studies. Although most of the techniques assembled here can be found in books of general plant anatomy or histology, they have rarely been brought together in one convenient place. The purpose of this chapter is to provide a source of techniques that will assist that more general study of wood anatomy.

In selecting techniques to be covered here, the emphasis has been on those that have been tried-and-tested by the contributors to this volume. However, the listing is neither exhaustive nor comprehensive; it includes detailed protocols for many techniques, and suggestions of others that are worth trying. Interested readers are also directed to the following publications, some of which have special 'wood sections', which may prove useful as sources of inspiration: Chamberlain, 1905; Brown, 1919; Averall, 1926; Johansen, 1940; Sass, 1958; Jensen, 1962; Purvis *et al.*, 1964; Feder and O'Brien, 1968; Berlyn and Mischke, 1976; Clark, 1981; O'Brien and McCully, 1981; Gahan, 1984; Ruzin, 1999.

For those who would like more background on wood biology/anatomy, the following texts, many of which are well illustrated, are suggested: IAWA Committee on Nomenclature, 1964; Meylan and Butterfield, 1972; Esau, 1977; Grosser, 1977; Schweingruber, 1978, 1990; Panshin and de Zeeuw, 1980; Harada and Côté, 1985; Carlquist, 1988; IAWA Committee, 1989; Fahn, 1990; Mauseth, 1991; Ilvessalo-Pfäffli, 1995; Siau, 1995; Desch and Dinwoodie, 1996; Haygreen and Bowyer, 1996; Higuchi, 1997; Kozlowski and Pallardy, 1997.

FRESH, HAND SECTIONS

Sections of unfixed material, sufficiently thin for viewing in the light microscope (LM) with brightfield (BF) or differential interference contrast (DIC, 'Nomarski'), can be cut by hand with a razor blade. Sections that prove too thick for BF may still be suitable for epifluorescence microscopy, when appropriately stained.

Iron-alum-staining of cell walls

This technique (Goffinett *et al.*, 1995) is effectively a reduced version of the tannic acid-iron alum-safranin-orange G stain of Sharman (1943) (see also Clark, 1981), and is extremely useful for emphasising cell walls, which become dark-stained. In the SVS it is particularly useful for enabling cell geometry and inter-relationships of cell-types to be observed, so that cell lineages can be discerned and patterns of cell division elucidated. It provides – quite literally – a most useful contrast to the calcofluor-stained image (see below), with the advantage that the staining pattern can be viewed in a conventional LM.

1   Cut thin sections of SVS (transverse sections (TS) are best) with a razor blade, and accumulate them in tap water.
2   Stain sections for 3 min in 5% (w/v) aqueous tannic acid.
3   Rinse sections briefly in tap water.
4   Stain in 1% (w/v) aqueous ferric ammonium sulphate (iron alum) for 1 min.
5   Rinse sections in tap water, mount in tap water, and view in a LM with BF optics.

Calcofluor-staining of cell walls

Note – this stain is also extremely useful for sectioned material embedded in LR White or butyl-methylmethacrylate resin (see section below).
For further details, see Hughes and McCully (1975), Clark (1981), and Galbraith (1981).

1   Cut thin sections of SVS (TS best) with a razor blade, and accumulate them in tap water.
2   Place sections in a few drops of 0.01% (w/v) aqueous Calcofluor White M2R New or Calcofluor White ST for approx. 2 min.
3   Rinse sections briefly with tap water, mount in tap water on a microscope slide.
4   View in an epifluorescence microscope; excite with UV light (in range 340–370 nm – Butt *et al.*, 1989), and view the blue-white emitted light (in range 420–530 nm – Butt *et al.*, 1989).

Although calcofluor has long been used to stain cellulose (α-1, 4-glucan) in cell walls, it will bind to a wide variety of α- and β-linked glucans (e.g. Maeda and Ishida, 1967), including callose (β-1,3-glucan – e.g. Hughes and McCully, 1975). In view of the great

interest in microfibril orientation and its influence on wood properties (e.g. Butterfield, 1998), it is noteworthy that Hughes and McCully (1975) mention calcofluor's fluorescence dichroic property, and suggest that it might be useful in studying cellulose microfibril orientation in cell walls. Other techniques for investigating microfibril angle are detailed in Huang *et al.* (1998).

In TS, calcofluor-staining is useful for demonstrating early stages of cambial development (Figure 2A), and overall cell arrangement within the SVS (see e.g. Figure 1 in Chaffey *et al.*, 1996). However, note that lignin also fluoresces strongly under the excitation and viewing conditions used for calcofluor. To differentiate between lignin autofluorescence and that due to bound calcofluor, it is necessary to view the section before and after application of calcofluor (e.g. Chaffey *et al.*, 1999).

Iodine test for starch

Note-this technique also works well for sections of material in butyl-methylmethacrylate, and in glycol methacrylate (e.g. Wetzel *et al.*, 1989).

Starch is a major reserve product of trees, and accumulates particularly in ray cells of the xylem and cambium (see e.g. Figure 2D in the transmission electron microscopy (TEM) chapter by Chaffey in this volume). A quick way of determining relative abundance and location of starch in woody tissues is to use I_2/KI solution (e.g. Ruzin, 1999).

Make up solution as: 1 g potassium iodide and 1 g iodine in 100 ml water. Add a drop of the solution to the hand-cut section, leave for approx. 5 min. Wash away excess, unbound solution, and view with BF.

Phloroglucinol/HCl

Lignin and lignification is one of the most studied aspects of secondary xylem development (e.g. Wardrop, 1981; see the chapter by Takabe in this volume). It can easily be visualised by eye using a solution of phloroglucinol (1,3,5-trihydroxybenzene) in HCl (the Wiesner reaction). There are several recipes for this stain (e.g. Ruzin, 1999). The following recipe works well: mix together 0.1 g phloroglucinol, 16 ml concentrated HCl, and 100 ml of 95% ethanol. Apply a drop to the sections for approx. one minute. The resulting red colouration can be quite dramatic, especially when contrasted with unstained tissues

(Figure 1C). Lignified elements in the xylem and the phloem/cortex stain with this procedure, and it is possible to follow the lignification of cells within a radial file by the increase in colour, from pink to red. Unfortunately, the colour is not permanent, and any photographic record of the reaction will need to be made soon after the colour develops. However, Speer (1987) describes a variant of the test that permits retention of the red colour for over a year.

Mäule reaction

Another colour test for lignin is the Mäule reaction (e.g. Krishnamurthy, 1999; see the chapter by Takabe in this volume for a detailed protocol), which uses aqueous potassium permanganate, followed by HCl and ammonia, to stain, primarily, syringyl-lignins (e.g. Wardrop, 1981; Takabe et al., 1992), and is therefore considered diagnostic for conifer lignin (Baas, 1999).

Other stains

It is worth mentioning here the seasonal study of cambial cytochemistry performed by Rao and Menon (1989). Although it is not clear from their methods section whether they used fixed or unfixed, embedded or hand sections, they performed a range of tests on 10–15 μm-thick radial longitudinal sections, including: toluidine blue (see below), and tannic acid-ferric chloride, for 'general histology, crystals and tannins'; I_2/KI for starch (see above); Sudan black for lipids; and mercuric bromophenol blue for proteins. Such studies are all too rare, but can provide a tremendous amount of information about the general biochemistry/physiology of the tissues and cells of the SVS during the seasonal cycle.

Differential interference contrast (DIC, 'Nomarski')

In order to ensure the correct osmolarity of fixative solutions (e.g. Rao and Catesson, 1987) or cryo-protectant media for freeze-fixation methods (see the chapter by Rensing in this volume), it is useful to be able to examine plasmolysis of cells of the SVS. DIC (e.g. O'Brien and McCully, 1981; Ruzin, 1999) of hand-cut sections is particularly instructive for this. It is also a good method for investigating cytoplasmic streaming in cambial cells and

their derivatives (tangential longitudinal sections (TLS) are best for this).

Polarised light

Polarised light (e.g. O'Brien and McCully, 1981; Ruzin, 1999) is an extremely useful technique for following the process of secondary wall formation. In particular, sequential cells in a secondary xylem radial cell file can be followed from their initiation in the cambium to final maturation. As those cells mature, the cellulose microfibrils in their walls become progressively more-ordered, as secondary wall deposition takes place, and exhibit bire-fringency when illuminated with polarised light. For references to the use of this technique, see e.g. Abe et al. (1995a, 1997); Itabashi et al. (1999); Murakami et al. (1999).

Scanning electron microscopy (SEM)

This technique can be employed on sections of woody material that have simply been excised, ethanol-dehydrated, dried and sputter-coated (e.g. Exley et al., 1974; Figure 15 in Chaffey et al., 1996), where it can be useful for corroborating details of structures seen in the LM. It can also be applied to sections on microscope slides, after removing the cover slip and any embedding resin (Carlquist, 1988). For more illustrations of the power – and beauty – of this technique, see the book by Meylan and Butterfield (1972).

Note – to avoid excessive distortion of material during drying, try covering the material with filter paper (to soak up water), placing a glass microscope slide either side of the filter paper, and holding the 'sandwich' together with an elastic band (see also Brown, 1919).

Tissue-printing

Although extremely useful, and relatively simple, tissue-printing (e.g. Cassab and Varner, 1989; Varner and Ye, 1995) has been little-used on trees so far (e.g. Roth et al., 1997). It can be used with great effect and accuracy to localise almost any material, but particularly proteins and polysac-charides, which can be transferred from a cut surface of the tissue to a nitro-cellulose membrane. For a detailed protocol and illustrations of its use, see the chapter by Micheli et al. in this volume.

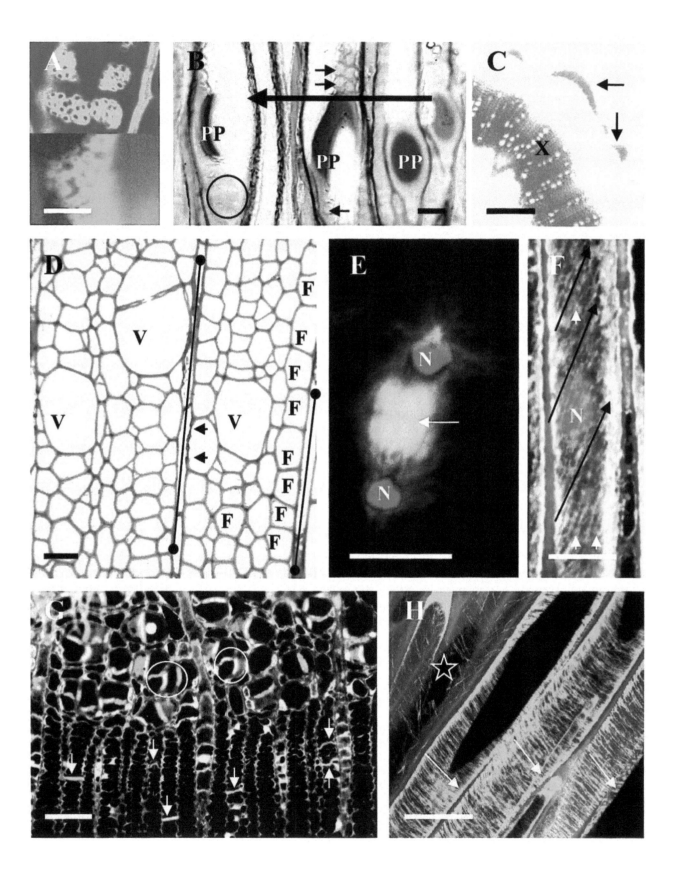

◄

Figure 1 Aspects of the anatomy and cell biology of the hardwood tree secondary vascular system illustrated with *Populus tremula* x *P. tremuloides* stem (A, D, E–H). and taproot (B) and stem (C) of *Aesculus hippocastanum*.
A Upper: calcofluor-stained sieve pores in a sieve element in resin-free radial longitudinal section of butyl-methylcrylate-embedded material; lower: calcofluor-aniline blue-double-stained sieve pores. Note displacement of the white-blue fluorescing calcofluor by the yellow-green fluorescing aniline blue.
B Methylene blue/azur A-stained radial longitudinal section of secondary xylem showing distinction between lignified walls of vessel elements (blue-green) and non-lignified perforation plate (red-purple). Note the increase in lignification with development of the vessel elements, and the development of the over-arching border of the bordered pits (arrows and encircled region) (direction of differentiation indicated by arrow).
C Hand-cut transverse section of epicotyl stained with phloroglucinol/HCl. Pronounced staining, but only of secondary and primary xylem and fibre-bundles (arrows) in the cortex.
D Transverse section of mature secondary xylem stained with toluidine blue. Note the differences in cell diameters of vessel elements and fibres, but also the variation within each cell-type. Parts of two rays (double-headed lines) are seen; note the contact pits between vessel elements and contact ray cells (arrows).
E CLSM image of mitotic spindle and daughter nuclei of a fusiform cambial derivative undergoing transverse cell division. Microtubules are localised with FITC/anti-tubulin antibody, nuclei are stained with propidium iodide. The unstained region (arrow) indicates the site where the division cell wall will form.
F CLSM image of radial longitudinal section of secondary xylem showing a differentiating fibre. Note the numerous, parallel-aligned microtubules (black arrows), the nucleus, and the red-fluorescent bodies (white arrows), which are in the same order of size as mitochondria. Staining as for Figure 1E.
G Transverse section of resin-free butyl-methylmethacrylate-embedded section viewed under epifluorescent illumination showing immunolocalisation of callose within the secondary vascular system with FITC/anti-callose antibody. Note the pronounced staining of the sieve plates in the phloem (encircled regions), and of the tangential walls of recently-divided cambial cells (arrows) (cf. Figure 4G in the transmission electron microscopy chapter by Chaffey in this volume).
H CLSM image of radial longitudinal section of secondary xylem showing differentiating fibres (staining as for Figure 1E, F). Note the close, parallel-aligned microtubules in fibres at late stages of differentiation (arrows) (in different orientation to those in Figure 1F), and the disorganised, sparse microtubules in fibres undergoing terminal stages of differentiation (☆). Note also the strong contrast between the lignified walls (red signal) and the FITC associated with the microtubules (green-yellow signal).
Key: CLSM, confocal laser scanning microscopy; F, fibre; FITC-fluorescein *iso*thiocyanate; N, nucleus; PP, perforation plate; V, vessel element; X, xylem. Scale bars represent: 10 μm (B, E, F), 20 μm (D, H), 25 μm (A), 70 μm (G), and 1 mm (C).

Pectin stains

A great deal of basic information about the chemical composition of cell walls can still be obtained by relatively simple techniques, such as ruthenium red or hydroxylamine-ferric chloride for pectins (e.g. O'Brien and McCully, 1981; Ruzin, 1999). Such stains can be used on a wide range of fresh and fixed and embedded material. For detailed protocols, and illustrations of the results, see the chapter by Micheli *et al.* in this volume.

MACERATIONS

The complicated 3-dimensional (3-D) arrangement of cells within the SVS (see e.g. the chapter entitled, An introduction to the problems of working with trees, in this volume) makes it extremely difficult to appreciate the structure of individual cells, and to measure them accurately in sectioned material. However, this problem is readily overcome by the technique of maceration, which separates the individual cells. Although principally used for the mature, robust cells of the secondary xylem (which

can withstand the harsh treatment involved), techniques also exist for macerating delicate phloem and the extremely fragile cambial cells.

Franklin's technique

This technique (after Franklin, 1945) works for a wide variety (all?) of hardwoods and softwoods. Proceed as follows:

1   Cut pieces of wood into lengths approx. 2 × 2 mm × 3 cm long ('matchsticks') from the particular area of interest.
    Note – cells differ in size, shape, etc. depending on their location, e.g. early-wood versus late-wood (Figure 3D); it is therefore important to be certain of the origin of your sample.
2   Accumulate pieces in a boiling tube.
3   Add approx. 30 ml of a mixture of equal parts glacial acetic acid and 20–30 volume hydrogen peroxide ('macerating solution').
    Note – the usual strength of H_2O_2 is 30%, which corresponds to '100 volume'; dilute

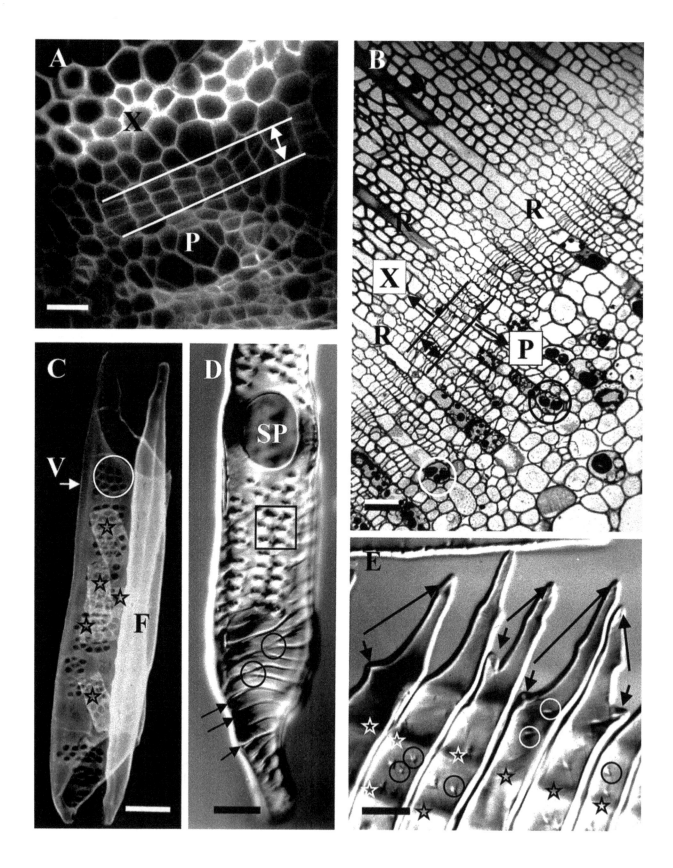

Figure 2 Aspects of the anatomy of the secondary vascular system of hardwood trees. All are from taproot of *Aesculus hippocastanum*, except for C, which is from stem of *Populus tremula* x *P. tremuloides*.
A Epifluorescent image of hand-cut transverse section stained with calcofluor white. Note the thin-walled developing cambium, whose limits are indicated by the parallel lines, and the bright fluorescence of thick-walled xylem cells.
B Transverse section of secondary vascular system prepared for transmission electron microscopy (aldehyde-fixation, osmium-post-fixation, embedded in Spurr resin) and stained with methylene blue/azur A. Note the narrow-diameter cells of the cambial zone (whose limits are indicated by the parallel lines), the radial files of cambial derivatives into the xylem and the phloem, and the dark tannin deposits in the phloem ray cells (encircled regions).
C Confocal laser scanning micrograph of macerated wood cells stained with acridine orange. Several different cell types are visible: five cuboid contact ray cells (☆), a large-diameter vessel element with simple perforation and contact pits (encircled), and a fibre with simple pits.
D Differential interference contrast image of unstained macerated vessel element. Note the tertiary thickenings (arrows), which are occasionally branched (encircled regions), the simple perforation, and the bordered pits (square region) (cf. Figure 2F in the cytoskeletal chapter by Chaffey in this volume).
E Differential interference contrast image of several secondary xylem fibres that have remained together during the maceration process indicating that they are products of the same cambial initial. Note the tapered, pointed tips, which are frequently branched (arrows), the simple pits (encircled regions), and the striations (☆), which are the impressions of ray cells that lay against the fibres within the tree.
Key: F, fibre; P, phloem; R, ray cell; SP, simple perforation; V, vessel element; X, xylem. Scale bars represent: 20 μm (D, E), 25 μm (A), 30 μm (C), and 40 μm (B).

with water as appropriate to get a 20–30 volume solution.

4 Stand the boiling tube in a bath of boiling water until the wood is bleached white, in fume cupboard.
Note – depending on the species and size of samples, this can take several hours. During that period it may be necessary to top up both the macerating solution and the water in the water bath, to replace that lost by evaporation. Note – alternative strategies are to leave the boiling tube and its contents at room temperature (RT) overnight (Baas, 1999); or to heat contents at 60 °C for 1–2 days (Kitin *et al.*, 1999).

5 When the wood is completely bleached (it will appear almost pure white), carefully decant the macerating solution.
Note – from this stage until step 10, perform operations gently, and avoid shaking the sample or premature disaggregation of the cells will result.

6 Gently add 30–40 ml of water to the boiling tube to wash out any remaining macerating solution, repeat 3 or 4 times.

7 Half-fill the tube with water.

8 Add Na_2CO_3 or $NaHCO_3$ to the water, a little at a time, to neutralise any remaining acetic acid.
Note – the solution will effervesce as (bi)carbonate is added; add it little-by-little to ensure that effervescence is not so vigorous that it breaks up the wood at this stage, or

froths over the top of the boiling tube taking your precious wood samples with it!

9 When the effervescence has stopped, decant the liquid and wash the wood 2 or 3 times to remove (bi)carbonate.

10 Finally, add approx. 15 ml of water to the tube, put your thumb over the end, and shake vigorously to break up the wood.

11 The wood-pulp mixture can be viewed immediately, or stored in the refrigerator (for many months).

Microscopic examination of macerate

The contents of the macerate can be viewed unstained, best observed with DIC (Figures 2D, E), or after staining with a variety of reagents for BF microscopy. Alternatively, it can be stained with a variety of fluorochromes and examined with either epifluorescence microscopy or confocal laser scanning microscopy (see below).

Other wood macerating techniques

Other maceration techniques exist in the literature, e.g. sodium hypochlorite (Spearin and Isenberg, 1947 – used by Tuominen *et al.*, 1997), the chromate method (particularly good for retaining clumps of cells so that their common lineage can be appreciated – Wenham, 1966; Wenham and Cusick, 1975), and Jeffrey's technique (e.g. Schmid, 1982). I have only used Franklin's method because it gives such good, clean macerates.

Maceration of cambial cells

A long-standing issue in wood anatomy has been the question of which of the axial cambial derivatives are longer than the fusiform initials (hence indicating extension growth), and by how much. Attempts to answer that question have made use of a variety of techniques, frequently comparing data from macerates of wood cells with measurements of cambial cells from sections. However, use of fixed, embedded material carries with it doubts over the possibilities of cell shrinkage/distortion during processing, and the problems of incomplete identification of whole cells. Cambial cells are far too delicate to withstand the wood maceration techniques considered above; however, Kitin et al. (1999) have recently demonstrated that this tissue can be macerated with gentle enzyme action. That facilitates truer comparisons between all cells, or, at least, the ability to compare like-with-like. This technique is also of interest because, potentially, it permits isolation of a purer cambial cell population for construction of DNA libraries than is usually used (e.g. Hertzberg and Olsson, 1998; Sterky et al., 1998; Label et al., 2000). The published procedure is as follows:

1   Fix excised blocks – approx. 20 mm square – of SVS in formalin: acetic acid: 50% (ethyl) alcohol (FAA, 1:1:18) overnight at RT.
2   Wash blocks in running tap water to remove all free FAA.
3   Cut blocks with a sharp razor blade to give slivers, 5–10 mm radially × 3 mm tangentially × approx. 2 cm long, containing tissues of SVS.
4   Treat slivers with pectinase (2 mg.ml^{-1}) (from stock solution dissolved in 0.1 M Sörensen's phosphate buffer) in an incubator at 35 °C for 2 h, then for 5 h on a shaking platform at RT.
     Note – Sörensen's phosphate buffer made up as follows:
     Solution A: 0.2 M Na$_2$HPO$_4$
     Solution B: 0.2 M NaH$_2$PO$_4$
     Mix 30 ml of Solution A with 70 ml of Solution B, add 100 ml distilled water; pH should be approx. 6.5
     Note – pectinase stock solution made up as follows:
     Dissolve powdered pectinase (0.01 units. mg^{-1}) in ice-cold 0.1 M Sörensen's phosphate buffer, pH 6.5, to give 50 mg. ml^{-1} solution.

5   Stop reaction by replacing enzyme solution with a mixture of 10% acetic acid: 25% ethanol (1:1 v/v).
     Note – samples can be stored temporarily in refrigerator at this stage.
6   Using dissecting needles, and a stereomicroscope, gently separate the phloem tissue from the cambial zone.
7   Gently scrape the cambial cells onto a glass slide and mount in a mixture of equal volumes of water and glycerin.
8   View macerated cambial cells with DIC.

Maceration of phloem

Generally, phloem cells are much neglected in the context of the SVS. However, as derivatives of the cambium they also warrant attention. For a method of macerating phloem cells using 'steam-pressure', see Peterson and Blais (1991).

SECTIONING FIXED MATERIAL

Although hand-cut sections (see above) can be extremely useful, for most detailed anatomical/ histochemical studies there is little substitute for semi-thin sections of fixed material. Whilst the 3-D nature of the SVS is largely lost with thin sections, the thinner the section, the less there will be inaccuracy from obliquely cut cell surfaces, and the better the resolution for detailed cell measurement and photomicroscopy. Over the years, considerable information has been accumulated concerning the anatomy of the SVS. However, most of it has concentrated on the largely dead secondary xylem, which is a remarkably robust material. With more recent focus on the cell biology of, for example, cambial activity and xylogenesis, new techniques have had to be developed in addition to the tried-and-trusted traditional techniques. This section presents an overview of some of the problems involved in fixing and sectioning the SVS, and offers some solutions (see also O'Brien and McCully, 1981; Ruzin, 1999 for more general comments and advice).

I   Fixation

Depending on the goal of the study, a wide range of fixation cocktails can be used for cells of the tree SVS (see the chapters by Chaffey, Funada, Hawkins *et al.*, Rensing, Samaj and Boudet in this volume for detailed protocols for a variety of investigations and formulations). Generally, all material fixed for TEM is suitable for semi-thin sectioning and LM viewing (although, depending on the resin used, the range of stains may be more or less limited).

Most fixatives considered in this volume use glutaraldehyde and/or *p*-formaldehyde (see Ruzin (1999) and Kiernan (2000) for useful reviews of the properties of these fixatives), which can introduce problems such as added fluorescence (e.g. Ruzin, 1999) or excessive denaturation of antigenic sites on proteins potentially compromising critical immunological work. Where fluorescence introduced by aldehydes is a problem, it can be reduced/eliminated with sodium borohydride (e.g. Christensen, 1998).

II Embedding media

A range of embedding media have been used for woody tissues. Old standards, such as paraffin wax, are still in use today. However, new techniques of investigation, such as immunocyto-chemistry, require new embedding media (or rediscovery of old, almost forgotten ones!). The major media used today for woody materials are briefly discussed below.

Tissue orientation

Before embedding material, or fixing it in the cryo-stat, it is necessary to consider the planes of section that are to be cut. Because of the 3-D nature of the SVS, it is necessary to cut transverse – or cross – (TS), radial longitudinal (RLS) and tangential longitudinal (TLS) sections in order to get an appreciation of cell geometry and arrangement in the different tissues (Figures 3A–C provide views of wood in, respectively, RLS, TLS, and TS). Some comments on achieving the desired tissue orientation are provided in the cytoskeleton chapter by Chaffey, and the TEM chapters by Chaffey, and Rensing in this volume.

Unembedded tissues

It is sometimes advantageous not to embed material. Frozen material of the tree SVS has been successfully cut using a cryostat by Funada *et al.* (50 μm thick) for immunolocalisation of cyto-skeleton (see the chapter by Funada in this volume), and for biochemical analysis (25 μm thick) by several workers (see the chapters by Uggla and Sundberg and references therein, and Micheli *et al.* in this volume).

Paraffin wax

This is the classic embedding medium (e.g. O'Brien and McCully, 1981; Ruzin, 1999), which has been used for many years in wood anatomy. It stains with a wide variety of stains for LM, cuts well with steel knives, and forms ribbons. However, although very good for examination of differentiated secondary vascular tissues, the fragile cells of the cambial zone can suffer considerable damage during sectioning (see also the chapter entitled An introduction to the problems of working with trees, in this volume). Furthermore, its relatively high melting point, >50 °C (e.g. Ruzin, 1999) renders it unsuitable for most immunological work, although it has been used successfully for β-glucuronidase (GUS) histo-chemistry of lignification enzymes in poplar stems (see the chapter by Hawkins *et al.* in this volume for a detailed protocol).

Steedman's wax

This wax, a mixture of polyethylene glycol 400 distearate and hexadecanol, is based on the 'polyester wax' developed by Steedman (1957), cuts well, forms ribbons, and stains with as wide a range of stains as paraffin wax. However, its lower melting point (approx. 37 °C), although useful for certain immunological work in the tree SVS (e.g. Chaffey *et al.*, 1996), makes it a softer embedding medium than paraffin. Consequently, preservation of the cambial zone and derivatives at early stages of dif-ferentiation is even more problematical (see Chaffey *et al.*, 1997c; and the cytoskeletal chapter by Chaffey in this volume).

LR white

This relatively low viscosity acrylic resin is liquid at RT, penetrates the cells of the SVS well, is avail-able in several grades of hardness, and stains readily with a wide range of LM stains. Although it does not form ribbons, the blocks of polymerised resin are much harder than wax, permitting cutting of much thinner sections, which can be used for TEM (e.g. Figures 4E–G in the TEM chapter by Chaffey, Figures 8, 10–13 in the chapter by Micheli *et al.*, and Figures 1, 2 in the chapter by Samaj and Boudet, in this volume), and which preserve the detailed anatomy of the cambial zone much better. It can be cured by heating, or at RT with added 'accelerator' (e.g. Ruzin, 1999), or by UV at low temperature, e.g. –20 °C (see the TEM chapter by Chaffey in this volume). Depending on the antigen being examined, heat-cured material may be suitable for immunolocalisation. However, because it is not easy to remove the embedding medium, immunolocalisation is largely limited to antigens that are exposed at the cut surface of sections, such as those associated with cell walls or the plasmalemma. LR White-embedded SVS of hybrid aspen has also been used successfully for high resolution in situ hybridisation studies (see the chapter by Regan and Sundberg in this volume and their Figure 3). A wide range of similar resins is now available, e.g. LR Gold and Unicryl (e.g. Ruzin, 1999), which are probably of greater value for immunogold work with the TEM; they largely remain to be explored with woody tissues.

Glycol methacrylate

This is a good medium for general wood anatomy (e.g. Wetzel et al., 1989); it stains with a wide variety of LM stains (Feder and O'Brien, 1968), and can readily be cut from 0.4 μm (Feder and O'Brien, 1968) up to 3 μm (Ruzin, 1999) thick. It is commercially available under such names as HEMA, JB-4, Historesin, and Technovit 7100 (Ruzin, 1999). Although it can be polymerised at RT (Feder and O'Brien, 1968), for immunological work it is of limited value since the resin cannot be removed from the sections without damaging the tissues (Ruzin, 1999), cf. butyl-methylmethacrylate (below). However, it has been used quite successfully in GUS histochemistry of lignification enzymes in woody stems (see the chapter by Hawkins et al. in this volume for a detailed protocol).

Butyl-methylmethacrylate (BMM)

This acrylic resin seems almost to be the ideal embedding medium for wood anatomical and immunological work at the LM level. Although it does not section thinly, or well, enough for TEM, it has properties similar to those of LR White and glycol methacrylate. However, the major advantage of BMM over those other two acrylic resins is that polymerised resin can be removed from sections with acetone (see the cytoskeleton chapter by Chaffey in this volume for further details), permitting access of antibodies to antigenic sites that

were previously inaccessible. For BF-LM, it cuts well at 1–3 μm; for immunolocalisation work, it cuts well at 6–10 μm, allowing an appreciation of the 3-D arrangement of such features as the cytoskeleton; and can be cut up to 30 μm thick – Berlyn, 1963). It has so far permitted immunolocalisation of α-(Figures 1E-H, Figures 1A, B, E, F in the cytoskeletal chapter by Chaffey in this volume) and β-tubulins, post-translationally modified tubulins, myosin, actin (Figures 1C, D in the cytoskeletal chapter by Chaffey in this volume), callose (Figure 1G) and various cell wall/plasmalemma-located antigens (Chaffey et al., 1997a, b, c, 1998, 1999, 2000a, b; Chaffey 2000; Chaffey and Barlow, 2000) in the tree SVS. Success reported by Evans et al. (1997) using a Golgi apparatus antibody in pine embryo tissues, and by Kronenberger et al. (1993) for in situ hybridisation in Arabidopsis, suggest that BMM will have wide, general applicability for cell biological work in trees.

Spurr resin

This epoxy resin is excellent for conventional TEM of woody tissues (see the TEM chapters by Rensing, and Chaffey in this volume), and for LM of charcoals and dry ancient woods (Smith and Gannon, 1973). However, problems of antibody-penetration into the resin and the need to polymerise by heating limit its use for immunolocalisation at the TEM level. Although it yields high-quality semi-thin sections for LM, its utility is limited by the restricted range of stains available. It can be formulated in several degrees of hardness, but when first promoted by Spurr (1969), it had to be made up in the laboratory, by weighing out the components. This is still performed in some laboratories today (see the chapter by Rensing in this volume); however, premix kits are marketed by such companies as TAAB in the UK. These kits require only that the three, pre-weighed liquid components are mixed together and the resin is ready for use. Kits are available in a variety of grades of hardness and have the advantage of being both more convenient than weighing out the ingredients each time, and of being of a consistent, reproducible formulation. However, this resin is a suspect carcinogen; Epon is an alternative that is more acceptable to many users.

III Sectioning

Note – although it is stressed more than once in the relevant chapters that follow, it is worth repeating the following words of caution here: woody tissues are extremely tough and will quickly blunt whatever knife is used to cut sections. Some cells of the SVS also contain crystal deposits (e.g. Bailey, 1910; Brown, 1919; Rao and Dave, 1984), which can cause serious damage to glass (and diamond!) knives.

There are many different ways to cut woody tissues, depending on the method of preparing the samples, which in turn depends upon the investigation being undertaken, e.g. from approx. 60 nm for TEM up to several hundred μm for confocal laser scanning microscopy (CLSM).

Razor or razor blade (hand sections)

Sections of relatively soft material can be cut with care and practice using a cut-throat razor or single- or double-edged razor blades. The main problems with this technique are that it is difficult to judge the thickness cut, to achieve reproducibility of thickness from section to section, and to cut sections thin enough to use transmitted light techniques. A razor blade is also recommended by Exley *et al.* (1974) for preparing wood blocks for SEM examination.

Vibrating microtome

This has the advantage of reproducibility of section thickness, but is limited to relatively soft material, for example, sectioning lightly-fixed, or fresh, material prior to immunolocalisation, *in vivo* examination, or reporter gene-histochemistry (e.g. the chapter by Hawkins *et al.* in this volume).

Sliding or sledge microtome

Traditionally used in wood anatomy laboratories for sectioning large pieces of wood (which, due to their inherent rigidity, can be sectioned without embedment) (see e.g. Garland, 1935), as an alternative to the rotary microtome (see next). It is still used today, e.g. in the recent CLSM study by Donaldson and Lausberg (1998), and for cryo-sectioning for immunolocalisation studies (see the chapter by Funada in this volume).

Rotary microtome

Using steel or glass knives, this is ideal for a variety of embedding media, particularly waxes such as paraffin or Steedman's (which has been used for immunolocalisation studies in the SVS – e.g. Chaffey *et al.*, 1996; see also the cytoskeleton

chapter by Chaffey in this volume). One of the problems with trees is that they can become rather large, which generally limits microscopical/ anatomical examination to small regions. An intriguing variation on the rotary microtome theme is the 'macrotome' (Lucansky, 1976), which can overcome some of the shortcomings of more conventional microtomes. This device is a modified meat slicer and has been used to cut whole sections of plant material up to 15.2 cm in diameter and up to 30.5 cm in length.

Ultratome

This is standard equipment in any TEM laboratory for cutting ultrathin sections, and is also extremely useful for cutting semi-thin sections for LM studies, or thicker sections (up to 10 μm) for immunological studies (see the cytoskeletal chapter by Chaffey in this volume) of resin-embedded tissues. Glass or diamond knives are needed.

Cryotome

Using either stainless steel or glass knives, this device has been used to considerable effect for longitudinal sectioning of the tree SVS for biochemical (see the biochemistry chapter by Magel, and the chapter by Micheli *et al.* in this volume), hormone (see the chapter by Uggla and Sundberg in this volume), and molecular-genetical (Hertzberg and Olsson, 1998) analyses.

IV Section adhesion

Sections can usually be successfully stuck onto untreated glass microscope slides by drying down on a drop of water over a hot plate at 60 °C for approx. 30 min. However, if subsequent prolonged washing or staining stages are anticipated, or if this sort of temperature is too high for the study being performed, it may be desirable to use slides that have been coated with an adhesive. However, the adhesive may contribute to 'background' staining, so should be chosen having regard to the particular investigation to be undertaken.

Note – background staining can also be contributed by the embedding medium; where possible it may be advantageous to use a removable embedding medium, such as BMM.

Adhesives and methods of adhesion used with success on sections of woody material include:

Mayer's egg albumen

This works very well for sections embedded in Steedman's wax or BMM, up to 10 μm thick, which are subsequently used for immunolocalisation of a wide range of proteins. Preparation details are provided in the cytoskeletal chapter by Chaffey in this volume (or see Clark, 1981).

Poly-L-lysine

This has not proved to be successful for sections embedded in BMM, but is suitable for LR White-embedded material (see the chapters by Hawkins *et al.*, and Micheli *et al.* in this volume for further details).

Gelatin

Although this is not generally used for woody tissues, it should be suitable (per O'Brien and McCully, 1981), and has been used by Kobayashi *et al.* (1994) for immunolocalisation of cytoskeletal proteins in BMM-sections of flax leaves (A. Hardham, personal communication).

Biobond

This has been successfully used for LR White-embedded sections (see the chapters by Hawkins *et al.*, and Samaj and Boudet in this volume for further details).

Silanisation

Silanised glass microscope slides have been used successfully for LR White-embedded sections for *in situ* hybridisation (see the chapter by Regan and Sundberg in this volume), a technique that has numerous lengthy aqueous washing steps.

Others

If none of the above is suited to your particular investigation, it might be worth consulting the articles by Fink (1987a, b, c) or the relevant section in the book by Ruzin (1999) for inspiration.

V Mountants

For viewing and recording purposes, sections are usually mounted in a liquid medium beneath coverslips, choice of both depends on the type of microscopic investigation pursued.

Non-fluorescence studies

For most studies using BF, DIC or polarising light, it may be sufficient to mount the sections in water for as long as viewing is required. The water can then be removed, the sections dried and stored for later reviewing. This can be advantageous for toluidine blue-stained sections that may require the colours to be 'resuscitated' by application of water (see O'Brien and McCully, 1981, who suggest breathing on the sections to achieve this).

If permanent mounts are desired, a range of media are available, e.g. DPX (see the chapters by Beardmore *et al.* and Hawkins *et al.* in this volume), Eukitt, and Canada balsam (see the chapter by Micheli *et al.* in this volume). Glycerol has been used for cryo-tangential sections (see the chapter by Uggla and Sundberg in this volume).

Fluorescence studies

A major problem in fluorescence work, particularly immunofluorescence studies, is the loss of fluorescence with time (e.g. also Johnson *et al.*, 1982; Longin *et al.*, 1993). This fading can be retarded by use of anti-fading mountants. A wide range is commercially available. One that has proved particularly effective is Vectashield® (see the cytoskeletal chapter by Chaffey, and the chapter by Micheli *et al.* in this volume). Laboratory-made *p*-phenylenediamine can also work well (see the cytoskeletal chapter by Chaffey in this volume). However, in choosing any mountant, and coverslip, it is necessary to be mindful of their relative refractive indices (see Table 1 in Bacallao *et al.*, 1995; see also comments in Jang *et al.*, 1992) in order to get the best out of the investigation.

LIGHT MICROSCOPY OF SEMI-THIN SECTIONS

This is useful – and essential! – for viewing semi-thin sections from block of material prior to ultra-thin sectioning for TEM, for assessing suitability of fixation, embedding, and location for TEM and LM. It is also useful for correlative LM and TEM work since sequential ultra-thin and semi-thin sections can be taken, thus permitting comparison of areas that are nearly the same with a variety of stains or treatments.

Unstained material

UV microscopy

This is particularly useful for examining progression of lignification during differentiation of wood cells (e.g. Murakami *et al.*, 1999), since lignin has a very strong absorbance in the UV (see the chapter by Takabe in this volume for further details).

DIC

This imaging technique can be particularly useful for semi-thin sections, as advocated by Isebrands and Larson (1973) for studying fusiform cambial cell division (they used 1–2 μm LSs of glutaraldehyde-osmium-fixed poplar internodes, embedded in Epon-araldite). See also Figure 5 A, C, D and F in the chapter by Beardmore *et al.* in this volume.

Polarised light

In both TS and LS, this is an extremely useful technique for examining stages in wall development during the differentiation of secondary vascular tissues. If used in conjunction with sections in which microtubules have been immunolocalised, it can be used in a correlative way to provide information relating microtubule orientation, or density, to stages of cell wall formation. For further information, see also fresh, hand sectioning above.

Note – DIC can also give a birefringent image similar to that obtained with polarised light (e.g. Riding and Little, 1984).

Stained material

Bright field LM comes into its own with stained semi-thin sections. A wide range of staining combinations have been used for such sections (e.g. O'Brien and McCully, 1981; Ruzin, 1999). Some that have proved particularly useful for examining the tree SVS are detailed below. It is stressed that it is well worth the trouble of experimenting with other stains; some suggestions are included at the end of this section.

Toluidine blue

This works well for sections embedded in Spurr epoxy resin, acrylic resins (LR White and BMM) (Figure 1D), and Steedman's wax. Although for 'critical work' one should ideally use 0.05% toluidine blue in benzoate buffer at pH 4.4 (Feder and O'Brien, 1968), a 0.01% (w/v) aqueous solution is suitable for general study and image recording (see also Graham *et al.*, 1987). It is, however, worth noting that Feder and O'Brien (1968) mention that use of other buffers at pH 4.4, or even distilled water, give the same results as the benzoate-solution.

Methylene blue/azur A

This works well with sections embedded in the same media as for toluidine blue, and gives the same colour reactions (Figures 1B, 2B). The recipe is derived from that of Humphrey and Pitman (1974) (see also Clark, 1981). The staining solution is made up as follows:

Methylene blue	0.13 g
Azur A (Sigma, No. A2918)	0.02 g
Glyerol	10 ml
Methanol	10 ml
Phosphate buffer, pH 6.9	30 ml
Distilled water	50 ml

The phosphate buffer is made as follows:

Anhydrous KH_2PO_4	9.078 g
Anhydrous Na_2HPO_4	11.876 g
Distilled water	1l

1   Collect semi-thin sections on a drop of water on an uncoated glass microscope slide.
2   Heat-dry sections for 30 min–1 h on a hot plate at approx. 60 °C.
3   Remove slide from hot plate, add a few drops of stain.
    Note – keep stain solution in a syringe fitted with a filter to prevent large particles of stain contaminating the section.
4   Replace slide on hot plate for approx. 20 sec (but do not let stain solution dry out!).
5   Wash excess stain from slide with gentle stream of water (use a pipette).
    Note – wash sections *thoroughly* to remove deposits of unbound stain.
6   Dry sections on hot plate.
7   Store stained slides unmounted in refrigerator at 4 °C until ready for viewing.
    Note – slides can be kept in good condition in a refrigerator for many months (probably for years).

8   To view, mount sections beneath cover slip
    with a drop of water.
9   After viewing, remove cover slip, dry slide,
    and store in refrigerator.
    Note – unlignified walls appear purple;
    nuclei appear blue, nucleoli appear purple/
    dark blue; tannins appear green; lignified
    walls appear blue-green.

Astra blue

Used on BMM-sections (without removal of resin).
Apply a drop of solution (made up as 0.5 g astra
blue in 100 ml of aqueous 2% (w/w) tartaric acid –
Harkes, 1973) to sections, heat for 10 sec at approx.
60 °C, wash off excess stain and view in BF. It has
some specificity for cellulose, and will stain non-
lignified cell walls blue (useful as a contrast to
safranin-staining, see below).

Safranin

Safranin is the classic wood stain (e.g. Garland,
1935). In the version described below it has been
used on BMM-sections (without removal of resin).
Apply a drop of 1% aqueous safranin solution to
sections, heat for 10 sec at approx. 60 °C, wash off
excess stain and view in BF. This will stain walls of
all cells of the SVS red. However, if such sections
are subsequently stained with astra blue (see
above), safranin will be displaced from non-
lignified walls, which will become blue. It requires
trial-and-error to get the timings of each stain
application sorted out, but the red-blue contrast
can be quite pleasing (and particularly good for
visualising the non-lignified, cellulosic gelatinous
(G)-layer in otherwise lignified tension fibres). See
the chapter by Hawkins *et al.* in this volume for
staining protocols for paraffin wax and glycol
methacrylate sections.

Iodine test for starch

Applied to BMM-sections (without removal of
resin), this test gives good staining of starch grains.
Add a drop of I₂/KI solution (see 'fresh, hand
sections' above) to sections, heat for 10 sec at approx.
60 °C, wash off excess stain and view in BF.

Feulgen-staining of nuclei

One of many interests in nuclei of cells of the SVS
is the question of nuclear genome size, whether

endopolyploidy accompanies xylem differentiation
(e.g. Dodds, 1981), or if DNA content varies in initials during the cambial seasonal cycle (e.g. Zhong
et al., 1995). To detect changes in DNA levels, the
nuclei can be appropriately stained and examined
in a microdensitometer. The following protocol
uses the Feulgen reaction to stain nuclei in cells of
the SVS. It has been developed for BMM-embedded material of *Aesculus hippocastanum* fixed in
p-formaldehyde.

1   Place 8–10 µm sections in a drop of water on
    Mayer's albumen-coated microscope slides
    (see the cytoskeletal chapter by Chaffey in
    this volume for formulation of adhesive),
    leave overnight at approx. 40 °C.
    Note – for microdensitometry, section thick-
    ness needs to be adjusted so as to include
    whole nuclei.
    Note – using poly-L-lysine as section-adhesive
    resulted in unacceptably high loss of sections
    from the slides during processing.
2   Remove BMM resin with acetone (try 15 min
    to begin with; extend time if required for full
    removal of resin).
    Note – if BMM resin was not removed, there
    was little nuclear staining, but substantial
    staining of lignified cell walls. Regardless of
    whether resin was removed or not, tannins
    (mainly present in phloem ray cells) stained
    red.
3   Wash sections in distilled water for 2 min.
4   Hydrolyse nucleic acids in nuclei with 5 M
    HCl for approx. 45 min at RT.
    Note – it may be necessary to experiment
    with hydrolysis times for your own material
    (see Fox (1969) for further information on this
    point).
5   Wash sections with distilled water for 2 × 2
    min.
6   Stain with Schiff's reagent (Sigma, No.
    395–2–016) for approx. 30 min at RT.
7   Wash sections in tap water for 3 × 1 min.
8   Stained sections can be stored in a
    refrigerator for later viewing.

A similar technique has been used by
Mellerowitz and co-workers to examine nuclear
genome size in softwoods (Mellerowicz *et al.*, 1989,
1992) and hardwoods (Zhong *et al.*, 1995).

Safranin-gentian violet-double-staining of epoxy resin sections

In general it is difficult to find satisfactory staining procedures for material embedded in epoxy resin. Both methylene blue/azur A and toluidine blue work well (see above), and a few other staining combinations that might be worth investigating are mentioned below. Another staining combination that has been successfully used with Spurr resin-embedded sections that have been prepared for TEM with glutaraldehyde-fixation and osmium-post-fixation is safranin followed by gentian (crystal) violet (Kitin *et al.*, 1999). The procedure is as follows:

1   Stain 1 μm-thick sections by floating on a 1% (w/v) aqueous solution of safranin for 30 min–1 h.
2   Wash with several changes of distilled water.
3   Float the sections on a 1% (w/v) aqueous solution of gentian violet for 30 min–1 h.
    Note – if high-contrast is required, stain at 45 °C.
4   Wash with several changes of distilled water.
5   Place sections on uncoated glass microscope slides, and dry down at 35–45 °C for 1 h.
6   Apply a drop of xylene to the sections.
7   Mount with a drop of Biolet mountant (Okenshoji Co.).
    Note – it is likely that several other mountants will be suitable for preparing permanent mounts of these sections.
    Note – gentian violet-staining tends to fade with time.

Although no clear selective (specific) staining of cell features is observed, cambial cell walls can be stained more intensely than with safranin alone, giving high contrast to these otherwise poorly contrasted cells.

Other stains worth trying for epoxy resin sections

Hayat (1993) provides details of the following staining combinations for epoxy-resin-embedded sections of plant tissues processed for TEM (i.e. aldehyde-fixed, osmium-post-fixed):
Ponceau 2R-periodic acid (after Gori, 1978) – for proteins;
Thionin-acridine orange (after Paul, 1980) – for a variety of cell and wall features;

Periodic acid-Schiff-toluidine blue O (after Yeung, 1990) – for lipids.

As far as I am aware, none of these three have yet been applied to woody material, although they all have promise, particularly for investigating seasonal fluctuations in amounts and distribution of reserve products.

Fineran (1997) gives a detailed protocol for staining lignin in TEM-prepared epoxy resin sections of woody tissues using the chlorine water/ethanolamine-silver nitrate method of Coppick and Fowler. Sections so stained can be viewed by LM or TEM.

Other stains for glycol methacrylate sections

Wetzel *et al.* (1989) stained proteins with 0.2% amido black in 7% acetic acid for 5 min. Proof of identity of protein bodies was obtained by elimination of amido black-staining of sections digested with pepsin.

Epifluorescence microscopy

Calcofluor

For references and staining details, see fresh, hand sections above. With calcofluor, nascent division cell walls in cambial zone appear to stain brighter than mature cell walls; sieve pores of sieve elements also stain well (Figure 1A). This fluorochrome is particularly useful for demonstrating early stages of wall deposition in association with wall elaboration in vessel elements (see Figure 2C in Chaffey *et al.*, 1997b). Another UV-excited fluorochrome that is worth investigating in this context is bisbenzimide (33258 Hoechst) (Hernández and Palmer, 1988), which also stains cell walls particularly well.

Aniline blue

The staining solution is made up as: 0.01% (w/v) aniline blue WS (water-soluble) in 0.0667 M (i.e. M/15) phosphate buffer, pH 10 (recipe from Clark, 1981). The solution remains usable for many months if kept in the refrigerator. In the SVS of hardwoods, this stain is particularly useful for identifying nascent division walls in the cambial zone and the sieve areas of sieve elements. In both instances, this is due to the callose present at those sites.

Note – although largely considered to be a UV-excited fluorochrome, it is advisable not to stain sections with aniline blue if you are using FITC for immunolocalisation studies since aniline blue exhibits similar fluorescence to FITC under FITC epifluorescent illumination conditions.

In view of differences in purity between batches, concerns over the specificity of aniline blue as a callose (β-1,3-linked glucan) stain have been expressed. The agent responsible for the staining reaction has been isolated and is marketed separately – sirofluor (e.g. Stone *et al.*, 1984; Nickle and Meinke, 1998). However, in the tree SVS the same staining image obtained with aniline blue is also seen with the anti-callose antibody (Figure 1G), supporting the view that aniline blue is a reliable guide to presence of callose. It is also worth mentioning the paper by Alché and Rodríguez-García (1997) which indicates that a wide range of fluorochromes – including acridine orange and DAPI – also have some affinity for callose.

Calcofluor/aniline blue

Sections are stained first with calcofluor (see above), then counter-stained with aniline blue (see above). The aniline blue displaces calcofluor from sites of callose (Jefferies and Belcher, 1974; Hughes and McCully, 1975); the yellow fluorescence of aniline blue being contrasted by the blue-white of calcofluor. This can be particularly useful for visualising sieve pores within sieve elements (Figure 1A).

DAPI-staining of DNA

DAPI (4',6-diamidino-2-phenylindole)-staining of nuclear DNA (e.g. Goodbody and Lloyd, 1994) works well in both BMM- and wax-embedded sections after removal of the embedding medium. Use at 1 μg.ml^{-1} in phosphate-buffered saline, stain for 2 min, wash off excess stain. It is a good stain for nuclear DNA, but not, apparently, for plastid DNA (Johansen *et al.*, 1995), cf. propidium iodide (see the confocal laser scanning microscopy section below). DAPI is a UV-excited fluorochrome, which can be examined under the same conditions as used for calcofluor viewing.

Propidium iodide-staining of nucleic acids

This stain has been used with BMM-embedded material, after removal of resin. Use at 2 μg.ml^{-1} in

phosphate-buffered saline, stain for 2 min, wash off excess stain. Staining of the nucleolus is intense, staining of the rest of nucleus less so. This is a visible wavelength-excited fluorochrome (cf. DAPI), and is suitable for use in confocal laser scanning microscopy (see section below).

Note – it is possible sequentially to 'triple-stain' resin-free BMM-embedded sections with DAPI, propidium iodide, and then calcofluor.

Immunofluorescence

This technique is becoming more used in cell biological study of the tree SVS. See the chapters by Chaffey, Funada, and Micheli *et al.* in this volume for examples and detailed protocols.

CONFOCAL LASER SCANNING MICROSCOPY

Over the past 10 years, the CLSM has taken its place as a valuable research tool for a wide variety of cell biological investigations (e.g. Running and Meyerowitz, 1996; Hepler and Gunning, 1998; see also the cytoskeletal chapters by Funada, and Chaffey in this volume for further references), particularly for fluorescence imaging. Although other techniques are available to cope with some of the limitations inherent in traditional fluorescence microscopy, such as wide field deconvolution microscopy (e.g. Scanalytics, Inc., 1997; Ruzin, 1999), CLSM appears to have captured the imagination and is the more widespread technique. For further details concerning the theory and practice of CLSM, interested readers are directed to the following books and articles: Shotton and White, 1989; Pawley, 1990, 1995; Knebel and Schnepf, 1991; Jang *et al.*, 1992; Matsumoto, 1993; Running and Meyerowitz, 1996; Sheppard and Shotton, 1997; Hepler and Gunning, 1998; Paddock, 1999; Ruzin, 1999; Wymer *et al.* 1999.

Although it will not replace the SEM, which has much better resolution, the optical sectioning ability of the CLSM can give images of relatively straightforward objects, such as transverse sections of wood (Figure 3C), which have a 3-D quality similar to those obtained by SEM. Furthermore, the optical sectioning ability of the CLSM, and the opportunity to see inside a structure, is a major advantage over the largely blind SEM (see discussion in Knebel and Schnepf, 1991).

Some examples of the application of the CLSM to tree biology are briefly considered below.

Cytoskeleton

A particularly good example in wood biology is the use of the CLSM in helping to elucidate the role of the cytoskeleton during secondary vascular differentiation of gymnosperms and angiosperm trees, using indirect immunofluorescence of microtubules and microfilaments (Abe *et al.*, 1995a, b; Chaffey *et al.*, 1997a, 1998, 2000a, b; Funada *et al.*, 1997, 2000; Furusawa *et al.*, 1998; Chaffey, 2000; Chaffey and Barlow, 2000; and the chapter by Funada in this volume and figures therein). The native autofluorescence of lignified wood cell walls can be used to good effect as a contrast to the fluorescence of the immunolocalisation system when 488 nm (for FITC) and 568 nm laser lines are used (see Figures 1F, H).

Cell wall composition

Ermel *et al.* (2000) have used CLSM for visualisation of the distribution of polysaccharide epitopes localised with the antibodies LM5 (galactan side-chains) and LM6 (arabinan side-chain of rhamnogalacturonan I) in relation to secondary vascular differentiation in poplar.

Wood-cell morphology

In a dramatic demonstration of the power of the CLSM technique, Knebel and Schnepf (1991) used sectioned material from pine and clematis, for studying secondary vascular tissues. The material was either stained with periodic acid-Schiff's reagent (if investigating lignified and non-lignified cells), or Schiff's reagent alone (for lignified cells only), and imaged with the 488 nm laser line. Donaldson and Lausberg (1998) compared CLSM and transmitted light images of wood for analysis of wood-cell dimensions, and found CLSM to be superior. They studied 20 μm-thick, sledge microtome-sections of radiata pine, stained with 0.1% aqueous safranin, and imaged with the 568 nm laser line.

A recent interesting use of CLSM is described by Kitin *et al.* (1999). Using acridine orange-stained, FAA-fixed, 60–80 μm thick cryo-tangential sections of *Kalopanax* cambium, they were able to measure the lengths of intact cambial initials *in situ*. A similar CLSM technique (Prior *et al.*, 1999) makes use of the fluorescence introduced by aldehyde-fixation to image up to 200 μm into resin-embedded blocks of plant material (at 488 nm). The fluorescence image was significantly enhanced with safranin-staining and excitation at 568 nm. Both these techniques should permit 3-D imaging of, and assist in the reconstruction and interpretation of the *in situ* cell-cell arrangements within, the tree SVS.

CLSM is also playing an important role in the analysis of pulped wood-fibres. Jang *et al.* (1992) studied unbleached kraft pulp of Western Canadian spruce, stained with 0.001% (w/v) aqueous solution of acridine orange, and imaged at 488 nm (see also Figure 2C). Martinez-Nistal *et al.* (1998) examined Texas red-stained eucalyptus and acacia pulp fibres. In view of the CLSM-imaging of sectioned wood stained with safranin (Figures 3A–C), it is likely that this dye will also be useful for staining and imaging macerated cells.

Miscellaneous applications

The CLSM has recently been used in a variety of studies concerned with wood degradation and microbiology:

Singh *et al.* (1997): glutaraldehyde-induced fluorescence of wood-degrading fungi (making use of the fact that fluorescence induced by glutaraldehyde fixation is primarily confined to living – or at least largely proteinaceous – cells (Ruzin, 1999), hence dead wood cells do not 'stain' with glutaraldehyde, but living fungal hyphae do);

Kim and Singh (1998): imaging degraded wood without additional staining (largely making use of native lignin autofluorescence);

Xiao *et al.* (1998a): immunofluorescent labelling of sapstain fungus in radiata pine;

Xiao *et al.* (1998b): visualisation of bacteria in wood using a fluorescent lipid probe.

The flow paths within radiata pine wood were studied by impregnating the material with fluorescein and imaging with a CLSM (Matsumura *et al.*, 1998).

Verbelen and Stickens (1995) demonstrated that it was possible to determine cellulose microfibril-orientation within walls of *living* cells using CLSM and polarised light-excitation of Congo red-stained walls. Although to my knowledge this has not yet been attempted in woody cells, in combination

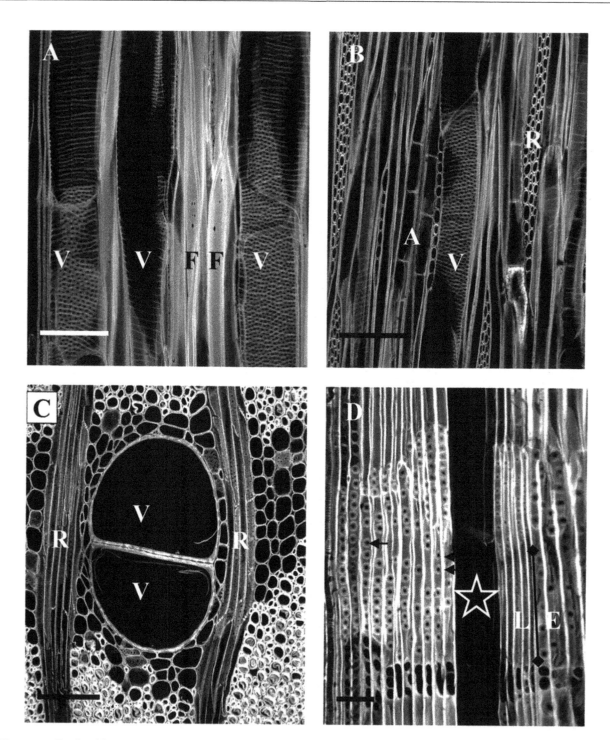

Figure 3   Confocal laser scanning microscopy images of wood. A, B and C are safranin-stained, D is a commercially-prepared teaching slide; all excited with the 568 nm laser line.
A Radial longitudinal section of lime, showing fibres and vessel elements.
B Tangential longitudinal section of lime, showing vessel element, axial parenchyma and cross-sectioned ray cells.
C Transverse section of iroko, showing two large-diameter vessel elements, surrounded by smaller-diameter fibres and parenchyma cells, and flanked by two xylem rays.
D Radial longitudinal section of pine. Note the resin duct (star), the tracheids with their prominent bordered pits (arrows) and the distinction between larger-diameter earlywood tracheids and narrower-diameter latewood tracheids.
Key: A, axial parenchyma; E, earlywood; F, fibre: L, latewood; R, xylem ray cell; V, vessel element. Scale bars represent

with immunolocalised tubulin it potentially offers a method for examining whether any correspondence exists between the orientation of cortical microtubules and that of recently-deposited microfibrils. Similarly, it should be possible to examine this using a UV laser to excite calcofluor-stained cells.

Whilst the usefulness of the CLSM in research applications is undeniable, it also has a valuable role to play in general education. In the context of wood cell biology, safranin-stained wood sections exhibit a strong fluorescence when excited with 568 nm light, which image well in the CLSM (Figures 3A–C). Commercially prepared teaching slides (staining combination unknown) can also provide quite dramatic images when imaged in the CLSM (Figure 3D). And sections stained with methylene blue/azur A also give a respectable fluorescence signal when irradiated at 568 nm.

Propidium iodide-staining of nucleic acids for CLSM

This stain also works well for epifluorescence visualisation of nucleic acids (see entry above for staining details); for CLSM it is viewed with 568 nm excitation. It is particularly useful in CLSM work of the tree SVS as a counter-stain to FITC-secondary antibodies (Figure 1E). Furthermore, unlike DAPI, it is possible that it also stains DNA-containing organelles. Figure 1F shows nuclear staining with propidium iodide, plus numerous red-fluorescing bodies. By reference to their size (approx. 1 μm), and their similarity to the image of carbocyanine dye-stained mitochondria in Olyslaegers and Verbelen (1998), it is suggested that these structures are mitochondria (or very small plastids). A range of other nuclear stains for CLSM is considered in Matsuzaki *et al.* (1997).

MISCELLANY

The majority of the techniques considered above concentrate on small areas of cells or tissues, and provide considerable detail about the micro-anatomy of the tree SVS. However, it is all too easy to miss the overall picture. Although largely 2-dimensional, a technique which can help provide information of cell geometry and cell-cell arrangements is the use of nail varnish to make impressions of exposed surfaces of different tissues of the tree SVS (Balasubramanian, 1979).

Such strips can be readily viewed in the LM and kept for reference and re-analysis. The SEM also has an important role in providing this bigger picture, especially when blocks are prepared to permit viewing of all three planes at the same time (Exley *et al.*, 1974).

Another technique which is enjoying something of a renaissance at present, and which gives invaluable insight into the 3-D nature of wood, when combined with SEM, is 'micro-casting' (Stieber, 1981; André, 1993, 2000; Fujii, 1993). Another solution to the problem of 3-D analysis of the SVS is the video-imaging system described by Lewis (1995) which accumulates images from serial sections of woody material, for subsequent reconstruction. However, although it represents a relatively cheap solution to the question of 3-D analysis, it has limitations which make it likely that it will be overtaken by use of CLSM of appropriately treated material.

CONCLUDING COMMENTS

There is probably a technique for every conceivable study of the tree SVS. If there isn't, it should be possible to adapt one that is more commonly used on herbaceous material. Experiment!

ACKNOWLEDGEMENTS

I thank Prof. J.R. Barnett for his version of Franklin's maceration technique; Ms Hazel Willott for drawing my attention to the iron alum-staining method; Dr P.W. Barlow for his version of the Feulgen reaction; Dr R. Funada for providing the details of the cambium maceration protocol and the safranin-gentian violet-staining of epoxy resin sections; and Dr M. Wenham for loan of his safranin-stained slides.

REFERENCES

Abe, H., Funada, R., Imaizumi, H., Ohtani, J., Fukuzawa, K. (1995a) Dynamic changes in the arrangement of cortical microtubules in conifer tracheids during differentiation. *Planta*, **197**, 418–421.

Abe, H., Funada, R., Ohtani, J. and Fukuzawa, K. (1995b) Changes in the arrangement of microtubules and microfibrils in differentiating conifer tracheids during the expansion of cells. *Annals of Botany*, **75**, 305–310.

Abe, H., Funada, R., Ohtani, J. and Fukuzawa, K. (1997) Changes in the arrangement of cellulose microfibrils associated with the cessation of cell expansion in tracheids. *Trees*, **11**, 328–332.

Alché, J.D. and Rodríguez-García, M.I. (1997) Fluorochromes for detection of callose in meiocytes of olive (*Olea europaea* L.). *Biotechnic and Histochemistry*, **72**, 285–90.

André, J.-P. (1993) Micromoulage des espaces vides intra- et intercellulaires dans les tissus végétaux. *Comptes Rendus de l' Academie des Sciences, Sciences de la Vie*, **316**, 1336–1341.

André, J.-P. (2000) Heterogeneous, branched, zigzag and circular vessels. In *Cell and Molecular Biology of Wood Formation*, edited by R. Savidge, J. Barnett and R. Napier, pp. 387–395. Oxford: Bios Scientific Publishers.

Averall, J.L. (1926) Suggestions to beginners on cutting and mounting wood sections for microscopic examination. *Journal of Forestry*, **24**, 791.

Baas, P. (1999) Book review. *IAWA Journal*, **20**, 394.

Bacallao, R., Kiai, K., Jesaitis, L. (1995) Guiding principles of specimen preservation for confocal fluorescence microscopy. In *Handbook of Biological Confocal Microscopy*, 2nd edn, edited by J.B. Pawley, pp. 311–325. New York, London: Plenum Press.

Bailey, I.W. (1910) Microtechnique for woody structures. *Botanical Gazette*, **49**, 57–58.

Balasubramanian, A. (1979) Improved imprinting technique for study of plant tissues. *Stain Technology*, **54**, 177–180.

Berlyn, G.P. (1963) Methacrylate as an embedding medium for woody tissues. *Stain Technology*, **38**, 23–28.

Berlyn, G.P. and Mischke, J.P. (1976) *Botanical Microtechnique and Cytochemistry*. Ames, Iowa: The Iowa State University Press.

Brown, F.B.H. (1919) The preparation and treatment of woods for microscopic study. *Bulletin of the Torrey Botanical Club*, **46**, 127–150.

Butt, T.M., Hoch, H.C., Staples, R.C. and St Leger, R.J. (1989) Use of fluorochromes in the study of fungal cytology and differentiation. *Experimental Mycology*, **13**, 303–320.

Butterfield, B.G. (ed.) (1998) *Microfibril Angle in Wood*. Canterbury, New Zealand: IAWA/IUFRO.

Carlquist, S. (1988) *Comparative Wood Anatomy*. Heidelberg: Springer.

Cassab, G.I. and Varner, J.E. (1989) Tissue printing on nitrocellulose paper: a new method for immunolocalisation of proteins, localization of enzyme activities and anatomical analysis. *Cell Biology International Reports*, **13**, 147–152.

Chaffey, N.J. (2000) Cytoskeleton, cell walls and cambium: new insights into secondary xylem differentiation. In *Cell and Molecular Biology of Wood Formation*, edited by R. Savidge, J. Barnett and R. Napier, pp. 31–42. Oxford: Bios Scientific Publishers.

Chaffey, N.J. and Barlow, P.W. (2000) Actin in the secondary vascular system. In *Actin: a Dynamic Framework for Multiple Plant Cell Functions*, edited by F. Baluška, P.W. Barlow, C. Staiger and D. Volkmann, pp. 587–600. Dordrecht, The Netherlands. Kluwer Academic Press.

Chaffey, N.J., Barlow, P.W. and Barnett, J.R. (1996) Microtubular cytoskeleton of vascular cambium and its derivatives in roots of *Aesculus hippocastanum* L. (Hippocastanaceae). In *Recent Advances in Wood Anatomy*, edited by L.A. Donaldson, B.G. Butterfield, P.A. Singh and L.J. Whitehouse, pp. 171–183. Rotorua, NZ: New Zealand Forest Research Institute.

Chaffey, N.J., Barlow, P.W. and Barnett, J.R. (1997a) Microtubules rearrange during differentiation of vascular cambial derivatives, microfilaments do not. *Trees*, **11**, 333–341.

Chaffey, N.J., Barlow, P.W. and Barnett, J.R. (1997b) Formation of bordered pits in secondary xylem vessel elements of *Aesculus hippocastanum* L.: an electron and immunofluorescent microscope study. *Protoplasma*, **197**, 64–75.

Chaffey, N.J., Barlow, P.W. and Barnett, J.R. (1998) A seasonal cycle of cell wall structure is accompanied by a cyclical rearrangement of cortical microtubules in fusiform cambial cells within taproots of *Aesculus hippocastanum* L. (Hippocastanaceae). *New Phytologist*, **139**, 623–635.

Chaffey, N.J., Barlow, P.W. and Barnett, J.R. (2000a) Structure-function relationships during secondary phloem development in *Aesculus hippocastanum*: microtubules and cell walls. *Tree Physiology*, **20**, 777–786.

Chaffey, N.J., Barlow, P.W. and Barnett, J.R. (2000b) A cytoskeletal basis for wood formation in angiosperm trees: the involvement of microfilaments. *Planta*, **210**, 890–896.

Chaffey, N.J., Barnett, J.R. and Barlow, P.W. (1997c) Visualization of the cytoskeleton within the secondary vascular system of hardwood species. *Journal of Microscopy*, **187**, 77–84.

Chaffey, N.J., Barnett, J.R. and Barlow, P.W. (1999) A cytoskeletal basis for wood formation in angiosperm trees: the involvement of cortical microtubules. *Planta*, **208**, 19–30.

Chamberlain, C.J. (1905) *Methods in Plant Histology*. Chicago: The University of Chicago Press.

Christensen, A.K. (1998) Using borohydride to quench autofluorescence of glutaraldehyde. *Microscopy Today*, **98–1**, 21.

Clark, G. (ed) (1981) *Staining procedures*, 4th edn. Baltimore, London: Williams and Wilkins.

Desch, H.E. and Dinwoodie, J.M. (1996) *Timber: Structure, Properties, Conversion and Use*, 7th edn. London: Macmillan Press Ltd.

Dodds, J.H. (1981) The role of the cell cycle and cell division in xylem differentiation. In *Xylem Cell Development*, edited by J.R. Barnett, pp. 153–167. Tunbridge Wells, Kent: Castle House Publications Ltd.

Donaldson, L.A. and Lausberg, M.J.F. (1998) Comparison of conventional transmitted light and confocal microscopy for measuring wood dimensions by image analysis. *IAWA Journal*, **19**, 321–336.

Ermel, F.E., Follet-Gueye, M.-L., Cibert, C., Vian, B., Morvan, C., Catesson, A.-M. and Goldberg, R. (2000) Differential localization of arabinan and galactan side chains of rhamnogalacturonan 1 in cambial derivatives. *Plant*, **210**, 732–740.

Esau, K. (1977) *Anatomy of Seed Plants*, 2nd edn. New York, London: John Wiley and Sons.

Evans, D.E., Clay, P.J., Attree, S. and Fowke, L.C. (1997) Visualization of Golgi apparatus in methacrylate embedded conifer embryo tissue using the monoclonal antibody JIM 84. *Cell Biology International*, **21**, 295–302.

Exley, R.R., Butterfield, B.G. and Meylan, B.A. (1974) Preparation of wood specimens for the scanning electron microscope. *Journal of Microscopy*, **101**, 21–30.

Fahn, A. (1990) *Plant Anatomy*, 4th edn. Oxford, New York: Pergamon Press.

Feder, N. and O'Brien, T.P. (1968) Plant microtechnique: some principles and new methods. *American Journal of Botany*, **55**, 123–142.

Fineran, B.A. (1997) Cyto- and histochemical demonstration of lignins in plant cell walls: an evaluation of the chlorine water/ethanolamine-silver nitrate method of Coppick and Fowler. *Protoplasma*, **198**, 186–201.

Fink, S. (1987a) Some new methods for affixing sections to glass slides. I. Aqueous adhesives. *Stain Technology*, **62**, 27–33.

Fink, S. (1987b) Some new methods for affixing sections to glass slides. II. Organicsolvent based adhesives. *Stain Technology*, **62**, 93–99.

Fink, S. (1987c) Some new methods for affixing sections to glass slides. III. Pressure-sensitive adhesives. *Stain Technology*, **62**, 349–354.

Fox, D.P. (1969) Some characteristics of the cold hydrolysis technique for staining plant tissues by the Feulgen reaction. *Journal of Histochemistry and Cytochemistry*, **17**, 266–272.

Franklin, G.L. (1945) Preparation of thin sections of synthetic resins and wood-resin composites, and a new macerating method for wood. *Nature*, **155**, 51.

Fujii, T. (1993) Application of a resin casting method to wood anatomy of some Japanese Fagaceae species. *IAWA Journal*, **14**, 273–288.

Funada, R., Abe, H., Furusawa, O., Imaizumi, H., Fukazawa, K. and Ohtani, J. (1997) The orientation and localization of cortical microtubules in differentiating conifer tracheids during cell expansion. *Plant & Cell Physiology*, **38**, 210–212.

Funada, R., Furusawa, O., Shibagaki, M., Miura, H., Miura, T., Abe, H. and Ohtani, J. (2000) The role of cytoskeleton in secondary xylem differentiation in conifers. In: *Cell and Molecular Biology of Wood Formation*, edited by R. Savidge, J. Barnett and R. Napier, pp. 225–264. Oxford: Bios Scientific Publishers.

Furusawa, O., Funada, R., Murakami, Y. and Ohtani, J. (1998) The arrangement of cortical microtubules in compression wood tracheids of *Taxus cuspidata* visualized by confocal laser microscopy. *Journal of Wood Science*, **44**, 230–233.

Gahan, P.B. (1984) *Plant Histochemistry and Cytochemistry*. London, New York, Tokyo: Academic Press.

Galbraith, D.W. (1981) Microfluorimetric quantitation of cellulose biosynthesis by plant protoplasts using Calcofluor White. *Physiologia Plantarum*, **53**, 111–116.

Garland, H. (1935) Notes on wood microtechnique. *Journal of Forestry*, **33**, 142–145.

Goffinet, M.C., Robinson, T.L. and Lakso, A.N. (1995) A comparison of 'Empire' apple fruit size and anatomy in unthinned and hand-thinned trees. *Journal of Horticultural Science*, **70**, 375–387.

Goodbody, K.C. and Lloyd, C.W. (1994) Immunofluorescence techniques for analysis of the cytoskeleton. In *Plant Cell Biology: a Practical Approach*, edited by N. Harris and K.J. Oparka, pp. 221–234. Oxford: Oxford University Press.

Gori, P. (1978) Ponceau 2R staining of proteins and periodic acid bleaching of osmicated subcellular structures on semi-thin section of tissues processed for electron microscopy: a simplified procedure. *Journal of Microscopy*, **113**, 111–113.

Grew, N. (1682) *The Anatomy of Plants, with an Idea of a Philosophical History of Plants and Several other Lectures read before the Royal Society*. New York, London: Johnson Reprint Corporation.

Grosser, D. (1977) *Die Hölzer Mitteleuropas: ein mikrophotographischer Lehratlas*. Berlin, Heidelberg, New York: Springer.

Harada, H. and Côté, W.A., Jr, (1985). Structure of wood. In *Biosynthesis and Biodegradation of Wood Components*, edited by T. Higuchi, pp. 1–42. Orlando: Academic Press.

Harkes, P.A.A. (1973) Structure and dynamics of the root cap of *Avena sativa* L. *Acta Botanica Neerlandica*, **22**, 321–328.

Hayat, M.A. (1993) *Stains and Cytochemical Methods*. New York, London: Plenum Press.

Haygreen, J.G. and Bowyer, J.L. (1996) *Forest Products and Wood Science*, 3rd edn. Ames, Iowa: Iowa State University Press.

Hepler, P.K. and Gunning, B.E.S. (1998) Confocal fluorescence microscopy of plant cells. *Protoplasma*, **201**, 121–157.

Hernández, L.F. and Palmer, J.H. (1988) Fluorescent staining of primary plant cell walls using bisbenzimide (33258 Hoechst) fluorochrome. *Stain Technology*, **63**, 190–192.

Hertzberg, M. and Olsson, O. (1998) Molecular characterisation of a novel plant homeobox gene expressed in the maturing xylem zone of *Populus tremula × tremuloides*. *Plant Journal*, **16**, 285–295.

Higuchi, T. (1997) *Biochemistry and Molecular Biology of Wood*. Berlin, Heidelberg: Springer.

Huang, C.-L., Kutscha, N.P., Leaf, G.J. and Megraw, R.A. (1998) Comparison of microfibril angle measurement techniques. In *Microfibril Angle in Wood*, edited by B.G. Butterfield, pp. 177–205. Canterbury, New Zealand: IAWA/IUFRO.

Hughes, J. and McCully, M.E. (1975) The use of an optical brightener in the study of plant structure. *Stain Technology*, **50**, 319–329.

Humphrey, C.D. and Pitman, F.E. (1974) A simple methylene blue-azure II-basic fuchsin stain for epoxy-embedded tissue sections. *Stain Technology*, **49**, 9–14.

IAWA Committee on Nomenclature (1964). *Multilingual Glossary of Terms used in Describing Wood Anatomy*. Winterthur: Konkordia.

IAWA Committee (1989) IAWA list of microscopic features for hardwood identification, *IAWA Bulletin, New series,* **10**, 219–332.

Ilvessalo-Pfäffli, M.-S. (1995) *Fiber Atlas: Identification of Papermaking Fibers.* Berlin, Heidelberg: Springer.

Isebrands, J.G. and Larson, P.R. (1973) Some observations on the cambial zone of cottonwood. *IAWA Bulletin,* **3**, 3–11.

Itabashi, T., Yokota, S. and Yoshizawa, N. (1999) The seasonal occurrence and histology of septate fibers in *Kalopanax pictus. IAWA Journal,* **20**, 395–404.

Jang, H.F., Robertson, A.G. and Seth, R.S. (1992) Transverse dimensions of wood pulp fibers by confocal laser scanning microscopy and image analysis. *Journal of Materials Science,* **27**, 6391–6400.

Jefferies, C.J. and Belcher, A.R. (1974) A fluorescent brightener used for pollen tube identification in vivo. *Stain Technology,* **49**, 199–202.

Jensen, W.A. (1962) *Botanical Histochemistry: Principles and Practice.* San Francisco, London: WH Freeman and Co.

Johansen, D.A. (1940) *Plant Microtechnique,* 1st edn, 6th impression. New York, London: McGraw-Hill Book Co. Ltd.

Johansen, B., Seberg, O., Petersen, G. and Arctander, P. (1995) Does DAPI detect cytoplasmic DNA? *American Journal of Botany,* **82**, 1215–1219.

Johnson, G.D., Davidson, R.S., McNamee, K.C., Russell, G., Goodwin, D. and Holborrow, E.J. (1982) Fading of immunofluorescence during microscopy: a study of the phenomenon and its remedy. *Journal of Immunological Methods,* **55**, 231–242.

Kiernan, J.A. (2000) Formaldehyde, paraformaldehyde and glutaraldehyde: what they are and what they do. *Microscopy Today,* **00–1**, 8, 10, 12.

Kim, Y.S. and Singh, A. (1998) Imaging degraded wood by confocal microscopy. *Microscopy Today,* **98–4**, 14.

Kitin, P., Funada, R., Sano, Y., Beeckman, H. and Ohtani, J. (1999) Variations in the lengths of fusiform cambial cells and vessel elements in *Kalopanax pictus. Annals of Botany,* **84**, 621–632.

Knebel, W. and Schnepf, E. (1991) Confocal laser scanning microscopy of fluorescently stained wood cells: a new method for three-dimensional imaging of xylem elements. *Trees,* **5**, 1–4.

Kobayashi, I., Kobayashi, Y. and Hardham, A.R. (1994) Dynamic reorganization of microtubules and microfilaments in flax cells during the resistance response to flax rust infection. *Planta,* **195**, 237–247.

Kozlowski, T.T. and Pallardy, S.G. (1997) *Physiology of Woody Plants,* 2nd edn. San Diego, London, Boston: Academic Press.

Krishnamurthy, K,V, (1999) *Methods in Cell Wall Cytochemistry.* Boca Raton, London, New York, Washington, DC: CRC Press.

Kronenberger, J., Desprez, T., Höfte, H., Caboche, M. and Traas, J. (1993). A methacrylate embedding procedure developed for immunolocalization on plant tissues is also compatible with in situ hybridization. *Cell Biology International Reports,* **17**, 1013–1021.

Label, P., Beritognolo, I., Burtin, P., Dehon, L., Couée, I., Bréton, C., Charpentier, J.-P., Jay-Allemand, C. (2000) Cambial activity and xylem differentiation in walnut (*Juglans* sp.). In *Cell and Molecular Biology of Wood Formation,* edited by R. Savidge, J. Barnett and R. Napier, pp. 209–221. Oxford: Bios Scientific Publishers.

Lewis, A.M. (1995) A video technique for imaging the three-dimensional architecture of wood. *IAWA Journal,* **16**, 81–86.

Longin, A., Souchier, C., Ffrench, M. and Bryon, P. (1993) Comparison of anti-fading agents used in fluorescence microscopy: image analysis and laser confocal microscopy study. *Journal of Histochemistry and Cytochemistry,* **41**, 1833–1840.

Lucansky, T.W. (1976) The macrotome: a new approach to sectioning large plant specimens. *Stain Technology,* **51**, 199–201.

Maeda, H. and Ishida, N. (1967) Specificity of binding of hexapyranosyl polysaccharides with fluorescent brighteners. *Journal of Biochemistry (Japan),* **62**, 276–278.

Martinez-Nistal, A., Alonso, M., González-Rio, F., Sampedro, A. and Astorga, R. (1998) 3D Reconstruction of wood fibres using confocal microscopy. *European Microscopy and Analysis,* **January**, 23–24.

Matsumoto, B. (ed.) (1993) *Methods in Cell Biology, Volume 38, Cell Biological Applications of Confocal Microscopy.* New York, London: Academic Press.

Matsumura, J., Booker, R.E., Donaldson, L.A. and Ridoutt, B.G. (1998) Impregnation of radiata pine by vacuum treatment: identification of flow paths using fluorescent dye and confocal microscopy. *IAWA Journal,* **19**, 25–33.

Matsuzaki, T., Suzuki, T., Fujikura, K. and Takata, K. (1997) Nuclear staining for laser confocal microscopy. *Acta Histochemica et Cytochemica,* **30**, 309–314.

Mauseth, J.D. (1991) *Botany: an Introduction to Plant Biology.* Philadelphia, Chicago, San Francisco, London: Saunders College Publishing.

Mellerowicz, E.J., Riding, R.T. and Little, C.H.A. (1989) Genomic variability in the vascular cambium of *Abies balsamea. Canadian Journal of Botany,* **67**, 990–996.

Mellerowicz, E.J., Riding, R.T. and Little, C.H.A. (1992) Periodicity of cambial activity in *Abies balsamea.* II. Effects of temperature and photoperiod on the size of the nuclear genome in fusiform cambial cells. *Physiologia Plantarum,* **85**, 526–530.

Meylan, B.A. and Butterfield, B.G. (1972) *Threedimensional Structure of Wood: A Scanning Electron Microscope Study.* London: Chapman and Hall Ltd.

Murakami, Y., Funada, R., Sano, Y. and Ohtani, J. (1999) The differentiation of contact cells and isolation cells in the xylem ray parenchyma of *Populus maximowiczii. Annals of Botany,* **84**, 429–435.

Nickle, T.C. and Meinke, D.W. (1998) A cytokinesis-defective mutant of *Arabidopsis* (*cyt1*) characterized by embryonic lethality, incomplete cell walls, and excessive callose accumulation. *Plant Journal,* **15**, 321–332.

O'Brien, T.P. and McCully, M.E. (1981) *The Study of Plant Structure: Principles and Selected Methods.* Melbourne: Termacarphi Pty Ltd.

Olyslaegers, G. and Verbelen, J.-P. (1998) Improved staining of F-actin and co-localization of mitochondria in plant cells. *Journal of Microscopy,* **192,** 73–77.

Paddock, S.W. (ed.) (1999) *Methods in Molecular Biology, Volume 122, Confocal Microscopy Methods and Protocols.* Totowa, NJ: Humana Press.

Panshin, A.J. and de Zeeuw, C. (1980) *Textbook of Wood Technology,* 4th edn. New York, San Francisco, London: McGraw-Hill, Inc.

Paul, R.N. (1980) The use of thionin and acridine orange in staining semithin sections of plant material embedded in epoxy resin. *Stain Technology,* **55,** 195–196.

Pawley, J.B. (ed.) (1990) *Handbook of Biological Confocal Microscopy,* 1st edn, revised. New York, London: Plenum Press.

Pawley, J.B. (ed.) (1995) *Handbook of Biological Confocal Microscopy,* 2nd edn. New York, London: Plenum Press.

Peterson, C.A. and Blais, M.A. (1991) A rapid method for macerating phloem. *Biotechnic & Histochemistry,* **66,** 242–245.

Prior, D.A.M., Oparka, K.J. and Roberts, I.M. (1999) *En bloc* optical sectioning of resin-embedded specimens using a confocal laser scanning microscope. *Journal of Microscopy,* **193,** 20–27.

Purvis, M.J., Collier, D.C. and Walls, D. (1964) *Laboratory Techniques in Botany.* London: Butterworths.

Rao, K.S. and Catesson, A.-M. (1987) Changes in the membrane component of nondividing cambial cells. *Canadian Journal of Botany,* **65,** 246–254.

Rao, K.S. and Dave, Y.S. (1984) Occurrence of crystals in vascular cambium. *Protoplasma,* **119,** 219–221.

Rao, K.S. and Menon, A.R.S. (1989) Histochemistry of cambium in *Holoptelea integrifolia* (Roxb.) Planch. and *Mangifera indica* L.: its seasonal significance. *Pakistan Journal of Botany,* **21,** 81–87.

Riding, T. and Little, C.H.A. (1984) Anatomy and histochemistry of *Abies balsamea* cambial zone cells during the onset and breaking of dormancy. *Canadian Journal of Botany,* **62,** 2570–2579.

Roth, R., Boudet, A.M. and Pont-Lezica, R. (1997) Lignification and cinnamyl alcohol dehydrogenase activity in developing stems of tomato and poplar: a spatial and kinetic study through tissue printing. *Journal of Experimental Botany,* **48,** 247–254.

Running, M.P. and Meyerowitz, E.M. (1996) Using confocal microscopy in the study of plant structure and development. *Aliso,* **14,** 263–270.

Ruzin, S.E. (1999) *Plant Microtechnique and Microscopy.* Oxford: Oxford University Press.

Sass, J.E. (1958) *Botanical Microtechnique,* 3rd edn. London: Constable and Company Ltd.

Scanalytics, Inc. (1997) High resolution 3-D fluorescence microscopy: a comparison of confocal laser scanning microscopy and a wide-field deconvolution technique. *Microscopy Today,* **97–6,** 10–12.

Schmid, R. (1982) Sonication and other improvements on Jeffrey's technique for macerating wood. *Stain Technology,* **57,** 293–299.

Schweingruber, F.H. (1978) *Mikroskopische Holzanatomie.* Zug, Switzerland: Zürcher AG.

Schweingruber, F.H. (1990). *Anatomie europäischer Hölzer.* Berne, Stuttgart: Paul Haupt.

Sharman, B.C. (1943) Tannic acid and iron alum with safranin and orange G in studies of the shoot apex. *Stain Technology,* **18,** 105–111.

Sheppard, C.J.R. and Shotton, D.M. (1997) *Confocal Laser Scanning Microscopy.* Oxford: Bios Scientific Publishers Ltd.

Shotton, D. and White, N. (1989) Confocal scanning microscopy: three-dimensional biological imaging. *Trends in Biochemical Sciences,* **14,** 435–439.

Siau, J.F. (1995) *Wood: Influence of Moisture on Physical Properties.* Virginia: Virginia Polytechnic Institute and State University.

Singh, A., Xiao, Y. and Wakeling, R. (1997) Glutaraldehyde fluorescence useful in confocal studies of fungi. *Microscopy Today,* **97–8,** 16.

Smith, F.H. and Gannon, B.L. (1972) Sectioning of charcoals and dry ancient woods. *American Antiquity,* **38,** 468–472.

Spearin, W.E. and Isenberg, I.H. (1947) The maceration of woody tissues with acetic acid and sodium hypochlorite. *Science,* **105,** 214.

Speer, E.O. (1987) A method of retaining phloroglucinol proof of lignin. *Stain Technology,* **62,** 279–280.

Spurr, A.R. (1969) A low viscosity epoxy resin embedding medium for electron microscopy. *Journal of Ultrastructural Research,* **26,** 31–43.

Steedman, H.F. (1957) Polyester wax: a new ribboning embedding medium for histology. *Nature,* **179,** 1345.

Sterky, F., Regan, S., Karlsson, J., Rohde, A., Holmberg, A., Amini, B., Bhalcrao, R., Larsson, M., Villarroel, R., Van Montagu, M., Sandberg, G., Olsson, O., Teeri, T.T., Boerjan, W., Gustafsson, P., Uhlén, M., Sundberg, B. and Lundeberg, J. (1998) Gene discovery in the wood-forming tissues of poplar: analysis of 5,692 expressed sequence tags. *Proceedings of the National Academy of Sciences of the United States of America,* **95,** 13330–13335.

Stieber, J. (1981) A new method of examining vessels. *Annals of Botany,* **48,** 411–414.

Stone, B.A., Evans, N.A., Bonig, I. and Clarke, A.E. (1984) The application of sirofluor, a chemically defined fluorochrome from aniline blue for the histochemical detection of callose. *Protoplasma,* **122,** 191–195.

Takabe, K., Miyauchi, S., Tsunoda, R. and Fukuzawa, K. (1992) Distribution of guaiacyl and syringyl lignins in Japanese beech (*Fagus crenata*): variation within an annual ring. *IAWA Bulletin, New series,* **13,** 105–112.

Tuominen, H., Puech, L., Fink, S. and Sundberg, B. (1997) A radial concentration gradient of indole-3-acetic acid is related to secondary xylem development in *Populus. Plant Physiology,* **115,** 577–585.

Varner, J.E. and Ye, Z.-H. (1995). Tissue printing to detect proteins and RNA in plant tissues. In *Methods in Plant Molecular Biology*, edited by P. Maliga, D.F. Klessig, A.R. Cashmore, W. Gruisem and J.E. Varner, pp. 79–94. New York: Cold Spring Harbor Laboratory Press.

Verbelen, J.-P. and Stickens, D. (1995) *In vivo* determination of fibril orientation in plant cell walls with polarization CSLM. *Journal of Microscopy*, **177**, 1–6.

Wardrop, A.B. (1981) Lignification and xylogenesis. In *Xylem Cell Development*, edited by J.R. Barnett, p. 115–152. Tunbridge Wells, Kent: Castle House Publications Ltd.

Wenham, M.E. (1966) An analysis of wood development in *Salix viminalis* L. *PhD Thesis*, University of Aberdeen (unpublished).

Wenham, M.W. and Cusick, F. (1975) The growth of secondary wood fibres. *New Phytologist*, **74**, 247–261.

Wetzel, S., Demmers, C. and Greenwood, J.S. (1989) Seasonally fluctuating bark protein are a potential form of nitrogen storage in 3 temperate hardwoods. *Planta*, **178**, 275–281.

Wymer, C.L., Beven, A.F., Boudonck, K. and Lloyd, C.W. (1999) Confocal microscopy of plant cells. In *Methods in Molecular Biology, Volume 122, Confocal Microscopy Methods and Protocols*, edited by S. Paddock, pp. 103–130. Totowa, NJ: Humana Press Inc.

Xiao, Y., Kreber, B. and Breuil, C. (1998a) Detection of *Ophiostoma piceae* in radiata pine using immuno-fluorescence labeling and confocal laser scanning microscopy. *Microscopy Today*, **98–3**, 24–25.

Xiao, Y., Singh, A. and Wakeling, R.N. (1998b) Detecting bacteria in wood using a fluorescent lipid probe. *Microscopy Today*, **98–7**, 16–17.

Yeung, E.C. (1990) A simple procedure to visualize osmicated storage lipids in semithin epoxy sections of plant tissues. *Stain Technology*, **65**, 45–47.

Zhong, Y., Mellerowicz, E.J., Lloyd, A.D., Leinhos, V., Riding, R.T. and Little, C.H.A. (1995) Seasonal variation in the nuclear genome of size of ray cells in the vascular cambium of *Fraxinus americana*. *Physiologia Plantarum*, **93**, 305–311.

Wood Formation in Trees ed. N. Chaffey
© 2002 Taylor & Francis
Taylor & Francis is an imprint of the
Taylor & Francis Group
Printed in Singapore.

4 Conventional (Chemical-fixation) Transmission Electron Microscopy and Cytochemistry of Angiosperm Trees

Nigel Chaffey

IACR – Long Ashton Research Station, Department of Agricultural Sciences, University of Bristol, Long Ashton, Bristol BS41 9AF, UK

ABSTRACT

The use of the transmission electron microscope in the study of wood cell biology is briefly reviewed. Detailed sampling, fixation and tissue-processing procedures for conventional (chemical-fixation) transmission electron microscopy are described for woody stems and roots of angiosperm trees. Procedures for zinc-iodide-osmium-tetroxide-, Thiéry- and phosphotungstic acid-staining, and immunolocalisation of cell components are provided. Some recent applications of transmission electron microscopy to the study of cambial cell biology and wood formation in angiosperm trees are summarised, and future directions indicated.

Key words: cambium, cytochemistry, cytoskeleton, plasmatubules, ultrastructural techniques, wood cells

INTRODUCTION

A brief history of the use of the transmission electron microscope in wood cell biology

Despite being widely scattered, there is a substantial body of published work that deals with application of the transmission electron microscope (TEM) to woody material. Therefore, this survey is of necessity brief and selective (biased towards angiosperm trees – for gymnosperms, see the chapter by Rensing in this volume), and is designed to give a flavour of the work that has been done in this area. This contribution extends, but does not supplant, the previous reviews of this aspect of wood research by Côté and Hanna (1985) and Singh *et al.* (1997).

The persistence and resilience of the wood cell wall means that it needs relatively little preparation for TEM, and for many years dead tracheary elements of angiosperm and gymnosperm trees have been much studied by this technique (e.g. Bosshard, 1951; Wardrop, 1954; Côté, 1958; Singh and Donaldson, 1999). Notwithstanding a long history of study, examination of wood cell walls continues to reveal new information, such as the interfibrillar bridges recently described from kraft pulp of radiata pine by Donaldson and Singh (1998). In addition to their cell biological relevance, such studies also generate information of practical significance, e.g. concerning the penetration of preservatives into wood (Thomas, 1976; Singh *et al.*, 1999). However, to see fine structural detail within differentiating cells, it is necessary to fix the cytoplasm.

Early attempts at fixation of woody cells followed the lead of those working with other biological material and used permanganate. As widely acknowledged, this fixative is excellent for enhancing membrane profiles within the cells but, at best, leaves much of the fine detail unrevealed, and, at worst, results in the loss of components from the cytoplasm (destruction of microtubules being a classic example of this).

As for many other areas of ultrastructural cell biology, the 'golden age' of wood TEM dawned with the promotion of aldehydes as fixatives, particularly the dialdehyde, glutaraldehyde, by Sabatini and co-workers (Sabatini *et al.*, 1963a, b).

A seminal publication from that era is the volume edited by Côté (1965) entitled, 'Cellular ultrastructure of woody plants'. That tome presents a survey of the state of the art of 'wood TEM' and includes several chapters on aspects of the cell biology of angiosperm trees (e.g. Cronshaw, 1965b – derivative differentiation; Harada, 1965 – vessels and ray parenchyma; Schmid, 1965 – pit structure; Wardrop, 1965 – xylem differentiation). It is also of interest because it captures that change-over period from use of permanganate- to glutaraldehyde-fixation in the images included by the contributors. Furthermore, use of glutaraldehyde-fixation by Ledbetter and Porter (1963) has contributed to much of the present day interest in the cytoskeleton of the tree SVS (see the cytoskeletal chapters by Chaffey, and Funada in this volume).

Ultrastructural observations of the tree SVS during the 1960s and 1970s provided a wealth of important cell-biological information about a wide range of phenomena, such as 'protective layers' (Czaninski, 1973; Chafe, 1974), tension wood and G-layers (Côté and Day, 1962; Mia, 1968; Côté et al., 1969; Fujita et al., 1974), vessel-associated cells (Czaninski, 1966, 1968, 1977), pit membranes (Machado and Schmid, 1964), microtubule function (Robards and Humpherson, 1967b; Robards and Kidwai, 1972; Nelmes et al., 1973), differentiation of secondary xylem (Robards, 1967, 1968), exocytosis (Robards and Kidwai, 1969a), and differentiation of ray parenchyma cells (Mia, 1972).

However, despite the fact that the major difficulties of fixation of the cambium for ultrastructural work were largely overcome in the 1960s (e.g. Srivastava, 1966; Evert and Deshpande, 1970), and notwithstanding the publication of a detailed protocol for TEM-processing of cambial tissue by Goosen-de Roo and van Spronsen (1978), relatively few workers have taken up the study of this remarkable tissue (see Table 1). And even today the perception still persists that such work is generally too difficult to attempt.

Table 1 Ultrastructural studies of hardwood cambia

Species	Study	Authors
Acer pseudoplatanus	general	Catesson (1974)
Aesculus hippocastanum	xylogenesis	Barnett (1981)
Ae. hippocastanum	nuclear microtubules	Barnett (1991)
Ae. hippocastanum	reactivation	Barnett (1992)
Ae. hippocastanum	seasonal	Rao (1985); Chaffey et al. (1998)
Ae. hippocastanum	tannins	Rao (1988)
Ae. hippocastanum	dormant	Rao and Catesson (1987)
Ae. hippocastanum	active	Chaffey et al. (1997c)
Ae. hippocastanum	cytoskeleton	Chaffey et al. (1996)
Fagus sylvatica	technic	Kidwai and Robards (1969a)
F. sylvatica	resting	Kidwai and Robards (1969b)
F. sylvatica	seasonal	Farooqui and Robards (1979)
Fraxinus americana	seasonal	Srivastava (1966)
Fr. excelsior	technic	Goosen-de Roo and van Spronsen (1978)
Fr. excelsior	cytokinesis	Goosen-de Roo et al. (1983)
Fr. excelsior	active	Goosen-de Roo et al. (1984)
Populus euramericana	tannins	Rao (1988)
Quercus rubra	technic	Kidwai and Robards (1969a)
Q. rubra	xylogenesis	Murmanis (1977)
Robinia pseudoacacia	seasonal	Farrar and Evert (1997a)
R. pseudoacacia	cell division	Farrar and Evert (1997b)
Salix fragilis	phytoferritin	Robards and Humpherson (1967a)
S. fragilis	technic	Kidwai and Robards (1969a)
S. fragilis	seasonal	Robards and Kidwai (1969b)
S. viminalis	seasonal	Sennerby-Forsse (1986)
S. dasyclados	seasonal	Sennerby-Forsse and von Fircks (1987)
Tectona grandis	plastids	Dave and Rao (1981)
Te. grandis	seasonal	Rao and Dave (1983)
Tilia americana	seasonal	Mia (1970)
Ti. americana	cell division	Evert and Deshpande (1970)
Ulmus americana	cell division	Evert and Deshpande (1970)

Although qualitative TEM is still an extremely valuable tool, a quantitative dimension to its use adds significantly to the value of such work. Two examples illustrate the benefit of this approach to trees. In the context of general tree physiological anatomy, it is important to understand the pathways of photosynthate transfer within the SVS. It has generally been assumed that the rays provide the main pathway for 2-way exchange of materials between the long-distance conduits of the xylem and the phloem, but it is necessary to have evidence for this view. Such support is found in the detailed ultrastructural analyses of the distribution and frequency of plasmodesmata within the walls of the cambium (Goosen-de Roo, 1981; Goosen-de Roo and Creyghton-Schouten, 1981), between the cambium and phloem (Farrar, 1995), and between xylem ray cells (Sauter and Kloth, 1986). The frequency of plasmodesmata was greatest in the tangential walls between adjacent ray cells, which is consistent with their being the major symplasmic route for xylem-phloem transport. Secondly, quantitative micromorphometric analysis of organelles and storage materials in poplar xylem ray cells by Sauter and van Cleve (1989) has proved to be a useful tool for the detection of cell-specific differences within this tissue, and provided a useful adjunct to the previously determined biochemical differences between the component cells.

One of the most powerful unions of recent times has been the marriage of TEM and immunolocalisation. Not only does this permit much needed correlative TEM and light microscopy, which provides corroborative evidence for the validity of the images obtained, but it extends immunolocalisation to a much finer level of resolution, aiding greater understanding. In woody tissues it has been used for a variety of studies, including analysis of storage proteins (e.g. Sauter and van Cleve, 1990; Clausen and Apel, 1991; Harms and Sauter, 1992; see also the chapter by Beardmore *et al.* in this volume), chemical characterisation of cell walls (e.g. Chaffey *et al.*, 1997b, c, 1998, 2000a; Guglielmino *et al.*, 1997; Awano *et al.*, 1998; Ermel *et al.*, 2000; see also the chapter by Micheli *et al.* in this volume), gum and tylose formation (Rioux *et al.*, 1998), and localisation of lignins (Ruel *et al.*, 1999) and enzymes of lignification (e.g. Samaj *et al.*, 1998; see also the chapter by Samaj and Boudet in this volume).

As will be clear from other chapters in this volume, the biology of the cell walls of the cambium and its derivatives is a subject of intense interest in the context of xylogenesis because they undergo such profound changes during cambial derivative differentiation (see e.g. Table 1 in the Introduction to this volume). From being cell features that were easy to prepare and examine in the early days of wood TEM, application of cytochemical and immunolocalisation techniques is now taking their study much further. Relatively early work, predominantly by French workers (e.g. Roland, 1978; Czaninski, 1979; Catesson and Roland, 1981; Catesson *et al.*, 1994), using a variety of subtractive techniques and well-established stains such as PATAg has succeeded in demonstrating the basics of cell wall chemistry, and of differences within and between the walls during xylogenesis. More recently, that work is being refined and extended with immunolocalisation techniques at the TEM level in examining the role of pectins and pectin methylesterases (Guglielmino *et al.*, 1997; see also chapter by Micheli *et al.* in this volume), and other cell wall components (Ermel *et al.*, 2000) during this process.

New techniques, or developments of existing ones, continue to keep TEM alive. This is seen clearly in the recent work which uses a quick-freezing, deep-etching technique to provide new information regarding the 3-dimensional (3-D) architecture of the wood cell wall during differentiation of gymnosperm and angiosperm trees (e.g. Fujino and Itoh, 1998; Hafrén *et al.*, 1999; see the chapter by Itoh in this volume). The value of that technique has itself been extended by the application of immunogold-labelling of polysaccharides to the deep-etched samples (see also the chapter by Itoh in this volume).

There is no chapter devoted to scanning electron microscopy (SEM) in this volume, so a few words concerning that technique are appropriate here. If tissue-preparation for TEM is difficult, preparing woody material for SEM can be as straightforward as simply drying slivers of wood, gold-coating them and viewing (e.g. Chaffey *et al.*, 1996). However, with a little more preparation, significantly more information can be obtained (Exley *et al.*, 1974; Kucera, 1986) and images obtained which dramatically reveal the intricate 3-D nature of woody tissues (e.g. Meylan and Butterfield, 1972) and of their various wall-elaborations, such as perforation plates (Butterfield, 1995). Refinements of SEM, such as the field emission-SEM (FE-SEM) (e.g. Vesk *et al.*, 1996), are now being applied to studies of wood cell development in angiosperm trees (*Fraxinus* – Prodhan *et al.*, 1995), gymnosperms (e.g. Sano *et al.*, 1999), and in bamboo (Crow *et al.*, 2000).

Considerations of successful sampling and fixation of woody material for TEM

These have already been covered in more detail in the chapter entitled 'An introduction to the problems of working with trees' in this volume. In summary, one must bear in mind the size of the cells of the SVS, their highly vacuolate nature, the delicacy of the cambial cells, and the fact that the SVS contains cells which represent a very wide range of stages of differentiation and physiological states. Fixation will generally be a compromise that attempts to deal with all these, sometimes conflicting, factors.

TECHNIQUES

Note – equipment and chemicals in bold type are considered in more detail in the SOLUTIONS, EQUIPMENT, ETC section.

Sampling strategy

Equipment needed:	*Chemicals needed:*
Hacksaw	Water
Bucket	
Glass Petri dish	
Single-edge razor blades (several!)	
Forceps	
Beaker (500 ml – 1 l)	

Duration of procedure: approx. 30 min (plus travelling time!)

This technique is suitable for saplings, with stem or root diameter up to approx. 3 cm (such as the one- to five-year-old *Aesculus hippocastanum* L. and *Populus tremula* × *P. tremuloides* for which the protocols detailed below were developed), and is illustrated in Figure 1. For larger trees, it will be necessary to remove material with a hammer and chisel (see the description in the chapter by Uggla and Sundberg in this volume).

1   To begin with, separate the stem/root of interest from the rest of the sapling using the hacksaw. Immediately place the cut end(s) under water (to minimise danger of embolism).

Note – I use a bucket of water, which is also useful for transporting the material back to the laboratory, if sampling the tree in the field. It also helps for washing soil, etc. from roots to permit their more ready sampling.

2   Cut stem/root into 3–4 cm lengths using the hacksaw. Accumulate these cylinders in a beaker of water.

3   To obtain slivers, stand one end of a 3–4 cm long cylinder in water in a glass Petri dish, and cut the cylinder in half – lengthwise – using a razor blade.

Note – this may require considerable force, and it is not unknown to resort to using a hammer to encourage the razor blade to cut all the way down through the cylinder.

4   Cover the exposed surfaces of the half cylinders with water (to prevent excess trauma to cells from water loss).

5   Repeat the lengthwise cut (fresh razor blade!) to obtain sliver of dimensions, approx. 3–4 cm long × approx. 2 mm thick × width of root/stem.

6   Then cut the sliver in half lengthwise, so that it is now half the width of root/stem. It is usually possible to get approx. 6–10 such half-slivers from a 2 cm diameter root/stem. However, if the cambium is active, bark-slippage may well frustrate your attempts to cut slivers; to avoid disappointing results, keep only those where no bark slippage is evident, i.e. with intact cambium. Unfortunately, there is no alternative but to keep cutting until you have sufficient 'good' slivers (and remember to allow for losses during subsequent processing, such as accidentally discarding material with solutions).

7   To avoid further damage to the tissues, fix the slivers (see next section) as soon as possible. Handle them, as little as possible, by their ends (which will subsequently be discarded) using forceps.

Note – any cylinders not used for TEM sampling can be kept for macerating, or for scanning electron microscope study, etc. (see the Wood microscopical techniques chapter in this volume). Slivers can also be excised and fixed in parallel for indirect immunofluorescence study (see the cytoskeletal chapter by Chaffey in this volume).

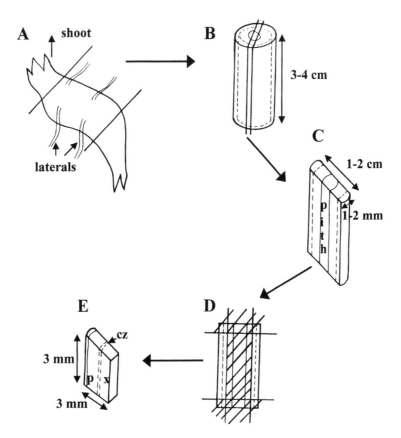

Figure 1 Diagrammatic representation of sampling procedure for transmission electron microscopy, illustrated using the taproot of *Aesculus hippocastanum*. A, taproot from which samples are to be taken; B, excised cylinder of taproot material; C, sliver removed from cylinder and ready for primary fixation; D, fixed sliver marked-out for trimming to final size – all of the pith, most of the secondary xylem, and the terminal few mm are removed and discarded; E, final trimmed block ready for osmium tetroxide-post-fixation.
Key: cz, cambial zone; p, phloem; x, secondary xylem; dashed line indicates location of cambial zone.

Processing for conventional transmission electron microscopy

Equipment needed:	*Chemicals needed:*
Rotator	**Basic TEM fix**
Glass Petri dish	Propylene oxide
Glass vials (approx. 15 ml capacity) and tops	Osmium tetroxide (OsO_4)
Embedding oven at approx. 70 °C	**Toluidine blue**
Transmission electron microscope	**Lead citrate**
Forceps	Uranyl acetate (saturated, aqueous)
Ultramicrotome and knives	Ethanol
Glass microscope slides	**Spurr resin**
Cover slips	**Methylene blue/azur A**
Single-edge razor blades (several)	Water
Light microscope, with bright field optics	
Embedding capsules	
Grids Cu/Pd 200 mesh, uncoated	
Hot plate at 70 °C	

Duration of procedure: approx. 4 days. To avoid problems inherent in working at weekends, I usually start sampling by mid-morning on a Monday, blocks are then ready to be cut on Friday morning.
Note – perform steps 1–2, 4–11 in stoppered glass vials on a rotator at room temperature (RT).

1 Fix the slivers for approx. 4.5 h in basic TEM fix.
Note – I have not found it necessary to vacuum-infiltrate the slivers at any stage during processing.

2 Wash the slivers thoroughly with deionised water (several changes in 30 min period), and transfer to a glass Petri dish containing water.

3 Trim the slivers further with a razor blade under deionised water in a glass Petri dish, so that all pith and much of the secondary xylem is removed (see Figure 1).
Note – do not be afraid to change the razor blade regularly: woody material is very tough and quickly blunts the blade! Spare the blade, spoil the block.

4 Post-fix in 1% (w/v) aqueous OsO$_4$ for approx. 1.5 h.

5 Wash in water for 3 × 5 min, keep in water overnight (approx. 16 h).

6 Dehydrate in a graded ethanol series: 25%, 50%, 75%, 90% for 30 min each, followed by 3 × 30 min in dry 100% ethanol.

7 Follow by 2 × 30 min in propylene oxide.
Note – don't forget to pierce the vial lids to avoid build-up of pressure in the vial as the propylene oxide evaporates. Lids can be forced off, taking your valuable samples with them!

8 Remove half of the volume of propylene oxide and replace with Spurr resin, leave overnight (approx. 19 h).

9 Replace half of the volume with fresh Spurr resin, leave for approx. 9 h.

10 Replace all of the volume with fresh Spurr resin, leave overnight (approx. 16 h).

11 Replace all of the volume with fresh Spurr resin, leave for approx. 4 h.

12 Replace all of the volume with fresh Spurr resin, leave for approx. 4 h.

13 Place tissue blocks in fresh Spurr resin in embedding capsules (no need to seal capsules), cure at 65–70 °C overnight (approx. 14 h).

Note – although it is necessary to orient the blocks in the capsules with the ultimate plane of sectioning in mind, I find it useful simply to allow the blocks to sink and settle with their radial face against the bottom of the capsule. That way, radial longitudinal sections can be cut easily by slicing across the diameter of the cylindrical resin block. Removing a 3 mm cylinder from the tissue-end of the resin block, and suitably orienting this short cylinder in the microtome, allows tangential longitudinal, or transverse, sections to be cut. Thus, one orientation in the resin block suits all planes of sectioning (and, although more wasteful of resin, is a lot easier than, e.g., embedding tissue blocks in the detached tops of the capsules).

14 Cut ultrathin sections and collect on uncoated 200 mesh Cu/Pd grids.
Note – for observation of series of cells within radial cell files, it is worth considering the use of coated slot-grids.
Note – before ultrathin-sectioning, it is necessary to check blocks for adequacy of fixation, infiltration of resin, etc. using thick sections and a light microscope. Sections stained with methylene blue/azur A or toluidine blue are suitable for this purpose. It is also a good idea to cut semi-thin sections (1–4 μm) for subsequent staining, etc. for light microscopy. Such sections can be placed on a drop of water on an uncoated microscope slide, stuck down at 60 °C overnight, and stored until needed. See the Wood microscopical techniques chapter in this volume for some suggested light microscopy techniques for these sections.

15 Contrast sections with uranyl acetate (can take up to 30 min), and

16 lead citrate (can also take up to 30 min), air-dry grids, and view in TEM.
Note – sections do take a long time to stain. It might be worth considering using a microwave oven to speed up the staining process (Cavusoglu et al., 1998).

Images of material processed in this way are shown in Figures 2B–D, 3A–D, 4C, H, and in Chaffey et al. (1996, 1997a, b, c, 1998, 1999, 2000a, b).

Note – this processing schedule also works satisfactorily for gymnosperms

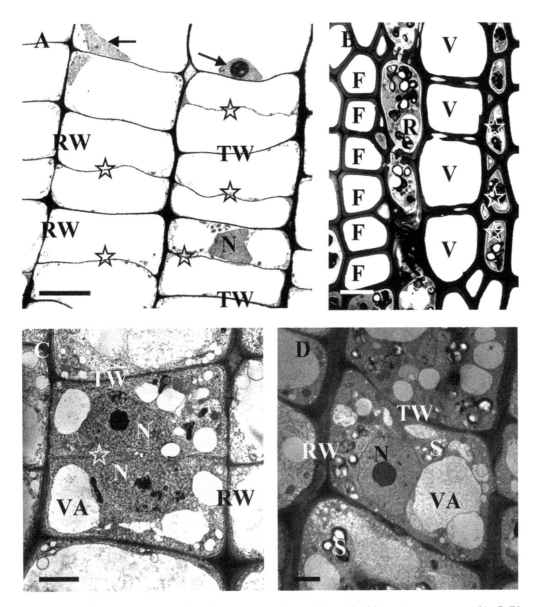

Figure 2 Transmission electron micrographs of transverse sections of *Aesculus hippocastanum* taproot (A, C, D) and stem (B).
A Low power CTEM view of fusiform cells of the cambial zone stained with PATAg for polysaccharides. Note the difference in cell wall thickness between tangential and radial walls, and the extremely thin nascent, tangential walls (☆) indicating cells that have recently divided. Note also the nucleus, which spans the radial width of the cambial cell, but which generally becomes localised to a tangential wall (arrows) as the derivative undergoes differentiation.
B Low power CTEM view of developing secondary xylem showing radial files of fibres, vessel elements and ray cells. Note the presence of cell contents in the ray cells, but their absence from most other cell types. Cells marked with ☆ may be the apices of ray cells or axial parenchyma cells, both of which retain cell contents long after they have been lost from other xylem cell types.
C CTEM of recently-divided ray cambial cell, indicated by the thin, lightly-stained tangential division wall (☆). Note the difference in thickness between the parental tangential and radial walls, the prominent nucleus and system of vacuoles.
D CTEM of ray cells from dormant cambium. Note the tangential and radial walls, which retain their relative difference in thickness (as in active cambia), but are much-thickened relative to those of ray cells from active cambia (cf. Figure 2C). A vacuole system is still present, but the numerous starch-containing plastids are a feature characteristic of the dormant state.
Key: CTEM, conventional transmission electron microscopy (aldehyde-fixation, osmium post-fixation, uranyl acetate-lead citrate double-staining); F, fibre; N, nucleus; R, ray cell; RW, radial cell wall; S, starch-containing plastids; TW, tangential cell wall; V, vessel element; VA, vacuole. Scale bars represent: 2 μm (C, D), 5 μm (A), and 10 μm (B).

Fixation with zinc-iodide-osmium-tetroxide

Although the staining reaction involved in this techniques is not fully understood, it is especially useful for staining endomembrane systems since electron-dense deposits accumulate within the nuclear envelope, endoplasmic reticulum, and dicytosome membranes (e.g. Hayat, 1993; Chaffey, 1995, 1996). The combination of this fixation technique, thick sections, and TEM using higher voltages than normal, permits a better appreciation of the 3-D structure of, or inter-relationships between, selected cell components. However, it is worth noting that cell features other than endomembrane components, such as mitochondrial envelopes and plasmodesmata, can become 'stained'. In this latter regard, the related technique, using osmium-tetroxide-potassium-ferricyanide and high voltage TEM of 2 μm sections, has been used by Barnett (1987) to study the role of plasmodesmata in xylogenesis.

Equipment needed:	Chemicals needed:
As for conventional TEM, plus watch glass	As for conventional TEM, plus **zinc-iodide-osmium-tetroxide solution**

Duration of procedure: approx. 4 days.
Note – steps 1, 3–5 performed on rotator at RT.

1   Fix excised slivers in basic TEM fix for approx. 4.5 h.
2   Wash in water and trim slivers to final block size.
3   Fix in zinc-iodide-osmium-tetroxide solution for approx. 2.5 h.
    Note – longer 'incubation' times will increase the degree of contrast and the types of cell components that stain.
4   Wash in water, and leave in water overnight (approx. 15 h).
5   Dehydrate and embed as for conventional TEM, steps 6–13 above.
6   Cut sections approx. 0.2 μm thick dry with glass knives.
7   Carefully pick up sections with fine-tipped forceps and transfer to approx. 5 ml of water in a watch glass.
8   Stretch the sections by placing watch glass on hot plate at approx. 70 °C, for approx. 60 sec.

9   Pick up sections on 200 mesh Cu/Pd grids.
10 Examine in TEM at 100 keV or more.

See Figure 3E for an example of material processed in this way (also Chaffey *et al.*, 1997c).

Processing for immunogold labelling at the ultrastructural level

Interest in immunolocalisation of proteins, etc. within cells of the SVS is increasing (see the Introduction above). The following procedure has so far been successfully used to immunolocalise pectins and callose within cambial cell walls (see below for immunolocalisation procedure). It is likely that it will also be suitable for immunolocalisation of other components of interest.

Note – although destined primarily for immunogold labelling, the LR White-embedded sections are also useful for a wide range of other cytochemical tests at the TEM level (see below re periodic acid-thiocarbohydrazide-silver proteinate- and phosphotungstic acid-staining).

Equipment needed:	Chemicals needed:
Rotator	Ethanol
Freezer at –20 °C	**JIM fix**
Embedding capsules (with lids)	LR White resin (Agar, hard grade, No. R1280)
Glass vials (approx. 15 ml capacity) plus caps	**PIPES buffer**
Refrigerator at 4 °C	Benzoin methyl ether (Sigma, No. B0279)
UV lamp	
Glass Petri dish	

Duration of procedure: approx. 4 days.
Note – steps 1, 3–8 performed at RT on a rotator in stoppered glass vials.

1   Fix slivers for approx. 4.5 h in JIM fix.
2   Following a PIPES buffer wash, for 3 × 5 min, cut slivers to final block size using razor blades under PIPES buffer in glass Petri dish.
3   Leave in PIPES buffer overnight (15–20 h).
4   Dehydrate in a graded water/ethanol series as for step 6 re conventional TEM above.
    Note – although it is possible to infiltrate LR White resin into material at the 70% ethanol

stage, this can result in unacceptable shrinkage of the tissues (Ruzin, 1999).

Note – polymerisation of LR White is inhibited by acetone (Ruzin, 1999), consequently, acetone should be avoided as a dehydrant.

5 Remove half of the final 100% ethanol volume and replace with LR White resin, and leave overnight (approx. 17 h).

6 Remove half of the resin/ethanol mixture, replace with fresh LR White resin, leave for approx. 9 h.

7 Replace all of the resin mixture with fresh LR White resin, leave overnight (15–20 h).

8 Replace all of the resin mixture with fresh LR White resin, leave for approx. 5.5 h.

9 Replace all of the resin mixture with fresh LR White resin containing 0.5% (w/v) benzoin methyl ether, transfer to a rotator in a refrigerator, and leave for approx. 4.5 h.

10 Place blocks in embedding capsules, fill with fresh LR White resin containing 0.5% (w/v) benzoin methyl ether, seal capsules. Orient blocks as appropriate for the sections that are to be taken. Cure with UV lamp at –20 °C overnight (approx. 14 h).

'Cytochemistry' techniques

Thiéry reaction (PATAg-staining)

This technique (Thiéry, 1967) is the TEM equivalent of the periodic acid-Schiff (PA-S) reaction used in light microscopy (e.g. Ruzin, 1999) and leads to deposition of silver at PA-S-positive sites. Although it has some specificity for α- and β-1,4-glucan linkages (Roland, 1974), it is used as a general stain for carbohydrates/polysaccharides. The reaction is usually performed on ultrathin sections that have been processed for TEM, but without an osmication-step.

Use of osmicated material is generally not recommended since any bound osmium that remains after the periodic-acid-oxidation step may cause non-specific deposition of silver (Lewis and Knight, 1977, but cf. Roland and Vian, 1991). However, provided one is careful to compare the staining pattern in non-osmicated material, it is possible to use osmium-treated tissues. I use osmication where the additional fixation is required to preserve/discern intracellular features such as coated vesicles, which are poorly preserved in fixation with aldehydes alone. Because of the possibility that osmium, or glutaraldehyde (Lewis and Knight,

1977) can interfere with the PATAg reaction, it is important to perform proper controls (see Courtoy and Simar, 1974).

PATAg-staining has been widely employed in wood cell biology (e.g. Freundlich and Robards, 1974; Czaninski, 1979; Parameswaran and Liese, 1981; Rao and Catesson, 1987; see also the chapter by Micheli et al. in this volume). A promising, recent extension of the technique is to combine it with immunolocalisation of cell wall components, such as pectins (see the chapter by Micheli et al. in this volume).

Equipment needed:	Chemicals needed:
Ultrathin Spurr resin sections on 200 mesh gold grids	Silver proteinate ('Roques' – TAAB, No. SO09)
Humid box	Acetic acid
Parafilm	Thiocarbohydrazide (Sigma, No. T-2137)
Forceps (to handle grids)	Periodic acid (Sigma, No. P-7875)

Duration of procedure: approx. 3–20 h.
Note – all steps performed at RT in a humid box.

1 Float sections for 30 min on 1% (v/v) aqueous periodic acid.

2 Wash thoroughly with water.

3 Float sections for 2–4 (up to 18) h on 0.2% (w/v) thiocarbohydrazide in 20% (v/v) acetic acid.
Note – increase in density of silver-staining with increased time in thiocarbohydrazide indicates a complex polysaccharide, such as a glycoprotein (e.g. Thiéry, 1969).

4 Wash sections briefly in aqueous acetic acid solutions of decreasing concentration (10, 8, 6, 4, 2, 0% (v/v) acetic acid), for 60 sec. in total.

5 Float sections for 30–60 min on 1% (w/v) aqueous silver proteinate solution, in darkness.

6 Examine sections in the TEM without further contrasting at 75 or 50 keV.

Controls

Omit either the periodic acid-oxidation, or the thiocarbohydrazide-incubation step (see also Courtoy and Simar, 1974).

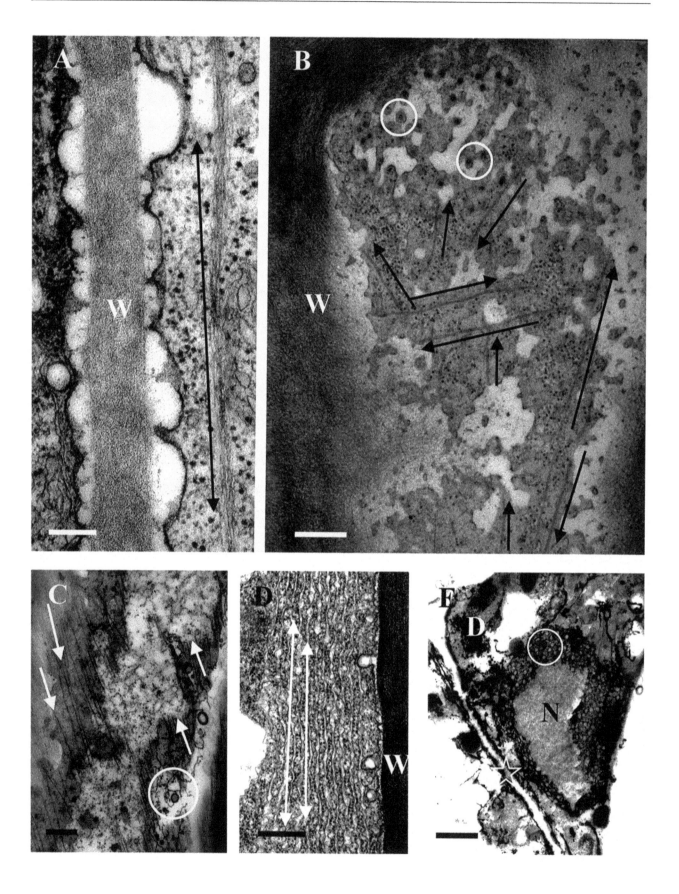

Figure 3 Ultrastructural aspects of the cytoskeletal and endomembrane systems of cells of the secondary vascular system of *Aesculus hippocastanum*. Radial longitudinal sections (A–D) and transverse section (E) of taproot material.
A CTEM of microfilament bundle (arrow) lying approx. parallel to the tangential wall of a fusiform cambial cell.
B CTEM of cell wall-plasmalemma-cytoplasm interface of a fusiform cambial cell. Note the microtubules (arrows), which are arranged at a variety of angles (i.e. randomly), and the plasmodesmata (encircled regions).
C CTEM of numerous, parallel-aligned microtubules (arrows) in a xylem fibre undergoing secondary wall deposition. The arrangement of microtubules approximates to a helix (cf. their random arrangement in cambial cells, Figure 3B; and cf. Figure 2B in the cytoskeletal chapter by Chaffey in this volume, and Figures 1F, H in the Wood microscopical techniques chapter in this volume). Note also the plasmatubules present in the paramural space (encircled region).
D CTEM of several profiles of rough endoplasmic reticulum (arrows) arranged approx. parallel to the tangential wall of a fusiform cambial cell in dormant cambium.
E Zinc-iodide-osmium-tetroxide-staining dramatically highlights the endomembrane system, as indicated in this ray cambial cell. Note the nuclear pores (encircled region) and the dictyosomes. Note also the dark-staining profiles of rough endoplasmic reticulum that lie close to the plasmalemma on each side of the unstained cell wall (☆).
Key: CTEM, conventional transmission electron microscopy (aldehyde-fixation, osmium post-fixation, uranyl acetate-lead citrate double-staining); D, dictyosome; N, nucleus; W, cell wall. Scale bars represent: 0.1 μm (A), 0.5 μm (B, D), and 1 μm (C, E).

In the SVS, a wide variety of cell features – walls, plasmalemma, starch grains, dictyosomes and associated vesicles, and paramural bodies – are stained with this procedure (e.g. Freundlich and Robards, 1974; Czaninski, 1979; Rao and Catesson, 1987; Chaffey *et al.*, 1997c, 1998, 2000a). It is particularly useful for identifying nascent cell walls in recently divided cells (Figure 4B), since it produces intense staining of the plasmalemma in those cells, and very little staining of the cell wall (possibly due to its callosic nature – see below and Figure 4G). Cell walls that are only slightly older are strongly stained with PATAg (Figures 2A, 4D). See also Figures 6, 7, 9, 13 in the chapter by Micheli *et al.* in this volume.

Equipment needed:	Chemicals needed:
Ultrathin Spurr resin sections on 200 mesh gold grids	Phosphotungstic acid
Humid box	Chromic acid
Parafilm	Water
Forceps (to handle grids)	

Phosphotungstic acid (PTA)-staining

This is performed on ultrathin sections of material processed for TEM, either with or without an osmication-step, collected on uncoated gold grids. Osmicated tissue can be used where it is important to discern small intracellular features that are otherwise poorly fixed, but it is necessary to remove the osmium prior to staining (Lewis and Knight, 1977).

Although there is still some debate over the specificity of the staining reaction (e.g. Lewis and Knight, 1977), which is markedly affected by pH (e.g. Hayat, 1993), it can stain PA-S-positive structures, hence it may appear similar to PATAg-staining (see preceding section). Since it is also seen as a guide to the presence of proteins or glyoproteins (e.g. Lewis and Knight, 1977), it is widely used as a plasmalemmal stain (e.g. Roland *et al.*, 1972; Morré, 1990). In the context of the tree SVS, it has been used by such workers as Freundlich and Robards (1974), Parameswaran and Liese (1981), and Rao and Catesson (1987).
Duration of procedure: 15–30 min.
Note – all steps performed at RT in humid box.

Proceed as follows (after Syrop and Beckett, 1972):

1 Float sections for 10–20 min on 1% (w/v) solution of phosphotungstic acid in 10% (w/v) aqueous chromic acid.
2 Wash sections thoroughly with water, air-dry grids and store till needed.
3 View in the TEM at 75 or 50 keV.

In the SVS of *Aesculus hippocastanum* there is markedly more staining associated with older, mature cell walls than in new or nascent cell walls. As for PATAg, the nascent wall-associated staining with PTA in such cells appears largely to be plasmalemmal. Although this can be useful as a method for distinguishing between nascent cell walls and those that are slightly older (Figure 4A), this staining can sometimes be obscured by general staining of the cytoplasm.

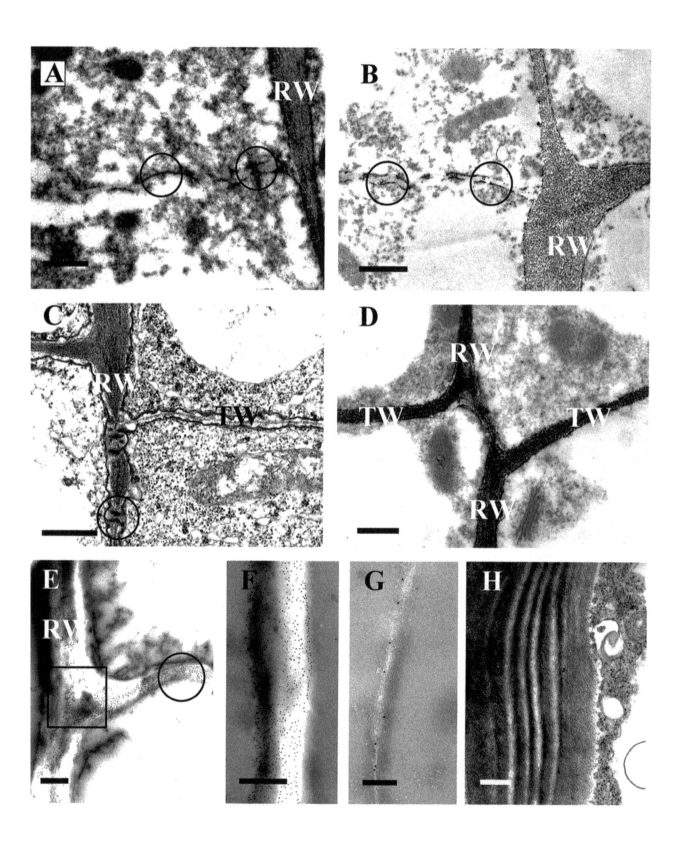

Figure 4 Aspects of cell wall ultrastructure and cytochemistry of cambial zone cells. All are of *Aesculus hippocastanum* taproot material, except G, which is from stem of *Populus tremula* x *P. tremuloides*; transverse (A–E), radial longitudinal (F–H) sections from active (A–G) and dormant (H) cambia.
A, B Phosphotungstic acid- (A) and PATAg- (B) staining of recently-divided fusiform cells. Note that the parental radial walls are heavily stained, but the nascent tangential wall (encircled regions) appears to be unstained, although the associated plasmalemma/wall interface is stained revealing the location of the division wall.
C Typical CTEM image of developing tangential wall between two radial cambial cells (more mature than those in Figures 4A, B), and mature radial and tangential walls. Note the plasmodesmata (encircled region).
D PATAg-staining of mature tangential and radial walls of fusiform cambial cells: staining throughout the walls (cf. nascent tangential wall in Figure 4B).
E Immunogold-labelling of radial and tangential walls of ray cambial cells with antibody, JIM 5, which indicates sites of unesterified pectins. Labelling is greatest in the middle lamella regions of walls (square region) (cf. Figure 4F) and the intercellular junction, but occurs throughout the developing tangential wall (encircled region).
F Immunogold-labelling of tangential wall between two adjacent fusiform cambial cells with antibody, JIM 7, which recognises methyl-esterified pectins. Labelling occurs throughout the cell wall (cf. Figure 4E).
G Immunogold-labelling of an extremely thin nascent tangential wall of a fusiform cambial cell with anti-callose antibody (cf. Figure 1G in the Wood microscopical techniques chapter in this volume).
H CTEM showing the multi-layered thickened tangential cell wall of a fusiform cambial cell from dormant cambium.
Key: CTEM, conventional transmission electron microscopy (aldehyde-fixation, osmium post-fixation, uranyl acetate-lead citrate double-staining); RW, radial cell wall; TW, tangential cell wall. Scale bars represent: 0.2 μm (G) and 0.5 μm (A–F, H).

Immunogold labelling of LR White-embedded tissue

To date I have used this procedure successfully with JIM antibodies (for cell wall/plasmalemmal epitopes – Chaffey *et al.*, 1997b, c, 1998, 2000a) and anti-callose antibody on LR White sections processed as described above. Other workers have used similar procedures for immunolocalisation of other cell wall components (see the chapter by Micheli *et al.* in this volume), and enzymes of lignification (see the chapter by Samaj and Boudet in this volume).

Equipment needed:	Chemicals needed:
Ultrathin LR White sections on 200 mesh gold grids	**Primary and secondary antibodies**
Humid box	**Blocking solution**
Parafilm	**Phosphate-buffered saline (PBS)**
Forceps (to handle grids)	**Lead citrate**
Filter paper	

Duration of procedure: approx. 4.5 h.
Note – sections used are those from material processed for immunogold labelling as described above.
Note – all steps performed at RT in humid box.

1   Float sections on blocking solution for approx. 45 min.

2   Drain excess solution from grid using filter paper, and incubate in primary antibody for approx. 2 h.

3   Wash thoroughly in PBS, 4 changes for 5 min each.

4   Drain excess solution from grid using filter paper, and incubate in secondary antibody for approx. 1 h.

5   Wash thoroughly in PBS, 4 changes for 5 min each.

6   Air-dry grids and store until ready to view.

7   View in TEM either with or without contrasting with lead citrate.

I have used this technique primarily for investigating the distribution of differently-esterified pectins with JIM5 (which recognises unesterified pectins), and JIM7 (which recognises methyl-esterified pectins) (Knox *et al.*, 1990). In cambial cell walls of the SVS of *Aesculus*, JIM5-staining is largely confined to the middle lamella (Figure 4E), whereas JIM7-staining is more evenly distributed throughout the cell wall (Figure 4F) (Chaffey *et al.*, 1997b, c, 1998). This staining pattern largely agrees with that previously described for these antibodies in primary tissues (Knox *et al.*, 1990), and has also been reported for poplar SVS (Guglielmino *et al.*, 1997). Recently, Ermel *et al.* (2000) used a similar technique to immunolocalise arabinans and galactans in poplar. Anti-callose antibody labels nascent tangential cell walls of the tree SVS (Figure 4G), which corroborates the immunofluorescent results with this antibody (see Figure 1G in the Wood

Table 2 A selection of fixation protocols recorded in the literature for conventional ultrastructural studies of angiosperm tree cambia

Ref.	Buffer	Primary fix	Secondary fix	Notes
1	P, pH 6.8–7.0	3% G, 0.25 M B, 2h @ RT, 22h @ 5 °C	2% Os, 0.05 M B + 0.15 M S, 2h @ RT	A-PO/'araldite' 4
2	NS	6% G + 0.15 M S, 18h @ 4 °C, pH 7.0	1% Os + 0.15 M S, 12h @ 4 °C, pH 7.4	TBA/E + 1% UA-PO/ Mollenhauer's 10
3, 4	0.05 M C or 0.1 M P, pH 7.2	3% G, 0.05 M C or 0.1 M P + 0.4 M S + 0.1 M 'chlorides', 4h @ RT or 4 °C	1% Os + 0.2 M B, +S to 0.2 or 0.4 M, 4h @ 4 °C	A + UA-PO/ Mollenhauer's 3, 10
5	0.05 M C, pH 7.5	6% G + B	2% Os	A/Araldite-Epon 14, 16
6	0.1 M NaC, pH 7.2	6.5% G + B, 24h @ RT	1% Os + B 24h @ RT	E-PO/Epon 5
7	0.1 M C, pH 7.5	2–4% G + B, 1–2h	2% Os + B	E/Araldite ± Epon 1, 6–8, 12, 15
8, 9	0.1 M NaC, pH 7.2	2.5% G + 2% F + B, >24h @ RT	1% Os, 24h @ RT	E-PO/Epon 4
10	0.05 M P, pH 7.0	5% G + B, 3–4h	2% Os, O/N @ 4 °C	A/Spurr 13
11	0.1 M P, pH 7.0–7.2	3% G + B ± 0.1 or 0.05 M S, 4h @ RT, O/N @ 4 °C	1–2% Os + B, 2h	A/Spurr 11
12	0.2 M C, pH 7.2	4 0 or 6.5% G + B + S, 2–3h	1% Os, 1–2h	E-PO/Araldite 2
13	0.01 M NaC, pH 7.4	4% G + B, 2–3h	1% Os, 1h	E-PO/Araldite 2, 7
14, 15	0.1 M C, pH 7.2	4% G + B, 1.5h @ RT	1% Os + B, 1h	E-PO/Araldite 1, 5, 7, 9, 12
16	S-P, 0.3 M PIPES, pH 7.0	4% G + B, 4h	1% aq. Os, 4h	A/TAAB 2
16, 17	S-P, pH 7	K + B, 4h	1% aq. Os, 4h	A/TAAB 2
18	0.05 M NaC, pH NS	5% G + B, 2–2.5h + 2.5–3h, ice-cold	2% Os + B O/N @ 4 °C	A (cold)/Spurr 9
19	0.05 M C, pH 7.0	6% G + B, 6h @ RT	2% Os + B 12h @4 °C	Spurr 3

Key:
Ref.: 1, Srivastava (1966); 2, Robards and Humpherson (1967a); 3, Robards and Kidwai (1969b); 4, Farooqui and Robards (1979); 5, Evert and Deshpande (1970); 6, Goosen de-Roo and van Spronsen (1978); 7, Catesson and Roland (1981); 8, Goosen de-Roo *et al.* (1983); 9, Goosen de-Roo *et al.* (1984); 10, Rao and Dave (1983); 11, Sennerby-Forsse (1986); 12, Rao and Catesson (1987); 13, Rao, 1988; 14, Funada and Catesson (1991); 15, Catesson *et al.* (1994); 16, Barnett (1991); 17, Barnett (1992); 18, Farrar and Evert (1997a); 19, Krabel (1998).
Buffer: C, unspecified cacodylate; NaC, sodium cacodylate; NS, not specified; P, phosphate; PIPES, Piperazine-*N,N'-bis*-[2-ethylsulphonic acid]; S-P, Sörenson's phosphate (see Glauert, 1986).
Primary fix/Secondary fix: aq., aqueous; B, buffer (defined in 'Buffer column'); F, formaldehyde; G, glutaraldehyde; K, 'Karnovsky's fixative' (see Karnovsky, 1965); O/N. overnight; Os, osmium tetroxide; RT, room temperature; S, sucrose; TA, tannic acid.
Notes: A, acetone-dehydration; E, ethanol-dehydration; PO, propylene oxide transfer step; TBA, tertiary butyl alcohol-dehydration; UA, uranyl acetate *en bloc* staining; 'Mollenhauer's (three recipes for Epon-araldite, or araldite epoxy resins, see Mollenhauer, 1964), Spurr, and TAAB refer to epoxy resins; 1, *Acer pseudoplatanus*; 2, *Aesculus hippocastanum*; 3, *Fagus sylvatica*; 4, *Fraxinus americana*; 5, *Fraxinus excelsior*; 6, *Pinus sylvestris*; 7, *Populus* x *euramericana*; 8, *Quercus macrocarpa*; 9, *Robinia pseudoacacia*; 10, *Salix fragilis*; 11, *Salix viminalis*; 12, *Sambucus nigra*; 13, *Tectona grandis*; 14, *Tilia americana*; 15, *Tilia platyphyllos*; 16, *Ulmus americana*.

microscopical techniques chapter in this volume), and is consistent with the absence of PATAg-staining of these walls (callose does not stain with PATAg – Roland, 1974).

Controls

Omit primary antibody, or secondary antibody-incubation steps. Replace primary antibody with pre-immune serum, if available. Another useful control is to compare the staining pattern observed with immunogold labelling with that seen in immunofluorescence staining (cf. Figure 4G in this chapter with Figure 1G in the Wood microscopical techniques chapter in this volume). See also the chapters by Micheli *et al.*, and Samaj and Boudet in this volume.

How difficult is it *really* to fix the cambium?

Table 2 summarises several published fixation protocols for the TEM study of cambia of angiosperm trees, and extends the comprehensive study undertaken by Kidwai and Robards (1969a). In brief, it shows that, for primary fixation, glutaraldehyde concentrations range from 2–6.5%, rarely is formaldehyde used, incubation times vary from 1.5 to more than 24 hours, at temperatures from 'ice-cold' to room temperature, in buffers whose strengths range form 0.05 to 0.25 M, and at pH from 6.8 to 7.5, often with addition of large amounts of sucrose (and sometimes with additional salts). Although not as variable, secondary fixations range from 1–2% osmium tetroxide, either buffered or aqueous (sometimes with added sucrose), at 4 °C or RT, for 1 h to overnight.

What can we conclude from this? Although some of the protocols may be species-specific, generally, it seems that these highly varied published protocols are testament to the resilience of the cambium in the face of our attempts to fix it. Which is good news, and should help to persuade others, who still consider the cambium to be too difficult to study, to have a go. However, and although the following may not find favour with editors of scientific journals (who want brevity in all things), it is worth repeating the plea by Kidwai and Robards (1969a) that, 'it is extremely important that … the fullest possible conditions of fixation are cited so that a realistic comparison can be made between the results of work done in different laboratories'.

SOLUTIONS, EQUIPMENT, ETC

Note – unless stated otherwise, 'water' means deionised water.

Antibodies

Primary

JIMs: rat-monoclonal antibodies, JIM5 or JIM7 (a kind gift from Dr JP Knox of the University of Leeds, UK). Used undiluted. Aliquots kept frozen at –20 °C.

Anti-callose (BioSupplies, No. 400–2). This mouse-derived IgG antibody is well characterised by Meikle *et al.* (1991) and specific for $(1 \rightarrow 3)$-β—glucan ('callose'). The immunofluorescence staining pattern in the SVS of *Aesculus* and hybrid aspen is the same as that of aniline blue (the standard fluorescent stain for callose – e.g. Clark, 1981). When reconstituted, according to the supplier's instructions, the solution has an antibody concentration of 1 mg. ml^{-1}. Used diluted 1:20 in **PBS**. Aliquots stored at –20 °C.

Secondary

10 nm gold-labelled anti-rat antibody (British BioCell, No. EM.GAT10), supplied as 0.25 ml solution with protein content 13 μg.ml^{-1}, or, 10 nm gold-labelled anti-mouse antibody (British BioCell, No. EM. GMHL10), supplied as 0.25 ml solution with protein content 15 μg.ml^{-1}.
Both secondary antibodies kept in refrigerator at +4 °C, and used diluted 1:100 in **PBSA**.

Basic TEM fix

2.5% (v/v) glutaraldehyde and 3.7% (w/v) *p*-formaldehyde in 25 mM **PIPES** buffer, pH 6.9. Use EM grade chemicals.

Blocking solution

6% (w/v) bovine serum albumen (BSA) (fraction V, Sigma, No. A2153), 0.1% (v/v) fish skin gelatin (Sigma, No. G-7765), 5% (v/v) normal goat serum (Sigma, No. S-2007), 0.05 M glycine in **PBS**. Filter through 0.45 μm filter, store aliquots at –20 °C.

Humid box

Plastic 'sandwich box', $11.5 \times 17.5 \times 3$ cm deep. The humid environment is created by placing a thoroughly wetted tissue in the base of the box and applying the lid. Place a piece of Parafilm on top of the tissue and apply drops of the solutions – approx. 20 μl – to this with a pipette. Ensure that the box is placed flat, otherwise solutions can run off the Parafilm or mix with other solutions! Where dark incubation conditions are specified, the humid box can simply be covered with a cardboard box, or similar, or placed in a cupboard.

Embedding capsules

Polypropylene, 8 mm diameter, flat-bottomed (TAAB, No. C095)
Polythene, 8 mm diameter, flat-bottomed (TAAB, No. C094)
Polythene capsules are routinely used for conventional TEM embedding. Although polypropylene capsules are recommended for LR White-embedding, polythene capsules can be used with no apparent detrimental effect on the tissue (and it is far easier to remove the resin block from these capsules).

JIM-fix

25 mM **PIPES buffer**, 0, 0.2 or 0.5% (v/v) glutaraldehyde, and 3.7% *p*-formaldehyde, pH 6.9.

Lead citrate

There are many formulations for this stain. The version used here is Fahmy's (after Fahmy, 1967; cited in Lewis and Knight, 1977; Hayat, 1993; Robards and Wilson, 1993).
To make:

1 Dissolve a single pellet of NaOH in 50 ml of water (which has been freshly boiled to expel all the air).
2 Add 0.25 g lead citrate and shake well to dissolve. The solution, pH of approx. 12, keeps for many months in a sealed tube in a refrigerator at +4 °C. Aliquots should be centrifuged before use.
3 Stain grids, taking the usual precautions to avoid precipitation of lead carbonate on the sections (e.g. Lewis and Knight, 1977).

Methylene blue/azur A

Used as a general stain for epoxy and acrylic resin sections to gauge adequacy of fixation, resin-infiltration, etc. (see the Wood microscopical techniques chapter in this volume for formulation of stain and detailed staining protocol).

PBSA

1% (w/v) bovine serum albumen BSA (fraction V, Sigma, No. A2153), 0.1% (v/v) fish skin gelatin (Sigma, No. G-7765) in **PBS**. Filter through 0.45 μm filter, store aliquots at –20 °C.

p-formaldehyde

Stock at 7.4% (w/v) in deionised water, either made up fresh, or freshly thawed from frozen aliquots stored at –20 °C (Doonan and Clayton, 1986).
To make:
Heat 100 ml of water in beaker to barely hand-holdable in microwave, add approx. 20 ml of this to 1.85 g *p*-formaldehyde in a 25 ml volumetric flask, shake to dissolve solid (which dissolves readily but not completely, resulting in a 'cloudy' solution). Add a few (approx. 5) drops of 1.0 M NaOH until the solution clears, and make up to 25 ml. Allow to cool before use.
Note – it is important to check the pH of the final solution, and to adjust to approx. 7. If this is not done, the resulting highly alkaline solution is likely to exceed the buffering capacity of the **Basic TEM fix** giving a fixative solution of unknown pH. [However, I should add that I have used formaldehyde solution whose pH was not adjusted; although the resulting fixative solutions had a pH in excess of 10.0, ultrastructurally the cells appeared 'normal'.]

Phosphate-buffered saline (PBS)

NaCl 8 g, KCl 0.2 g, Na_2HPO_4 1.15 g, KH_2PO_4 0.2 g, NaN_3 0.2 g.l^{-1} of water, pH approx. 7.3–7.6.

PIPES buffer

Piperazine-*N,N'-bis*-[2-ethylsulphonic acid] (Sigma, No. P6757). Stock at 0.2 M, pH 6.9, stored in refrigerator. To make a litre of solution, add sufficient solid to approx. 750 ml of water, check the pH

Table 3 Seasonal changes in ultrastructure of fusiform cambial cells within taproots of *Aesculus hippocastanum* L.

Feature	Active cambium	Dormant cambium
Cell walls	thin, amorphous	thick, lamellate
Nucleus[1]	single	single
Mitochondria	++ (unbranched)	++ (unbranched)
Plastids	++ (little starch)	++ (more starch)
Microtubules[2]	random	helical
Microfilaments[2]	axial	axial
Vacuome	single, large vacuole	dispersed, small vacuoles
Endoplasmic reticulum	++ (rough only)	++ (rough mainly, some smooth)
Dictyosomes	++ (active)	++ (less active)
Secretory vesicles	++	+(+)
Coated vesicles	++	+(+)
PCR/TGN[3]	+	+
Plasmalemma	undulated	undulated
Oleosomes	+	++
Microbodies	+	+ (+)
Ribosomes[4]	++	++

Key: [1], occasionally containing microtubule-like structures; [2], cortical elements; [3], partially-coated reticulum/*trans* Golgi network-like structure (Chaffey *et al.*, 1997c); [4], frequently in polysomal configurations; +, +(+), ++, increasing abundance.

(which will be very low), and use 1.0 M KOH to increase pH. Take pH to approx. 6.5, leave solution to stir for approx. 30 min, before increasing the pH to 6.9, and making up to 1L.

Note – this procedure will use a lot of KOH, so be patient.

Spurr resin

Hard grade, supplied as a pre-mix kit (TAAB, No. S031). When made up, store at –20 °C; but allow to reach RT before using. Although this is my resin of choice, some laboratories forbid its use (on health

Table 4 Seasonal changes in ultrastructure of ray cambial cells within taproots of *Aesculus hippocastanum* L.

Feature	Active cambium	Dormant cambium
Cell walls	thin, amorphous	thick, lamellate
Nucleus[1]	single, centrally-located	single, centrally-located
Mitochondria	++ (unbranched)	++ (unbranched)
Plastids	++ (little starch)	++ (more starch)
Microtubules	random CMTs; EMTs	helical CMTs; EMTs
Microfilaments	CMFs rare; EMFs	CMFs rare; EMFs
Vacuome	single, large vacuole	numerous, small vacuoles
Endoplasmic reticulum	++ rough only	++ rough only
Dictyosomes	++ (active)	++ (less active)
Secretory vesicles	++	+
Coated vesicles	++	+(+)
PCR/TGN[2]	+	+
Plasmalemma	much undulated	undulated
Oleosomes	+	++
Microbodies[3]	+	+(+)
Ribosomes	++	++

Key: [1], occasionally containing microtubule-like structures; [2], partially-coated reticulum/*trans* Golgi network-like structure (Chaffey *et al.*, 1997c); [3], occasionally with a crystalline inclusion; +, +(+), ++, increasing abundance; CMFs, cortical microfilaments; CMTs, cortical microtubules; EMFs, endoplasmic microfilaments; EMTs, endoplasmic microtubules.

grounds); a satisfactory alternative is Epon or TAAB resin.

Toluidine blue

For staining of resin-embedded sections to gauge adequacy of fixation, resin-infiltration, etc., use a 0.01% (w/v) aqueous solution. Staining colours as for methylene blue/azur A above (see also the entry in the Wood microscopical techniques chapter in this volume).

Zinc-iodide-osmium-tetroxide solution (ZIO)

Prepared according to the recipe of Harris (1978). Add equal volumes of 2% (w/v) aqueous osmium tetroxide and zinc iodide solution together, immediately prior to use. Zinc iodide solution is made as follows: Add 1.5 g powdered zinc to 0.5 g resublimed iodine in 10 ml water. Sonicate the solution for approx. 1 min, stir for a further 5 min, filter, use.

APPLICATIONS

The introduction to this chapter covers some of the many contributions of TEM to wood biology; some applications of the procedures described above have been mentioned in the relevant technical section. General aspects of cambial ultrastructure have been thoroughly dealt with in recent publications, Catesson (1994), Iqbal (1994, 1995), Larson (1994), and Lachaud et al. (1999) and it is neither necessary nor appropriate to repeat that work here. Rather, this section will highlight a few of the more recent TEM observations of relevance to wood cell biology.

Root versus shoot

It is implicit, if not explicit, in the majority of the published literature that the ultrastructure of root cambial cells is the same as those of the shoot. In order to test that assumption I have studied the cambium of the taproots of *Aesculus hippocastanum* and compared it with the cambium of the epicotyl and shoot (Chaffey et al., 1997c, 1998). A summary of the results of root cambial ultrastructure is presented in Table 3 (for fusiform cells) and Table 4 (for ray cells) and illustrated in the accompanying figures. Encouragingly (for nobody likes to over-

turn long-held beliefs!), active cambia in root and shoot are alike. Furthermore, as for shoots, there are many ultrastructural similarities between ray and fusiform cambial cells of roots.

It is well established that shoot cambial cells of temperate trees undergo a seasonal cycle of activity. Associated with that is a cycle of ultrastructural changes (e.g. Fig. 5.19 in Larson, 1994). In *Aesculus*, a cycle of ultrastructural changes, similar to that in shoots, also occurs in root cambial cells (Tables 3 and 4).

Plasmatubules

Plasmatubules (PTs) (Harris et al., 1982; Chaffey and Harris, 1985a, b) are small-diameter, tubular evaginations of the plasmalemma found at sites where there is an inferred, but relatively short-term, high flux between the apoplasm and the symplasm (Harris and Chaffey, 1985, 1986). Although greeted with some scepticism when they were first described, the validity of the PTs has since been demonstrated by use of a wide variety of fixation procedures (Chaffey and Harris, 1985a). Plasmatubules have subsequently been described in several sites, e.g., growing pollen tubes (Kandasamy et al. 1988; Dawkins and Owens, 1993), the host-parasitic angiosperm interface (Coetzee and Fineran, 1989; Condon and Kuut, 1994), infection hyphae of *Colletotricum* (Pring et al., 1995), and lead-grown onion roots (Wierzbicka, 1998).

Plasmatubules are seen in differentiating xylem cells in *Aesculus hippocastanum* (Figure 3C) (Chaffey et al., 1999); but I have not seen them in differentiating phloem cells. The plasmalemma-bounded symplasmic compartment of an individual differentiating xylem cell is surrounded by a large apoplasmic space, which consists of its own cell wall, the walls of other cells, and the lumina of dead, 'empty' xylem cells. The observed distribution of PTs within secondary xylem cells can be related to their inferred role in symplasm-apoplasm exchange, in this instance facilitating the resorption of the products of xylem cell-lysis from the extracellular medium. This suggested role for PTs is attractive since it would permit recycling of valuable materials, and receives experimental support from the study of Coetzee and Fineran (1989) who examined uptake of radioactively labelled lysine by dwarf mistletoe from its host. Furthermore, absence of PTs from differentiating phloem cells is consistent with this suggestion for xylem, since here symplasmic routes between adjacent cells via plasmodesmata are maintained. PTs are illustrated and discussed in the study of tracheid differentiation in *Cryptomeria* (Takabe and Harada, 1986), differentiating tracheary elements of *Salix dasyclados* (Sennerby-Forsse and von Fircks, 1987), and differentiating bast fibres in flax (Sal'nikov *et al.*, 1993). PTs are also evident in the published work of Cronshaw (1965a) and Barnett (1981) on xylem fibres of *Acer* and *Aesculus*, respectively.

Orientation of microtubules in cambial cells: a cautionary tale

The majority of early TEM studies of active cambial cells of the tree SVS concluded that the MTs were present in an ordered arrangement (see discussion in Chaffey *et al.*, 1996). My initial immuno-fluorescent studies of MTs in active cambial cells of *Aesculus* showed only random arrays (Chaffey *et al.*, 1996; see the cytoskeletal chapter by Chaffey in this volume). Since almost every published account of immunofluorescent localisation of MT orientation in interphase cells – although none of them dealing with cambium – showed them as being parallel transverse, longitudinal or oblique, there was some uncertainty over the validity of the image seen in the cambium. However, the reproducibility of that random array in different fixations for indirect immunofluorescence (IIF) study (Chaffey *et al.*, 1996, 1997d), *and* demonstration of the same array in TEM (Figure 3B; Chaffey *et al.*, 1996, 1997a, b), *and* its presence in several other woody taxa (Chaffey *et al.*, 1997d), *and* its different arrangement in dormant cambia (Chaffey *et al.*, 1998), all attested to its validity (see also discussion in Chaffey *et al.*, 1996).

That study carries several important messages. One, when examining cell types with techniques that have not previously been tried, believe what you see, it may be correct (subject to it being reproducible and corroborated!). Two, TEM is 'selective' to the extent that it is at best a 2-dimensional representation of a 3-D array, which may well appear different when the full picture is revealed. Three, the undoubted value of using correlative microscopy, in this case immunolocalisation at the light microscope level and TEM. Although this is not strict correlative microscopy, in the sense that serial or same sections are viewed with LM and TEM (e.g. Hayat, 1987), tissue from the same

sampling is subsequently processed for TEM or for IIF. In that way, the fact that different processing methods are used on a common structure, e.g. cyto-skeleton, and both give similar images, strengthens the validity of the observed image.

FUTURE PROSPECTS

As with most new techniques, when TEM first became commercially available, there was initially a flood of work that attempted to exploit the new technology. Thereafter, newer techniques come along and, to some extent, the older ones go out of fashion. However, because of the vastly increased resolution of TEM over light microscopy (LM), it seems likely that TEM – in one guise or another – is going to be around for many years to come. Although study of ultrastructure 'for its own sake' is not readily fundable in today's research environment (and is often frowned upon!), it is still an extremely rewarding pursuit, and continues to provide new insights into the workings of the cell. Further, because of the technical difficulties in performing TEM on woody material (see 'An introduction to the problems of working with trees', in this volume), the study of this material lags many years behind that performed on more amenable tissues: there is still much to do!

However, it is acknowledged that to get the most out of any investigation it is necessary to use several techniques, which often embrace several specialisms. Thus, nowadays, TEM increasingly takes its place alongside LM and fluorescence microscopy in unlocking the secrets of the cell.

Continued use of correlative TEM and other observational techniques

It is difficult to overstress the importance of getting information from as many different techniques as possible in order to arrive at a consensus view of a particular structure. To that end, TEM will continue to play an important role in all future 'anatomical' studies. Increasingly, TEM will also have application to 'molecular' studies that require, for example, precise information about the localisation of products of gene expression. In that regard, the combination of SEM and immunogold localisation techniques (e.g. Reichelt *et al.*, 1995) is a promising development which should be applied to study of the tree SVS.

Improvements in fixation of the tree SVS

In order to make fuller use of the improved resolution of TEM over LM, it is necessary to ensure that the best – i.e. most life-like – fixation of the cells of the tree SVS is obtained. In particular, it will be necessary to experiment with 'fast-fixation' procedures. Of these, freeze-fixing is the most widespread. Although freeze-etching, freeze-fixation and/or freeze-substitution procedures have been successfully applied to trees (e.g. Parish, 1974; Inomata *et al.*, 1992; Ristic and Ashworth, 1993; Fujikawa *et al.*, 1999; see the chapter by Rensing in this volume), the accessibility of the cells of the SVS, and the wide variety of cell types (especially in angiosperms – see Table 1 in the Introduction to this volume), present considerable technical difficulties of adequate fixation of all cells with avoidance of ice-crystal damage. The comparatively recent development of microwave-fixation may offer more hope for 'tree work'. Such fixation can be fast and give ultrastructural images comparable to those from the best freeze-fixation procedures (Login and Dvorak, 1994), and antigenicity can be satisfactorily preserved for immunological work (Benhamou *et al.*, 1991).

ACKNOWLEDGEMENTS

I thank Prof. Nick Harris for my early introduction to the joys of TEM, and Prof. John Barnett for my more recent induction to the fascinating world of wood ultrastructure. I am also grateful to Dr J.P. Knox for generous gifts of antibodies.

REFERENCES

Awano, T., Takabe, K. and Fujita, M. (1998) Localization of glucuronoxylans in Japanese beech visualized by immunogold labelling. *Protoplasma*, **202**, 213–222.

Barnett, J.R. (1981) Secondary xylem cell development. In *Xylem Cell Development*, edited by J.R. Barnett, pp. 47–95. Tunbridge Wells, Kent, UK: Castle House Publications Ltd.

Barnett, J.R. (1987) Changes in the distribution of plasmodesmata in developing fibre-tracheid pit membranes of *Sorbus aucuparia* L. *Annals of Botany*, **59**, 269–279.

Barnett, J.R. (1991) Microtubules in interphase nuclei of *Aesculus hippocastanum* L. *Annals of Botany*, **68**, 159–165.

Barnett, J.R. (1992) Reactivation of the cambium in *Aesculus hippocastanum* L.: a transmission electron microscope study. *Annals of Botany*, **70**, 169–177.

Benhamou, N., Noel, S., Grenier, J. and Asselin, A. (1991) Microwave energy fixation of plant tissue: an alternative approach that provides excellent preservation of ultrastructure and antigenicity. *Electron Microscopy Techniques*, **17**, 81–94.

Bosshard, H.H. (1951) Elektronmikroskopische Untersuchungen im Holz von *Fraxinus excelsior* L. *Berichte Schweizerischen Botanischen Gesellschaft*, **62**, 482–508.

Butterfield, B.G. (1995) Vessel element differentiation. In *The Cambial Derivatives*, edited by M. Iqbal, pp. 93–106. Berlin: Gebrüder Borntraeger.

Catesson, A.-M. (1974) Cambial cells. In *Dynamic Aspects of Plant Ultrastructure*, edited by A.W. Robards, pp. 358–390. London: McGraw-Hill Book Co.

Catesson, A.-M. (1994) Cambial ultrastructure and biochemistry: changes in relation to vascular tissue differentiation and the seasonal cycle. *International Journal of Plant Sciences*, **155**, 251–261.

Catesson, A.M., Funada, R., Robert-Baby, D., Quinet-Szély, M., Chu-Bâ, J. and Goldberg, R. (1994) Biochemical and cytochemical cell wall changes across the cambial zone. *IAWA Journal*, **15**, 91–101.

Catesson, A.-M. and Roland, J.-C. (1981) Sequential changes associated with cell wall formation and fusion in the vascular cambium. *IAWA Bulletin, New series*, **2**, 151–162.

Cavusoglu, I., Kahveci, Z. and Sirmali, S.A. (1998) Rapid staining of ultrathin sections with the use of a microwave oven. *Journal of Microscopy*, **192**, 212–216.

Chafe, S.C. (1974) Cell wall formation and 'protective layer' development in the xylem parenchyma of trembling aspen. *Protoplasma*, **80**, 335–354.

Chaffey, N.J. (1995) Structure and function in the grass ligule: the endomembrane system of the adaxial epidermis of the membranous ligule of *Lolium temulentum* L. (Poaceae). *Annals of Botany*, **76**, 103–112.

Chaffey, N.J. (1996) Structure and function of the root cap of Lolium temulentum L. (Poaceae): parallels with the ligule. Annals of Botany, 78, 3–13.

Chaffey, N.J., Barlow, P.W. and Barnett, J.R. (1996) Microtubular cytoskeleton of vascular cambium and its derivatives in roots of *Aesculus hippocastanum* L. (Hippocastanaceae). In *Recent Advances in Wood Anatomy*, edited by L.A. Donaldson, B.G. Butterfield, P.A. Singh and L.J. Whitehouse, pp. 171–183. Rotorua, NZ: New Zealand Forest Research Institute.

Chaffey, N.J., Barlow, P.W. and Barnett, J.R. (1997a) Cortical microtubules rearrange during differentiation of vascular cambial derivatives, microfilaments do not. *Trees*, **11**, 333–341.

Chaffey, N.J., Barlow, P.W. and Barnett, J.R. (1998) A seasonal cycle of cell wall structure is accompanied by a cyclical rearrangement of cortical microtubules in fusiform cambial cells within taproots of *Aesculus hippocastanum* (Hippocastanaceae). *New Phytologist*, **139**, 623–635.

Chaffey, N.J., Barlow, P.W. and Barnett, J.R. (2000a) Structure-function relationships during secondary phloem development in *Aesculus hippocastanum*: microtubules and cell walls. *Tree Physiology*, **20**, 777–786.

Chaffey, N.J., Barlow, P.W. and Barnett, J.R. (2000b) A cytoskeletal basis for wood formation in angiosperm trees: the involvement of microfilaments. *Planta*, **210**, 890–896.

Chaffey, N.J., Barnett, J.R. and Barlow, P.W. (1997b) Cortical microtubule involvement in bordered pit formation in secondary xylem vessel elements of *Aesculus hippocastanum* L. (Hippocastanaceae): a correlative study using electron microscopy and indirect immunofluorescence microscopy. *Protoplasma*, **197**, 64–75.

Chaffey, N.J., Barnett, J.R. and Barlow, P.W. (1997c) Endomembranes, cytoskeleton and cell walls: aspects of the ultrastructure of the vascular cambium of taproots of *Aesculus hippocastanum* L. (Hippocastanaceae). *International Journal of Plant Sciences*, **158**, 97–109.

Chaffey, N.J., Barnett, J.R. and Barlow, P.W. (1997d) Visualization of the cytoskeleton within the secondary vascular system of hardwood species. *Journal of Microscopy*, **187**, 77–84.

Chaffey, N.J., Barnett, J.R. and Barlow, P.W. (1999) A cytoskeletal basis for wood formation in angiosperm trees: the involvement of cortical microtubules. *Planta*, **208**, 19–30.

Chaffey, N.J. and Harris, N. (1985a) Plasmatubules: fact or artefact? *Planta*, **165**, 185–190.

Chaffey, N.J. and Harris, N. (1985b) Localisation of ATPase activity on the plasmalemma of scutellar epithelial cells of germinating barley (*Hordeum vulgare* L.). *Journal of Experimental Botany*, **36**, 1612–1619.

Clark, G. (ed) (1981) *Staining procedures*, 4th edn. Baltimore, London: Williams and Wilkins.

Clausen, S. and Apel, K. (1991) Seasonal changes in the concentration of the major storage protein and its mRNA in xylem ray cells of poplar trees. *Plant Molecular Biology*, **17**, 669–678.

Coetzee, J. and Fineran, B.A. (1989) Translocation of lysine from the host *Melicope simplex* to the parasitic dwarf mistletoe *Korthalsella lindsayi* (Viscaceae). *New Phytologist*, **112**, 377–381.

Condon, J. and Kuut, J. (1994) Anatomy and ultrastructure of the primary endophyte of *Ileostylus micranthus* (Loranthaceae). *International Journal of Plant Sciences*, **155**, 350–364.

Côté, W.A., Jr (1958) Electron microscope studies of pit membrane structure: implications in seasoning and preservation of wood. *Forest Products Journal*, **8**, 296–301.

Côté, W.A., Jr (ed) (1965). *Cellular ultrastructure of woody plants*. Syracuse: Syracuse University Press.

Côté, W.A., Jr and Day, A.C. (1962) The G layer in gelatinous fibers – electron microscopic studies. *Forest Products Journal*, **12**, 333–338.

Côté, W.A., Jr, Day, A.C. and Timell, T.E. (1969) A contribution to the ultrastructure of tension wood fibers. *Wood Science and Technology*, **3**, 257–271.

Côté, W.A. and Hanna, R.B. (1985) Trends in application of electron microscopy to wood research. In *Xylorama*, edited by L.J. Kucera, pp. 42–50. Basel, Boston, Stuttgart: Birkhäuser Verlag.

Courtoy, R. and Simar, L.J. (1974) Importance of controls for the demonstration of carbohydrates in electron microscopy with the silver methenamine or the thiocarbohydrazide-silver proteinate methods. *Journal of Microscopy*, **100**, 199–211.

Cronshaw, J. (1965a) The organization of cytoplasmic components during the phase of cell wall thickening in differentiating cambial derivatives of *Acer rubrum*. *Canadian Journal of Botany*, **43**, 1401–1407.

Cronshaw, J. (1965b) Cytoplasmic fine structure and cell wall development in differentiating xylem elements. In *Cellular Ultrastructure of Woody Plants*, edited by W.A. Côté, Jr, pp. 99–124. Syracuse: Syracuse University Press.

Crow, E., Murphy, R.J. and Cutler, D.F. (2000) Development of fibre and parenchyma cells in culms of bamboo (*Phyllostachys viridi-glauescens* (Carr.) Riv. & Riv.) with emphasis on the cytoskeleton during cell wall deposition. In: *Cell and Molecular Biology of Wood Formation*, edited by R. Savidge, J. Barnett and R. Napier, pp. 265–276. Oxford: Bios Scientific Publishers.

Czaninski, Y. (1966) Aspects infrastructuraux de cellules contiguës aux vaiseaux dans la xylème de *Robinia pseudo-acacia*. *Comptes Rendus de l'Academie des Sciences, Série D*, **262**, 2336–2339.

Czaninski, Y. (1968) Étude du parenchyme ligneux du Robinier (parenchyme a réserves et cellules associées aux vaisseaux) au cours du cycle annuel. *Journal de Microscopie*, **76**, 145–164.

Czaninski, Y. (1973) Observations sur une nouvelle couche pariétale dans les cellules associées aux vaisseaux du Robinier et du Sycomore. *Protoplasma*, **77**, 211–219.

Czaninski, Y. (1977) Vessel-associated cells. *IAWA Bulletin*, **1977/3**, 51–55.

Czaninski, Y. (1979) Cytochimie ultrastructurale des parois du xylème secondaire. *Biologie Cellulaire*, **35**, 97–102.

Dave, Y.S. and Rao, K.S. (1981) Plastid ultrastructure in the cambium of teak (*Tectona grandis* L.f.). *Annals of Botany*, **49**, 425–427.

Dawkins, M.D. and Owens, J.N. (1993) *In vitro* and *in vivo* pollen hydration, germination, and pollen-tube growth in white spruce, *Picea glauca* (Moench) Voss. *International Journal of Plant Sciences*, **154**, 506–521.

Donaldson, L.A. and Singh, A.P. (1998) Bridge-like structures between cellulose microfibrils in radiata pine (*Pinus radiata* D. Don) kraft pulp and holocellulose. *Holzforschung*, **52**, 449–454.

Doonan, J.H. and Clayton, L. (1986) Immunofluorescence studies on the plant cytoskeleton. In *Immunology in Plant Science*, edited by T.L. Wang, pp. 111–136. Cambridge, London: Cambridge University Press.

Ermel, F.E., Follet-Gueye, M.-L., Cibert, C., Vian, B., Morvan, C., Catesson, A.-M. and Goldberg, R. (2000) Differential localization of arabinan and galactan side chains of rhamnogalacturonan 1 in cambial derivatives. *Planta*, **210**, 732–740.

Evert, R.F. and Deshpande, B.P. (1970) An ultrastructural study of cell division in the cambium. *American Journal of Botany*, **57**, 942–961.

Exley, R.R., Butterfield, B.G. and Meylan, B.A. (1974) Preparation of wood specimens for the scanning electron microscope. *Journal of Microscopy*, **101**, 21–30.

Fahmy, A. (1967) An extemporaneous lead citrate stain for electron microscopy. *Proceedings of the 25th EMSA Conference* (cited in Robards and Wilson, 1993).

Farooqui, P. and Robards, A.W. (1979) Seasonal changes in the ultrastructure of cambium of *Fagus sylvatica* L. *Proceedings of the Indian Academy of Sciences*, **88B**, 463–472.

Farrar, J.J. (1995) Ultrastructure of the vascular cambium of *Robinia pseudoacacia*. *PhD Thesis*, University of Wisconsin, Madison, USA (unpublished).

Farrar, J.J. and Evert, R.F. (1997a) Seasonal changes in the ultrastructure of the vascular cambium of *Robinia pseudoacacia*. *Trees*, **11**, 191–202.

Farrar, J.J. and Evert, R.F. (1997b) Ultrastructure of cell division in the fusiform cells of the vascular cambium of *Robinia pseudoacacia*. *Trees*, **11**, 203–215.

Freundlich, A. and robards, A.W. (1974) Cytochemistry of differentiating plant vascular cell walls with special reference to cellulose. *Cytobiologie*, **8**, 355–370.

Fujikawa, S., Kuroda, K., Jitsuyama, Y., Sano, Y. and Ohtani, J. (1999) Freezing behavior of xylem ray parenchyma cells in softwood species with differences in the organization of cell walls. *Protoplasma*, **206**, 31–40.

Fujino, T. and Itoh, T. (1998) Changes in the three-dimensional architecture of the cell wall during lignification of xylem cells in *Eucalyptus tereticornis*. *Holzforschung*, **52**, 111–116.

Fujita, M., Saiki, H. and Harada, H. (1974) Electron microscopy of microtubules and cellulose microfibrils in secondary wall formation of poplar tension wood fibres. *Mokuzai Gakkaishi*, **20**, 147–156.

Funada, R. and Catesson, A.-M. (1991) Partial cell wall lysis and the resumption of meristematic activity in *Fraxinus excelsior* cambium. *IAWA Bulletin, New series*, **12**, 439–444.

Glauert, A.M. (1986) *Fixation, Dehydration and Embedding of Biological Specimens*. Amsterdam: North Holland Publishing Company.

Goosen-de Roo, L. (1981) Plasmodesmata in the cambial zone of *Fraxinus excelsior* L. *Acta Botanica Neerlandica*, **30**, 156.

Goosen-de Roo, L., Bakhuizen, R., van Spronsen, P.C. and Libbenga, K.R. (1984) The presence of extended phragmosomes containing cytoskeletal elements in fusiform cambial cells of *Fraxinus excelsior* L. *Protoplasma*, **122**, 145–152.

Goosen-de Roo, L., Burggraaf, P.D. and Libbenga, K.R. (1983) Microfilament bundles associated with tubular endoplasmic reticulum in fusiform cells in the active cambial zone of *Fraxinus excelsior* L. *Protoplasma*, **116**, 204–208.

Goosen-de Roo, L. and Creyghton-Schouten, E.A.M. (1981) Plasmodesmata in cell walls in the cambial zone of *Fraxinus excelsior* L. *Abstract from the 2nd Cell Wall Meeting, Göttingen*.

Goosen-de Roo, L. and van Spronsen, P.C. (1978) Electron microscopy of the active cambial zone of *Fraxinus excelsior* L. *IAWA Bulletin*, **1978/4**, 59–64.

Guglielmino, N., Liberman, M., Jauneau, A., Vian, B., Catesson, A.M. and Goldberg, R. (1997) Pectin immunolocalisation and calcium visualization in differentiating derivatives from poplar cambium. *Protoplasma*, **199**, 151–160.

Hafrén, J., Fujino, T. and Itoh, T. (1999) Changes in cell wall architecture of differentiating tracheids of *Pinus thunbergii* during lignification. *Plant & Cell Physiology*, **40**, 532–541.

Harada, H. (1965) Ultrastructure of angiosperm vessels and ray parenchyma. In *Cellular Ultrastructure of Woody Plants*, edited by W.A. Côté, Jr, pp. 235–249. Syracuse: Syracuse University Press.

Harms, U. and Sauter, J.J. (1992) Localization of a storage protein in the wood ray parenchyma cells of *Taxodium distichum* (L.) L.C. Rich. by immunogold labelling. *Trees*, **6**, 37–40.

Harris, N. (1978) Nuclear pore distribution and relation to adjacent cytoplasmic organelles in cotyledon cells of developing *Vicia faba*. *Planta*, **141**, 121–128.

Harris, N. and Chaffey, N.J. (1985) Plasmatubules in transfer cells of pea (*Pisum sativum* L.). *Planta*, **165**, 191–196.

Harris, N. and Chaffey, N.J. (1986) Plasmatubules – real modifications of the plasmalemma. *Nordic Journal of Botany*, **6**, 599–607.

Harris, N., Oparka, K.J. and Walker-Smith, D.W. (1982) Plasmatubules: an alternative to transfer cells? *Planta*, **156**, 461–465.

Hayat, M.A. (ed.) (1987) *Correlative Microscopy in Biology: Instrumentation and Methods*. London, New York: Academic Press.

Hayat, M.A. (1993) *Stains and Cytochemical Methods*. New York, London: Plenum Press.

Inomata, F., Takabe, K. and Saiki, H. (1992) Cell wall formation of conifer tracheid as revealed by rapid-freeze and substitution method. *Journal of Electron Microscopy*, **41**, 369–374.

Iqbal, M. (1994) Structural and operational specializations of the vascular cambium of seed plants. In *Growth Patterns in Vascular Plants*, edited by M. Iqbal, pp. 211–271. Portland, Oregon: Dioscorides Press.

Iqbal, M. (1995) Structure and behaviour of vascular cambium and the mechanism and control of cambial growth. In *The Cambial Derivatives*, edited by M. Iqbal, pp. 1–67. Berlin: Gebrüder Borntraeger.

Kandasamy, M.K., Kappler, R. and Kristen, U. (1988) Plasmatubules in the pollen tubes of *Nicotiana sylvestris*. *Planta*, **173**, 35–41.

Karnovsky, M.J. (1965) A formaldehyde-glutaraldehyde fixative of high osmolality for use in electron microscopy. *Journal of Cell Biology*, **27**, 137A (cited in Barnett, 1991).

Kidwai, P. and Robards, A.W. (1969a) The appearance of differentiating vascular cells after fixation in different solutions. *Journal of Experimental Botany*, **20**, 664–670.

Kidwai, P. and Robards, A.W. (1969b) On the ultrastructure of resting cambium of *Fagus sylvatica* L. *Planta*, **89**, 361–368.

Knox, J.P., Linstead, P.J., King, J., Cooper, C. and Roberts, K. (1990) Pectin esterification is spatially regulated both within cell walls and between developing tissues of root apices. *Planta*, **181**, 512–521.

Krabel, D. (1998) Mikroanalytische Untersuchungen zur Physiologie des Baumkambiums am Beispiel von *Thuja occidentalis* L. und *Fagus sylvatica* L.

Contributions to Forest Sciences No. 2. Dresden: Technische Universität.

Kucera, L.J. (1986) Splitting wood specimens for observation in the scanning electron microscope. *Journal of Microscopy*, **142**, 71–77.

Lachaud, S., Catesson, A.-M. and Bonnemain, J.-L. (1999) Structure and functions of the vascular cambium. *Comptes Rendus de l'Academie des Sciences, Série III, Sciences de la vie*, **322**, 633–650.

Larson, P.R. (1994) *The Vascular Cambium: Development and Structure.* Berlin: Springer.

Ledbetter, M.C. and Porter, K.R. (1963) A "microtubule" in plant cell fine structure. *Journal of Cell Biology*, **19**, 239–250.

Lewis, P.R. and Knight, D.P. (1977) *Staining Methods for Sectioned Material.* Amsterdam: North Holland Publishing Company.

Login, G.R. and Dvorak, A.M. (1994) *Methods of Microwave Fixation for Microscopy: a Review of Research and Clinical Applications 1970–1992.* Stuttgart, New York: Gustav Fischer Verlag.

Machado, R.D. and Schmid, R. (1964) Observations on the structure of pit membranes in hardwood cells. *Proceedings of the Third European Regional Conference on Electron Microscopy*, 163–164.

Meikle, P.J., Bonig, I., Hoogenraad, N.J., Clarke, A.E. and Stone, B.A. (1991) The location of $(1\rightarrow3)$-β-glucans in the walls of pollen tubes of *Nicotiana alata* using a $(1\rightarrow3)$-β-glucan-specific monoclonal antibody. *Planta*, **185**, 1–8.

Meylan, B.A. and Butterfield, B.G. (1972) *Three-dimensional Structure of Wood: A Scanning Electron Microscope Study.* London: Chapman and Hall Ltd.

Mia, A.J. (1968) Organization of tension wood fibres with special reference to the gelatinous layer in *Populus tremuloides* Michx. *Wood Science*, **1**, 105–115.

Mia, A.J. (1970) Fine structure of active, dormant, and aging cambial cells in *Tilia americana*. *Wood Science*, **3**, 34–42.

Mia, A.J. (1972) Fine structure of the ray parenchyma cells in *Populus tremuloides* in relation to senescence and seasonal changes. *Texas Journal of Science*, **24**, 245–260.

Mollenahuer, H.H. (1964) Plastic embedding mixtures for use in electron microscopy. *Stain Technology*, **39**, 111–114.

Morré, D.J. (1990) Plasma membrane cytochemistry. In *The Plant Plasma Membrane*, edited by C. Larsson and I.M. Møller, pp. 76–92. Berlin: Springer.

Murmanis, L. (1977) Development of vascular cambium into secondary tissue of *Quercus rubra* L. *Annals of Botany*, **41**, 617–620.

Nelmes, B.J., Preston, R.D. and Ashworth, D. (1973) A possible function of microtubules suggested by their abnormal distribution in rubbery wood. *Journal of Cell Science*, **13**, 741–751.

Parameswaran, N. and Liese, W. (1981) Ultrastructural localization of wall components in wood cells. In *The Ekman-Days 1981 Morphological Distribution of Wood Components: New Understanding of the Chemical Structure of Wood Components, Volume 1*, pp. I:16–I:31, International Symposium on Wood and Pulping Chemistry, Stockholm, 9–12 June, 1981.

Parish, G.R. (1974) Seasonal variation in the membrane structure of differentiating shoot cambial-zone cells demonstrated by freeze-etching. *Cytobiologie*, **9**, 1131–143.

Pring, R.J., Nash, C., Zakaria, M. and Bailey, J.A. (1995) Infection process and host range of *Collectotrichum capsici*. *Physiological and Molecular Plant Pathology*, **46**, 137–152.

Prodhan, A.K.M.A., Funada, R., Ohtani, J., Abe, H. and Fukuzawa, K. (1995) Orientation of microfibrils and microtubules in developing tension-wood fibres of Japanese ash (*Fraxinus mandshurica* var. *japonica*). *Planta*, **196**, 577–585.

Rao, K.S. (1985) Seasonal ultrastructural changes in the cambium of *Aesculus hippocastanum* L. *Annales des Sciences Naturelles, Botanie, 13ème série*, **7**, 213–228.

Rao, K.S. (1988) Fine structure details of tannin accumulations in non-dividing cambial cells. *Annals of Botany*, **62**, 575–581.

Rao, K.S. and Catesson, A.-M. (1987) Changes in the membrane components of nondividing cambial cells. *Canadian Journal of Botany*, **65**, 246–254.

Rao, K.S. and Dave, Y.S. (1983) Ultrastructure of active and dormant cambial cells in teak (*Tectona grandis* L. f.). *New Phytologist*, **93**, 447–456.

Reichelt, S., Ensikat, H.-J., Barthlott, W. and Volkmann, D. (1995) Visualization of immunogold-labeled cytoskeletal proteins by scanning electron microscopy. *European Journal of Cell Biology*, **67**, 89–93.

Rioux, D., Nicole, M., Simard, M. and Ouellette, G.B. (1998) Immunocytochemical evidence that secretion of pectin occurs during gel (gum) and tylosis formation in trees. *Phytopathology*, **88**, 494–505.

Ristic, Z. and Ashworth, E.N. (1993) New infiltration method permits use of freeze-substitution for preparation of wood tissues for transmission electron microscopy. *Journal of Microscopy*, **171**, 137–142.

Robards, A.W. (1967) The xylem fibres of *Salix fragilis* L. *Journal of the Royal Microscopical Society*, **87**, 329–352.

Robards, A.W. (1968) On the ultrastructure of differentiating secondary xylem in willow. *Protoplasma*, **65**, 449–464.

Robards, A.W. and Humpherson, P.G. (1967a) Phytoferritin in plastids of the cambial zone of willow. *Planta*, **76**, 169–178.

Robards, A.W. and Humpherson, P.G. (1967b) Microtubules and angiosperm bordered pit formation. *Planta*, **77**, 233–238.

Robards, A.W. and Kidwai, P. (1969a) Vesicular involvement in differentiating plant vascular cells. *New Phytologist*, **68**, 343–349.

Robards, A.W. and Kidwai, P. (1969b) A comparative study of the ultrastructure of resting and active cambium of *Salix fragilis* L. *Planta*, **84**, 239–249.

Robards, A.W. and Kidwai, P. (1972) Microtubules and microfibrils in xylem fibres during secondary cell wall formation. *Cytobiologie*, **6**, 1–21.

Robards, A.W. and Wilson, A.J. (eds) 1993. *Procedures in Electron Microscopy.* New York, Brisbane, Toronto: John Wiley & Sons.

Roland, J.-C. (1974) Cytochimie des polysaccharides végé-taux: détection et extraction sélectives. *Journal de Microscopie*, **21**, 233–244.

Roland, J.-C. (1978) Early differences between radial walls and tangential walls of actively growing cambial zone. *IAWA Bulletin*, **1978/1**, 7–10.

Roland, J.-C., Lembi, C.A. and Morré, D.J. (1972) Phosphotungstic acid-chromic acid as a selective electron-dense stain for plasma membranes of plant cells. *Stain Technology*, **47**, 195–200.

Roland, J.C. and Vian, B. (1991) General preparation and staining of thin sections. In *Electron Microscopy of Plant Cells*, edited by J.L. Hall and C.R. Hawes, pp. 1–66. London, New York, Toronto: Academic Press.

Ruel, K., Burlat, V. and Joseleau, J.-P. (1999) Relationship between ultrastructural topochemistry of lignin and wood properties. *IAWA Journal*, **20**, 203–211.

Ruzin, S.E. (1999) *Plant Microtechnique and Microscopy*. Oxford, New York: Oxford University Press.

Sabatini, D.D., Bensch, K. and Barrnett, R.J. (1963a) Cytochemistry and electron microscopy: the preservation of cellular ultrastructure and enzymatic activity by aldehyde fixation. *Journal of Cell Biology*, **17**, 19–58.

Sabatini, D.D., Miller, F. and Barrnett, R.J. (1963b) Aldehyde fixation for morphological and enzyme histochemical studies with the electron microscope. *Journal of Histochemistry and Cytochemistry*, **12**, 57–71.

Sal'nikov, V.V., Ageeva, M.V., Yumashev, V.N. and Lozovaya, V.V. (1993) Ultrastructural analysis of bast fibers. *Russian Plant Physiology*, **40**, 416–421.

Samaj, J., Hawkins, S., Lauvergeat, V., Grima-Pettenati, J. and Boudet, A. (1998) Immunolocalization of cinnamyl alcohol dehydrogenase 2 (CAD 2) indicates a good correlation with cell-specific activity of CAD 2 promoter in transgenic poplar shoots. *Planta*, **204**, 437–443.

Sano, Y., Kawakami, Y. and Ohtani, J. (1999) Variation in the structure of intertracheary pit membranes in *Abies sachalensis*, as observed by field-emission scanning electron microscopy. *IAWA Journal*, **20**, 375–388.

Sauter, J.J. and Kloth, S. (1986) Plasmodesmatal frequency and radial translocation rates in ray cells of poplar (*Populus* × *canadensis* Moench 'robusta'). *Planta*, **168**, 377–380.

Sauter, J.J. and van Cleve, B. (1989) Micromorphometric determination of organelles and of storage material in wood ray cells – a useful method for detecting differentiation within a tissue. *IAWA Bulletin, New series*, **10**, 395–403.

Sauter, J.J. and van Cleve, B. (1990) Biochemical, immunochemical, and ultrastructural studies of protein storage in poplar (*Populus* × *canadensis* 'robusta') wood. *Planta*, **183**, 92–100.

Schmid, R. (1965) The fine structure of pits in hardwoods. In *Cellular Ultrastructure of Woody Plants*, edited by W.A.

Côté, Jr, pp. 291–304. Syracuse: Syracuse University Press.

Sennerby-Forsse, L. (1986) Seasonal variation in the ultrastructure of the cambium in young stems of willow (*Salix viminalis*) in relation to phenology. *Physiologia Plantarum*, **67**, 529–537.

Sennerby-Forsse, L. and von Fircks, H.A. (1987) Ultrastructure of cells in the cambial region during winter hardening and spring dehardening in *Salix dasyclados* Wim. grown at two nutrient levels. *Trees*, **1**, 151–163.

Singh, A., Dawson, B., Franich, R., Cowan, F. and Warnes, J. (1999) The relationship between pit membrane ultrastructure and chemical impregnability of wood. *Holzforschung*, **53**, 341–346.

Singh, A.P. and Donaldson, L.A. (1999) Ultrastructure of tracheid cell walls in radiata pine (*Pinus radiata*) mild compression wood. *Canadian Journal of Botany*, **77**, 32–40.

Singh, A.P., Kim, Y.S. and Dawson, B.S. (1997) Application of electron microscopy in wood formation and utilisation. *Proceedings of The First ASEAN Microscopy Conference*, edited by A.R. Razak, M. Omar, S. Tay and T.Y. How, pp. 13–17. Malaysia: Senai Johor.

Srivastava, L.M. (1966) On the fine structure of the cambium of *Fraxinus americana* L. *Journal of Cell Biology*, **31**, 79–93.

Syrop, M.J. and Beckett, A. (1972) The origin of ascospore-delimiting membranes of *Taphrina deformans*. *Archives of Microbiology*, **86**, 185–191.

Takabe, K. and Harada, H. (1986) Polysaccharide deposition during tracheid wall formation in *Cryptomeria*. *Mokuzai Gakkaishi*, **32**, 763–769.

Thiéry, J.-P. (1967) Mise en évidence des polysaccharides sur coupes fines en microscopie électronique. *Journal de Microscopie*, **6**, 987–1018.

Thiéry, J.-P. (1969) Rôle d'appareil de Golgi dans la synthèse des mucopolysaccharides étude cytochimique. I Mise en évidence de mucopolysaccharides dans les vésicules de transition entre l'ergastoplasme et l'appareil de Golgi. *Journal de Microscopie*, **6**, 689–708.

Thomas, R.J. (1976) Anatomical features affecting liquid penetrability in three hardwood species. *Wood and Fiber*, **7**, 256–263.

Vesk, P.A., Vesk, M. and Gunning, B.E.S. (1996) Field emission scanning electron microscopy of microtubule arrays in higher plant cells. *Protoplasma*, **195**, 168–182.

Wardrop, A.B. (1954) The mechanism of surface growth involved in the differentiation of fibres and tracheids. *Australian Journal of Botany*, **2**, 165–175.

Wardrop, A.B. (1965) Cellular differentiation in xylem. In *Cellular Ultrastructure of Woody Plants*, edited by W.A. Côté, Jr, pp. 61–97. Syracuse: Syracuse University Press.

Wierzbicka, M. (1998) Lead in the apoplast of *Allium cepa* L. root tips – ultrastructural studies. *Plant Science*, **133**, 105–119.

5 Chemical and Cryo-fixation for Transmission Electron Microscopy of Gymnosperm Cambial Cells

Wood Formation in Trees ed. N. Chaffey
© 2002 Taylor & Francis
Taylor & Francis is an imprint of the
Taylor & Francis Group
Printed in Singapore.

Kim H. Rensing

Department of Botany, University of British Colombia, Vancouver, British Columbia, V6T 1Z4, Canada

ABSTRACT

Some considerations of chemical fixation, and cryo-fixation and -substitution (CFS) of cambial cells are briefly discussed within the more general context of plant cell fixation. Techniques are provided for collection, chemical fixation, and resin-embedding of cambial material from stems and branches of gymnosperms. Detailed protocols for cryo-fixation (by plunge-freezing), and cryo-substitution (both with, and without dedicated equipment) are presented with some discussion towards avoiding or resolving technical problems. A short discussion on fixation artefacts and the benefits of CFS is presented.

Key words: cambium, cryo-fixation, cryo-substitution, gymnosperms, TEM, ultrastructure

INTRODUCTION

Wood is the product of cell divisions in the lateral secondary meristem known as the cambium. The cambium takes the form of a thin cylinder of primary-walled cells between the xylem to the inside and the phloem to the outside, and produces the cells that mature into both these tissues. Within the cambium, the meristematic 'fusiform initials' produce the axial elements, while the 'ray initials' produce the transverse elements. Fusiform initials are greatly elongated while ray initials are roughly isodiametric, reflecting the respective shapes of the cells they produce. In gymnosperms, the predominant axial elements of the xylem are tracheids, while those of the phloem are sieve cells and axial parenchyma. The rays are composed of both ray tracheids and ray parenchyma.

In gymnosperms, each fusiform initial divides to form a radial file of daughter cells. Both tracheids and sieve cells in a single radial file develop from the same initial. Thus, a single file of cells containing all the stages of tracheid and sieve cell maturation can be isolated in section and examined by light and electron microscopy. Careful chemical fixation of the cambium can yield sections that show the ultrastructure of the initials and their derivatives.

The primary obstacle to studying cellular systems by transmission electron microscopy (TEM) is that chemical fixation distorts the living form of cells. Thus, it is necessary to recognise that the structures seen by TEM are highly modified representations of the living system. Chemical fixation is a gradual process whereby cell components are stabilised for observation by light microscopy (LM) or TEM. Many processes using chemical fixation are feasible for electron microscopy (Willingham, 1983; Bullock, 1984; Hopwood 1985), but the most common is a two-stage process using aldehyde (i.e., glutaraldehyde, formaldehyde) followed by osmium tetroxide. Aldehydes stabilise mainly proteins, while osmium tetroxide stabilises lipids and adds contrast to the image.

Similar chemicals are used for preparing animal and plant tissues for TEM, but for plants, greatly extended processing times are needed to allow penetration of aqueous chemical fixatives and liquid resins. This is because cell walls, plastids and a vacuolar apparatus are exclusive to plants. The cell wall is a significant obstacle to the passage of many substances and thereby slows the movement of fixatives into plant cells. Fixatives can disrupt cell processes long before their concentration is high enough to immobilise the cell structure. Long periods of dehydration are necessary, for as the cell wall loses water, its permeability decreases and residual water becomes exponentially more difficult to extract. Plastids often contain substances that further slow fixative penetration as well as materials that fix poorly with chemical preservatives. The long fixation times necessary for preservation can lead to significant extraction of cell contents. Vacuoles introduce another difficulty since some contain materials, such as polyphenols or tannins, which react adversely with fixatives to produce precipitates. Other vacuoles have largely aqueous contents that can be at a different pH and osmotic potential from the cytoplasm. In cells such as those of the cambium, the cytoplasm lies in a very thin peripheral layer around an aqueous vacuole that occupies nearly all of the cell volume. Here, even minute osmotic pressure differences between vacuole and fixative can induce substantial distortions of ultrastructure or even its complete destruction.

Other features of cambial anatomy conspire to make its study notably arduous. Fusiform cells, being axially very long, are easily damaged during preliminary dissection and are difficult to fix in their entirety. The delicate composition of their thin primary cell walls and their predominantly aqueous contents sandwiched between the all but inflexible secondary cell walls of the phloem and xylem makes sectioning very difficult (see also the chapter entitled 'An introduction to the problems of working with trees' by Chaffey, in this volume). Traditional paraffin-embedding and sectioning for LM (Johansen, 1940) yields poor results with woody tissues. For this reason, embedding in media regularly used for TEM preparation is necessary for LM of the cambium (Barnett, 1971; see also the 'Wood microscopical techniques' chapter in this volume). Such resins or plastics minimise the difference in hardness between the tissues. However, the time investment per section is about ten times that of the paraffin technique and the cost is greater.

Many researchers have studied the chemically fixed ultrastructure of the cambium and its derivatives in woody plants. A large proportion of the work has been done on angiosperm species (see TEM chapter by Chaffey in this volume), but there is also a substantial body of work on gymnosperms (Table 1). Our current knowledge of cambial ultrastructure is the result of careful use of chemical fixation, and there is a great deal yet to discover, particularly in combination with other techniques (see e.g. the TEM chapter by Chaffey, and the chapter by Micheli et al. in this volume). Advances in biotechnology and particularly immunolabelling show much promise for learning a great deal more about tree cell processes from the chemically preserved cambium.

The cryo-fixation/substitution (CFS) method for preparation of plant specimens for electron microscopy is considered to produce results superior to those from conventional methods (McCully and Canny, 1985; Kiss et al., 1990; Sluiman, 1991). Shibaoka (1993) reported that the 'freeze-substitution method does not give satisfactory results ... in situ' but this is no longer the case. It is extremely well suited to preserving individual cells such as tobacco BY-2 protoplasts (Samuels et al., 1995) and maple suspension culture cells (Zhang and Staehelin, 1992). CFS has also been successfully applied to: stamen hairs, and germinating moss spores (Lancelle et al., 1986), pollen tubes (Lancelle and Hepler, 1989), root tips (Kiss et al., 1990, Staehelin et al., 1990, Ding et al., 1991, Samuels et al., 1995), leaf tissues (Ding et al., 1991) and pollen (Runions et al., 1999).

In the gymnosperms, CFS has been used to preserve portions of developing tracheids (Inomata et al., 1992), but little else has been published. Cryotechniques are relatively recent, as well as more expensive, time consuming and difficult to achieve than chemical fixation. However, as a progression from chemical fixation, CFS has the potential to produce valuable new information. It can more accurately preserve native biochemical conformation and ultrastructural detail, as it is virtually instantaneous. Also, the cells are not exposed to fixative chemicals, osmotic gradients or pH stress, which are all detrimental to immunolocalisation. Considering these factors and the unique structure of cambial cells, CFS is particularly suited to the study the cambium by electron microscopy.

Table 1 Comparison of published fixation procedures on the ultrastructure of cambial cells and their derivatives in gymnosperms

Author	Species	Primary fixative	Membranes	Other
Srivastava and O'Brien (1966)	*Pinus strobus*	3% glutaraldehyde in 0.025M phosphate buffer	plasmalemma folded inwards tonoplast collapsed organelle membranes wrinkled	paramural bodies gaps in the cytoplasm cytoplasm condensed
Mahmood (1968)	*Pinus radiata* *Pinus patula*	2% potassium permanganate	no membranes remained	cell contents not preserved
Murmanis and Sachs (1969)	*Pinus strobus*	3% glutaraldehyde in 0.02M phosphate buffer	membranous formations	cell contents aggregated
Murmanis (1971a, b)	*Pinus strobus*	3% glutaraldehyde in 0.02M phosphate buffer	plasmalemma separated from the wall vacuoles wrinkled	structures obscured "particles" resemble osmium precipitate
Wodzicki and Brown (1973)	*Pinus echinata*	2% glutaraldehyde in 0.1M cacodylate buffer	broken and irregular membranes membranous formations	
Barnett (1973, 1977)	*Pinus radiata*	4% glutaraldehyde in Sörenson's phosphate buffer with 5% (0.15M) sucrose	plasmalemma undulated and folded inwards tonoplast collapsed	electron opaque material at the plasmalemma uneven shaped nuclei "paramural bodies"
Timell (1980)	*Picea abies*	Karnovsky's	plasmalemma folded inwards tonoplast collapsed	magnification too low to interpret ultrastructure
Riding and Little (1984)	*Abies balsamea*	Karnovsky's formaldehyde and glutaraldehyde	plasmalemma folded inwards tonoplast collapsed membranous formations	
Inomata *et al.* (1992)	*Cryptomeria japonica* *Pinus densiflora*	cryo-fixation/substitution	smooth and continuous	homogenous cell wall uniform cytoplasm
Savidge and Barnett (1993)	*Pinus contorta*	Karnovsky's in Sörenson's buffer	plasmalemma separated from the wall membranous formations	electron opaque material in ER, vacuoles and outside the plasmalemma
Rensing and Owens (1994)	*Pseudotsuga menziesii*	3% glutaraldehyde in 0.075M and 0.05M phosphate buffer with 3mM CaCl$_2$	plasmalemma undulated and folded inwards vacuoles collapsed	

Techniques

<table>
<tr><td>

Equipment needed:

Hammer and chisel
Sharp woodworking saw
Wet paper towels
Plastic bags

Labelling markers

Single-edge razor blades (several)

Double-edged razor blades (several)

10 cm glass Petri dishes

No. 5 fine forceps

3 ml glass vials with caps
pH meter

Vacuum pump
Rotating mixer
Aluminium foil
40 mm diameter aluminium weighing boats
Embedding oven at 60 °C
Jeweller's saw
8 mm diameter mounting stubs (Pelco Ltd)
Hot plate at 60 °C
Ultramicrotome
Light microscope
Glass slides
75-mesh hexagonal copper grids
Glass or diamond ultramicrotome knife

</td><td>

Chemicals needed:

Double distilled water (ddH$_2$O)
Na$_2$HPO$_4$.2H$_2$0
NaH$_2$PO$_4$.2H$_2$0
8% Glutaraldehyde (EM grade)
(Marivac Ltd, No. CG006-0)
Osmium tetroxide
(Marivac Ltd, No. CO004-3)
*ERL 4206 (vinyl cyclo hexane dioxide)
(SPI Supplies, No. 02815-AF)
*DER 736 (diglycidyl ether of polypropylene glycol)
(SPI Supplies, No. 02830-AF)
*NSA (nonenyl succinic anhydride)
(SPI Supplies, No. 02829-AF)
*DMAE (dimethyl aminoethanol)
(SPI Supplies, No. 02824-AB)
Silica Gel
Cyanomethacrylate glue
(e.g. Superglue™ or Crazy Glue™)
Toluidine Blue O (C.I. No. 52040)
NaOH
Lead citrate
Lead nitrate
Lead acetate
Dental wax
Ethanol
Propylene oxide
Methanol
Acetone

</td></tr>
</table>

* Kits that include all the components of ERL (Spurr's) resin are also available (see TEM chapter by Chaffey in this volume).

Sample Preparation

Duration of procedure: approx. 30 to 60 min

1   Small material such as seedling stems, tree branches or roots can be collected and brought whole to the laboratory.
2   Material from mature trees can be collected from freshly cut trees or by removing smaller pieces from the stem of living trees. Chisel out a 3 to 5 cm square from the stem then take sections from within this sample (see also chapter by Uggla and Sundberg in this volume).
    Note – when the cambium is actively dividing, it may be better to cut out a wedge of tissue with a very sharp wood-working saw. Although more damage is suffered by the tree, the cambium is not torn by flexion of the sample as with the method above.
3   Wrap the tissues in soaking wet paper towels, seal them in a plastic bag and transport to the laboratory.

4   Prepare the primary fixative as outlined below.

5   Remove any needles or lateral branches from the collected material, and using a single-edged razor blade, trim away the ends that were cut in the field.

6   Cut small material (e.g. branches etc.) into short (approx. 2.5 cm) lengths with a razor blade, and radially bisect each piece just prior to cutting sections.

7   Carefully trim off the bark, cortex and most of the xylem if present, as these are mostly impermeable to fixatives.

8   With a double-edged razor blade (broken in half) cut thin radial sections containing the cambium and a portion of the adjacent xylem and phloem. Change blades frequently.

9   Immediately immerse each section in the primary fixative or a cryo-protectant depending on the chosen technique (see below).

10   To minimise the effects of evaporation from the cut surface, quickly cut a few sections then prepare or expose a new face. This is particularly important for cryo-fixation and -substitution, as thin sections will be made from just below the surface.

11   For cryo-fixation (see below), trim larger sections to 1–2 mm wide and up to 3 mm long in a glass Petri dish with cryo-protectant. The hand-cut sections should be thin enough to observe individual cells by LM, so any that are greater than approx. 0.5 mm thick should be discarded.
     Note – for cambium of mature trees, longer sections may be necessary so as to include entire fusiform cells, but the return of well-fixed material will be lower.

Chemical fixation for light and transmission electron microscopy

Many protocols are available for chemical fixation of wood and cambial tissue. The following procedure is simple and has produced satisfactory results for the cambium of gymnosperms. Make all solutions from distilled water in immaculately clean glassware, and use 'EM grade' chemicals. Phosphate buffer is recommended because it is non-toxic, readily available and inexpensive, but other buffers (e.g., sodium cacodylate, HEPES (N-2-hydroxyethylpiperazine-N'-2-ethanesulfonic acid), PIPES (piperazine-N,N'-bis-[2-ethyl-

sulphonic acid])) may work equally well. The working buffer is best made just prior to use from pre-made stock solutions:

Stock A: 8.9 g – $Na_2HPO_4 \cdot 2H_2O$ in 500 ml ddH_2O (= 0.1 M)

Stock B: 7.8 g – $NaH_2PO_4 \cdot 2H_2O$ in 500 ml ddH_2O (= 0.1 M)

Tissues can be processed through to embedding in small (e.g. 3 ml) glass vials with plastic caps.

FIXATION

Duration of procedure: approx. 7 h

1   Prepare the primary fixative.
a.   Add stock B to stock A until pH 6.8 is reached
     Note – if a pH meter is not readily available, borrow one each time new stock is made then record the required quantities for future reference.
b.   Combine with glutaraldehyde then dilute with double distilled water to the desired concentration.
     For 4 ml of 2.5% glutaraldehyde primary fixative:

	For 0.075 M	*For 0.05 M*
0.1 M. buffer pH 6.8	0.3 ml	0.2 ml
8% glutaraldehyde	1.25 ml	1.25 ml
ddH_2O	2.45 ml	2.55 ml

Note – low molarity buffers are best for fixing cambium because of the low solute concentration in the vacuole. It may be necessary to adjust the buffer molarity with seasonal changes in cambial morphology. It was found that 0.075 M phosphate buffer worked well for fixing dormant or late-season cambium, while 0.05 M buffer worked better during the peak time of cell division (Rensing and Owens, 1994).

2   Fix the dissected tissue pieces for 2–3 h in 2.5% buffered glutaraldehyde at room temperature (RT). Fixation times should be extended for colder temperatures.
     Note – an alternative primary fixative is a mixture of formaldehyde and glutaraldehyde ('Karnovsky's'). Proportions range from 2% formaldehyde with 2% glutaraldehyde to

4% formaldehyde with 1% glutaraldehyde. Make a formaldehyde concentrate just prior to use by heating p-formaldehyde in distilled water until it is just dissolved then allowing the solution to cool. Mix this with glutaraldehyde and buffer to the desired concentration.

3   Two 5-min exposures to vacuum are usually necessary to remove air from intercellular spaces. Even if the tissues sink, there are often small gaps in the prepared tissues from tiny air pockets. Raise the vacuum to 15 mm mercury slowly to reduce rapid, destructive expansion of air bubbles in the tissues.

4   Following the primary fixation, remove the excess aldehydes with three 10-min immersions in buffer alone.

For 12 ml of buffer rinse:

	For 0.075 M	For 0.05 M
0.1 M buffer	0.9 ml	0.6 ml
ddH$_2$O	11.1 ml	11.4 ml

5   Post-fix (i.e. secondary fixation) for 2–3 h in 1% osmium tetroxide in buffer.

For 4 ml of 1% OsO$_4$:

	For 0.075 M	For 0.05 M
0.1 M buffer	0.3 ml	0.2 ml
4% OsO$_4$	1 ml	1 ml
ddH$_2$O	2.7 ml	2.8 ml

6   Follow with a 5-min distilled water wash.

DEHYDRATION

Duration of procedure: approx. 1.5 h.
Water is not miscible with most embedding resins, so it is replaced gradually by resin-miscible organic solvents.

1   Immerse the fixed tissues in increasing concentrations of alcohol (most commonly ethanol or methanol) or acetone for 10 min at each concentration.

2   A useful series is: 10%, 20%, 40%, then 80% acetone followed with three changes of 100% acetone.

3   Resin infiltration can be started from pure acetone, as residual acetone ultimately will evaporate.
Note – if ethanol is used for dehydration, an intermediate step of three 10-minute changes of propylene oxide is recommended.

INFILTRATION

Duration of procedure: approx. 1 h for steps 1 and 2, and at least 2.5 days for the remainder of infiltration and polymerisation.

The initial stages of infiltration with embedding medium are the most critical (Mersey and McCully, 1985) so resin must be added slowly to prevent damage. Use a hard grade resin mixture to reduce the compressibility differences between the well-developed xylem cell walls and the delicate cambium. This difference usually produces uneven section thickness and can cause tearing between the tissues. ERL resin (Spurr, 1969) works well for plant tissues because of its low viscosity. The following produces 40.4 grams of hard grade resin, about the amount needed for processing one batch of samples:

ERL 4206	10.0 g
DER 736	4.0 g
NSA	26.0 g
DMAE	0.4 g

The resin is made by weighing the components then gently, but thoroughly, mixing them in a beaker. The complete mix can be stored for about 2 months in 10 ml plastic syringes sealed in plastic and covered with aluminium foil in a freezer. Individual syringes of resin can be removed from the freezer and used after they warm to room temperature. Condensation will prevent polymerisation if the cold resin is used directly from the freezer.

If immunochemistry is to be performed, use the hard grade of LR White (SPI Supplies, No. 02645-AB), but polymerisation must be done in an oxygen-free environment (see e.g. TEM chapter by Chaffey in this volume for a detailed protocol).

1   Add one drop of resin mixture to the tissues in 1 ml dehydrant, mix by rocking gently, and then leave the samples capped on a rotator for 5 min.

2 Continue adding drops in the same manner until the mixture is approx. 25% resin (i.e. about 1.3 ml total) then leave the samples on a rotator for at least 1 h.

3 Increase the resin component to 75% by discarding 1 ml of the resin mixture and adding 0.75 ml of pure resin. Mix thoroughly by rocking, then place the open vials (i.e. with caps off) on a rotator in a fume hood.

4 Place some silica gel on the rotator with the vials to avoid moisture absorption by the resin. Moisture prevents the resin from polymerising fully; the blocks will be soft and sticky.

5 Cover the rotator head with aluminium foil to exclude light and leave for 12 h (overnight). The acetone will evaporate from the mixture leaving almost pure resin.
Note – light causes the resin to increase in viscosity so covering the samples greatly extends the functional infiltration time.

6 Completely exchange the solutions with fresh 100% resin a minimum of three times over 24 h. The infiltration time can be extended indefinitely for difficult tissue if the resin is replaced at least once a day.

7 Arrange the infiltrated samples in 40 mm diameter aluminium weighing boats and cover them to a depth of 3 mm with fresh hard-formulation resin. The radial surface should lie flat on the bottom of the dish.
Note – abelling markings scribed into the aluminium with a dissecting needle will transfer to the polymerised resin.

8 Polymerise for 18 h at 60 °C.

9 Remove the foil from the embedded material, and with a jeweller's saw, cut out small blocks containing one piece of tissue.

10 Mount blocks on 8 mm diameter aluminium stubs with cyanomethacrylate glue ('Superglue') or a drop of ERL resin. These mounting stubs fit into the chucks of Reichert, SPI and LKB ultramicrotomes.
Note – pre-roughen the top of the mounting stub with coarse sandpaper to increase adhesion.

11 Allow the glue or resin to set overnight at 60 °C before sectioning.

This method of mounting embedded material allows great flexibility for precisely orienting the tissues for sectioning. The smooth face (from the bottom of the embedding mould) produces radial longitudinal sections. Using these sections as a reference, the block can be precisely trimmed on the ultramicrotome, removed from the stub with a razor blade, then remounted to cut tangential or transverse sections. Thus, a single block can be sectioned in all three planes (see TEM chapter by Chaffey in this volume for an alternative orientation method).

SECTIONING

1 Cut reference sections 0.6–0.8 μm thick with an ultramicrotome. If the material is particularly woody, a cheap or well-used diamond knife is a great help. One or two sections through woody material will generally render a glass knife unusable.

2 Transfer the sections to a drop of water on a glass slide and heat it on a hot plate at 60 °C until the water has evaporated (do not allow to boil).

3 Stain the sections with 0.5% Toluidine Blue (C.I. No.52040) in 0.1% aqueous NaOH (O'Brien and McCully, 1981).
Cover the sections with a drop of stain, heat on the 60 °C hot plate for 5 to 10 seconds (but do not allow solution to dry out), then wash off the excess from the slide with a gentle stream of water.

4 Evaluate the sections with a LM (a cover slip is only needed for high magnification or photography).

5 Cut thin sections (at 70 nm) only from the best blocks.
Note – if the sections tend to fall apart, mount them on formvar-coated grids. However, image contrast will be lower than with uncoated grids.
Note – 75-mesh hexagonal grids provide the best balance between support and viewing space. Hexagonal mesh was preferred to square mesh because the long cambial cells seemed to always line up behind the parallel bars of square mesh grids.

6 Stain grids with 1% uranyl acetate in 50% ethanol and Sato's lead stain (Sato, 1967). This produces sufficiently high contrast in a short time even with ERL resin.

a) Immerse each grid in a drop of uranyl acetate for 10 min, rinse in 50% ethanol and then in each of three beakers of distilled water.

Note – do not allow the sections to dry.
b) Immerse each grid in a drop of Sato's lead stain (1967) for 5 min, rinse in three beakers of distilled water and allow to dry.

Sato's Lead Stain

1.0 g lead citrate	
1.0 g lead nitrate	
1.0 g lead acetate	Mix all for 2 min, then
2.0 g sodium citrate	add 18 ml 4% NaOH
82 ml pre-boiled dH$_2$O	

CRYO-FIXATION AND CRYO-SUBSTITUTION

When freezing is uniform and sufficiently rapid, aqueous solutions will freeze in a vitreous (glass-like) state with no movement of material. If water freezes slowly, ice crystals grow and displace the cellular layout. Cryo-protectants reduce ice crystallisation by decreasing the freezing point of the extracellular matrix. However, if the extracellular matrix freezes before the intracellular matrix, an osmotic gradient is established and water moves out of the cell to crystallise in the extracellular matrix.

The best cryo-protectant was found to be 0.2 M sucrose in distilled water. Ding (1991) found that 0.2 M sucrose was not harmful to plant cell metabolism and did not cause plasmolysis. This cryo-protectant has been used for several types of plant material and worked well for cambial tissue of gymnosperms. For cambial cells, buffering the cryo-protectant proved to be detrimental to cryo-fixation, most likely due to the additional osmotic stress from the buffer salts.

Many other cryo-protectants are possible if sucrose is found to be ineffective for other material. The least suitable can be eliminated from further consideration with the following technique:

1 Immerse hand-cut tissue pieces in potential cryo-protectants on a microscope slide then observe them with differential interference contrast (DIC) microscopy.
2 Solutions that reduce or halt cytoplasmic streaming or cause plasmolysis need not be considered further. For instance, sucrose solutions up to 0.4 M had no observable effect on cytoplasmic streaming in cambial cells over more than 4 h of observations, but salt solutions (e.g., NaCl, KCl) plasmolysed the cells.

Note – do not include wetting agents, such as Tween, in the cryo-protectant and ensure that all glassware is immaculately clean. Any residual detergents could induce plasmolysis in the fusiform cells during later processing. Salt solutions at low concentrations that did not affect cytoplasmic streaming, were not effective cryo-protectants.

If cryo-fixation/substitution is not achieved, it may be difficult to determine which part of the technique is responsible. Ray cells are much easier to fix than fusiform cells, so observation of these should provide the first measure of success. Once some level of preservation is achieved, each step leading to embedded material can be modified independently to fine-tune the process.

Experimentation indicated that maximal cryo-protection was achieved after about 1 h in the cryo-protectant. Longer times did not appear to affect the samples, but to minimise potential extraction of material, cryo-fixation should be started 30 min following dissection. With experience, it is possible to cryo-fix about 50 samples in approx. 60 minutes. Be aware that the process of CFS is one of attrition; up to 75% of the sections will be damaged during hand-sectioning, lost during plunge-freezing, and simply poorly fixed or substituted. Thus, although 10 to 15 tissue blocks out of 50 may have worthwhile sections, every block must be carefully checked.

Processing damage can be checked in Toluidine Blue-stained 60 μm-thick sections cut on the ultramicrotome. Use care to check each of the first sections from the block. The outer cells will be damaged during dissection, but the cells immediately beneath these will be the best-fixed. Those greater than three to four layers deep are usually damaged by ice crystal growth. Plasmolysed cell contents appear as dark staining material in the cell lumens. Lesser degrees of damage may not be visible by LM; some sectioning damage, ice crystallisation, or incomplete fixation will only be apparent by TEM.

Poorly fixed cells are immediately apparent by TEM. In the worst cases, fusiform cells contain electron opaque masses of coagulated cell contents similar to poorly chemically fixed cells (Figure 1). Occasionally, only distorted nuclei remain

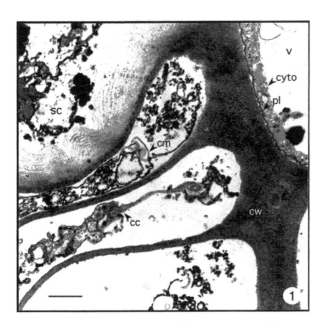

Figure 1 Transverse section of the inactive cambium and phloem of *Pseudotsuga menziesii* showing varied degrees of plasmolysis due to poor chemical fixation. The cell wall retains its shape, and the primary cell wall and the secondary sieve cell wall can be distinguished from the staining density. However, the cell contents range from relatively well fixed to the upper right to completely extracted and coagulated on the lower left. Key: cc, condensed cytoplasm; cm, condensed membrane; cw, cell wall; cyto, cytoplasm; pl, plasmalemma; sc, sieve cell; v, vacuole. Scale bar = 1 μm.

duced by the crystal growth do not become apparent until over 25,000 × magnification. However, this small amount of distortion rarely detracts from the quality of the ultrastructural detail at lower magnification. Where no damage has occurred, all membranes will be smooth and continuous with no wrinkles or folds, the cytoplasm will be uniform with no gaps and the cell walls will be fine-grained and homogenous. In sections made within (parallel with) the wall–membrane interface, the plasmalemma is only apparent as darker areas of little detail. Overall, sections cut from just beneath the outer layer yield the best results.

The following technique was established for the Reichert KF80 immersion cryo-fixation system and the Reichert CSauto cryo-substitution apparatus (Leica Inc.), but the protocol can be adapted for other types of cryo-fixation equipment. In the KF80, liquid nitrogen (LN$_2$) surrounds a chamber that is filled with the cryogen, liquid propane. Tissue holders are held in a pair of fine forceps on a spring-loaded rod. The tissues are plunged downward into the cryogen by releasing the spring tension. With the CSauto, an internal LN$_2$ Dewar maintains the substitution temperature and processing is fully controlled by the machine. Substitution time, temperature and re-warming rates can be set by the operator and the substitution fluid can be changed during processing. A protocol is also given below for cryo-substitution without dedicated equipment.

identifiable. These cells are usually in the first sections cut from a block that were damaged during processing. Scattered osmiophilic and membranous material in the otherwise vacant spaces of the fusiform cells indicate severe plasmolysis. In less damaged cells, the plasmalemma is separated from the cell wall, and the cytoplasmic detail is largely distorted. Certain features may be recognisable, but the structure is thick and granular with a condensed appearance. Damage in this case is most likely from ice crystallisation in the cytoplasm, and from water moving out of the cell and then crystallising between the plasmalemma and the cell wall. For a number of reasons, all sections may contain cells with different amounts of damage. Sections often cut through cells where the freezing front has passed unevenly, leaving only portions well-fixed. Other cells can appear to be well-fixed at low magnifications with the TEM, but at higher magnifications show varied levels of crystallisation damage. In some cases, spaces pro-

Equipment needed:	Chemicals needed:
Wooden applicator sticks	Sucrose
Fine-tipped forceps	Dental wax
32 gauge (or similar) wire	Formvar
5 cm diameter filter paper	1, 2-dichloroethane
10 cm glass Petri dish	Liquid nitroge (LN$_2$)
Reichert KF80 (or equivalent)	Propane
Reichert CSauto (or equivalent)	Acetone
	Crystalline osmium tetroxide
	ERL or LR White resin
	Molecular sieve

PREPARATION OF THE REICHERT KF80 AND CSAUTO

Duration of procedure: approx. 1 hour.
Several techniques for holding the sections during plunge-freezing may be tried. However, a balance must be attained between supporting the specimen and minimising the thermal inertia. Unsupported sections break or deform on passing through the cryogen surface, and the freezing time increases with the mass of the tissue and holder. Copper TEM grids and light gauge copper strands often bend when they pass through the cryogen surface, either dislodging or bending the sections. The following method (after Lancelle, 1986) was found to work well:

1 Make support loops from 32-gauge silver motor-winding wire or similar material.

a) Strip the insulation from 2 cm long pieces and bend them in half around a wooden applicator/swab stick. Twist the ends to make a loop around the stick then slide it from the stick.

b) Trim the ends to leave a 0.5 cm handle.

c) Plunge each loop 3 times into 5 by 10 mm rectangles of formvar (0.6% in dichloroethane) floating on water so that the film folds around the loop holes. Three layers are stronger than one of equivalent thickness. Uncoated loops do not retain sections during plunging.

d) Affix the coated loops to a sheet of dental wax by their handles until they are dry.

e) Used loops can be cleaned with dichloroethane and recoated with formvar.

2 Turn on the cryo-fixation system (Reichert KF80) and switch it to heating mode to drive off any moisture. Ensure that the transfer holders and sample baskets are placed in the chamber so that they are also moisture-free.

3 Obtain sufficient LN_2 in a Dewar for the entire cooling and fixation process. Between 5 and 10 litres will be necessary for 1 to 2 hours of continuous fixation. Allow for maximum use of LN_2 during the first few runs then re-estimate the amount needed. If the Reichert CSauto is being used, an additional 8 litres are needed to pre-cool and fill its internal Dewar.

4 Turn on the cryo-substitution machine, open the chamber and ensure the correct sample basket holder is in place.

5 Fill the internal Dewar (this takes about 0.5 h). The cryo-substitution apparatus can be filled automatically if the equipment is available or manually by pouring LN_2 through a special funnel.

6 Assemble the necessary equipment such as wooden applicator sticks, fine forceps, small filter papers, and the coated plunging loops.

7 Dissect the material for freezing into the cryo-protectant medium (e.g. 0.2 M unbuffered sucrose) as outlined above. Maximum cryo-protection should be achieved while completing the next step.

8 Switch the KF80 to cooling mode and connect the external LN_2 Dewar (or pour LN_2 directly into the chamber).

9 When the cryogen chamber has cooled to approx. −140 °C, slowly add propane to plunge vessel taking care to keep the chamber colder than −40 °C, the vapourising temperature of propane.

10 Allow the cryo-fixation machine to cool to −180 °C.

CRYO-FIXATION (WITH A REICHERT KF80)

Duration of procedure: approx. 1 hour.

1 Place a small filter paper on the closed lid of the cryogen chamber. Keep the cooling chamber closed as much as possible to reduce LN_2 loss and minimise frost build-up.

2 Place a coated loop in the plunging forceps so it is in line with the plunging axis, then place a sample for freezing on the loop and remove the excess fluid with filter paper.
Note − the positioning is important to produce a laminar rather than a turbulent transit through the cryogen and so that the specimen is first to contact the cryogen. Turbulence causes more insulating bubbles to form around the specimen and often dislodges it from the support.

3 Move the plunge unit over the chamber and bring it upright. Open the plunging chamber and the cryogen chamber, then push the release button to plunge-freeze the specimen.

4 Wait until the propane stops bubbling, then lift the specimen to just above the cryogen chamber.

5   Close the lid of the cryogen chamber and release the forceps dropping the loop on to the filter paper. The filter paper absorbs the excess liquid propane from the specimen and loop. Rotate the plunge unit out of the way.

6   Transfer the frozen samples still on the loops to specimen baskets in the cryo-fixation chamber. Repeat the process until sufficient samples are frozen.

7   Place the specimen baskets in a liquid nitrogen-chilled transfer unit that can be moved to the pre-cooled cryo-substitution machine.

CRYO-SUBSTITUTION (WITH A REICHERT CSAUTO)

Duration of procedure: approx. 7 days

1   Pre-warm the metal cover for the chamber head of the Reichert CSauto.

2   Open the cryo-substitution chamber and place the warmed cover over the head to prevent frost accumulation.

3   Place a basket of oven-dried molecular sieve at the base of each sample column.

4   Transfer specimens from the KF80 to cryo-substitution chamber and distribute the specimen baskets among the columns.

5   To prevent the specimens from floating away in the substitution medium, cap each column with a basket filled with molecular sieve. Free-space can be filled with baskets of molecular sieve for additional water scavenging.

6   Remove the transfer units and the head cover, then close the machine. Slowly inject the substitution medium through the appropriate port.
    Note – the best substitution medium for preservation and contrast was found to be 1.5% osmium tetroxide dissolved from crystals in fresh dry acetone. Acetone was chosen over alcohols because it is miscible with epoxy resins and does not require the use of propylene oxide as an intermediate step. Acetone alone can be used for cryo-substitution, but the TEM images lack the contrast provided by the osmium. More volatile media, such as tetrahydrofuran evaporated during the

lengthy substitution, leaving the tissues completely dry (the CSauto is an open system).

7   Substitute at –90 °C for 90 h, then warm at 5 °C per hour to 15 °C (for 21 h).

8   Remove the tissues from the substitution apparatus and place them in Petri dishes of fresh acetone at 15 °C.

9   Separate the tissues from the plunging loops, and set the loops aside for cleaning and re-use.

10 Transfer the samples into capped vials with 1 ml of fresh, dry acetone. The tissues can then be infiltrated with ERL resin or LR White, and sectioned as outlined above.

CRYO-SUBSTITUTION (WITHOUT DEDICATED SUBSTITUTION EQUIPMENT)

Equipment needed:	Chemicals needed:
Small plastic cooler (e.g. Coleman™ L'il Oscar™)	Dry ice
10 ml screw cap vials	Acetone
Stainless steel ball bearings	Molecular sieve
Lidded screen baskets	Crystalline osmium tetroxide
Large forceps	
Cold resistant gloves	
Freezer at –20 °C or colder	
*Reichert KF80	

*   This procedure does not eliminate the need for cryo-fixation equipment.

Duration of procedure: approx. 10 days
During preparation:

1   Obtain small lidded screen baskets (such as those used for critical-point drying) that will fit into 10 ml screw-cap vials.

2   Place one or two stainless-steel ball bearings into each basket for ballast.

3   Place these in the cryo-fixation chamber of the KF80 to pre-cool at the start of the cryo-fixation process, and transfer the cryo-fixed specimens directly into them.

4   Half-fill a small plastic cooler (e.g. Coleman™ Li'l Oscar™) with a slurry of dry ice and

acetone (Styrofoam coolers will melt from the acetone fumes).

5   Prepare one 10 ml screw-cap vial for each group of cryo-fixed samples:
a) Add oven-dried molecular sieves until the vial is 1/3 full.
b) Top up the vial with prepared substitution fluid.
c) Tightly cap and submerge the vials in the acetone slurry to pre-cool.

After cryo-fixation:

1   Remove a chilled substitution vial from the dry ice slurry using forceps (work with one at a time).
2   Using gloves, open the cap (it can freeze on).
3   Transfer the specimen baskets into the substitution fluid.
4   Ensure the samples are completely immersed and the cap is tightly closed.
5   Replace the vial in the slurry.
6   Top up the cooler with dry ice and place it in the coldest freezer available.
7   Top up the dry ice every day (if a –20 °C freezer is used, it will last a little less than 2 days).
8   After 1 week, allow the dry ice to evaporate so that the cooler warms to –20 °C.
9   After 2 days, move the cooler to a refrigerator at 4 °C for 24 h.
10 Bring to RT in a fume hood.
11 The tissues can be processed further as outlined above.

APPLICATIONS

Light Microscopy

During the preparation of TEM sections, reference sections are produced for light microscopy. As mentioned, using TEM preparation techniques for LM of the cambium yields arguably the best LM sections, so these should be observed with consideration to collecting data. LM can produce useful information that is not apparent with TEM. Transmission electron microscopy shows only a small cell area and, in general, is prohibitively labour intensive to gain quantitative data. Even at low magnification (i.e. 40 × objective), LM clearly shows the cambial zone organisation, the cell wall

thickness, and the state of the cytoplasm. Cell numbers can be assessed from radial or transverse sections, and the difference between cell wall structure in dormant versus active cells, or in early- versus late-wood tracheids is readily apparent. In addition, the cytoplasm of dormant cells is visible surrounding numerous unstained vacuoles, and nuclei are easily seen. However, in comparison, the active cells appear to contain little more than some nondescript material and the nuclei.

At high magnification (i.e. with a 100 × oil-immersion objective), the multi-vacuolate nature of the dormant cytoplasm is more readily apparent. The darkly staining nuclei and darker staining nucleoli are apparent. In the active cambium, division can be observed in many sections and thus large areas can be assessed. If captured at the right time, chromosomes can be seen and the stage of mitosis can be determined. The phragmoplast as well as the new membrane and wall forming between them are easily distinguished. The cytoplasm can be seen around the nuclei, at the cell tips, in the phragmoplasts and along the newly formed membrane. However, even at this magnification, many areas of the cells do not appear to have a peripheral layer of cytoplasm. Numerous darkly staining organelles, visible as small darkly staining dots, are present in the phragmoplasts and the visible cytoplasm. However, identification of these is only possible by electron microscopy.

Transmission Electron Microscopy

It is known that fixatives can cause drastic changes in membranous structures of the living cell (Mersey and McCully, 1978; Sluiman, 1991). These changes become preserved and examined by TEM, and if the images are considered accurate representations of the living cell, data can be misinterpreted. However, Kellenberger (1992) noted that controlled distortions induced by careful chemical fixation may yield useful information because it can make certain features more distinct. Precipitation can make cellular substances visible by TEM (Rao 1988; Savidge and Barnett, 1993), but interpretation of their significance must be made with care. The origin of the substances is unknown and they bind stain only when condensed. The characteristics of such substances could be difficult to resolve since they have undergone chemical and conformational changes.

During chemical fixation, contact with the fixative in buffer causes water to leave a cell by mass flow (Taylor, 1988). This can cause moderate damage as the cell membrane shrinks from the wall then moves back towards its original state as water re-enters the cell with the fixative. The plasmalemma is stabilised to an extent by its association with the cell wall but, full turgor is not achieved as the osmotic potential equilibrates during the inward movement of the fixative (Taylor, 1988), and the cell becomes fixed with space between the wall and the plasmalemma. The plasmalemma thereby appears wrinkled in electron micrographs because its area is greater than the surface area of the cytoplasm it surrounds (Figure 2).

The greatest obstacle to preservation of active cambial cells is the large central vacuole. Because there is relatively little cytoplasm in active cambial cells, mass flow (Taylor, 1988) occurs from the vacuole during fixation. Chemical fixation relies on the chemical cross-linkage of cellular components, yet the largely aqueous vacuoles contain little material that can be chemically bound leaving it essentially unfixed. If the tonoplast is ruptured during fixation, the vacuole will not regain turgor and it will appear in micrographs as folded membrane in a vacant space. If enough water is extracted, the tonoplast can condense on itself possibly in association with portions of the cytoplasm. These remnants are visible in sectioned material as electron-dense, membranous masses in the cell lumen (Figure 1). Any remaining cytoplasm associated with the plasmalemma may be fixed in place, or, without the tonoplast confining it, could move and be fixed elsewhere. With extreme plasmolysis, the entire cell contents, including the plasmalemma, coagulate leaving the cell lumen vacant except for large deeply staining masses (Figure 1). All these artefacts can be seen in electron micrographs of chemically fixed cambial cells of gymnosperms (Table 1) and angiosperms (e.g. Kidwai and Robards, 1969; Murmanis, 1977; Rao and Dave, 1983; Sennerby-Forsse, 1986; Rao and Catesson, 1987; Barnett, 1992; Chaffey *et al.*, 1997; Farrar and Evert, 1997). Thus, even with careful fixation, an ideal balance between the osmotic potentials of the fixative solutions and the large aqueous vacuole has not been fully achieved. Wrinkles, undulations (Figure 2) and inward folding of the plasmalemma or tonoplast were reported or visible in all chemically fixed cambial cells except those that were completely destroyed.

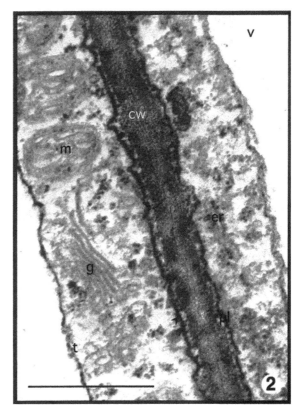

Figure 2 Radial section from chemically fixed active fusiform cells of *Pinus ponderosa* showing a portion of the peripheral cytoplasm. The cytoplasm has gaps and is unevenly distributed while the plasmalemma and tonoplast are wrinkled or infolded. Note the spaces between the cell wall and the plasmalemma (arrowheads), an indication of water movement from the cell during fixation. The thin primary cell wall, plus the presence of mitochondria and of Golgi with prominent budding vesicles indicate a high level of activity in these cells. Key: cw, cell wall; er, endoplasmic reticulum; g, Golgi; m, mitochondrion; pl, plasmalemma; t, tonoplast; v, vacuole.
Scale bar = 0.5 μm.

The factors that cause these artefacts also alter the organisation between the cell wall, the plasmalemma and the cytoskeleton. Sections cut parallel to and within this interface often show little detail (Figure 4).

Dormant cells are easier to preserve chemically as they do not have large aqueous central vacuoles and maintain higher solute concentrations than active cells. The cells contain numerous small vacuoles and large amounts of membranous material. Membranes separated by even a thin aqueous layer will remain distinct, but reduction in volume

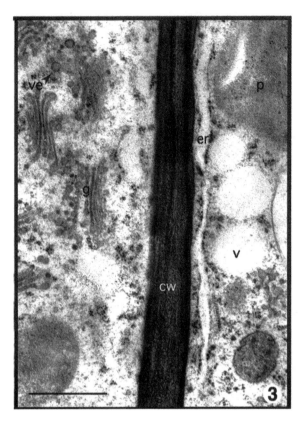

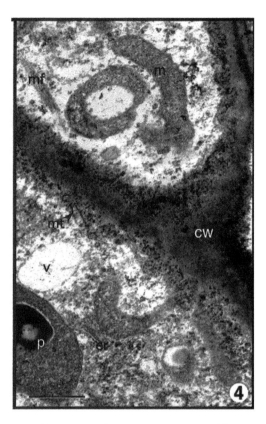

Figure 3 Radial sections of cryo-fixed and substituted (CFS) fusiform initials of *Pinus ponderosa* during spring reactivation. This section is similar to that shown in Figure 2. Occasional clusters of ribosomes are distributed within the uniform cytoplasm. The plasmalemma is barely visible to the right side of the cell wall because it is so closely associated with the wall. The Golgi and large number of associated vesicles indicate a significant amount of membrane trafficking and are much more prominent than in chemically fixed cells (cf. Figure 2). Key: cw, cell wall; er, endoplasmic reticulum; g, Golgi; p, plastid; v, vacuole; ve, vesicle. Scale bar = 0.5 μm.

Figure 4 Chemically fixed radial sections of *Pinus ponderosa* cambium showing the cytoplasm near the cell periphery. In these areas the mitochondria are often greatly elongated or cresent shaped and microfilament bundles and microtubules are usually visible. Because the section is nearly within the plane of the cell wall, very little detail is discernible at the wall-cytoplasm interface. However, the microtubules are clearly intimately associated with the cell wall. A starch grain within the plastid is virtually electron opaque. Key: cw, cell wall; er, endoplasmic reticulum; m, mitochondrion; mf, microfilament bundle; mt, microtubule; v, vacuole. Scale bar = 0.5 μm.

puts membranes in closer contact with each other. This increases the likelihood that membranous material will fuse to produce artefactual vacuoles or fuse with the plasmalemma to form apparent invaginations or 'plasmalemma invaginations' (Rao and Catesson, 1987; Rensing and Owens, 1994).

Cryo-fixation and substitution has limited use as a preparatory technique for light microscopy alone. The procedure is costly, time consuming and requires a high level of operator skill and specialised equipment. However, it is an exceptionally good method for preserving the unique ultrastructure of fusiform initials, and thus for electron microscopy, the results are well worth the extra cost and effort (compare Figures 2 and 3). CFS is particularly beneficial for preserving membranous structures and retention of much of the cambial cell anatomy depends on the integrity of membranes. When the tonoplast is preserved, the cytoplasm is more likely to be retained, particularly in active cells with thin peripheral cytoplasm (Figure 3). Lancelle *et al.* (1986) and Ding *et al.* (1991) noted that successfully cryo-fixed and substituted plant cells have a characteristically smooth plasmalemma that is tightly appressed to the cell wall. With successful CFS of cambial cells, no wrinkles, undulations or folds were present in the plasmalemma or tonoplast, nor was undefined electron opaque material present between wall and membrane or associated with any membrane (Figure 3). Because CFS induces little distortion at the interface between the cell wall and the plasmalemma (Figure 5), this technique can be useful in the study of cell wall production and the cytoskeleton. Microtubules clearly extend from the cell wall while microfibril bundles are often associated with coated vesicles.

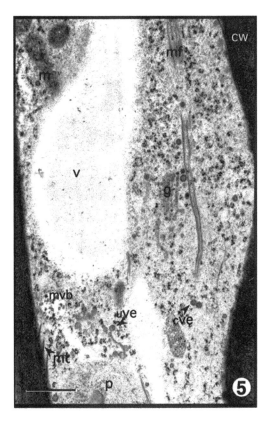

Figure 5 CFS fusiform cell of *Pinus ponderosa* in radial section. This is from close to the cell periphery and shows many of the same features as Figure 4 but these appear more clearly defined using CFS. Microtubules arch into the cytoplasm from the cell wall and the individual microfilaments within the microfilament bundles are visible. Multivesicular bodies and vacuoles are preserved with full rounded membranes, coated and uncoated vesicles can be distinguished, and the cytoplasm is uniform and gap free. Key: cve, coated vesicle; cw, cell wall; g, Golgi; m, mitochondrion; mf, microfilament bundle; mt, microtubule; mvb, multivesicular body; p, plastid; uve, uncoated vesicle; v, vacuole. Scale bar = 0.5 μm.

FUTURE PROSPECTS

Until recently, chemical fixation was the only option for ultrastructural research. Development of equipment and techniques for routine cryo-fixation and substitution of plant material has provided electron microscopists with several new avenues for research. Transient cellular events are preserved because cryo-fixation is virtually instantaneous. High pressure cryo-fixation with cryo-substitution has preserved novel features of plant Golgi stacks and transient membrane-recycling related plasmalemma configurations (Staehelin *et al.*, 1990). Cellular processes initiated by experimental treatments can be preserved within seconds of the treatment so exact time-related studies can be undertaken. Using a combination of immunogold localisation and CFS, the electron microscopist has the capacity to make great advances in our knowledge of biochemical processes within the plant cell. For instance, immuno-cytochemical analysis of cryo-fixed Golgi has shown that biosynthesis of pectins involves all types of Golgi cisternae while assembly of xyloglucan was confined to the *trans* Golgi cisternae and the *trans* Golgi network (Zhang and Staehelin, 1992). Because cambial cells can be preserved with remarkable fidelity using CFS, it is now possible to undertake similar studies to elucidate the biochemistry of processes such as division and differentiation that are critical to tree growth (and wood production). It should also be possible to gain new knowledge of the plant cytoskeleton (eg. microtubules) and the movement of cellular substances (eg. via microfibril bundles) from cryo-fixed and substituted cambial cells. Cryo-techniques can also be applied to the xylem and phloem so similar progress should be achievable for these tissues.

ACKNOWLEDGEMENTS

Research funding was provided by the Canadian Department of Natural Resources, Maritimes Division, and the National Science and Engineering Research Council of Canada. I would like to thank my PhD supervisor, Dr. R.A. Savidge in the Faculty of Forestry and Environmental Management, University of New Brunswick. I am also greatly indebted to Dr. J.N. Owens and the Department of Biology, University of Victoria for use of the cryo-fixation equipment and the TEM.

REFERENCES

Barnett, J.R. (1971) Electron-microscopy preparation techniques applied to the light microscopy of the cambium and its derivatives in *Pinus radiata* D. Don. *Journal of Electron Microscopy* **94**, 175–180.

Barnett, J.R. (1973) Seasonal variation in the ultrastructure of the cambium in New Zealand grown *Pinus radiata* D. Don. Annals of Botany. **37**, 1005–1115.

Barnett, J.R. (1977) Tracheid differentiation in *Pinus radiata*. *Wood Science and Technology* **11**, 83–92.

Barnett, J.R. (1992) Reactivation of the cambium in *Aesculus hippocastanum* L.: a transmission electron microscope study. *Annals of Botany* **70**, 169–177.

Bullock, G.R. (1984) The current status of fixation for electron microscopy: a review. *Journal of Microscopy* **133**, 1–15.

Chaffey, N., Barnett, J. and Barlow, P. (1997) Endomembranes, cytoskeleton, and cell walls: aspects of the ultrastructure of the vascular cambium of taproots of *Aesculus hippocastanum* L. (Hippocastanaceae). *International Journal of Plant Science* **158**, 97–109.

Ding, B., Turgeon, R. and Parthasarathy, M.V. (1991) Routine cryo-fixation of plant tissue by propane jet freezing for freeze-substitution. *Journal of Electron Microscopy Technique* **19**, 107–117.

Farrar, J.J. and Evert, R.F. (1997) Seasonal changes in the ultrastructure of the vascular cambium of *Robinia pseudoacacia*. *Trees Structure and Function*. **11**, 191–202.

Hopwood, D. (1985) Cell and tissue fixation, 1972–1982. *Histochemistry Journal* **17**, 389–442.

Inomata, F., Takabe, K. and Saiki, H. (1992) Cell wall formation of conifer tracheid as revealed by rapid-freeze and substitution method. *Journal of Electron Microscopy* **41**, 369–374.

Johansen, D.A. (1940) *Plant Microtechnique*, New York: McGraw-Hill Book Company.

Kellenberger, E., Johansen, R., Maeder, M., Bohrmann, B., Stauffer, E., and Villiger, W. (1992) Artefacts and morphological changes during chemical fixation. *Journal of Microscopy* **168**, 181–201.

Kidwai, P. and Robards, A.W. (1969) On the ultrastructure of resting cambium of *Fagus Sylvatica* L. *Planta* **89**, 361–368.

Kiss, J.Z., Giddings, T.H., Staehelin, L.A. and Sack, F.D. (1990) Comparison of the ultrastructure of conventionally fixed and high pressure frozen/freeze substituted root tips of *Nicotiana* and *Arabidopsis*. *Protoplasma* **157**, 64–74.

Lancelle, S.A., Callahan, D.A. and Hepler, P.K. (1986) A method for rapid freeze fixation of plant cells. *Protoplasma* **131**, 153–165.

Lancelle, S.A. and Hepler, P.K. (1989) Immunogold labelling of actin on sections of freeze-substituted plant cells. *Protoplasma* **150**, 72–74.

Mahmood, A. (1968) Cell grouping and primary wall generations in the cambial zone, xylem, and phloem in *Pinus*. *Australian Journal of Botany* **16**, 177–195.

McCully, M.E. and Canny, M.J. (1985) The stabilization of labile configurations of plant cytoplasm by freeze-substitution. *Journal of Microscopy* 27–33.

Mersey, B. and McCully, M.E. (1978) Monitoring the course of fixation of plant cells. *Journal of Microscopy* **114**, 49–76.

Murmanis, L. (1971) Particles and microtubules in vascular cells of *Pinus strobus* L. during cell wall formation. *New Phytologist* **70**, 1089–1093.

Murmanis, L. (1971) Structural changes in the vascular cambium of *Pinus strobus* L. during an annual cycle. *Annals of Botany* **35**, 133–141.

Murmanis, L. (1977) Development of vascular cambium into secondary tissue of *Quercus rubra* L. *Annals of Botany* **41**, 617–620.

Murmanis, L. and Sachs, I.B. (1969) Seasonal development of secondary xylem in *Pinus strobus* L. *Wood Science and Technology* **3**, 177–193.

O'Brien, T.P. and McCully, M.E. (1981) The study of plant structure: principles and selected methods. Termarcarphi Pty. Ltd., Melbourne.

Rao, K.S. (1988) Fine structural details of tannin accumulations in non-dividing cambial cells. *Annals of Botany* **62**, 575–581.

Rao, K.S. and Catesson, A.-M. 1987) Changes in the membrane components of nondividing cambial cells. *Canadian Journal of Botany* **65**, 246–254.

Rao, K.S. and Dave, Y.S. (1983) Ultrastructure of active and dormant cambial cells in Teak (*Tectona grandis* L. f). *New Phytologist* **93**, 447–456.

Rensing, K.H. (2000) Ultrastructure of the vascular cambium of *Pinus ponderosa* seedlings preserved by cryofixation and cryosubstitution. PhD thesis. University of New Brunswick, Fredericton, Canada.

Rensing, K.H. and Owens, J.N. (1994) Bud and cambial phenology of lateral branches from Douglas-fir (*Pseudotsuga menziesii*) seedlings. *Canadian Journal of Forest Research* **24**, 286–296.

Riding, R.T. and C.H.A. Little (1984) Anatomy and histochemistry of *Abies balsamea* cambial zone cells during the onset and breaking of dormancy. *Canadian Journal of Botany* **62**, 2570–2579.

Runions, C.J., Rensing, K.H., Takaso, T. and Owens, J.N. (1999) Pollination of *Picea orientalis* (Pinaceae): saccus morphology governs pollen buoyancy. *American Journal of Botany* **86**, 190–197.

Samuels, A.L., Giddings, T.H. and Staehelin, L.A. (1995) Cytokinesis in Tobacco BY-2 and root tip cells: a new model of cell plate formation in higher plants. *Journal of Cell Biology* **130**, 1345–1357.

Sato, T. (1967) A modified method for lead staining of thin sections. *Journal of Electron Microscropy* **16**, 133.

Savidge, R.A. and Barnett, J.R. (1993) Protoplasmic changes in cambial cells induced by a tracheid differentiation factor from pine needles. *Journal of Experimental Botany* **44**, 395–405.

Sennerby-Forsse, L. (1986) Seasonal variation in the ultrastructure of the cambium in young stems of willow (*Salix viminalis*) in relation to phenology. *Physiologia Plantarum* **67**, 529–537.

Shibaoka, H. (1993) The use of tobacco BY-2 cells for the studies of the plant cytoskeleton. *Journal of Plant Research* (Special Issue) **3**, 3–15.

Sluiman, H.J. (1991) Cell division in *Gloeotilopsis planctonica*, a newly identified ulvophycean alga (chlorophyta) studied by freeze fixation and freeze substitution. *Journal of Phycology* **27**, 291–298.

Srivastava, L.M. and O'Brien, T.P. (1966) On the ultrastructure of cambium and its derivatives. I. Cambium of *Pinus strobus* L. *Protoplasma* **61**, 257–276.

Staehelin, L.A., Giddings, T.H., Kiss, J.Z. and Sack, F.D. (1990) Macromolecular differentiation of golgi stacks in root tips of *Arabidopsis* and *Nicotiana* seedlings as visualized by high pressure frozen and freeze-substituted samples. *Protoplasma* **157**, 75–91.

Taylor, D.P. (1988) Direct measurement of the osmotic effects of buffers and fixatives in *Nitella flexilis*. *Journal of Microscopy* **150**, 71–80.

Timell, T.E. (1980) Organization and ultrastructure of the dormant cambial zone in compression wood of *Picea abies*. *Wood Science Technology* **14**, 161–179.

Willingham, M.C. (1983) An alternative fixation processing method for pre-embedding ultrastructural immunocytochemistry of cytoplasmic antigens: the GBS (glutaraldehyde-borohydride-saponin) procedure. *Journal of Histochemistry and Cytochemistry* **31**, 791–798.

Wodzicki, T.M. and Brown, C.L. (1973) Organization and breakdown of the protoplast during maturation of pine tracheids. *American Journal of Botany* **60**, 631–640.

Zhang, G.F. and Staehelin, L.A. (1992) Functional compartmentation of the golgi apparatus of plant cells: Immunocytochemical analysis of high-pressure frozen and freeze-substituted sycamore maple suspension culture cells. *Plant Physiology* **99**, 1070–1083.

Deep-etching Electron Microscopy and 3-dimensional Cell Wall Architecture

Wood Formation in Trees ed. N. Chaffey
© 2002 Taylor & Francis
Taylor & Francis is an imprint of the Taylor & Francis
Group
Printed in Singapore.

Takao Itoh

Wood Research Institute, Kyoto University, Uji, Kyoto 611–0011, Japan

ABSTRACT

This chapter describes a deep-etching technique for the study of cell wall formation. Deep-etching electron microscopy allows one to visualise the 3-dimensional cell wall architecture undisturbed by embedding media. The highly porous structure of primary as well as unlignified cell wall was first revealed by this technique. This novel method coupled with immunogold-labelling will provide information on the mutual linkages of cell wall components at the molecular level.

Key words: deep-etching, *Eucalyptus tereticornis*, immunogold-labelling, *Pinus thunbergii*, *Populus alba*

INTRODUCTION

New attempts to visualise directly the molecular organisation of the cell wall were made after the development of techniques for high-resolution electron microscopy, such as deep-etching in combination with quick-freezing. This technique was first applied by Heuser and Salpeter (1979) in an attempt to observe the organisation of acetylcholine receptors in animal cells. Goodenough and Heuser (1985) were the first to apply the technique to plant cells for examining the cell wall of a green alga, *Chlamydomonas*, which is rich in glycoproteins.

McCann *et al.* (1990) applied the technique for an examination of the complicated molecular structure of the cell wall in tissue of the onion (*Allium cepa*). Further extensive work on the distribution of polysaccharides of the plant cell wall matrix and its relevance to the architecture of the cell wall in onion tissue has been done by McCann *et al.* (1992). Based on results obtained by deep-etching electron microscopy, Bruce *et al.* (1994) presented a new cell wall model in place of the long-held one of Albersheim (1975). Satiat-Jeunemaitre *et al.* (1992) investigated three types of cells from different species for a comparison of random, polylamellate and helicoidal cell walls. We have applied the deep-etching technique to suspension cultured cells and protoplasts of *Populus*, epidermal cells in mung bean, and xylem cells in *Eucalyptus* and *Pinus* to visualise their cell wall intact (Itoh and Ogawa, 1993; Fujino and Itoh, 1998a, b; Suzuki *et al.*, 1998a, b; Hafrén *et al.*, 1999). We believe that the data collected by deep-etching will help to reconstruct the three-dimensional network of cell wall components.

THE LIMITATIONS OF CONVENTIONAL TRANSMISSION ELECTRON MICROSCOPY

The cell wall has many functions in living plants. It has long been investigated from chemical as well as structural aspects. To investigate the cell wall at the sub-cellular and molecular level, it is necessary to observe it under the transmission electron microscope (TEM) to know the relationship between its structure and chemical linkages during cell differentiation. The fine structure of the cell wall has been investigated by conventional TEM such as thin sectioning and replica techniques. The former is suitable for revealing the thickness and layering of the cell wall. However, before making thin sections, the cell wall materials are dehydrated with alcohol, then embedded in resins (see e.g. chapters by Chaffey and Rensing in this volume). So, one normally observes the cell wall while it is embedded in polymerised resins. The pores in the cell wall, if they exist, are also embedded with resins, making it difficult to visualise them by thin-sectioning. Nevertheless, using TEM we have learned that most xylem cells are composed of three layers, S_1, S_2 and S_3 (Harada, 1965a, b). We can also easily identify tension and compression wood cells by making thin sections of them. However, thin sections give rather vague images of the cell wall structure with a faint contrast making it difficult to see individual cellulose microfibrils. Furthermore, the identity and localisation of cell wall substances cannot be achieved without using immuno-gold labelling.

The replica technique is superior for revealing the details of the surface structure. When we apply the technique to xylem cells, the orientation and arrangement of individual microfibrils are disclosed clearly. Although the lamellation of cellulose microfibrils can be seen by the conventional replica technique, the successive structures of the cell wall from the surface to the interior cannot. Furthermore, before applying the technique, the materials must be dried, which may damage or modify the fine structure of the cell wall.

These weak points of thin sectioning as well as replica techniques have been overcome by the invention of the novel technique, "deep-etching".

TECHNIQUES

Deep-etching techniques

A Sampling

Equipment needed:	Chemicals needed:
Chisel	Water
Hammer	
PVC plate (approx. 10 cm square, 5 mm thick)	
Razor blades (double-edged)	
Forceps	

Dissect fresh tissue, including the cambial zone, on a PVC plate with a double-headed razor blade into small pieces (approx. $2 \times 2 \times 2$ mm) without prior chemical fixation.

B Quick-freezing

Equipment needed:	Chemicals needed:
Quick-freezing apparatus and accessories	Instant adhesive
Filter paper (4 mm square, Advantec No. 2)	Liquid helium
Aluminium plate (commercially available and specially designed for quick-freezing apparatus, diameter: 13 mm, thickness: 0.2 mm)	Baker's yeast (used for shock absorber and mounting agent)
Double-sided tape	Helium gas
Forceps	Liquid nitrogen
Plastic ring (commercially available and specially designed for quick-freezing apparatus, diameter: 15 mm, thickness: 0.6 mm, ring width: 1.5 mm) small container for liquid nitrogen	

Duration of procedure: approx. 2 h
A series of deep-etching steps, from text headings "B" to "F", is schematically illustrated in Figure 1.

1   Prepare the aluminium plate by fixing a filter paper on to it with double-sided tape; the plastic ring is fixed to the aluminium plate with instant adhesive.
2   Load a small amount (approx. 1 mm^3) of yeast paste onto the filter paper.
3   Plug the aluminium plate, with specimen, upside down on top of the freezing head of the quick-freezing apparatus (Meiwa Co. Ltd, No. QF-5000).
4   Drop the specimens onto an ultrapure copper block, which has been pre-cooled to -269 °C by a spray of liquid helium from a storage Dewar mounted beneath the apparatus. Monitor the block temperature with a thermocouple plugged into it.
5   Separate the plunger from the freezing head and transfer it to a small container filled with liquid nitrogen.
6   Remove the frozen specimen with the aluminium plate from the plunger.

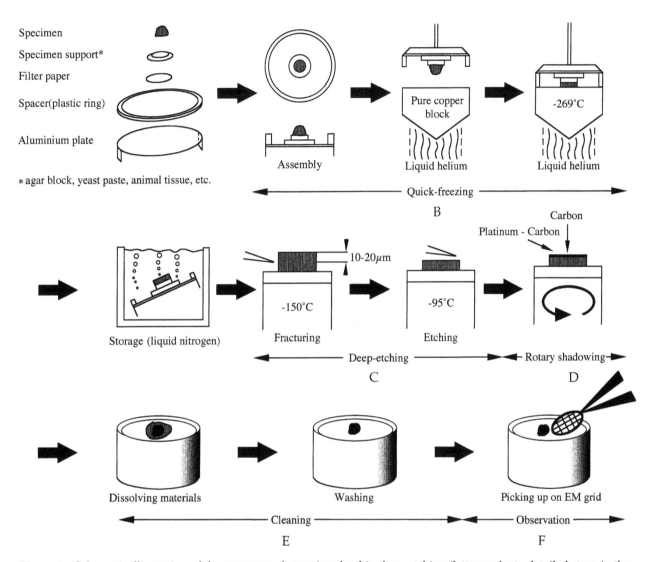

Specimen
Specimen support*
Filter paper
Spacer(plastic ring)
Aluminium plate

*agar block, yeast paste, animal tissue, etc.

Assembly

Pure copper block

Liquid helium

-269°C

Liquid helium

Quick-freezing

B

Storage (liquid nitrogen)

Fracturing

10-20μm

-150°C

Etching

-95°C

Carbon

Platinum - Carbon

Deep-etching

Rotary shadowing

C

D

Dissolving materials

Washing

Picking up on EM grid

Cleaning

Observation

E

F

Figure 1   Schematic illustration of the sequence of steps involved in deep-etching (letters refer to detailed steps in the text). The figure was modified from Fig.1 drawn by Ikeuchi (1990).

7   Separate the aluminium disc from the freezing-head, under liquid nitrogen, and pull the plastic ring off of the disc.

C   Fracturing and etching

Equipment needed:	Chemicals needed:
Freeze-etching apparatus	Liquid nitrogen
Forceps	Nitrogen gas
Single-edge "SCHICK" razor blade (Warner-Lambert Co. Ltd)	

Duration of procedure: approx. 2h

1   Quickly load the frozen specimen with the aluminium plate on the specimen holder of a freeze-etching apparatus (Balzers, No. BAF 400D).

2   After obtaining a vacuum better than 2×10^{-4} Pa and reaching an appropriate specimen holder temperature of approx. –150 °C, fracture each specimen very superficially, with great care, using a frozen knife which is capable of being operated from outside the bell jar.

      Note – the specimen is barely grazed with a razor blade (fresh blade each time)

3   Increase the temperature to –95 °C and allow the fractured specimen to etch for 15 min.

D   Rotary shadowing

Equipment needed:

Carbon rod (to produce carbon specimen replicas)

Tungsten cathode

Quartz crystal

Carbon rod (to produce carbon-platinum coatings)

Platinum insert

Thickness monitor

Immediately after etching, rotary-shadow the specimen at 25 °C by evaporating the platinum insert with high power (1,900 V and 90 mA) for approx. 7 sec, and coat by carbon-evaporation at 2,400 V and 100 mA. During shadowing and carbon-coating, the thickness is kept at a uniform 2 nm by a thickness monitor.

E   Cleaning the replica

Equipment needed:	*Chemicals needed:*
Forceps	Potassium dichromate
Formvar-coated copper grids (50 mesh or greater)	Sulphuric acid
Porcelain or glass vessel	Distilled water

Duration of procedure: 2h–overnight

After replication, the tissue is dissolved by floating the specimen on 50% sulphuric acid containing 5% potassium dichromate, for 2h–overnight at room temperature.

The replicas are washed three times in distilled water, then collected on formvar-coated copper grids.

F   Observation and photography

Equipment needed:

Transmission electron microscope

Electron microscopy film

A transmission electron microscope (JEOL 2000 EXII) operated at 100 kV is used to observe the replicas. Micrographs are taken on Fuji electron microscopic film (FG, orthocromatic) and reversed to positives before printing.

APPLICATIONS

The technique enables us to visualise the three-dimensional structure of the cell wall in different cells, tissues, organs and plants. In other words, the technique gives an image of replication after freeze-drying of the materials at very low temperature, approx. –100 °C. Thus, most of the free water has been removed from the outer layer of the specimens before replication. In that sense, the image does not show the hydrated state; however, it is closer to the native state than that obtained by conventional techniques.

Visualisation of the primary wall

Strictly speaking, it is difficult to observe the real primary wall in the cambial zone of tree species by the deep-etching technique. This is because differentiating xylem cells, with accompanying deposition of secondary wall, may be observed. However, three examples of deep-etched images obtained from real primary wall will now be described.

Primary wall in intact tissue

The primary wall architecture of elongating epidermal cells in pea epicotyls is visualised by the technique.

In deep-etching, plant tissues are collapsed by the slamming into a cold metal block during the quick-freezing step and it is very difficult to identify an individual tissue owing to the shadowing image. However, the cell wall of the outer epidermis is easily distinguished from that of other tissues because of its cuticle, very thick walls, and helicoidal pattern, in which the change of the microfibrillar orientation is regular and progressive from one lamella to the next.

Figure 2 shows deep-etched images of tangential sections of outer epidermal walls in elongating and elongated regions of pea (*Pisum sativum*) stem. The cell wall architecture in the elongating region is composed of cellulose microfibrils and granular substances that completely infiltrate the spaces between the cellulose microfibrils (Figure 2a). The

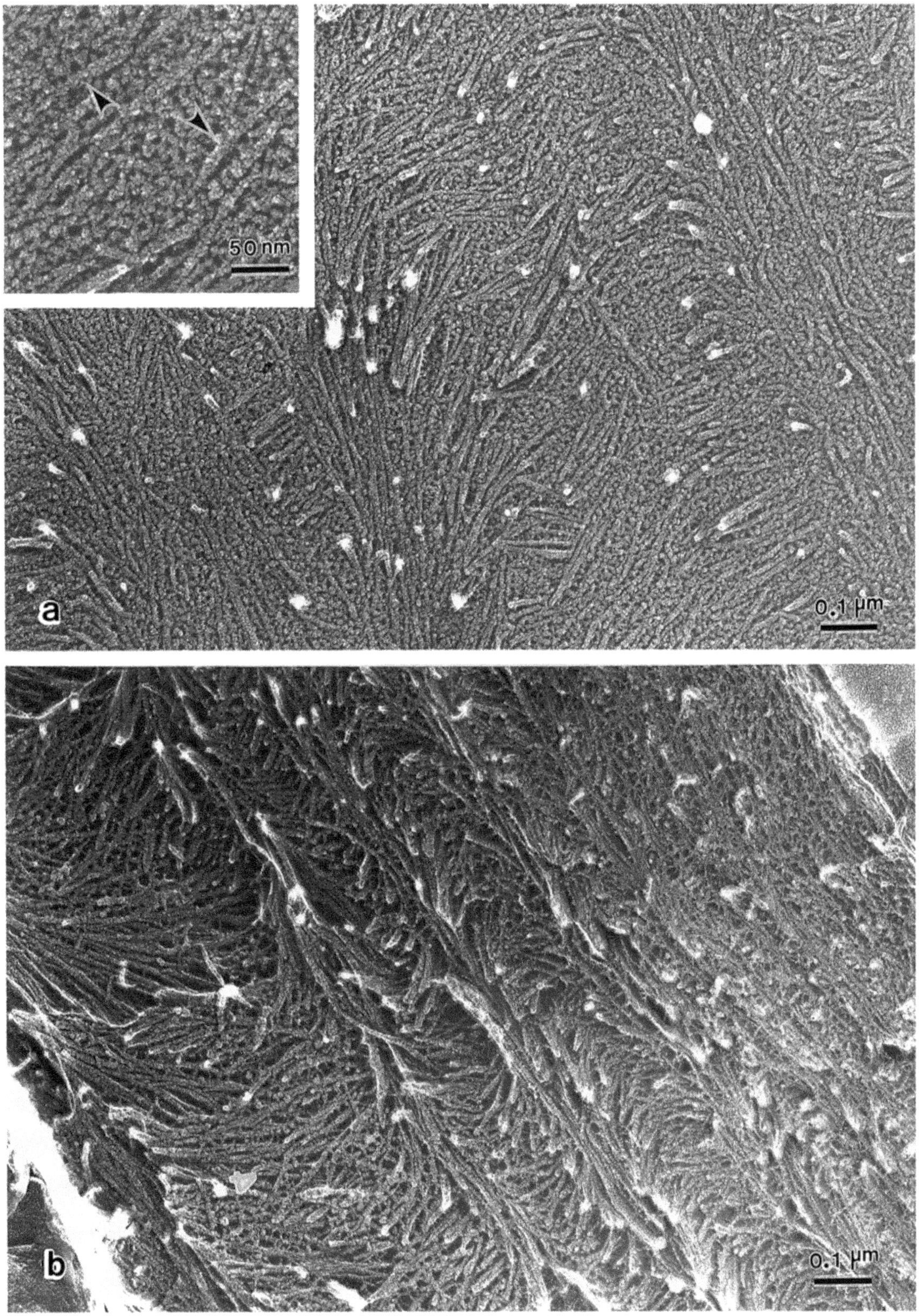

Figure 2 Deep-etched images of outer epidermal walls in elongating (Figure 2a) and elongated (Figure 2b) regions of pea (*Pisum sativum*).

pore size within the cell wall is approx. 5.5 nm. It seems that several granular substances are connected to each other, rather than existing independently (Figure 2a inset). This cell wall architecture of microfibrils and granular substances is continuous from the innermost wall layer just above the plasma membrane to the outermost wall layer below the cuticle. The cellulose microfibrils and granular substances are the only structural features observed.

A substantial decrease in the density of the granular substances between cellulose microfibrils is observed in the walls of elongated epidermal cells (Figure 2b). Larger and less numerous spaces are formed between the cellulose microfibrils after the cessation of elongation growth and fewer granular substances are detected between the cellulose microfibrils in the cell wall. The average pore size within the elongated cell wall is approx. 13 nm. There are no apparent differences in cell wall architecture between the inner and outer layers in the cell wall. However, the helicoidal orientation of cellulose microfibrils is not distinct in the outer layer of the wall.

After extraction of pectins by ethylenediamine-tetraacetic acid (EDTA), the outer epidermal walls in the elongating region are observed by the deep-etching technique. The granular substances completely disappear and only the network of cellulose microfibrils is left in the wall (Figure 3a). The surfaces of the cellulose microfibrils show smooth contours. After EDTA treatment, the space between cellulose microfibrils becomes distinct and has a width of approx. 30 nm. The pore size also increases significantly after EDTA treatment.

After pectin extraction by EDTA in the elongated region of the epidermal wall, followed by observation with the deep-etching technique, cellulose microfibrils are revealed more clearly than in the non-extracted material and the average pore size within the cell wall increases to 20 nm (Figure 3b). However, no substantial differences of cell wall architecture are noted in this region between extracted and non-extracted wall. An abundant network structure composed of granular substances disappears from the cell wall during elongation. The granular substances are demonstrated to be pectic polysaccharides by their disappearance upon EDTA treatment and by chemical analysis of the EDTA-extractable substances (Fujino and Itoh, 1998b). It is proposed that the association of the granular substances is involved in the swelling of the cell

walls in the elongating region. Two observations suggest that the formation of the pectic gel itself is not involved in controlling the wall porosity (Fujino and Itoh, 1998b). Labelling with the monoclonal antibody JIM5, which recognises unesterified pectins (Knox *et al.*, 1990), is much more extensive in the cell walls of the elongated region compared to those in the elongating region. The pore size of the cell wall is greater in the nonelongated than elongating region (Fujino and Itoh, 1998b).

Primary wall in tissue culture

The structure of the cell wall regenerated from leaf mesophyll-derived protoplasts of white poplar (*Populus alba* L.) is visible by conventional thin-sectioning TEM. The freshly isolated protoplasts, as well as 3-day-old cells are spherical and no deposition of cell wall materials is observed (see figure 2a and 2b in Suzuki *et al.*, 1998b). At 10 days of protoplast culture, deposits of wall materials are scattered unevenly over the outside of the plasma membrane (see figure 3a and 3b in Suzuki *et al.*, 1998b). At 20 days of protoplast culture, some cells have an enlarged nucleus (data not shown), and the wall thickness has increased relative to 10-day-old cells. At higher magnification the cells at this stage show a loosened wall with many spaces, possibly because of the low density of cell wall materials (see figure 4 in Suzuki *et al.* 1998b). At 30 and 45 days of protoplast culture, the cells have formed large colonies with asymmetric division (data not shown). The cell wall materials are densely deposited compared to those of 20 day-old cells (see figure 5 in Suzuki *et al.* 1998b). However, the thickness of the cell walls varies little between 20- and 30- or 45-day-old cells.

Quick-freezing and deep-etching electron microscopy disclosed the deposition of cellulose microfibrils in 3-day-old cells (Figure 4a), contrary to the result obtained by conventional thin-sectioning TEM. The microfibrils are oriented randomly in close contact with the plasma membrane and the surface of the plasma membrane is not covered completely with microfibrils. This implies that the plasma membrane is supported partially by microfibril deposition. No microfibrils extend outside the cell beyond the plasma membrane. Figure 4b shows the wall architecture of 10-day-old cells. The right side of the figure shows the PF-face (that is, protoplasmic fracture face) of the plasma membrane. A dense deposition of micro-

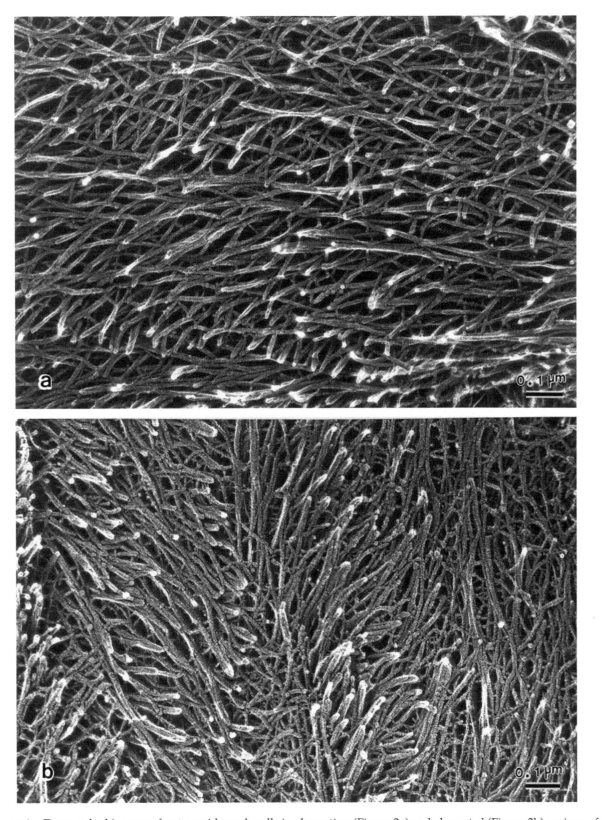

Figure 3 Deep-etched images of outer epidermal walls in elongating (Figure 3a) and elongated (Figure 3b) regions of pea after extraction of pectins with EDTA.

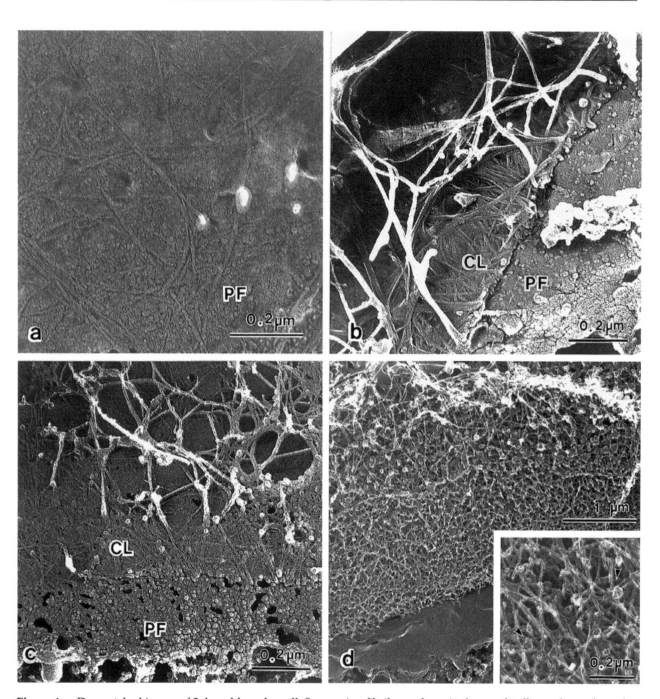

Figure 4a   Deep-etched image of 3 day-old poplar cell. Some microfibrils are deposited sporadically on the surface of the plasma membrane (compare with figure 2 in Suzuki *et al.*, 1998b which shows no deposition of microfibrils). PF: Protoplasmic face of the plasma membrane

Figure 4b   Deep-etched image of poplar 10-day-old cell. The plasma membrane is completely covered with thin microfibrillar lamella. All microfibrils outside the lamellae are extended and loosened, resulting in the formation of a network architecture with large meshes. PF: Protoplasmic face of the plasma membrane, CL: Cellulosic lamellae

Figure 4c   Deep-etched image of poplar 20-day-old cell. The diameter of the meshes in the network structure decreased because of the dense and increased deposition of microfibrils. All microfibrils outside the lamellae are extended and loosened, resulting in the formation of a network architecture with small meshes. PF: Protoplasmic face of the plasma membrane, CL: Cellulosic lamellae

Figure 4d   Deep-etched image of poplar 30-day-old cell. The fine network architecture is seen throughout the cell wall. Some granular substances are also observed in the cell wall (arrowhead in the inset). Thin lamellae cannot be observed because the cell is covered by a thick cell wall layer.

fibrillar lamellae can be seen in the middle of the figure. The plasma membrane is completely covered with a cellulosic lamella (CL) at least two microfibrils thick, each layer with a different orientation. The left side of the figure shows the microfibrils deposited previously as lamellae. These microfibrils seem to be loosened and markedly extended during cell swelling, thus making large pores, often more than 300 nm in size, in the cell wall. The mean diameter of the pores at this stage is 101 nm. The plasma membrane is always covered with new lamellae. It is suggested that the new cell wall lamellae which are deposited tightly on the surface of the plasma membrane are extremely difficult to observe by ultra-thin sectioning, because they are too thin to be visualised.

After 20 days of protoplast culture, the deposition of cell wall materials increases, with microfibrillar networks of mean diameter 33 nm seen in the entire wall, and the size of the mesh decreases gradually because of the dense and increased deposition of microfibrils (Figure 4c). The plasma membrane at this stage is also covered with thin wall lamellae. The microfibrils protruding from these thin wall lamellae increase in number as compared with the protoplast culture for 10 days. After 30 days of protoplast culture, the deposition of the cell wall increases more than that of 20-day-old cells, leaving the network architecture of the cell wall with a mean diameter of 25 nm, which is smaller than that of 20-day-old cells (Figure 4d). The fine networks are constructed by the deposition of microfibrils approx. 6–10 nm in width at this stage. Some granular substances are also observed in the cell wall (arrowheads in the inset of Figure 4d).

Primary wall of cambial and/or enlarging cells

As cell wall development proceeds, an early primary wall in the differentiating xylem cells of *Eucalyptus tereticornis* is deposited with randomly oriented microfibrils of 11–14 nm in diameter forming a fibril meshwork with a pore size of 8–28 nm. Abundant cross-links, 2.2 nm in diameter, can be seen between the microfibrils. Based on deep-etching images of sequentially extracted poplar cells in suspension culture, cross-links with similar features to those observed here are suggested to consist of pectin and hemicelluloses.

The cell corner and middle lamellae of *Eucalyptus tereticornis* show highly porous and meshwork structures (Figures 5a and 5b). The fibrous structure of the meshwork shows branching and is 11.4 nm in mean diameter. The mesh in the meshwork on the cross-fracture is approx. 18 nm in mean diameter. Highly ordered microfibrils are absent from both cell corners and middle lamellae.

A more mature primary wall is shown in Figures 6a and 6b. The cell wall structure displays an increasingly more lateral alignment of microfibrils with considerably greater association of fibrils running almost unidirectionally, mostly perpendicular to the cell axis. The fibrils are aggregated into larger bundles of two or more microfibrils with less spacing in between them, but the microfibrils and the cross-links have the same features as in the earlier stage of primary cell wall formation. The positioning of these structures is determined using stereo images; the meshwork structure is located over the lamellae with ordered microfibrils, which are located over the plasma membrane. The meshwork structure is successive to the middle lamella. A meshwork structure with overlapping thin microfibrillar lamellae with ordered microfibrils is shown in Figure 6a. The mesh of the meshwork on the tangential fracture is approx. 24 nm in diameter. The lamella with ordered microfibrils show a number of empty spaces of dimension approx. 18 nm in the shorter axis. The individual microfibrils are 14 nm in mean diameter. Cross-links (17 nm) between microfibrils are often observed in the walls during primary wall formation (Figure 6b). These observations indicated that the cell corners, middle lamellae and primary walls are all highly porous.

Visualisation of secondary wall

Unlignified wall

The appearance of the unlignified secondary wall can be seen in Figure 7, with a wavy, rather porous, microfibrillar meshwork, in the secondary xylem cells of *Pinus thunbergii*. The microfibrils aggregate in bundles of different sizes, with an individual fibril 6.2 nm wide. A distinct feature of the secondary wall seems to be a system of "slit-like" pores scattered in the cell wall between the microfibrils. These structures, 8–40 nm in diameter, are of various lengths with narrowing ends. Numerous cross-links between the microfibrils can

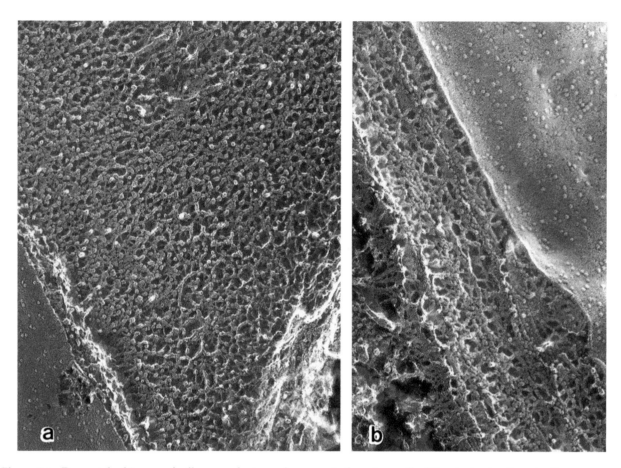

Figure 5a Deep-etched image of cell corner during enlargement of a xylem cell of *Eucalyptus tereticornis* after cross-fracture. (magnification: 93,000 ×)
Figure 5b Deep-etched image of middle lamella during enlargement of wood fibres of *Eucalyptus tereticornis* after cross-fracture. Note numerous pores in the network of amorphous materials. The primary wall has a loose microfibrillar structure. (magnification: 75,000 ×)

be seen in the slit-like pores. The cross-links are 1.6 nm in diameter.

Lignifying and lignified wall

It is quite difficult in the deep-etching replica images to distinguish lignifying from lignified wall during the differentiation of xylem cells. Figures 7 and 8 show the secondary wall with different degrees of lignification. In Figure 7, the microfibrils are unidirectional, but still many pores and cross-links can be seen. Therefore, the micrograph is likely to show a secondary cell wall under development with only slight lignin encrustation. In the final stage of development when cell wall deposition has stopped and the mature fibre is fully lignified, as in Figure 8, the S_2 wall is highly compacted and the encrustation of

lignin seems to seal the "pore system" in the cell structure. Hence, slit-like pores or cross-links are difficult to distinguish in the xylem. The lignin seems to be deposited around the microfibrils. The microfibrils are, on average, 9.3 nm in diameter. Since a microfibril diameter of 6.2 nm is measured on the unlignified cell wall, the microfibril structure is about 3.1 nm thicker *after* cell wall maturation. The surface of the microfibrils in Figure 8 is somewhat more irregular and less smooth than before lignification (cf. Figure 7). Therefore, the remaining deposition of cell wall constituents seems not to be completely evenly distributed on the polysaccharide meshwork. Chemical analyses of native mature xylem indicated 28.2 % Klason lignin (+ 0.2% acid soluble), which is normal in *Pinus thunbergii* (28.3%, Sarkanen and Hergert, 1971).

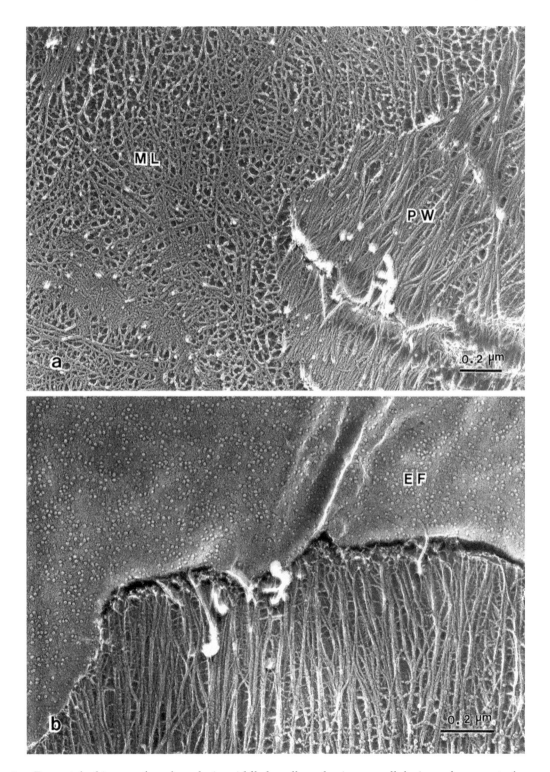

Figure 6a Deep-etched image of meshworks in middle lamella and primary wall during enlargement of a wood fibre of *Eucalyptus tereticornis*. The horizontal direction of the figure corresponds to the longitudinal axis of the wood fibre. PW: Primary wall, ML: Middle lamella

Figure 6b Deep-etched image of the cell wall during enlargement of a wood fibre of *Eucalyptus tereticornis*. Many slit-like spaces and cross-links can be seen between ordered cellulose microfibrils. The horizontal direction of the figure corresponds to the longitudinal axis of the wood fibre. EF: E-fracture face.

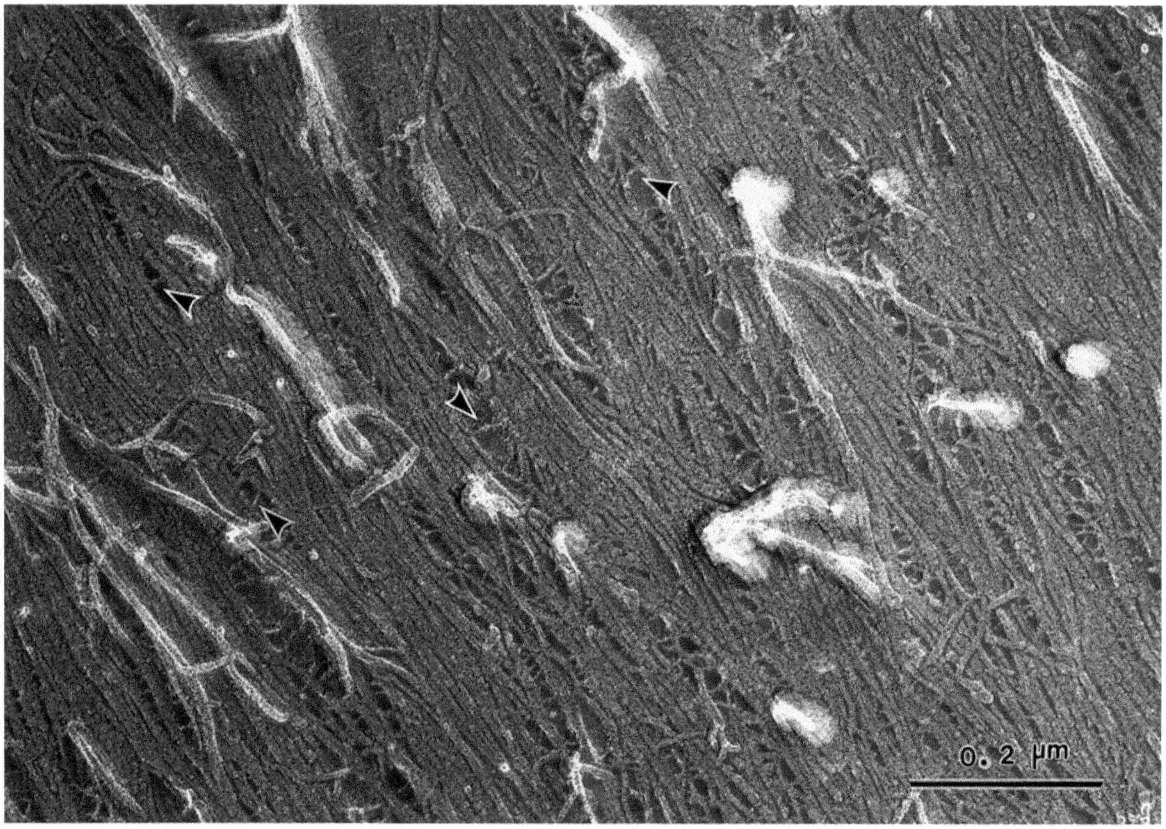

Figure 7 Deep-etched image of an unlignified S₂ wall in the differentiation of wood fibres of *Pinus thunbergii*. A dense structure with scattered oval pores and cross-links is shown (cross-links are indicated by arrowheads).

Delignified wall

Delignification of wood by sodium chlorite is fairly selective for lignin, although some polysaccharides are also dissolved. The effect of delignification of the secondary cell wall in mature xylem is shown in Figure 9. The general appearance seems to be similar to that of the unlignified S₂ wall, as shown in Figure 7. The slit-like pores are again apparent, with explicit cross-links and slit-like pores between the microfibrils. The microfibril size, 7.7 nm, and porosity pattern are somewhat altered. The delignified microfibrils in Figure 9 are significantly larger in diameter than the corresponding microfibrils in the unlignified S₂ wall (Figure 7). The cross-links are again visible after the removal of lignin and they are of about the same size, 1.9 nm, as before lignification. The chemical analysis of the delignified wood samples indicates an almost complete delignification. The carbohydrate analyses show that arabinose and

galactose are degraded to a greater extent than glucose and xylose by the sodium chlorite treatment (Hafrén *et al.*, 1999).

In conclusion, the deep-etching technique shows us the highly porous structure of the cell wall. The pores may be involved in transport of water and dissolved solutes in the cell wall, which is diminished by the deposition of lignin.

FUTURE PROSPECTS

It has been shown in the past several years that cellulose as well as hemicellulose and pectin constitute the three-dimensional networks in the cell wall of higher plants (McCann and Roberts, 1991; Satiat-Jeunemaitre *et al.*, 1992; Itoh and Ogawa, 1993). These reports were mainly performed by using deep-etching techniques. Using these techniques, it is relatively easy to determine the distribution of cellulose because of its fibrillar form; it is not

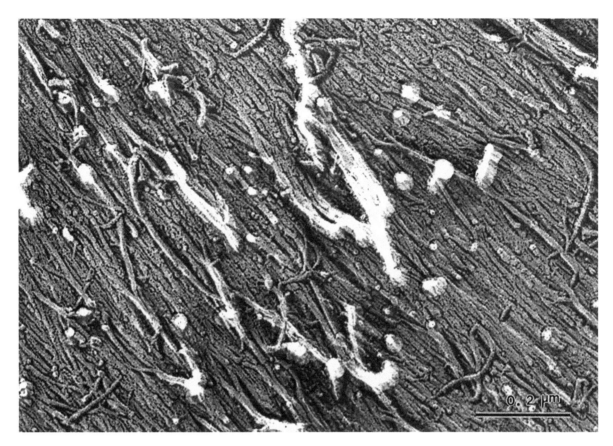

Figure 8 The closed structure of a fully lignified S_2 wall in the mature wood fibre of *Pinus thunbergii*. Neither pores nor cross-links are visible. The deep-etched image shows a radial view of the wood sample.

Table 1 Distribution of anti-xyloglucan-linked gold particles bound to microfibrils

Localisation	Culture time (h)	
	3	120
On single mf[1]	9^2 $(31.0)^3$	6 (4.3)
In intersection of 2 mfs	6 (20.7)	19 (13.7)
On associated mfs	13 (44.8)	114 (82.0)
On cross-linked mfs	1 (3.5)	0 (0)

Key: [1], mf, microfibril; [2], number of gold particles; [3], number of gold particles as a % of total.

so easy to know the localisation of hemicellulose and pectin in the cell wall. Immunogold electron microscopy applied to ultra-thin sections gives us a novel idea of the distribution of hemicellulose and pectin. The localisation of xyloglucan, which may actively be involved in cell elongation, has been extensively investigated by the labelling of gold particles on macerated and thin-sectioned cell wall using antibodies (Baba *et al.*, 1994). However, it is difficult to solve the three-dimensional relationship between cellulose and hemicellulose by these conventional techniques.

Deep-etching electron microscopy gives new insight into the three-dimensional aspect of cell wall organisation, but does not give any information concerning the chemical nature of the cell wall constituents. However, deep-etching electron microscopy coupled with immuno-gold labelling allows one to understand better the chemical nature of specific structures of the cell wall in its three-dimensional organisation.

The distribution of xyloglucans was investigated by this novel technique during the regeneration of cell wall in *Nicotiana tabacum*. Neither cell wall materials nor gold particles can be observed in freshly isolated protoplasts at all (data not shown). The gold particles, however, are found in

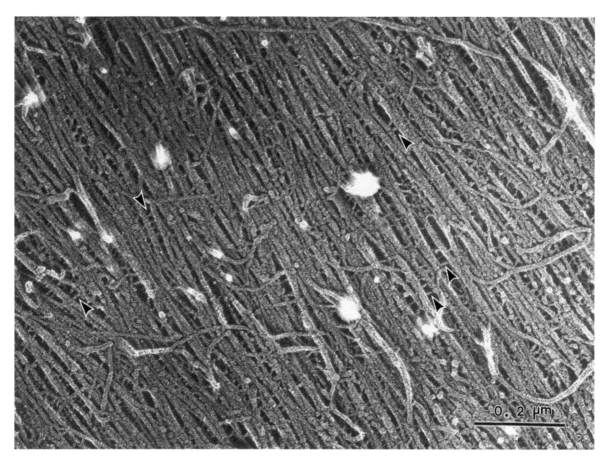

Figure 9   Delignified S$_2$ wall from the mature wood fibre of *Pinus thunbergii* as shown by a deep-etched image. Between the microfibrils cross-links are visible, and pores formed like gaps (cross-links are indicated by arrowheads).

the cell wall structure of 3-hour-old cells and some cell wall material is visible. These gold particles are mostly bound to cellulose microfibrils, but few, only 0–2 particles, are bound to a single microfibril. At a more advanced stage of cell development, the number of gold particles attached to single microfibrils increases. Gold particles in the cell wall of 120-hour-old cells are more abundant than at the former stage. Most of these particles exist on single or laterally-associated microfibrils and at the intersection between two crossed microfibrils (Figure 10). They are distributed randomly, as also seen in ultra-thin sections (data not shown). Although there are not so many gold particles in the thin lamellae adjacent to the plasma membrane, a number of gold particles are observed in the loose network architecture of the cell wall which is formed outside the lamellae (Figure 10). Most of the particles are localised at the intersection between two

crossed microfibrils and on the associated microfibrils (Figure 10). The distribution of the gold particles bound to microfibrils is shown in Table 1. The table indicates that the particles are observed along the associated microfibril in both 3- and 120-hour-old cells. Comparing 3- and 120-hour-old cells, the amount of gold particles associating with a single microfibril decreases, but that on the associated microfibrils increases (Table 1). As a control in these experiments, the specimens were incubated with the rabbit-derived normal serum instead of the polyclonal anti-xyloglucan antibodies; no labelling was noted (data not shown).

This novel technique enables us to understand the structure and mutual linkages of cell wall components in the three-dimensional organisation, and its application to the wood cell wall will lead to deeper insights into secondary vascular differentiation.

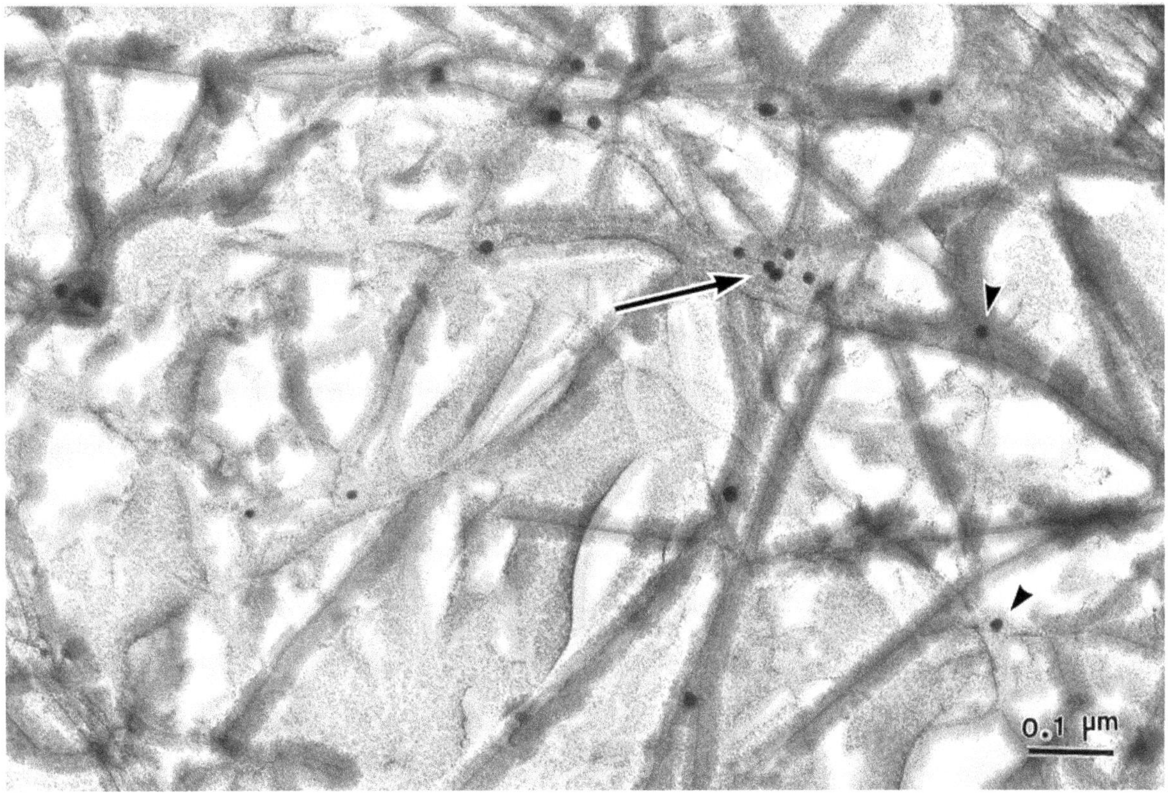

Figure 10 Deep-etched image of 120-hour-old cells of *Nicotiana tabaccum* labelled by the anti-xyloglucan antibodies. The number of gold particles is much higher in 120-hour-old cells than in 3-hour-old cells. Gold particles largely appear on the associated cellulose microfibrils (arrow) and at intersections between crossed microfibrils (arrowheads).

ACKNOWLEDGEMENTS

The author expresses his thanks to his graduate students, Dr T. Fujino, Dr K. Suzuki, and Dr T. Ogawa and a visiting student, Dr J. Hafrén, from STFI, Stockholm, Sweden for assisting with the elegant, but time-consuming deep-etching electron microscopy to investigate cell wall architecture. The author also expresses his thanks to Dr T. Fujino for preparing the drawing of a series of deep-etching techniques.

REFERENCES

Albersheim, P. (1975) The walls of growing plant cells. *Sci. Amer.* **232**, 80–95.

Baba, K., Sone, Y., Misaki, A. and Hayashi, T. (1994) Localisation of xyloglucan in the macromolecular complex composed of xyloglucan and cellulose in the pea stems. *Plant Cell Physiol.* **35**, 439–444.

Bruce, A., Bray, D., Lewis, J., Raff, M., Roberts, K. and Watson, J. D. (1994) *Molecular Biology of Plant Cell.* Garland Publishing Inc. 1002

Fujino, T. and Itoh, T. (1998a) Changes in the three-dimensional architecture of the cell wall during lignification of xylem cells in *Eucalyptus tereticornis.* *Holzforsch.* **52**, 111–116.

Fujino, T. and Itoh, T. (1998b) Changes in pectin structure during epidermal cell elongation in pea (*Pisum sativum*) and its implications for cell wall architecture. *Plant Cell Physiol.* **39**, 1315–1323.

Goodenough, U.W. and Heuser, J.E. (1985) The *Chlamydomonas* cell wall and its constituent glycoproteins analyzed by the quick-freeze, deep-etch technique. *J. Cell Biol.* **101**, 1550–1568.

Hafrén, J., Fujino, T. and Itoh, T. (1999) Changes in cell wall architecture of differentiating tracheids of *Pinus thunbergii* during lignification. *Plant Cell Physiol.* **40**, 532–541.

Harada, H. (1965a) Ultrastructure of gymnosperm cell walls. In *Cellular Ultrastructure of Woody Plants*, edited by W.A. Côté, Jr., pp. 235–249. Syracuse University Press.

Harada, H. (1965b) Ultrastructure of angiosperm vessels and ray parenchyma. In *Cellular Ultrastructure of Woody Plants*, edited by W.A. Côté, Jr., pp. 215–233. Syracuse University Press.

Heuser, J.E. and Salpeter S.R. (1979) Organisation of acetylcholine receptors in quick-frozen, deep-etched, and rotary-replicated torpedo postsynaptic membrane. *J. Cell Biol.* **82,** 150–173.

Ikeuchi, Y. (1990) Deep-etching technique. *Jikken-igaku* **8**(**5**), 160 (In Japanese).

Knox, J.P., Linstead, P.J. King J., Cooper C. and Roberts, K. (1990) Pectin esterification is spatially regulated both within cell walls and between developing tissues of root apices. *Planta* **181,** 512–521.

McCann, M.C. and Roberts, K. (1991) Architecture of the primary cell wall. In *The Cytoskeletal Basis of Plant Growth and Form*, ed. C.W. Lloyd, pp. 109–129. Academic Press.

McCann, M.C., Wells, B. and Roberts, K. (1990) Direct visualisation of cross-links in the primary plant cell wall. *J. Cell Sci.* **96**, 323–334.

McCann, M.C., Wells, B. and Roberts, K. (1992) Complexity in the spatial localisation and length distribution of plant cell-wall matrix polysaccharides. *J. Microscopy* **166**, 123–136.

Satiat-Jeunemaitre, B., Martin, B. and Hawes, C. (1992) Plant cell wall architecture is revealed by rapid-freezing and deep-etching. *Protoplasma* **167**, 33–42.

Suzuki, K., Baba, K., Itoh, T. and Sone, Y. (1998) Localisation of the xyloglucan in cell walls in a suspension culture of tobacco by rapid-freezing and deep-etching techniques coupled with immunogold labelling. *Plant Cell Physiol.* **39**, 1003–1009.

Suzuki, K., Itoh, T. and Sasamoto, H. (1998) Cell wall architecture prerequisite for the cell division in the protoplasts of white poplar, *Populus alba* L. *Plant Cell Physiol.* **39**, 632–638.

Secondary Ion Mass Spectrometry Microscopy: Application to the Analysis of Woody Samples

Wood Formation in Trees ed. N. Chaffey
© 2002 Taylor & Francis
Taylor & Francis is an imprint of the
Taylor & Francis Group
Printed in Singapore.

Marie-Laure Follet-Gueye

SCUEOR, Faculté des sciences de l'université de Rouen, F-76821 Mont-Saint-Aignan Cedex, France

ABSTRACT

Secondary ion mass spectrometry (SIMS) microscopy was used to investigate ion distributions in the cambial regions of one-year-old beech branches. This required a new method of wood sample preparation which entails fixing and dehydrating in the vapour phase. This new method is compared with the pyroantimonate precipitation method.

Key words: beech cambium; chemical fixation; ion; SIMS microscopy

INTRODUCTION

Mineral nutrient uptake is probably involved in cambial activity and vascular differentiation in trees (Larson, 1967; Wardrop, 1981). Pollutions that modify the mineral soil composition bring about disturbances in xylogenesis (Ulrich, 1981; Zöttl et al., 1989). Of all the factors involved in cambial activity or in vascular differentiation, the effects of environmental or hormonal (Avery, 1937; Sachs, 1981) parameters have been relatively well studied; the role of minerals is less clearly understood. Magnesium has a positive effect on vascular production via its role in photosynthesis (Küppers et al., 1985). Potassium is implicated in the growth of meristematic cells and of differentiating cambial cells (Pitman et al., 1971). The involvement of calcium in cell wall thickening has been especially studied by Albersheim (1978). Calcium allows the formation of intermolecular bonds through pectins that play a fundamental role in the regulation of cell wall extensibility. For example in poplar, it was shown that the thickening walls of phloem derivatives contained acidic pectins and calcium throughout the wall. In contrast, in cell walls of xylem derivatives acidic pectins were restricted to cell junction and middle lamellae and rarely accompanied by calcium ions (Guglielmino et al., 1997). Finally, calcium ions are involved in many processes: e.g. polarised growth (Weisenseel et al., 1981), mitosis (Hepler et al., 1981), cytoplasmic streaming (Kamiya, 1981). In the light of the above, we have used secondary ion mass spectrometry microscopy to study the involvement of the major mineral elements (Ca^{2+}, K^+, Na^+ and Mg^{2+}) in cambial activity and vascular differentiation in beech trees.

Secondary ion mass spectrometry (SIMS) microscopy, developed by Castaing and Slodzian in 1962, combines secondary ion emission with mass spectrometry. When coupled to analytical imaging, SIMS is one of the few microscopical methods that, simultaneously, both detects and localises in situ the isotopes of virtually all elements of the periodic classification. The detection limit of the SIMS method is of the order of a few ppm, even of ppb in some cases. The first applications of SIMS to biological samples were initiated by Galle (1970). However, although SIMS has been used in toxicology (Chassard-Bouchaud and Galle, 1986; Berry et al., 1987) and pharmacology (Larras-Regard and Mony, 1995), it has rarely been applied to plant tissues (Campbell *et al.*, 1979; Spurr and Galle, 1980; Grignon *et al.*, 1997).

Hydrated biological samples, in their natural state, are not directly usable in the vacuum conditions of the ion microscope. They should be stabilised by fixation and dehydration methods, which must both preserve ultrastructures and the distribution of the ions. The fixed and dehydrated samples must then be embedded in resin in order to obtain a flat surface after microtomy. In the particular case of woody samples, which contain many types of cells, this compromise between good preservation of structure and retention of diffusible ions is very hard to find.

There are two fixation processes that ensure ion retention in biological tissues: the precipitation and the cryo-fixation techniques. Precipitation methods are only performed when the ions of interest are solidly associated with structural polymers, such as cell walls (Jauneau *et al.*, 1992; Ripoll *et al.*, 1992). The cryo-fixation methods, based on rapid sample freezing (Chandler *et al.*, 1984) and subsequent dehydration by cryo-substitution, do not usually preserve ion *distribution*. In order to study the distribution of diffusible ions (Na^+, K^+) in woody samples, we have adapted a new method previously described for pollen preparation by Lhuissier *et al.* (1997), which uses fixation and dehydration in the vapour phase. In our procedure, woody material is fixed by aldehydic vapour of acrolein and dehydrated with acidified dimethoxy, 2–2 propane (DMP) vapour. We have compared the characteristics of this new method with the potassium pyro-antimonate precipitation method (Mentré and Escaig, 1988) in cambial zone of beech sample.

TECHNIQUES

Fixation and dehydration in the vapour phase

Principle

For fixation-dehydration in the vapour phase, it is necessary to maintain the woody sample in an hermetically sealed enclosure so that the atmosphere remains saturated with vapour of the fixative reagent as long as necessary for a total tissue fixation. After this fixation step, the woody sample is kept in the enclosure and the fixative vapours are progressively replaced with vapours of dehydrating reagent. The choice of fixative reagent depends on its chemical properties. The reagent must have a vapour density greater than that of air in order to

rapidly saturate the enclosure, and be sufficiently hydrophobic in order to limit the movement of diffusible ions (which might otherwise occur if the water vapour pressure is too high). Acrolein, a three-carbon monoaldehyde (C_3H_4O), is one of a few fixative reagents that satisfy both of these criteria. Only a few millilitres of this reagent are sufficient to saturate quickly a small hermetic enclosure, and the vapours of reagent penetrate tissues in a few seconds (Luft, 1959). Although this aldehyde molecule is particularly dangerous, use of a less dangerous alternative, e.g. glutaraldehyde, was not possible because of its lower vapour density.

Moreover, to preserve ultrastructure, the use of aqueous osmium tetroxide, a common post-fixative, has not been possible because of the high level of water vapour. And, we have observed that after acrolein fixation, it is necessary to wash the sample environment with water vapours in order to prevent the reduction of osmium tetroxide. So this step isn't compatible with the study of soluble ion distribution. Finally, osmium generates electrostatic charges at the surface of sections that perturb the secondary ion beam during SIMS analysis (Edelman, 1991).

The dehydrating agent used is acidified DMP which eliminates water from the tissue by converting it to acetone and methanol. Addition of a few drops of hydrogen chloride catalyses the action of the DMP (Müller and Jacks, 1975).

Materials and methods

Beech samples: One- or two-year-old branches were cut with pruning shears from a single beech tree (*Fagus sylvatica* L.) which was approx. 50 years old, located in a park near the campus of the University of Rouen. The branches were sampled in mid-December and in May, corresponding, respectively, to the rest and active meristematic period of the cambial seasonal cycle. The first internodes were excised with a razor blade and used for SIMS microscopy analysis.

Equipment needed:	Chemicals needed:
Hermetic enclosure (Figure 1)	Acrolein 90% (Aldrich, No. 11,022-1)
Razor blades	2,2-Dimethoxypropane 98% (DMP) (Aldrich, No. D 13,660-8)

Petri dishes with lids	Hydrochloric acid fuming 37% (HCl) (Aldrich, No. 25,814-8)
10ml glass pipettes	Propylene oxide 99% (PO) (Aldrich, No. 11,020-5)
Eppendorf tubes (capacity: 1ml)	Spurr resin kit (Agar, No. R 1032); medium grade (ERL 4206, 10g; DER 736, 6g; NSA, 26g)
Embedding oven at 60 °C Flat embedding mould (Agar, No. G 369)	

Duration of procedure: approx. 50–120 h.

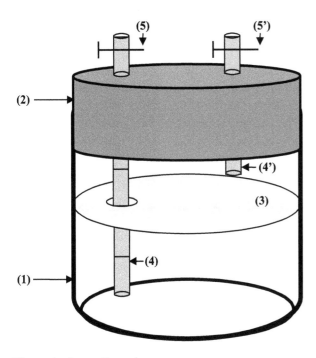

Figure 1 hermetic enclosure
(1) glass container
(2) silicone stopper on top chamber
(3) plastic grid
(4) longer and (4') shorter pipe
(5) (5'): on-off systems

The hermetic enclosure (Figure 1) is a glass container 6 cm wide and up to 10 cm high. It is hermetically sealed by means of a silicone top which is pierced by two resistant plastic pipes of different lengths which are equipped with an on-off system to allow exchange of gases or liquids where desired. The

longer pipe (15 cm) is used to introduce or remove the reagents, whereas the shorter pipe (3 cm) is opened during introduction or removal of reagent to preserve the internal vapour pressure inside the class container. A plastic grid, pierced for the passage of the longer pipe, is fixed in the middle of the glass container in order to prevent the pipe falling to the bottom of the container. This assembly is kept under a fume hood.

1   Cut excised fragments perpendicular to the axis of beech branches with a razor blade so that they are half an internode, approx. 3 cm, long.
    Note – where present, cork is carefully removed with a razor blade.
2   Place 6–10 fragments on the plastic grid of the hermetic enclosure.
3   With both taps in the 'on' position, generate acrolein vapours in the enclosure by adding 10 ml of a 90% (w/v) acrolein solution via a 10 ml glass pipette, then maintain both taps in 'off' position.
    Note – acrolein has a high vapour pressure and is extremely toxic (it is a component of tear gas). It is therefore very important always to work carefully in a fume hood, and use gloves.
4   Fix for at least 5 h (and up to 72 h) at room temperature (RT).
    Note – the duration of the fixative step is a function of the 'meristematic state' of the cambium, and of the quantity of xylem tissue present in the sample. We recommend: 5 h for active cambium: 72 h for dormant cambium.
5   After fixation, pump out the acrolein with a 10 ml glass pipette.
    Note – one pumping is usually not sufficient to remove all the acrolein solution; in order to eliminate any trace of acrolein inside the container, we add 10 ml of DMP acidified with a few drops of concentrated (37%, w/w) HCl.
6   Dehydrate in the same container used for fixation, by exposing the fragments to the vapours of acidified DMP for 5 min.
7   Replace acidified DMP, and dehydrate for a further 1 h.
    Note – dehydrated samples became brittle and hard.

8 Place a few ml of PO into each of two glass Petri dishes. Replace the lids rapidly to minimise loss of PO by evaporation. Also place 1 ml of PO in each of a series of Eppendorf tubes (and seal the tubes).

9 After dehydration, open the silicone top of the container and carefully, but rapidly (only few seconds), remove the fragments from the container by means of forceps, and immerse in PO in a Petri dish.

   Note – it is very important that the fragments remain totally covered by the PO. It is therefore important to monitor the level of PO, and top up as necessary.

10 Transfer fragments to the second Petri dish for dissection in a fume hood.

11 Using a razor blade, remove and discard the extreme ends of each fragment and section the remainder so that small pieces, approx. 3 mm^3, containing cambial tissue with adjacent phloem and xylem cells, are isolated.

   Note – wear gloves and perform the dissection in a fume hood.

12 Using forceps, accumulate the small pieces in an Eppendorf filled with PO.

   Note – if the fragments aren't totally immersed in PO their subsequent embedding becomes impossible. Such fragments are discarded and not processed further.

   Note – at this step if material was insufficiently fixed, the sample texture will change rapidly after the transfer to PO (within 5 min) and we can observe any retraction of the tissue.

13 Well-fixed and dehydrated samples are sequentially infiltrated with Spurr resin:PO mixtures:1:3 (v/v) for 1 h, 1:1 (v/v) for 1 h, and 3:1 (v/v) overnight. Finally, the samples are infiltrated twice, for 1 h and then for 2 h, in pure resin in an Eppendorf.

14 Transfer samples to fresh resin in embedding moulds, and polymerise at 60 °C for 24 h.

Fixation by precipitation technique

Principle

The precipitation methods immobilises the diffusible elements by precipitating them *in situ* during the fixation step. The choice of precipitation agent, which is added to the fixative mixture, depends on the ions to be studied. We use potassium pyro-antimonate (KSb(OH)$_6$) because of its large spectrum of ion precipitation (Klein *et al.*, 1972). For those authors, Na, Mg and Ca ions are precipitated at, respectively, critical concentrations of 10^{-2}, 10^{-4} and 10^{-6} M. We use a pyro-antimonate method that has been modified from that described by Mentré and Escaig (1988) in which the fixative reagent is a *p*-formaldehyde solution.

Equipment needed:	Chemicals needed:
Glass vials	Acetic acid 99.8% (Aldrich, No. 10,908-8)
Razor blades	Kaliumhexahydroxo-antimonat K[Sb(OH)$_6$] (potassium pyro-anti monate) (Merck, No. R:20/22) Phenol 99 % (Aldrich, No. 18,545-0)
Eppendorf tubes (capacity: 1ml) Embedding oven at 60 °C	Ultra-pure water (Milli-Q, Millipore) *p*-formaldehyde (CH$_2$O)n (Prolabo, No. 88320)
Flat embedding mould (Agar, No. G 369)	Ethyl alcohol (ethanol) (Aldrich, No. 27,074-1) Propylene oxide 99% (PO); (Aldrich, No. 11, 020-5)
Glass Petri dish	Spurr resin kit (Agar, No. R 1032); medium grade (ERL 4206, 10g; DER 736, 6g; NSA, 26g)

Materials and methods
Duration of procedure: approx. 46 h.

1 Dissect small pieces in a Petri dish containing fixative solution by means of a razor blade, approximately 3 mm^3 in volume, from the cambial area of the collected internodes.

2 Fix for 3 h at 4 °C, in a mixture of 4 ml of 5% (w/v) potassium pyro-antimonate solution, 1 ml of 10% (w/v) *p*-formaldehyde solution and 0.05g of phenol. Adjust the pH of the mixture to 7.6 with 0.02 M acetic acid.

3 Rapidly (few seconds) wash samples in ultra-pure water.

4　Dehydrate in 5, 10, 20, 40 and 60% aqueous ethanol (5 min in each), and eventually for 3 x 30 min in anhydrous ethanol.

5　Finally embed samples in Spurr resin using the protocol described above for samples fixed and dehdydrated in the vapour phase.

Sectioning for SIMS microscopy

Sectioning was performed using a diamond knife mounted on an ultramicrotome (Ultracut S, Leica). For SIMS imaging, 2.5 μm thick sections were mounted in tight contact with the surface of finely polished stainless steel stubs. Two models of assembly were tested: dry collage and water collage.

In dry collage, the section is transferred onto the surface of a carbon adhesive stuck on the stub. The stubs with their adherent sections are stored under vacuum until used for SIMS imaging.

In water collage, the section is transferred onto the surface of a micro-drop of ultra pure water deposited on the stub. The water drop is immediately dried (for a few sec) using a heat plate at 60 °C.

SIMS microscopy

Principle

SIMS imaging was carried out using a Cameca IMS 4f ion microscope (Cameca) operating in the scanning mode. The principle of SIMS imaging has already been described (e.g. Thellier *et al.*, 1993). Briefly, when a beam of ions of a few keV (the primary ions) is finely focused onto the surface of a solid sample, it vapourises the most superficial atomic layers of the sample. Part of that sputtered matter is in the form of monoatomic or polyatomic ions (secondary ions). These secondary ions are collected and dispersed using a high performance mass spectrometer which allows the selection of the secondary ions of the element(s) under consideration. The selected secondary ions issuing from the bombarded area (less than 1 μm^2 in the scanning mode) of the sample surface are counted using an appropriate device (electron multiplier or Faraday cup) and the value obtained (i.e. the signal) is stored in the memory of the computer. The beam is rastered over the surface of the sample, thus producing a 256×256 pixels scanning digital image.

Finally, the images are edited using a grey-level or a false-colour scale. The detection limit of the SIMS method is of the order of a few ppm for any isotope of most elements. This method is also characterised by an excellent mass resolution, which makes it possible to remove most of the problems of mass interference.

In our present study, we have set the mass resolution at the value of 2000 (which means that two secondary ions with masses m' and m, differing by only m/2000, were discriminated). Under such conditions, although there was an intense emission of organic polyatomic ions with masses close to 23 and 40, it was possible to map specifically the distributions of the isotopes ^{23}Na, ^{24}Mg, ^{39}K and ^{40}Ca in the plant samples. The primary ions were O^{2+} ions, 15 keV in energy. The lateral resolution of the ion images was approx. 0.5 μm.

Measurement of normalised ion signals

Because of matrix effects (Thellier *et al.*, 1993), the SIMS method is not well suited to absolute quantitation of ions/elements. The reason is that a given elemental concentration may produce signals differing by orders of magnitude, depending on the chemical composition of the matrix. The usual approach to overcome this difficulty consists of considering ratios of signals instead of the signals themselves. In our present study, we determine the ion signals (^{23}Na, ^{24}Mg, ^{39}K and ^{40}Ca) by considering the ratio (termed the 'normalised signal') of the signal of each element under study to that of a well-chosen normalising element (^{12}C in our present case). The reason why it is preferable to take only signal ratios into account is that the matrix effects have an almost identical influence on the signals of both elements involved in the ratio, therefore, they have practically no effect on the value of the ratio. Moreover, the normalised signals do not depend on the analytical conditions and they are not very sensitive to the topographic irregularities of the sample surface, thus allowing quantitative comparison between different images (Thellier *et al.*, 1993).

For the analysis of the SIMS images, we have developed a new 32-bit image processing program which can be used under Windows 95 or Windows 4.0 environments. It is a raster format image editing program (Bitmap editor). The program can perform standard functions, e.g. filtering or segmentation, and new dedicated functions for SIMS image

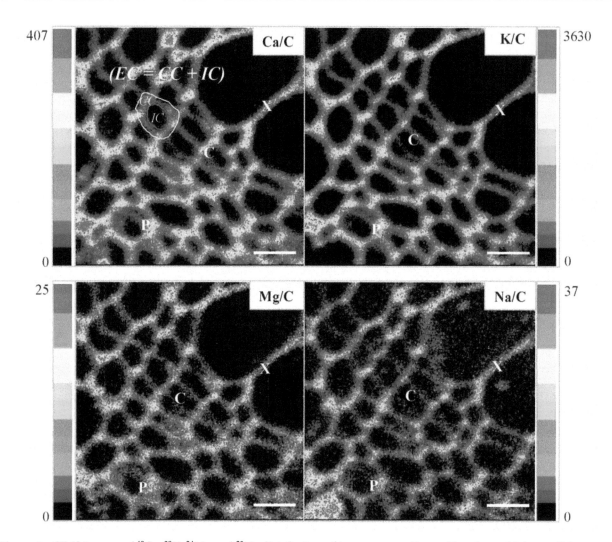

Figure 2   SIMS images of ^{40}Ca, ^{39}K, ^{24}Mg and ^{23}Na distribution of transverse sections of beech cambial zone (May fixation) fixed by the vapour phase technique. The false colour scale indicates the ion normalised signal (ion signal intensity for 10 min acquisition time / ^{12}C signal intensity for 10 min acquisition time).
Contours delimited respectively by the yellow line correspond to the entire cell (EC), the red line to the internal cell compartment (IC). The pixels situated between the two closed lines correspond to the cell contour (CC). C, cambial cells; P, phloem derivative cell; X, xylem derivative cell. Bars = 10 μm.

processing. We have used this program to quantify, on a cell-by-cell basis, the normalised Na, Mg, K and Ca signals. This was done by calculating the ratio of the sums of the Na (or Mg, K and Ca) and of the C pixel values over the image area corresponding to each particular plant cell (Figure 2).

In practice, using the computer mouse, the contours of each plant cell are drawn on the SIMS image of ^{40}Ca in the form of two closed lines (Figure 2). The program accumulates the ^{40}Ca count values corresponding, first, to the pixels within the contour delimited by the yellow line (the entire cell), secondly, to the pixels within the contour delimited by the red line (the internal cell compartment: IC), and thirdly, to the pixels situated between the two closed lines (the cell contour: CC). Then, on the corresponding ^{23}Na, ^{24}Mg, ^{39}K, and ^{12}C (which may be acquired simultaneously with the ^{40}Ca image), it automatically draws contours identical to those on the ^{40}Ca image, and adds up the ^{23}Na, ^{24}Mg, ^{39}K and the ^{12}C count values for the pixels within each contour.

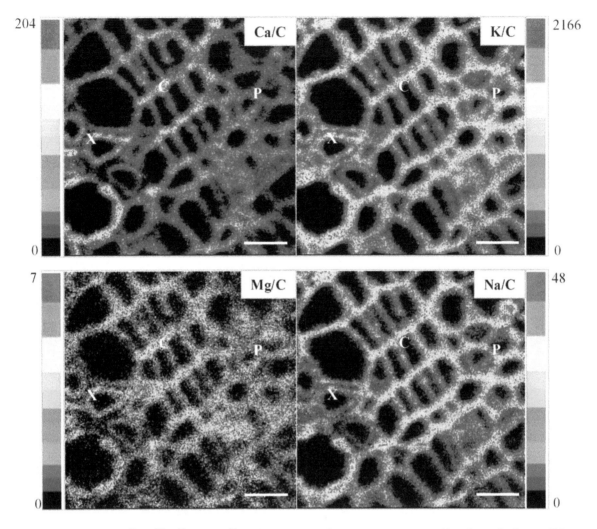

Figure 3 SIMS images of ^{40}Ca, ^{39}K, ^{24}Mg and ^{23}Na distribution of transverse sections of beech cambial zone (May fixation) fixed by the precipitation method. The false colour scale indicates the ion normalised signal (ion signal intensity for 10 min acquisition time / ^{12}C signal intensity for 10 min acquisition time). *C*, cambial cells; *P*, phloem derivative cell; *X*, xylem derivative cell. Bars = 10 μm.

EVALUATION OF SAMPLE PREPARATION PROCEDURES

Ultrastructural preservation

Fixation by acrolein vapours and dehydration by acidified DMP vapours

The degree of ultrastructural preservation obtained in the cambial area of beech branches depends on the stage of vascular differentiation. The more secondarily thickened the cell wall (phloem fibres and mature xylem cells), the more difficult it becomes to preserve cell ultrastructure. In fact, the limiting point of our new technique seems to reside in the final sample transition from the vapour phase to the liquid medium. At this step, possibly the physical tensions applied during the dissection or the first step of sample inclusion. This consideration is much more important than the heterogeneous pattern of cells in the samples. So, resting cambial area, which contains a thin layer of cambial cells enclosed between mature xylem and phloem cells, becomes more brittle than active cambial cells. We can optimise the ultrastructural preservation of cambial cells by increasing the duration of the fixation step. However, the best results were obtained within the active cambial area of one-year-old beech branches.

Comparison with the precipitation method

For determination of ultrastructural preservation, transmission electron microscopy (TEM) was carried out in order to compare the precipitation method with the new fixation technique in vapour phase. Ultrathin sections (80 nm thick), corresponding to each preparation procedure were collected on uncoated copper grids and stained with a mixture of 7% uranyl acetate and 7% lead citrate solution. Imaging was performed using a Zeiss EM109 operating at 80 keV.

Fixation of the active cambial area of young beech branches using acrolein vapours is better than that with *p*-formaldehyde in the presence of

pyro-antimonate. Figure 4 shows the ultrastructural patterns of young active cambial area as a function of the fixation procedure. In the case of the precipitation method, the ultrastructural patterns of cambial cells are not well preserved. For example, in the active cambial cell, the cytoplasmic material is very disturbed. Moreover, thylakoid membranes of chloroplasts are very deformed. Only the cell wall apears not to be seriously affected by this procedure of sample preparation.

By contrast, with the new method, fixation and dehydration in the vapour phase, the ultrastructure of the cambial area is not very disturbed. Active cambial cells are much better preserved

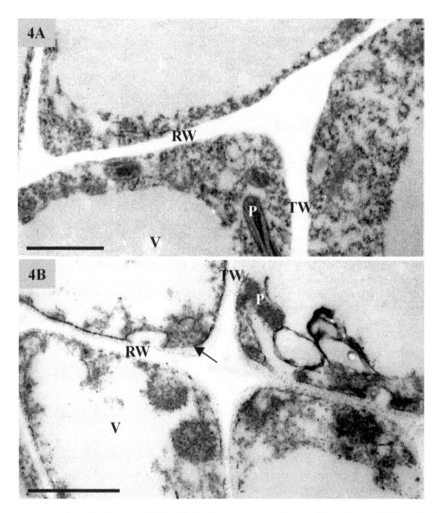

Figure 4   Electron micrographs of active cambial cells in transverse sections of beech cambial region (May fixation) fixed by the vapour phase technique (A) or with the precipitation method (B). P, plastid; RW, radial wall; TW, tangential wall; V, vacuole.
Note plasmalemmal detachment (black arrow in Figure 4B) in the case of the precipitation technique.
Bars = 1 μm.

compared to that observed with the precipitation method. Several internal membrane systems were preserved: chloroplasts, dictyosomes and mitochondria. With the new technique, the ultrastructure is closer to that achieved by classical methods of TEM preparation (not shown, but see e.g. the TEM chapter by Chaffey in this volume).

It is widely recognised that the cambial area of the tree, which contains different cell types (highly vacuolated cambial cells; differentiating derivatives; xylem cells with secondary wall deposits), is particularly difficult to prepare for microscopical analysis (Kidwai and Robards, 1969). Those authors showed that it is necessary to use two fixative solutions (glutaraldehyde/osmic acid) to obtain good cambial preservation. Thus, use of a single fixative in the precipitation method here is less satisfactory than the 'classical method'. Although the cells do not appear to be plasmolysed, the endomembrane system and organelles such as plastids were rarely preserved. It is noteworthy, however, that Kidwai and Robards (1969) observed only mediocre cambial cell preservation in willow when fixed for 4 h with an aqueous acrolein solution. Here, by exposure of young beech branches to acrolein *vapour* for 5 h, we obtained acceptable preservation of the active cambial zone (Figure 4).

Ion retention

Plant samples contain two types of ions. One is directly associated with the cellular structure (cell wall), the other is free (in solution in the cytosol or in the vacuole). Because Ca and Mg, and particularly Na and K are diffusible ions, they can be disturbed or loosened by use of 'classical fixation techniques' (Virk and Cleland, 1988). We have already said that there are two fixation methods which allow ion retention: the precipitation techniques and 'cryo' methods. This last technique uses ultrarapid freezing (at −193 °C) of biological samples in order to solidify tissue water, stop enzyme reactivity, and immobilise cellular constituents (Chandler *et al.*, 1984; Moor, 1987; Kaeser *et al.*, 1989; Ryan, 1992). The dehydration step can consist of a lyophilisation process or cryo-substitution with organic solvents. Even though this process decreases artefacts concerned with retention of diffusible constituents caused by classical chemical fixation techniques, several difficulties, such as the formation of ice crystals and the problems in obtaining good embedment of cryo-fixed samples, must still be surmounted. Prevention of the formation of intracellular vitreous ice crystals, which can affect ultrastructural preservation, necessitates the use of cryo-protectants (Ding *et al.*, 1991), which may also effect the retention of diffusible ions in the cells.

Using soybean leaf, Grignon *et al.* (1997) showed that both ultrastructural and ion retention patterns were satisfactorily preserved by use of ion-inert solvents, such as DMP or acetone, and resin-infiltration at low temperature. Unfortunately, when we applied this technique (propane jet-freezing, cryo-substitution with acetone, and embedding in Spurr resin at low temperature) to beech it did not preserve the cambial area satisfactorily (results not shown). However, Ristic and Ashworth (1993) have described a specific protocol of ligneous cryo-fixation which achieved good sample embedding and ultrastructural preservation. They rehydrated the wood sample after the cryo-substitution step for 48 h in a 'buffer bath' in order to increase membrane permeability. Clearly, this technique is not suitable for SIMS analysis of diffusible ions. For this reason we have only compared the ion retention characteristics between a precipitation method and our new method.

We have used atomic absorption spectrometry (AAS) to determine ion contents in one-year-old beech branches.

Equipment needed:	Chemicals needed:
Glass balloon (capacity 50 ml) (Prolabo, No. 03 370.348)	Lanthanum nitrate 99.9% (Aldrich, No. 23,855-4)
Propylene tubes (capacity 50ml) (Prolabo, No. 01142.633)	Sodium atomic absorption standard solution 1g/ml (Aldrich, No. 20,750-0)
	Magnesium atomic absorption standard solution 1g/ml (Aldrich, No. 20,727-6)
Spectrophotometer	Calcium atomic absorption standard solution 1g/ml (Aldrich, No. 30,590-1)

Potassium atomic
absorption standard
solution 1g/ml
(Aldrich, No. 31,694-6)
Nitric acid, fuming
HNO_3 37% (w/v)
(Aldrich, No. 30,907-9)

Duration of procedure: approx. 76 h plus spec-
trophotemeter time

1   Dry samples for 72 h at 60 °C. The mean
     dry weight for each experiment was about
     130 mg (any more prevents good wood
     mineralisation).
2   Place dried samples on a glass balloon con-
     taining 10 ml of ultrapure, concentrated (37%
     w/v) HNO_3.
     Note – the glass balloon was previously
     washed with concentrated HNO_3 for removal
     of any adhering minerals.
3   Perform sample mineralisation by backward-
     boiling in HNO_3 for 4 h at 20 °C, and recover
     the mineralisation solution in propylene
     tubes.
4   Dilute the mineralisation solution 1:10 in
     HNO_3 (1 mol/l) solution (w/v), then at 1:20,
     1:50 or 1:100 (respectively, for Na, Mg, Ca
     and K estimation) in the HNO_3 1 M solution
     containing 1% of lanthanum nitrate (w/v).
     Note – standard solutions, at 1g/ml, of Ca,
     Mg, Na, and K were also diluted using the
     lanthanum-HNO_3 solution.
5   Measure the absorbancy of the mineralisation
     solutions and the standard solutions in a
     spectrophotometer.

In this way, ion contents were established for 10
first internodes of young beech branches harvested
in May. Each internode was divided into two equal
fragments. Ion contents (weight ppm) of one fragment which had been fixed, dehydrated (sample
was treated by either the new method or by the precipitation technique), and finally bathed for 5 min in
PO, were compared with those of the control half,
which was neither fixed nor dehydrated. The
weight % variation in ion content between the
control and assay halves were calculated [(ion
control content-ion assay content) × 100] and the
mean % presented in Table 1.

Table 1 Weight % of variation in ion content between
control and experimental samples fixed, dehydrated,
and bathed for 5 min in propylene oxide.

	% Ca	% K	% Mg	% Na
[a] precipitation	−1 ± 14	+58 ± 54	−11 ± 12	+7.0 ± 10
[b] vapour phase	−3.4 ± 15	−8 ± 7	−4.5 ± 12	+0.2 ± 20

Each value represents the mean of 10 measurements ± standard
error.
(a) Fixation by p-formaldehyde with presence of potassium
    pyro-antimonate; dehydration in ethanol series
(b) Fixation by acrolein vapours; dehydration by acidified DMP
    vapours.

Except for K, and sometimes Na, content in the
case of the precipitation procedure, we usually
noted a small diminution of ion content between
the control and the assay. The precipitation technique, which uses potassium pyro-antimonate,
contaminated the K ion content in the respective
assay in the order of +58%.

With our new method of sample preparation,
ion content in beech fragments fixed by acrolein
vapour and dehydrated by acidified DMP vapours
were not significantly different from those in the
control fragments (Table 1). Even with the high
level of ion content variability between each
fragment, measurement by AAS shows that the
differences in % of ion content between assay and
control were smaller in our new method than in
the precipitation technique.

With the precipitation method, it is probable that
some diffusible ions are lost into the aqueous solution. In its original form, the pyro-antimonate
method for sample preparation (Mentré and Escaig,
1988) consisted of precipitating cell calcium as tiny
crystals of calcium pyro-antimonate and observing
the subcellular distribution of the crystals by TEM.
We have used TEM to estimate the significance of
any internal ion movement in the cambial area by
observing the crystal size (results not shows). In
resting or active cambial region of beech, crystals
remain small, corresponding to little movement of
precipitated ions. Using fixation and dehydration in
the vapour phase, estimation of intracellular ion
movement in the cambial zone is more difficult than
for the precipitation method. Although we might
think that all metabolic reactions or active intercellular transport cease as soon as the sample is exposed
to acrolein vapour, the high vapour density of the
acrolein may cause an immediate saturation of
sample environment that helps to prevent redistrib-

ution of ions. However, we can exclude the persistence of a passive intercellular transport, which must be low because of the low level of aqueous vapours and solutions.

It is possible that diffusible ions, either free in the cytoplasm or in the vacuole, moved, respectively, to the plasmalemma or the tonoplast. This is almost certainly important in the case of highly vacuolate, active cambial cells. It is probably during the dehydration step that major ion movement occurs, either to the tonoplast or inside the vacuole to form crystals. In the new procedure, the samples were fixed and dehydrated *in toto* without moving. Moreover, it is likely that use of acidified DMP vapour as dehydration reagent, which eliminates tissue water by converting it to acetone end methanol (Müller and Jacks, 1975), effectively limits movement of diffusible ions.

SIMS imaging

Figures 2 and 3 show the secondary normalised signal of ^{40}Ca, ^{39}K, ^{24}Mg and ^{23}Na in the active cambial area of young beech branches. Regardless of the method of sample preparation and the ion considered, the secondary emission ion signal represents a negative gradient between the phloem and the xylem tissues, being maximal in the walls of differentiating phloem cells. In this type of plant sample, where the active cambial cells are highly vacuolated, the principal source of secondary ion signal was concentrated near the cell wall. Because there is only a thin parietal layer of cytoplasm in these cells, it is difficult to distinguish between the cell wall and cytoplasm. For that reason, we designate the whole (wall + cytoplasmic layer) by the term 'cell contour' (*CC*), and the vacuole as the 'internal cell' (*IC*). Thus, the entire cell (*EC*) value is the sum of these two different areas.

In order to estimate any ion movements which may occur during the collage process, the normalised ion signals from semi-thin sections were studied as a function of collage processing (Figures 5, 6). We compared water collage with dry collage. In Figure 5, we have represented the normalised ion signal in the total area of active beech cambial cell measured after the precipitation method or fixation in vapour phase. Regardless of the technique of fixation used, the divalent ions Ca^{2+} and Mg^{2+} appear less sensitive to the collage process than the monovalent ions, Na^+ and K^+. Moreover, the analy-

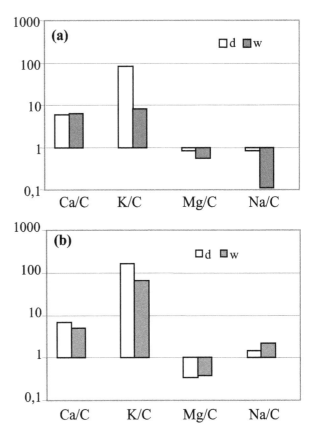

Figure 5 Normalised ion (^{40}Ca, ^{39}K, ^{24}Mg and ^{23}Na) signal in active beech cambial cells (*EC*: entire cell) prepared by fixation and dehydration in vapour phase (a) or by the precipitation method (b), as a function of the collage semithin section procedure (d: dry collage; w: water collage). Each value is the mean of 15 measurements carried out using 5 different images corresponding to the different cambial area of three samples. Logarithmic scale.

sis of means by Student's *t* test (Table 2) shows that the samples fixed and dehydrated in the vapour phase are particularly sensitive to the collage protocols. For this method, the water collage cause a significative decrease of the K and Na normalised signal compared with the dry collage process. In Figure 6 we represent the ratio of ion normalised ion signal in the cell contour to the internal part of the cell (*CC/EC*) as a function of collage process. Generally, we noted that this ratio is increased by water collage. Student's *t* test (Table 3) reveals that this increase affected the normalised signal of monovalent ion in the case of fixation in vapour phase more than the signal of divalent ions.

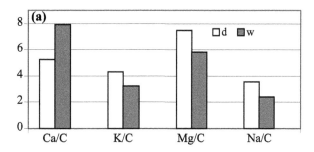

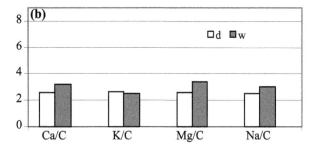

Figure 6 Normalised ion (^{40}Ca, ^{39}K, ^{24}Mg and ^{23}Na) signal ratios in the contour (CC : cell contour) to the internal part (IC : internal cell compartment) of the cambial cell. The cambial area of beech branch was prepared by fixation and dehydration in vapour phase (a) or by the precipitation method (b), as a function of the collage semithin section procedure (d: dry collage; w: water collage). Each value is the mean of 15 measurements carried out using 5 different images corresponding to the different cambial area of three samples. Logarithmic scale.

Our study concerning the effects of collage process in SIMS microscopy demonstrates that dry collage is essential to preserve the whole ion present in the sample, after fixation. In SIMS imaging, it is important to analyse a solid sample with a regular surface, in order to prevent artefactual disturbance of secondary ion emission which is enhanced by an irregular surface. Having observed the surfaces of semi-thin sections after dry collage or water collage by beam electron microscopy (results not shown), we noted that the better, more regular surface was obtained with the water collage process. However, study of the normalised ion signal as a function of the collage process shows that this collage technique was more damaging to the preservation of ion distribution (Tables 2 and 3). Generally, that study also revealed that in the two techniques of sample preparation, some ions were solidly fixed, whereas for others, although they were conserved by the fixation process, their association to cellular structures was much weaker.

Table 2 Mean analysis of the data represented in Figure 5
Student's t values[a] for the different normalised ion signals

Collage procedure compared	^{40}Ca	^{39}K	^{24}Mg	^{23}Na
Vapour phase				
d/w	0.5 (2.3/8)	**16 (2.3/8)**	1.1 (2.3/8)	**12 (2.3/8)**
Precipitation				
d/w	3.0 (3.1/14)	**10.0 (2.1/14)**	1.7 (2.1/14)	**4.3 (2.1/14)**

[a] Calculated t value (theoretical t value for 0.05 confidence level/degree of freedom)
Values in bold indicate a significant difference between the two means that are compared. d: dry collage; w: water collage.

In the case of pyro-antimonate precipitation, SIMS analysis allows localisation not only of calcium pyro-antimonate crystals, but also of small amounts of Ca, Na or Mg, which are not precipitated by the pyro-antimonate but are easily removed by collage by aqueous or organic solutions. In our new technique of fixation and dehydration in vapour phase in which the bound ions were probably more brittle, only the dry collage process is possible.

Table 3 Mean analysis of the data represented in Figure 6
Student's t values[a] for the different normalised ion signals

Collage procedure Compared	^{40}Ca	^{39}K	^{24}Mg	^{23}Na
Vapour phase				
d/w	1.8 (2.3/8)	**3.0 (2.3/8)**	0.9 (2.3/8)	**3.6 (2.3/8)**
Precipitation				
d/w	0.9 (2.1/14)	0.3 (2.1/14)	1.1 (2.1/14)	0.8 (2.1/14)

[a] Calculated t value (theoretical t value for 0.05 confidence level/degree of freedom).
Values in bold indicate a significant difference between the two means that are compared. d: dry collage; w: water collage.

Applications

Using this new method of tissue preparation and SIMS microscopy, we have investigated the involvement of mineral nutrients in cambial activity and vascular differentiation on beech seedlings (Follet-Gueye *et al.*, 1999). In this work, an unexpected ionic interaction between calcium and sodium has been characterised within the walls of beech seedlings: whereas a decrease of the Na signal in the cell walls was always paralleled by an increase of the calcium signal at the same site, unexpectedly an increase of the normalised Na signal in the cell walls of the Na-enriched seedlings was paralleled by an increase of the Ca signal. We have also shown that this interaction was involved in the differentiation of the secondary walls. In fact, in the hypocotyl of sodium-enriched seedlings the cambium is functional but the number of xylem and phloem cells was slightly smaller than in the control or Na-depleted seedlings. Moreover in the Na-enriched seedlings, secondary wall thickening of the cells of the secondary xylem was not visible.

FUTURE PROSPECTS

The SIMS analysis of mature xylem cells has been practised for several years. In fact, Martin *et al.* (1994) showed that this technique may have a wide application in dendrochronology. The recent study of Brabander *et al.* (1999) on old red oak xylem shows the ability of SIMS analysis to explore metal pollution. Our future work will extend the study of the cambial area to another tree that can be well preserved by acrolein fixation in the vapour phase.

CONCLUSION

Application of SIMS imaging to plant samples requires a fixation preparation that permits, simultaneously, good quality ultrastructure preservation and satisfactory ion retention at the cellular level. The new preparation method we have devised has proved to be particularly suitable for SIMS analysis. An added bonus of our method is its good ultrastructural preservation, even of woody material. This new method provides an alternative technique that will allow a comparative study for plant sample analysis by SIMS microscopy.

ACKNOWLEDGMENTS

I gratefully acknowledge F. Lhuissier for his assistance in this new sample preparation method and D. Gibouin and F. Lefebvre for the realisation of SIMS images. This work was supported by a doctoral fellowship of the 'ministère de l'enseignement supérieur et de la recherche' and was carried out in the Processus Ioniques Cellulaires laboratory of Rouen University. I thank M. Demarty for his contribution to images analysis. I am grateful to Prof. C. Ripoll, my PhD supervisor, for his support.

REFERENCES

Albersheim, P. (1978) Concerning the structure and biosynthesis of the primary walls of plants. *Int. Rev. Biochemistry.* II, **16**, 127–150.

Avery, G.S., Burkholder, P.R. and Creighton, H.B. (1937) Production and distribution of growth hormone in shoots of *Aesculus* and *Malus*, and its probable role in stimulating cambial activity. *Am. J. Bot*, **24**, 51–58.

Berry, J.P., Escaig, F. and Galle, P. (1987) Etude de la localisation intracellulaire du béryllium par microscopie ionique analytique. *C.R. Acad. Sci.* Paris, **304**, III, 239–243.

Brabander, D.J., Keon N., Stanley, R.H.R. and Hemond, H.F. (1999) Intra-ring variability of Cr, As, Cd, and Pb in red oak revealed by secondary ion mass spectrometry: Implications for environmental biomonitoring. *Proc. Natl Acad Sci.*, USA, **96**, 14635–14640.

Campbell, N.A., Stika, K.M. and Morisson, G.H. (1979) Calcium and potassium in the motor organ of the sensitive plant: localization by ion microscopy. *Science*, **204**, 185–186.

Castaing, R. and Slodzian, G. (1962) Microanalyse par émission ionique secondaire. *J. de microscopie*, Paris, **1**, 395–414.

Chandler, D.E. (1984) Comparaison of quick-frozen and chemically fixed sea-urchin eggs: structural evidence that cortical granule exocytosis is preceded by a local increase in membrane mobility. *J. Cell Sci*, **72**, 23–36.

Chassard-Bouchaud, C. and Galle, P. (1986) Bioaccumulation de l'aluminium par les organismes marins. Mise en évidence par microscopie corpusculaire analytique. *C.R. Acad. Sci. Paris*, **302**, III, 55–61.

Ding, B., Turgeon, R. and Parthasarathy, M.V. (1991) Routine cryo-fixation of plant tissue by propane jet freezing for freeze substitution. *J. Electron Microsc. Tech*, **19**, 107–117.

Dünisch, O. and Bauch, J. (1994) Influence of mineral elements on wood formation of old growth Spruce (*Picea abies* [L.] Karst.). *Holzforschung*, **48**, 5–14.

Edelman, L. (1991) Freeze substitution and the preservation of diffusible ions. *J. of Microscopy*, **161**, 217–228.

Follet-Gueye, M.L., Demarty, M., Verdus, M.C. and Ripoll, C. (1999) SIMS imaging of Ca and Na, and effect of the ionic interaction Ca/Na on wall maturation of phloem fibres and xylem cells of hypocotyls of beech seedlings. *C.R. Acad. Sci. Paris*, **322**, 383–392.

Follet-Gueye, M.L., Verdus, M.C., Demarty, M., Thellier, M. and Ripoll, C. (1998) Cambium preactivation in beech correlates with a strong temporary increase of calcium in cambium and phloem but not in xylem cells. *Cell Calcium*, **24** (3), 205–211.

Galle, P. (1970) Sur une nouvelle méthode d'analyse cellulaire utilisant le phénomène d'émission ionique secondaire. *Ann. Phys. Biol. Med.*, **42**, 83–94.

Grignon, N., Halpern, S., Jeusset, J., Briancon, C. and Fragu, P. (1997) Localization of chemical elements and isotopes in the leaf of soybean (*Glycine max*) by secondary ion mass spectrometry microscopy: critical choice of sample preparation procedure. *J. of Microscopy*, **186**, 51–56.

Guglielmino, N., Liberman, M., Jauneau, A., Vian, B., Catesson, A.M. and Goldberg R. (1997) Pectin immunolocalization and calcium visualization in differentiating derivatives from poplar cambium. *Protoplasma*, **199**, 151–160.

Hepler, P.K., Wick, S.M. and Wolniak, S.M. (1981) The structure and role of membranes in the mitotic apparatus. In *International Cell Biology*, 1980–1981, edited by. H.G. Scheiger, pp. 673–686. Berlin: Springer-Verlag.

Jauneau, A., Ripoll, C., Rihouey, C., Demarty, M., Thoiron, A., Martini, F. and Thellier, M. (1992) Localisation de Ca et Mg par microscopie ionique analytique dans des plantules de lin: utilisation d'une méthode de précipitation au pyroantimonate de potassium. *C.R. Acad. Sci. Paris*, **315** (III), 179–188.

Kaeser, W., Koyo, H.W. and Moor, H. (1989) Cryofixation of plant tissues without pretreatment. *J. of Microscopy*, **154**, 279–288.

Kamiya, N. (1981) Physical and chemical basis of cytoplasmic streaming. *Annu. Rev. Plant Physiol*, **32**, 205–36.

Kidwai, P. and Robards, A.W. (1969) The appearance of differentiating vascular cells after fixation in different solutions. *J. Exp. Bot*, **20**, 664–670.

Klein, R.L., Yen, S. and Athureson Klein (1972) Critique of the K-pyroantimonate method for semi quantitative estimation of cations in conjunction with electron microscopy. *J. Histochem. Cytochem.*, **20**, 65–78.

Küppers, M., Zechw, Schulze, E.D. and Beck, E. (1985) CO_2 Assimilation transpiration und wachstum von *Pinus sylvestris* L. bei unterschiedlicher magnesiumversorgung. *Forstw. Cbl*, **104**, 23–26.

Larras-Regard, E. and Mony, M.C. (1995) Scanning ion images; analysis of pharmaceutical drugs at organelle levels. *Int. J. Mass Spectrometry Ion Processes*, **143**, 147–160.

Larson, P.R. (1967) Assessing wood quality of fertilized coniferous trees. *Forest Fertilization Symposium*, Grainville, 275–280.

Luft, J.H. (1959) The use of acrolein as a fixative for light and elctron microscopy. *Anat. Record*, **133**, 305.

Martin, R.R., Sylvester T. and Biesinger, M.C. (1994) Secondary ion mass spectrometry (SIMS) in the analysis of elemental micropatterns in tre rings. *Can. J. For. Res*, **24**, 2312–2313.

Mentré, P. and Escaig, F. (1988) Localization by pyroantimoniate. I. Influence of the fixation on the distribution for preparation of calcium and sodium. An approach by analytical ion microscopy. *J. Histochem. Cytochem*, **31**, 48–54.

Moor, H. (1987) Theory and practice of high pressure freezing. *In: Cryotechniques of Biological Electron Microscopy*, edited by R.A. Steinbrecht and K. Zierold, pp. 175–191. Springer Verlag, Berlin, Heidelberg.

Muller, L.L. and Jacks, T.J. (1975) Rapid chemical dehydration of samples for electron microscopic examinations. *J. Histochem. Cytochem*, **23**, 107–110.

Pitman, M.G., Mowat, J. and Nair, H. (1971) Interactions of processes for accumulation of salt and sugar in barley plant. *Aust. J. Biol. Sci*, **24**, 619–631.

Ripoll, C., Jauneau, A., Lefebvre, F., Demarty, M. and Thellier, M. (1992) SIMS determination of the distribution of the main mineral cations in the depth of the cuticle and the pectocellulosic wall of epidermal cells of flax stems: problems encountered with SIMS depth profiling. *Biol. Cell*, **74**, 135–148.

Ripoll, C., Pariot, C., Jauneau, A., Verdus, M.C., Catesson, A.M., Morvan, C., Demarty, M. and Thellier, M. (1993) Involvement of sodium in a process of cell wall differentiation in plants. *C.R. Acad Sci*. Paris, **316**, 1433–1437.

Ristic, Z., Ashworth, E.N. (1993) New infiltration method permits use of freeze substitution for preparation of wood tissues for transmission electron microscopy. *J. of Microscopy*, **171**, 137–142.

Ryan, K.P. (1992) Cryofixation of tissues for electron microscopy, a review of plunge cooling methods. *Scanning Microscopy*, **6**, 715–743.

Sachs, T. (1981) The control of the patterned differentiation of vascular tissues. *Adv. Bot. Res*, **9**, 151–262.

Spurr, A.R. and Galle, P. (1980) Localisation of elements in botanical materials by secondary ion mass spectrometry. In *Secondary Ion Mass Spectrometry, SIMS 2*, edited by Springer-Verlag (Springer Ser. Chem. Phys., 9), pp. 252–255. New York.

Thellier, M., Ripoll, C., Quintana, C., Sommer, F., Chevalier, P. and Dainty, J. (1993) Physical methods to locate metal atoms in biological systems. *Methods Enzymol*, **227**, 535–586.

Ulrich, B. (1981) Destabilisierung von waldökosystemen durch akkumulation von luftverunreinigungen. *Der Forst. und Holtzwirt*, **36**, 526–532.

Virk, S.S. and Cleland, R.E. (1988) Calcium and the mechanical proporties of soybean hypocotyl cell-wall loosening. *Planta*, **176**, 60–37.

Wardrop, A.B. (1981) Lignification and xylogenesis. In *Xylem cell development*, edited by J.R. Barnett. Castle Publ. Ltd., pp. 115–152. London.

Weisenseel, M.H. and Kicherer, R.M. (1981) Ionic currents as control mechanisms in cytomorphogenesis. In *Cytomorphogenesis in Plants*, edited by. O. Kiermayer, pp. 379–99. Wein: Springer -Verlag.

Zöttl, H.W., Hüttl, R., Fink S., Tomlinson, G.H. and Wismiewski, J. (1989) Nutritional disturbances and histological changes in declining forests. *Water, Air and Soil Pollution*, **48**, 87–109.

Immunolocalisation of the Cytoskeleton in the Secondary Vascular System of Angiosperm Trees and its Visualisation Using Epifluorescence Microscopy

Wood Formation in Trees ed. N. Chaffey
© 2002 Taylor & Francis
Taylor & Francis is an imprint of the
Taylor & Francis Group
Printed in Singapore.

Nigel Chaffey

IACR – Long Ashton Research Station, Department of Agricultural Sciences, University of Bristol, Long Ashton, Bristol BS41 9AF, UK

ABSTRACT

Aspects of the biology of the cytoskeleton of plant cells are briefly summarised, and techniques for its study introduced. Detailed protocols are provided for chemical- and freeze-fixation, processing, embedding, immunolocalisation, and epifluorescent visualisation of microtubules and microfilaments in the secondary vascular system of angiosperm trees embedded in either low melting point Steedman's wax or butyl-methylmethacrylate. Results of epifluorescence visualisation of the microtubule and microfilament cytoskeleton during wood formation in horse-chestnut (*Aesculus hippocastanum* L.) are summarised, and future directions of cytoskeletal studies on angiosperm trees considered.

Key words: actin, cambium, cytoskeleton, indirect immunofluorescence, microfilaments, microtubules, tubulin, wood cells

INTRODUCTION

What is the cytoskeleton?

The cytoskeleton is a 'dynamic three-dimensional array of protein filaments found in all eukaryotic cells' (Seagull, 1994). In plants it embraces microtubules (MTs) (Cyr, 1994), microfilaments (MFs) (McCurdy and Williamson, 1991), intermediate filaments (IFs) (Shaw *et al.*, 1991; Menzel, 1993), microtubule-organising centres (MTOCs) (Lambert, 1993;

Marc, 1997), and an ever expanding collection of MT- and MF-associated proteins, such as kinesins (e.g. Liu *et al.*, 1996) and myosins (Reichelt *et al.*, 1999), respectively. Although considerably more is known about the cytoskeleton of animal cells, improvements in techniques, particularly the use of immunofluorescence procedures (Lloyd, 1987), have heralded a great increase in our understanding of the plant cytoskeleton since MTs were first reported in higher plants by Ledbetter and Porter (1963).

However, the majority of work on higher plants to date has concentrated on understanding the role of these cytoplasmic components in single-cell systems, such as pollen tubes (e.g. Pierson *et al.*, 1986; Li *et al.*, 1997) or root hairs (e.g. Bibikova *et al.*, 1999; Miller *et al.*, 1999), or cells and tissues of the primary plant body (e.g. Baluška *et al.*, 1992; Shibaoka *et al.*, 1996; Sonobe, 1997; Barlow and Baluška, 2000). Apart from a few pioneering transmission electron microscope (TEM) studies of the MT cytoskeleton in differentiating wood cells (e.g. Robards and Humpherson, 1967), cells and tissues of the secondary plant body have largely been ignored. As with all other work on the cell biology of the secondary vascular system (SVS), this neglect has probably been occasioned more by perceived difficulties in adequately fixing, etc. this material (see 'An introduction to the problems of working with trees' in this volume) rather than any intrinsic assumption that they are not interesting.

What is known about the plant cytoskeleton? To date, the major components recognised are the MTs and the MFs. Microtubules are comprised of helical arrays of heterodimers of the proteins, α- and β-tubulin, and are approx. 25 nm in external diameter, with a hollow centre of approx. 7 nm. They exist in four major arrays; three are characteristic of the different phases of the cell cycle, the fourth characterises non-dividing, interphase cells:

1   At pre-prophase, a dense cortical band of MTs (the pre-prophase band, or PPB) (e.g. Wick, 1991) develops near the wall and marks the future plane of cell division. Although this array subsequently disappears, it also seems to mark the sites of attachment of the new division wall to the parental wall.

2   The spindle that separates the daughter chromosomes, during both mitosis and meiosis (e.g. Franklin and Cande, 1999), is composed principally of MTs. Although the mechanism by which chromosome-separation occurs is still a matter of debate, success of the process is crucial to cell division and organised growth.

3   Microtubules are also associated with formation of the new cell wall during the succeeding phase of cytokinesis (e.g. Wick, 1991). Here they are seen at the sites of wall formation, the phragmoplasts, but are apparently absent from the intervening cell plate. The phragmoplast MTs appear to be involved in cell wall formation, possibly by providing guides for the accumulation of vesicles at these sites.

4   Finally, interphase cells – and non-dividing cells generally – tend to have an ordered arrangement of MTs, either transverse, longitudinal or oblique (e.g. Shibaoka, 1991) to the cell's long axis (although cambial cells are different – see below). Since the early reports of co-alignment of cytoplasmic MTs and cellulose microfibrils (Ledbetter and Porter, 1963), the principal view of the role of MTs in interphase cells concerns their apparent ability to influence the orientation of nascent microfibrils in the cell wall (reviewed by Giddings and Staehelin, 1991). It should be noted that this view is not accepted by all workers for all sites of cell wall synthesis (e.g. Emons et al., 1992; Emons and Mulder, 2000), and it is possible that a third factor may serve to co-orient both the MTs and the microfibrils (e.g. Preston, 1988), and there may be 2-way exchange of information between the MTs and the microfibrils (Fisher and Cyr, 1998). However, the widely acknowledged role of the relatively inflexible cellulose microfibrils in determining the direction of cell growth (e.g. Green, 1969), implicates interphase MTs in cell differentiation. Interphase MTs can be further sub-divided into cortical (present near the plasmalemma), and endoplasmic (extending from the cell nucleus into the cortex) (e.g. Baluška et al., 1992). Cell wall formation is associated with the CMTs; the EMTs are implicated in communication between the nucleus and cell cortex, and positioning of the nucleus (Barlow and Baluška, 2000).

Microtubules are formed from soluble pools of the tubulin heterodimers that polymerise, with accompanying hydrolysis of GTP, when a critical concentration has been met. Generally, this is believed to happen at specific sites in the cell, possibly at the nuclear surface or other sites at the periphery of the cell, and is associated with MTOCs which contain a third type of tubulin, γ-tubulin (e.g. Lambert, 1995). Microtubules may be separate or linked to each other, or other cellular components, by MT-associated proteins (MAPs) (Cyr, 1991; Schellenbaum et al., 1992).

Although individual MFs are only approx. 7 nm in diameter, they are relatively easy to visualise in the TEM because they are frequently aggregated into bundles, of many individual MFs, and can be several hundred μm long. Bundling may be facilitated by MF-associated proteins. Microfilaments are homopolymers of F (filamentous)-actin, which polymerise from the soluble form, G (globular)-actin, with concomitant hydrolysis of ATP. The most widely acknowledged role for MFs in plant cells is in cytoplasmic streaming (e.g. Shimmen and Yokota, 1994). However, a recent elegant study by Boevink et al. (1998) in epidermal cells of tobacco suggests a role for MFs in movement of Golgi stacks around the cell.

As more is becoming known about the distribution of MFs in plant cells, it is becoming clear that they are often co-localised with MTs (see e.g. review by Seagull, 1989). Whether this means that both cytoskeletal components are actively involved in the associated cell biological process, or one component is stabilising the other so that it can perform its role alone, is not yet clear. However, their frequent close-association signals the need to understand more fully the interactions between these cytoskeletal components, and to extend this study to other components, such as IFs (Shaw *et al.*, 1991), MTOCs (Lambert, 1995), and the wide range of MT- and MF-associated proteins, particularly the motor proteins such as myosin (Yokota *et al.*, 1995; Reichelt *et al.*, 1999), kinesin (Liu *et al.*, 1996) and dynein (Moscatelli *et al.*, 1995).

Tubulins and actin are coded for by multi-gene families in plants (e.g. Fosket, 1989; Meagher and McLean, 1990; Hussey *et al.*, 1991; Meagher, 1991; Fosket *et al.*, 1993). This may indicate that different isoforms participate in different MT and MF arrays, which may in turn be expressed in different cells, tissues or organs at specific stages of the life cycle of the plan (e.g. Silflow *et al.*, 1987; Hussey *et al.*, 1988; Janßen *et al.*, 1996; Chu *et al.*, 1998; Eun and Wick, 1998; Uribe *et al.*, 1998). Tubulins, particularly α-tubulin can be modified post-translationally, e.g. acetylated, tyrosinated, phosphorylated (e.g. Hussey *et al.*, 1991; Smertenko *et al.*, 1997; Gilmer *et al.*, 1999), further expanding the complexity of this family of proteins.

The cytoskeleton is an extremely dynamic organelle (e.g. Lloyd, 1994). Apart from the four arrays mentioned above, the orientation of MTs is responsive to a variety of stimuli, such as hormones (Shibaoka, 1991; Baluška *et al.*, 1999), electric fields (Hush and Overall, 1996), physical strain (Fischer and Schopfer, 1998), and gravity (Himmelspach *et al.*, 1999), (see also the table in Fischer and Schopfer, 1997). Too little is currently known about the factors that may affect MF orientation.

Although it is easy to consider the cytoskeleton in isolation from the rest of the cell machinery, it must be borne in mind that is an integral component of the cell and interacts with other cell organelles and compartments, such as the plasmalemma (Lloyd *et al.*, 1996). Recent study supports the idea of a cell wall-plasmalemma-cytoskeleton continuum (e.g. Miller *et al.* 1997), and accords the cytoskeleton a place in the ever more complex signal transduction pathways that are becoming identified in plant cells (see Fig. 3 in Trewavas and Malhó, 1997).

The marked degree of conservatism between animal and plant cytoskeletal proteins (e.g. Burns and Surridge, 1994), dramatically verified by the fact that animal tubulins can be incorporated into plant MTs (e.g. Zhang *et al.*, 1990; Wasteneys *et al.*, 1993), permits the use of antibodies raised against animal proteins to probe the plant cytoskeleton. However, caution is needed in the interpretation of results from use of such antibodies (e.g. de Ruijter and Emons, 1999). As more plant-derived antibodies become available (e.g. Andersland *et al.*, 1994; Koropp and Volkmann, 1994; Goddard *et al.*, 1998; Kandasamy *et al.*, 1999) – if not yet commercially – our knowledge of plant-specific cytoskeletal arrays will increase. However, until that time, use of animal-derived antibodies will continue to have an important place in botanical research allowing us to identify more homologous proteins in the plant cytoskeleton.

In addition to the works cited above, and for further general background, interested readers are directed to the following articles which deal specifically with elements of the plant cytoskeleton: Lloyd (1982, 1986, 1991); Tiwari *et al.* (1984); Clayton (1985); Hepler (1985); Roberts *et al.* (1985); Derksen *et al.* (1990); Emons *et al.* (1991); Staiger and Lloyd (1991); Fosket and Morejohn (1992); Goddard *et al.* (1994); Lambert and Lloyd (1994); Davies *et al.* (1996); Nick *et al.* (1997); Nick (1998, 1999); Kost *et al.* (1999); Volkmann and Baluška (1999); Staiger (2000); Staiger *et al.* (2000).

Relevance of the cytoskeleton to the secondary vascular system of trees

Some of the most fundamental changes that take place when cambial derivatives differentiate during xylogenesis and phloem formation relate to cell size/shape and wall structure, i.e. morphogenesis (see Table 1 in the Introduction to this volume). As mentioned above, of great importance to, and a major constraint upon, plant-cell morphogenesis is the orientation of relatively inflexible cellulose microfibrils within their walls, a process in which MTs have been much implicated. Observations to date certainly lend credence to such an interpretation of the role of MTs in xylogenesis (see the chapter by Funada in this volume, and below), and for an involvement of MFs (see below).

Techniques for study of the cytoskeleton

Transmission electron microscopy

Classically – if we can talk of the 1960s in such terms – the transmission electron microscope (TEM) was the only way to visualise the cytoskeleton (see the TEM chapter by Chaffey in this volume). Even today, particularly in combination with immunolocalisation procedures, it continues to play an important role in providing fine-structural detail of cytoskeletal components, and of the inter-relationships between them and other cytoplasmic elements. It also has an important role in providing images of the cytoskeleton derived from a fixation and processing method different to that used for indirect immunofluorescence (IIF) (e.g. Traas, 1990). The similarity of images from TEM and IIF technique is good evidence for the existence of different cytoskeletal arrays, and of their validity (e.g. Chaffey et al., 1996; see also the TEM chapter by Chaffey in this volume).

Immunofluorescence

Notwithstanding the ability to serially section and reconstruct images (e.g. Kristen et al., 1989), the ultrathin sections required for TEM are a considerable obstacle to understanding the 3-dimensional arrangement of the cytoskeleton. The small area of each ultrathin section is also a hurdle to an appreciation of the arrangement of cytoskeletal arrays between cells, e.g. in a radial file of secondary xylem cells in wood. The advent of immunofluorescent procedures for the plant cytoskeleton (reviewed in Lloyd, 1987) and the concomitant use of sections two orders of magnitude thicker than those used for TEM has generated a remarkable amount of data in a relatively short time.

For many years it was necessary to be satisfied with the static images revealed by IIF of the plant cytoskeleton. However, relatively recent improvements in procedures for introducing fluorescently labelled proteins into cells (e.g. Gilroy, 1997), the use of the confocal laser scanning microscope (CLSM) (e.g. Hepler and Gunning, 1998; see also the chapter by Funada in this volume), and development of green fluorescent protein (GFP) technology (e.g. Ludin and Matus, 1998; Tsien, 1998; Conn, 1999; Sullivan and Kay, 1999) have allowed direct visualisation of MTs (Zhang et al., 1990; Hepler et al., 1993; Wasteneys et al., 1993; Hush et al., 1994b; Yuan et al., 1994, 1995; Hepler and Hush, 1996; Wymer et al., 1997; Marc et al., 1998;

Himmelspach et al., 1999) and MFs (Hepler et al., 1993; Kost et al., 1998) in living plant cells. This ability to follow cytoskeletal dynamics in vivo is providing even greater insight into the many and varied roles of this cell component.

Biochemistry

Of equal interest to the documentation of cytoskeletal arrays in cells is identification of the proteins concerned. Extraction of cell proteins and their separation using 2-dimensional gel electrophoresis (e.g. Williamson et al., 1989) permits their being probed for different isoforms (e.g. of tubulins – Hussey et al., 1988) using antibodies. If the material is sampled carefully, e.g. using the cryo-sectioning technique (see the chapter by Uggla and Sundberg in this volume), such an approach will also give information about the cell-specificity of cytoskeletal protein isoforms. Raising antibodies against those different isoforms allows the tissues/cells themselves to be probed, providing valuable information about array-specific isoforms (e.g. Goddard et al., 1998).

Molecular

Allied to an appreciation of the pattern of cytoskeletal protein isoform usage in cells/arrays is knowledge of the expression of the genes that code for those proteins. Using well-established techniques of molecular biology and genetics, multi-gene families of actin and tubulin proteins have been identified in plants (e.g. Fosket, 1989; Meagher, 1991), and the patterns of the expression of the genes studied. In general, it is found that different cells/tissues/organs of the plant often express a selected range of isoforms, which may be different at different stages of their development (e.g. Hussey et al., 1988; Meagher, 1991; Chu et al., 1998; Uribe et al., 1998).

Combining gene expression studies, using promoter-GUS technology (see the chapter by Hawkins et al. in this volume), with messenger RNA in situ hybridisation (see the chapter by Regan and Sundberg in this volume), and the use of specific antibodies against protein isoforms (e.g. the chapters by Micheli et al., and Samaj and Boudet in this volume) directs a formidable arsenal of modern 'weapons' at the target of understanding cytoskeletal dynamics within the plant cell. Furthermore, fusion of traditional gene expression techniques with relatively recently developed GFP technology and methods of in vivo observation permits study of a new level of complexity/organisation within the plant cytoskeleton (e.g. Boevink *et al.*, 1998; Nebenführ *et al.*, 1999), and of its dynamics (Granger and Cyr, 2000).

Brief history of study of the role of the cytoskeleton in wood formation

Until 1993, the limited information that was available concerning the cytoskeleton in wood formation in trees was provided solely by TEM, generally as part of a broader ultrastructural study, only occasionally as the main focus of the investigation (e.g. Robards and Humpherson, 1967) (for more information, see Chaffey *et al.*, 1996). The paper by Uehara and Hogetsu (1993), which deals principally with a correlative TEM and IIF study of the involvement of MTs in bordered pit formation in tracheids of the gymnosperm, *Taxus cuspidata*, is probably the first to document the use of IIF techniques to study the cytoskeleton in wood formation. In the subsequent 7 years, papers on this topic have appeared from two groups: Funada and colleagues in Japan, working principally on gymnosperms (see Funada *et al.*, 2000; and the chapter by Funada in this volume); and that of Chaffey and collaborators in the UK, mainly studying angiosperms (e.g. Chaffey, 2000; summarised below).

Considerations of successful sampling and fixation of woody material for immunofluorescence work on the cytoskeleton

The general criteria of relevance for IIF studies of the cytoskeleton are essentially the same as those for other microscopical work on the tree SVS, and readers are directed to the chapter, An introduction to the problems of working with trees, in this volume. Specifically for IIF, the main problems are autofluorescence of native cell components, such as lignin and chlorophyll, and fluorescence induced by use of glutaraldehyde as fixative. The protocols described below have been designed with these considerations in mind.

Why epifluorescence?

With the advent of CLSM for IIF studies of the plant cytoskeleton, and the advantages of its use over traditional fluorescence microscopy (see e.g. the chapter by Funada in this volume), why use epifluorescence? The major reason for use of epifluorescence is the cost of a CLSM, which is considerable,

and for many laboratories is prohibitive. Furthermore, CLSMs are still relatively scarce, and so booking time on such an instrument is not always easy, even if the equipment is relatively close to your laboratory. Without doubt, CLSMs will become cheaper and a time can be envisaged when they will be standard equipment in every cell biological laboratory. Until that time, epifluorescence continues to be an acceptable, and cheap, alternative to CLSM, which is within the reach of almost all investigators, and which is capable of revealing much about the 3-dimensional arrangement of the plant cytoskeleton. It is also gratifying to read that IIF 'on ultrathin cryosections provided convincing images with higher resolution than confocal LSM' (Ishiko *et al.*, 1998).

TECHNIQUES

Techniques and procedures for IIF of the cytoskeleton have been presented in books many times. However, in contrast to the number of citations for animal cytoskeleton, few chapters (and no books that I am aware of) have been devoted specifically to plant material. Techniques-oriented articles which give a good introduction to IIF of the plant cytoskeleton are those of Lloyd *et al.* (1979); Doonan and Clayton (1986); Wick and Duniec (1986); Gubler (1989); Wick *et al.* (1989); Williamson *et al.* (1989), Baskin *et al.* (1992, 1996); Goodbody and Lloyd (1994); and Hush *et al.* (1994a). Two that are highly recommended, particularly from the point of view of trouble-shooting, are the comprehensive reviews by Wick (1993) and Brown and Lemmon (1995).

It is important to stress at the outset that the procedures reproduced here are not the last word on the subject of IIF of the SVS. I am sure that further improvements can be made, and it is possible that they may not work perfectly for material other than those angiosperm trees (*Aesculus hippocastanum* (horse-chestnut), *Populus tremula* × *P. tremuloides* (hybrid aspen), and taxa listed in Table 1) for which they were developed. So, do not be afraid to experiment, using the protocols in this chapter as a starting point, and the above-listed articles as sources of inspiration.

Note – equipment and chemicals in bold type are considered in more detail in the CHEMICALS, EQUIPMENT, ETC section.

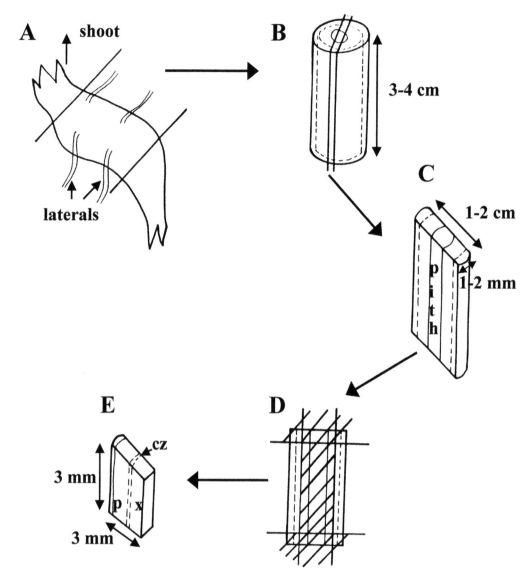

Figure 1 Diagrammatic representation of sampling procedure for indirect immunofluorescence studies, illustrated using the taproot of *Aesculus hippocastanum* L. A, taproot from which samples are to be taken; B, excised cylinder of taproot material; C, sliver removed from cylinder and ready for pre-fixation; D, fixed sliver marked-out for trimming to final size – all of the pith, most of the secondary xylem, and the terminal few mm are removed and discarded; E, final trimmed block ready for overnight buffer-storage.
Key: cz, cambial zone; p, phloem; x, secondary xylem; dashed line indicates location of cambial zone.

Sampling strategy

Equipment needed:	Chemicals needed:
Single-edge razor blades (several)	Water
Hacksaw	
Glass Petri dish	
Bucket	
Beaker (0.5–1L)	
First-aid box (cuts are all too common)	

Duration of procedure: approx. 30 min (plus travelling time!)

Sampling for IIF is the same as that used for TEM and is diagrammatically illustrated in Figure 1. The legend to that figure should be consulted for details of the sampling procedure. Excised radial longitudinal slivers of material are 3–4 cm long × 1–2 mm thick × 1–2 cm wide. Usually, the sliver is then cut in half longitudinally. However, whatever size sliver you use, ensure that the entire length of the sliver will fit into your vial and still be covered by solution!

1   To begin with, separate the stem/root of interest from the rest of the sapling using the hacksaw. Immediately place the cut end(s) under water (to minimise danger of embolism).
    Note – I use a bucket of water, which is also useful for transporting the material back to the laboratory, if sampling the tree in the field. It is also useful for washing soil, etc. away from the roots to permit their more ready sampling.
2   Cut stem/root into 3–4 cm lengths using the hacksaw. Accumulate these cylinders in a beaker of water.
3   To obtain slivers, stand one end of a 3–4 cm long cylinder in water in a glass Petri dish, and cut the cylinder in half – lengthwise – using a razor blade.
    Note – this will require considerable force, and it is not unknown to resort to using a hammer to encourage the razor blade to cut all the way down through the cylinder.
4   Cover the exposed surfaces of the half cylinders with water (to prevent excess trauma to cells from water loss).
5   Repeat lengthwise cut (fresh razor blade!) to obtain sliver of dimension, approx. 3–4 cm long × approx. 2 mm thick × width of root/stem.
6   Then cut sliver in half lengthwise, so that it is now half the width of the root/stem. It is usually possible to get approx. 6–10 such half-slivers from a 2 cm diameter root/stem. However, if the cambium is active, bark-slippage may well frustrate your attempts to cut slivers with intact cambium: unfortunately, there is no alternative but to keep cutting until you have sufficient 'good' slivers (and remember to allow for losses during subsequent processing, such as accidentally discarding material with solutions).
7   To avoid further damage to the tissues, fix slivers (see next section) as soon as possible. Handle them, as little as possible, by their ends (which will subsequently be discarded) using forceps.
    Note – cylinders not used for IIF sampling can be kept for macerating, or for scanning electron microscope study, etc. (see the Wood microscopical techniques chapter in this

volume). Slivers can also be excised and fixed in parallel for TEM study (see the TEM chapter by Chaffey in this volume)

Fixation and embedding in Steedman's wax

The procedure detailed here is modified from a protocol originally developed for tomato and maize roots (e.g. Baluška *et al.*, 1992). See Brown *et al.* (1989), Brown and Lemmon (1995), and Vitha *et al.* (2000b) for additional background information on this technique.

Equipment needed:	*Chemicals needed:*
Embedding moulds	**'Wax fixative'**
Refrigerator at 4 °C	Ethanol
Glass vials (capacity approx. 15 ml) and tops	**Toluidine blue**
Rotator	**Steedman's wax**
Embedding oven at 37 °C	**Microtubule-stabilising buffer (MTSB)**
Single-edge razor blades (several)	Acid fuchsin
Glass Petri dishes	

Duration of procedure: 4 days (blocks can be cut late on fourth day, but it is probably better to leave them overnight in the refrigerator until the 5th day).
Note – steps 1, 2, 4-penultimate 100% ethanol stage in step 5 are performed at room temperature (RT) in sealed glass vials on a rotator; the final 100% ethanol step and subsequent stages to embedment are carried out at 37 °C in an embedding oven (with occasional shaking by hand to mix contents).

1   Fix slivers for approx. 4.5 h in Wax fixative.
    Note – vacuum-infiltration has not been found to be necessary.
2   Wash in MTSB, several changes in 15 min.
3   Trim slivers to final block size (approx. 2–3 × 2–3 × 1–2 mm) with razor blade in Petri dish under MTSB.
4   Leave in MTSB overnight (15–20 h).
5   Dehydrate in a graded ethanol/water series: 30%, 50%, 70%, 80%, 90% for 30 min at each stage, followed by 2 × 30 min in dry 100% ethanol.

Note – when set, the wax is opaque which makes it difficult to see the tissue and ensure its correct orientation for cutting. To overcome this problem, it is recommended to surface-stain the tissue by adding a few drops of a saturated solution of acid fuchsin in 95% ethanol to the 90% ethanol stage. Any excess stain will be removed during subsequent dehydration steps.

6   After 30 min in the final 100% ethanol step, add an equal volume of molten Steedman's wax, gently shake the tube to mix the contents, and leave overnight (15–20 h).

7   Replace half of the volume of the wax/ethanol mixture with molten Steedman's wax, leave for approx. 8 h.

8   Replace all of the volume with molten Steedman's wax and leave overnight (15–20 h).

9   Replace all of the volume with fresh molten Steedman's wax, leave for approx. 3.5 h.

10   Replace all of the volume with fresh molten Steedman's wax, leave for approx. 3.5 h.
Note – before embedding the tissue blocks, consider the planes of section that are to be cut, and orient the blocks appropriately. Although radial longitudinal sections (RLS) are probably the most useful for IIF studies, both tangential longitudinal (TLS) and transverse sections (TS) are also important in gaining an appreciation of the 3-dimensional structure of cytoskeletal arrays. Further, oblique TSs can be extremely useful in providing views of the radial and tangential walls *of the same cell* in the same tissue section.

11   Embed tissue in fresh molten Steedman's wax in pre-warmed silicone moulds.
Note – To prevent premature hardening of the wax during the time it takes to fill all the wells, I usually keep the mould in the oven until all wells are filled.

12   Allow the wax to cool on the bench, remove the wax blocks from the wells, and store in the refrigerator until ready for cutting.

Fixation/embedding using butyl-methylmethacrylate

Steedman's wax-embedment is a satisfactory procedure for examining the SVS in *Aesculus*

(Chaffey *et al.*, 1996; 1997a) *during the early stages of secondary thickening*. However, as soon as secondary thickening is appreciable – after only the first few weeks of seedling growth – problems are encountered with wax-embedded tissue. In particular, the blocks section poorly, frequently destroying the delicate cambial tissue and the valuable early stages in the sequence of development of cambial derivatives. These problems are substantially reduced by use of butyl-methyl-methacrylate (BMM), a virtue of resin over wax for routine anatomical study previously emphasised by both Berlyn (1963) and Baskin *et al.* (1992). Further, in view of the correlative study that is advocated (Chaffey *et al.*, 1996), the BMM embedding protocol fits in conveniently alongside the conventional TEM processing (see the TEM chapter by Chaffey in this volume) that is performed on parallel samples of tissue, thereby eliminating the inconvenience of alternating between wax- and resin-processing procedures. For complementary bright-field microscopy, BMM-sections readily stain with a wide range of stains traditionally used for wax sections, where full use can be made of the superior anatomical preservation achieved compared with its wax counterpart. For some detailed staining procedures for BMM-sections, see the Wood microscopical techniques chapter in this volume.

Cutting qualities of BMM are such that sections can be cut at 6 to 30 μm (Berlyn, 1963), and 22 μm sections were used by Kobayashi *et al.* (1994) in their IIF study of MTs and actin in rust-infected flax. I routinely cut sections at 6–10 μm and have not experienced the collapse of the cytoplasm upon resin-removal that was predicted by Baskin *et al.* (1992). Rather, the thickness of such sections allows a better appreciation of the 3-dimensional nature of the cytoskeletal arrays.

The fixation procedure described here permits visualisation of a range of cytoskeletal proteins – α-tubulin and actin (see below), β-tubulin, post-translationally modified variants of α-tubulin, and myosin (Chaffey and Barlow, unpublished) – and other cell wall and plasmalemmal components, such as callose (Figure 1G in the 'Wood microscopical techniques' chapter in this volume), pectins and proteoglycans (Chaffey and Barlow, unpublished). In view of this success, it is also likely that material processed by this fixation protocol can be used for IIF detection of other cytoskeletal proteins, such as MTOCs and IFs.

Equipment needed:	Chemicals needed:
Single-edge razor blades (several)	**3-maleimidobenzoic acid *N*-hydroxysuccinimide ester (MBS)**
Freezer at –20 °C	**Butyl-methylmethacrylate (BMM)**
Refrigerator at 2–5 °C	Benzoin methyl ether (Sigma, No. B-0279)
UV lamp	Ethanol
Embedding capsules	**Basic cytoskeleton fixative**
Glass vials (approx. 15 ml capacity) and tops	**PIPES buffer**
Rotator	

Duration of procedure: approx. 4.5 days.

Note – don't forget to allow sufficient time for frozen solutions to thaw prior to use.

Note – steps 1, 2, 4–9 performed at RT on rotator in stoppered glass vials.

1 Pre-fix excised slivers in MBS for 1.5 h.
2 Fix in basic cytoskeleton fixative for approx. 3 h.
3 Transfer to 25 mM PIPES buffer, trim material to final block size.
4 Store in 25 mM PIPES buffer overnight (15–20 h).
5 Dehydrate in ethanol-water series, 30 min each at 30, 50, 70, 80, 90% and 3 × 30 min in 100%.
6 To the final ethanol step, add equal volume of BMM, and leave overnight (approx. 20 h).
7 Replace half of the volume in vial with pure BMM, leave for approx. 9 h.
8 Replace all of the volume with pure BMM and leave overnight (15–20 h).
9 Replace all of the volume with pure BMM and leave for approx. 4.5 h.
10 Replace all of the volume with BMM containing 0.5% (w/v) benzoin methyl ether, and leave rotating in the dark at approx. +5 °C (refrigerator) (to prevent premature polymerisation of resin), for approx. 5 h.
11 Place blocks in embedding capsules, in appropriate **orientation**, add fresh BMM containing 0.5% (w/v) benzoin methyl ether until nearly full, replace top, and perform **resin-curing** overnight with a UV lamp at –20 °C. Blocks

can be stored at RT and antigenicity retained for many months (up to a year in some cases). Note – this processing schedule works equally well for gymnosperms.

Freeze-fixation

There continues to be debate as to whether chemical-fixation or freeze-fixation preserves cellular structure better (e.g. Kiss *et al.*, 1990; Robards, 1991, Bourrett *et al.*, 1999). Even though a wide variety of fixation and incubation cocktails have been employed for visualisation of MTs in plant cells (e.g. Wick *et al.*, 1981; Simmonds *et al.*, 1985; Goodbody and Lloyd, 1994), the images obtained are not immune from the criticism that they are 'artefactual' (e.g. aldehydes might induce bending and breaking of MTs – Cross and Williams, 1991). Although a wide variety of chemical fixations for IIF and TEM give the same images of the MT and MF cytoskeleton in angiosperm trees (Chaffey *et al.*, 1996, 1997c), in order to provide further corroborative evidence of the validity of these chemical-fixed images, two freeze-fixation, freeze-substitution procedures have also been tried on stems of *Aesculus hippocastanum* and hybrid aspen.

Unfortunately, the location of the SVS within the tree organ does not render it suitable for the plunge-freezing method used by Baskin *et al.* (1996) on whole *Arabidopsis* roots, or the freeze-shattering technique of Wasteneys *et al.* (1997). However, the hardwood SVS appears amenable to freezing (e.g. Regan *et al.*, 1999a; see also the chapter by Regan and Sundberg in this volume), at least for observations at the light microscope level (where it seems likely that any ice crystals formed are below the limits of resolution of that imaging technique – Baskin *et al.*, 1996). Whether the freezing techniques described below will also be suitable for TEM work on the SVS remains to be established. Preliminary investigation of this material with IIF of α-tubulin suggests that the images obtained with freeze-fixation are the same as those from chemical-fixed material.

Encouragingly, Wasteneys *et al.* (1998), studying epidermal cells of *Arabidopsis* root, found no qualitative difference in MT preservation between chemical-fixed and cryo-fixed, freeze-substituted samples. Even more encouraging is the work of Wymer *et al.* (1997), which showed that images of chemically fixed MTs appeared the same as those seen using an *in vivo* visualisation technique.

Equipment needed:	Chemicals needed:
Freezer at –80 °C	Methanol
Freezer at –20 °C	Acetone
Dewar (insulated flask) for liquid nitrogen	**Butyl-methylmethacrylate (BMM)**
Vials with tops which will withstand –80 °C	Glutaraldehyde
Glass Petri dish	Liquid nitrogen
Single-edge razor blades (several)	Benzoin methyl ether (Sigma, No. B-0279)
Forceps	

Method 1: methanolic glutaraldehyde

Duration of procedure: approx. 45.5 days.
Note – ensure necessary solutions are prepared and pre-cooled prior to starting sampling!

This method is modified from the procedure used by Regan *et al.* (1999a):

1 Excise slivers (see Sampling Strategy above).
2 Using forceps, transfer slivers to, and accumulate in, liquid nitrogen.
3 Transfer frozen slivers to vials containing 0.2% (v/v) glutaraldehyde in methanol (pre-cooled to –80 °C), and leave at –80 °C for 42 days.
    Note – it is worthwhile experimenting with different lengths of time at –80 °C to see what suits your material best; use 42 days as a guide.
4 Transfer vials to bench at RT, leave for approx. 2 h.
5 Transfer slivers to methanol in glass Petri dish, and trim to final block size using a razor blade.
    Note – perform steps 6–10 at RT on a rotator.
6 Transfer slivers to vial of fresh methanol at RT, leave for 1 h.
7 Add an equal volume of BMM, leave overnight (approx. 22 h).
8 Replace half of the volume of resin mixture with fresh BMM, leave for approx. 8 h.
9 Replace all of the resin mixture with fresh BMM, leave overnight (approx. 16 h).
10 Replace all of the resin mixture with fresh BMM, leave for approx. 4 h.
11 Replace all of the resin mixture with BMM containing 0.5% (w/v) benzoin methyl ether,

and leave rotating for approx. 4 h in the refrigerator.
12 Place blocks in embedding capsules, with appropriate **orientation**, fill to the top with BMM containing 0.5% (w/v) benzoin methyl ether, and perform **resin-curing** at –20 °C with UV lamp overnight (approx. 16 h). Store blocks at RT.

Method 2: acetone-fixation

Duration of procedure: approx. 30 days.
Note – ensure necessary solutions are pre-cooled prior to starting sampling!

1 Excise slivers (see Sampling Strategy above).
2 Transfer slivers to, and accumulate in, liquid nitrogen.
3 Transfer frozen slivers to acetone (pre-cooled to –80 °C), and leave at –80 °C for 25 days.
    Note – it is worthwhile experimenting with different lengths of time at –80 °C to find the most suitable time for your own material: use 25 days as a guide.
4 Transfer vials to bench at RT, leave for approx. 1 h.
5 Trim slivers to final size in acetone in glass Petri dish, return to vial.
    Note – perform steps 6–12 at RT on a rotator.
6 Add an equal volume of BMM to vial, leave over-night (approx. 15 h).
7 Replace half of the volume of resin mixture with fresh BMM, leave for approx. 8 h.
8 Replace all of the resin mixture with fresh BMM, leave overnight (approx. 16 h).
9 Replace all of the resin mixture with fresh BMM, leave for approx. 8 h.
10 Replace all of the resin mixture with fresh BMM, leave overnight (approx. 16 h).
11 Replace all of the resin mixture with fresh BMM, leave for approx. 5 h.
12 Replace all of the resin mixture with fresh BMM, containing 0.5% (w/v) benzoin methyl ether, leave for approx. 5 h rotating in refrigerator.
13 Place blocks in embedding capsules, with appropriate **orientation**, fill to top with BMM containing 0.5% (w/v) benzoin methyl ether, and perform **resin-curing** at –20 °C with UV lamp overnight (approx. 16 h). Store blocks at RT.

Immunolocalisation of α-tubulin in Steedman's wax-embedded material

Equipment needed:	Chemicals needed:
Spirit lamp to melt wax	Ethanol
Microtome and knives	**Enzyme digest**
Incubator at 37 °C	**Phosphate-buffered saline (PBS)**
Epifluorescence micro-	**Microtubule**
scope with camera,	**stabilising**
film,	**buffer (MTSB)**
appropriate filters,	
and objectives	
Coplin jars (or	Triton X-100
equivalent)	
Slide storage box	**Toluidine blue**
Glass cover slips	**Antibodies**
Humid box	**Anti-fade mountant**
Glass microscope slides	**Counter-stains**
coated with **Mayer's**	
egg albumen	
Refrigerator at 5 °C	
Forceps	
Pasteur pipettes	

Duration of procedure: approx. 4–4.5 h (excluding sectioning time and overnight drying-down).

Note – don't forget to allow sufficient time for frozen solutions to thaw prior to use.

Note – perform steps 1–6 and 9 at RT; steps 7 and 8 at 37 °C in the dark in a humid box (in incubator).

1 Cut sections at 6 μm, mount on microscope slides, and dry down overnight at RT.

Note – to assess orientation of block, position of section in SVS, quality of fixation, infiltration of wax, etc., sections will need to be cut for examination in the light microscope. Basic staining of these sections with toluidine blue, whilst still in the wax ribbon, is sufficient for these purposes. However, it is usually a good idea to cut some additional sections at 2 μm, and dry them down on uncoated glass microscope slides for subsequent staining, etc. If cutting many sections from a given block, it may be necessary periodically to chill the block in the refrigerator to avoid excessive warming of the block (remember, melting point of Steedman's wax is 37 °C, so section-cutting on very warm days may be problematical).

2 Dewax sections with 100% ethanol for 3 × 10 min.

3 Rehydrate in 80%, 50%, and 30% ethanol-water 5 min each.

4 Stand in MTSB in Coplin jar for 35–50 min.

5 Treat sections with enzyme digest for 15 min.

6 Rinse sections with MTSB (use Pasteur pipette), then stand in 1% (w/v) Triton X-100 in MTSB for 30 min.

7 Rinse with MTSB, then incubate with the primary antibody for 1 h at 37 °C in the dark (in incubator).

8 Rinse sections with MTSB, then incubate with the secondary antibody for 1 h at 37 °C in the dark (in incubator).

9 Rinse with MTSB, apply counter-stains as appropriate (e.g. DAPI and/or calcofluor), then a final rinse with PBS.

10 Minimise autofluorescence of tissue by staining sections with toluidine blue, until pale blue, and wash off excess stain with PBS.

11 Finally, air-dry the sections, apply anti-fade mountant and cover slip, and store in refrigerator at 4 °C until ready to view.

Note – the fluorescence image significantly improves with time after mounting in anti-fade mountant (see also Wymer et al., 1999); therefore, sections are critically studied *only* after allowing the FITC image to 'develop' overnight in the refrigerator.

12 View sections with appropriate filters in an epifluorescence microscope and record images.

Immunolocalisation in butyl-methylmethacrylate-embedded material

α-tubulin

Equipment needed:	Chemicals needed:
Humid box	**Phosphate-buffered saline (PBS)**
Glass cover slips	**Counter-stains**
Pipettes (Pasteur and analytical)	**Blocking solution**

<table>
<tr><td>
Epifluorescence micro-
scope fitted with
camera, **film,**
appropriate filters, and
objectives
Microtome and knives
Slide-drying rack at
40 °C
Refrigerator at 4 °C
Coplin jars (or
equivalent)
Hot plate at 60–70 °C
Multi-well slides coated
with **Mayer's egg albumen**
</td><td>
Antibodies

Acetone
Toluidine blue

PBSA
Anti-fade mountant
</td></tr>
</table>

Duration of procedure: approx. 6 h (excluding sectioning time and overnight drying-down).
Note – don't forget to allow sufficient time for frozen solutions to thaw prior to use.
Note – steps 3–11 performed at RT, in humid box for steps 5, 6 and 8.
Note – do not allow sections to dry out during steps 3–11.

1 Cut sections dry using glass or stainless steel knives at 6–10 μm, and transfer singly to a drop of water in a well on a Mayer's egg albumen-coated multi-well glass microscope slide.
   Note – during cutting, the sections curl up. To prevent this, carefully hold the edge of the section with a pair of fine-tipped forceps as it is being cut. This takes practice to perfect, but will significantly assist the subsequent flattening of the sections. It is good practice to cut some sections at 2–4 μm for light microscopy, to gauge position in SVS, quality of fixation/resin-embedding, and for subsequent staining, etc. Collect these sections on a drop of water on an uncoated microscope slide and dry down overnight at approx. 60 °C.
2 Stretch sections at approx. 60 °C for 10 sec, and dry down overnight at approx. 40 °C.
3 Remove resin by standing the slides in 100% acetone, with *gentle* agitation (to minimise danger of sections coming unstuck), for 45–60 min, drain off excess acetone, and,
4 immediately (to prevent drying out) immerse sections in PBS, stand for 2 min. then repeat with fresh PBS.

5 Drain off excess PBS, apply blocking solution (approx. 13 μl per well), and incubate for approx. 1 h (in humid box).
   Note – ensure box and slides kept horizontal throughout incubation period, to prevent drying-out and undesirable well–well mixing of solutions.
6 Drain liquid from slide, apply primary antibody (approx. 13 μl per well), and incubate for approx. 2 h (in humid box).
7 Wash off antibody solution with gentle stream of approx. 1.5 ml of PBS, then stand slides for 3×5 min in Coplin jar of PBS.
8 Drain PBS from slide, apply secondary antibody (approx. 13 μl per well), and incubate for approx. 1 h (in humid box).
9 Following a 3×5 min PBS-wash, counterstains may be applied, if desired.
10 Minimise autofluorescence of tissue by staining sections with toluidine blue until pale blue, rinse finally in PBS.
11 Remove excess PBS (but do not allow sections to dry out) before mounting in anti-fade mountant and applying cover slip.
12 Store slides in refrigerator until ready to view.
   Note – as for Steedman's-wax-embedded sections, the fluorescence image significantly improved with time after mounting in anti-fade mountant. Sections are critically studied *only* after allowing the FITC image to 'develop' overnight in the refrigerator. Comfortingly, the FITC fluorescence appears to last for several months when stored under these conditions, but I would recommend viewing sections as soon as is practicable (lest *your* sections fade).
13 View sections in an epifluorescence microscope with appropriate filters.
   Note – this processing schedule works equally well for gymnosperms.

Actin

Generally, there are two methods for localising actin MFs in plant cells, using an antibody (e.g. Vitha *et al.*, 1997) or with a fluorescent phallatoxin, such as rhodamine-labelled phalloidin (e.g. Tewinkel *et al.*, 1989). However, I have been unable to stain MFs in fixed material of *Aesculus* and hybrid aspen SVS with fluorescent-labelled phalloidin. Similar lack of staining with gymnosperms has also been reported by Funada (personal communication; see also his chapter in this volume). Success has, however, been achieved in these two angiosperms with anti-actin antibody. Although I have not tried the antibody on Steedman's wax-embedded angiosperm SVS, Vitha *et al.* (1997, 2000a, b) have shown that it can work for maize and cress roots embedded in this wax; and I see no reason why it should not also work in the tree SVS. Note, however, that Vitha *et al.* (1997) were unable to obtain good preservation of MFs using dimethyl sulphoxide (DMSO) in their procedure.

> *Equipment needed:*   *Chemicals needed:*
> As for α-tubulin above   As for α-tubulin above

Duration of procedure: approx. 6 h (excluding sectioning time and overnight drying-down).
Note – this processing schedule works equally well for gymnosperms.

1   All steps as for α-tubulin localisation, except using anti-actin as the primary antibody.

Controls

Controls for tubulin and actin localisation are rarely mentioned in research papers these days, possibly because the antibodies used are so well established and reliable. However, controls are necessary to satisfy oneself – and others! – of the validity of the staining reaction (see also Burry, 2000), particularly in new taxa or cell systems. Controls, which I routinely employ, are:

1   omission of primary antibody
    Incubate sections with blocking solution instead of primary antibody: no staining noted.
2   omission of secondary antibody
    Incubate with PBSA solution instead of secondary antibody: no staining noted.
3   use of primary antibody of no relevance to tissue being studied
    Substitution of UBIM22 (a monoclonal antibody against rat bone marrow cells (Pain *et al.*, 1992)) for the primary antibody resulted in a very low level of antibody binding, but did not stain the α-tubulin-containing structures (Chaffey *et al.*, 1996).

4   use of alternative antibody
    Where two antibodies have well-documented and distinct distributions, the one can be used as a control for the other (e.g. Samaj *et al.*, 1998; Vitha *et al.*, 2000b). For instance, in fusiform cambial cells and xylem fibres, I have used anti-actin and anti-α-tubulin in this way (e.g. Chaffey *et al.*, 1999, 2000b).

Although it is widely recommended to use 'normal' serum, i.e. serum from the pre-immunised animal that was subsequently used for antibody production, as a control, this is practically impossible where commercially obtained antibodies are concerned. As a compromise, and since the secondary antibody used here is produced in goats, I use normal goat serum as a component of both the blocking solution and the PBSA.

Appropriate use of controls notwithstanding, there is no substitute for repetition of immunostaining until you get a consistent, reproducible pattern of staining.

A few words of caution concerning 'carry-over' of solutions are pertinent. If slides are placed in the humid box so that they touch a neighbour, there is the possibility that solution will be carried over from one slide to the next. Similarly, do not mix treatments on a slide since carry-over from well to well is very common. Another possible source of 'carry-over' occurs during the washing steps in the Coplin jars: even though antibodies – and stains – are tremendously diluted here, there is still the possibility that tissue may be exposed to chemicals other than those intended! (see also comments in Wick, 1993).

CHEMICALS, EQUIPMENT, ETC.

Note – unless stated otherwise, 'water' means deionised water.

Antibodies

Primary

Anti-α-tubulin (Amersham, No. N356; supplied as 0.5 ml solution, roughly 150 μg. ml^{-1}), 20μl aliquots stored at –20 °C. Used diluted at 1:10 with **PBS**.
Anti-actin (ICN, No. 69100; supplied as 0.1 or 0.5 ml solutions), stored in refrigerator.
Used diluted at 1:100 with **PBS**.
Both antibodies are IgG fractions raised in mouse.

Secondary

Fluorescein *iso*thiocyanate (FITC)-labelled anti-mouse IgG (Sigma, No. F-0257; supplied as 0.5 ml solution), 20 μl aliquots stored at –20 °C. Used diluted at 1:30 with **PBSA**. Raised in goat.

Note – FITC can be bleached by daylight. Therefore, it is advisable to try to perform this antibody incubation step in a darkened room. Once the sections have been processed for IIF, storing the slides in a slide box in the refrigerator overcomes that problem.

Anti-fade mountant

For Steedman's wax sections, a solution of *p*-phenylenediamine in glycerol (Johnson and Araujo 1981; Johnson *et al.*, 1982) was used as the anti-fade mountant. To make: dissolve 50 mg of *p*-phenylenediamine in 5 ml **PBS**. Add 45 ml of glycerol and adjust to approx. pH 8.0 with 0.5 M carbonate: bicarbonate buffer, pH 9.0. The buffer is made by mixing appropriate volumes of 0.5 M Na_2CO_3 and $NaHCO_3$ together to get pH 9.0. Rapid browning of the solution at RT renders it unsuitable for critical IIF; but this can be largely avoided by storing aliquots at –20 °C. Although I have used the mountant as made up above for successful IIF of MTs (and obtained the same images as with Vectashield® as anti-fade mountant), note that Swartz and Santi (1996) stress the importance of ensuring that the pH of the mountant is between 8.5 and 10.5 to avoid possible artefacts which affect interpretation of the IIF staining pattern.

However, in view of problems experienced with reliability of the glycerol and the self-made mountant, I now routinely use the commercially available Vectashield® (Vector Laboratories, No. H-1000). This is an aqueous solution (containing *p*-phenylenediamine – Longin *et al.*, 1993), supplied as 10 ml volumes in a dark bottle, which needs to be kept in a refrigerator. Although sections mounted in Vectashield® and stored in the refrigerator retain immunofluorescence for several months, it is best to observe sections as soon as possible after staining. If desired, cover slips can be sealed with nail varnish to make the mounts more permanent.

Appropriate filters

Since the choice of filter sets may well be dictated by the epifluorescence microscope available, users will have to decide for themselves which to use. To assist in that choice, the following table gives some relevant data concerning the three fluorochromes mentioned in this chapter.

Fluorochrome	Absorption (nm)	Emission (nm)
FITC[a]	490	520
Calcofluor[b]	365	435
DAPI[a]	350	470

[a] Tsien and Waggoner (1990); [b] Herman (1998)

Basic cytoskeleton fixative

25 mM **PIPES buffer**, pH 6.9, 3.7% (w/v) *p*-**formaldehyde,** 0.2% (v/v) glutaraldehyde, 10% (v/v) **DMSO**.

Blocking solution

6% (w/v) bovine serum albumen (BSA) (fraction V, Sigma, No. A2153), 0.1% (v/v) fish skin gelatin (Sigma, No. G-7765), 5% (v/v) normal goat serum (Sigma, No. S-2007), 0.05 M glycine in **PBS**. Filter through 0.45 μm filter, store aliquots at –20 °C.

Butyl-methylmethacrylate (BMM)

A mixture of 4 parts of n-butyl-methacrylate(TAAB, No. BO14): 1 part methylmethacrylate (TAAB, No. MO08) with 5 mM dithiothreitol (DTT) (Sigma, No. D-0632) (after Baskin *et al.*, 1992).

Note – this resin does not form ribbons during cutting (cf. **Steedman's wax**). However, the single sections that are obtained are ideally suited to use with **multi-well slides**.

Counter-stains

Sections are routinely counter-stained, sequentially, after immunolocalisation with the following:
DAPI (4',6-diamidino-2-phenylindole), 1 μg. ml^{-1} in **PBS**, for nuclear DNA (e.g. Goodbody and Lloyd 1994); 2 min, then,
Calcofluor White M2R New, or Calcofluor White ST, 0.01% (w/v) in **PBS**, for 'cell walls' (e.g. Chaffey, 1994, 1996); 2 min.
Remove excess stain with a PBS-rinse. But note, 'carry-over' of calcofluor can occur when slides are held in PBS for washing sufficient to stain sections that did not have the stain applied directly to them!

Dimethyl sulphoxide (DMSO)

DMSO is widely used in IIF studies of the MT cytoskeleton (e.g. Schroeder *et al.*, 1985; Wick *et al.*, 1989; Baluška *et al.*, 1992; Hush *et al.*, 1994a; and the chapter by Funada in this volume), commonly at a 10% (v/v) concentration. Although the same MT arrangements have been seen in material fixed without DMSO (Chaffey *et al.*, 1997c), addition of DMSO is used to facilitate comparison with the published literature. Contrary to the recent study by Vitha *et al.* (1997), excellent preservation of MFs in the angiosperm tree SVS is obtained using DMSO in the fixative.

Embedding capsules

Two types of capsule are routinely used for BMM-embedding:
Polypropylene, 8 mm diameter, flat-bottomed (TAAB, No. C095), and
Polythene, 8 mm diameter, flat-bottomed (TAAB, No. C094).

Although the literature recommends the use of polypropylene for acrylic resin-embedding, polythene appears equally suitable, and has the major advantage that it is much easier to release the resin block.

Enzyme digest

This is made up as: 1% (w/v) hemicellulase (Sigma, No. H-2125) containing 5 mM ethylene glycol-*bis* (*p*-aminoethyl ether)*N,N,N',N'*-tetraacetic acid (EGTA), 0.4 M mannitol, 1% (v/v) Triton X-100, 1% **MTSB**. Prior to use, filter the solution and add 0.1% (v/v) of a saturated solution of phenylmethyl-sulphonyl fluoride (PMSF – a general protease inhibitor) in *iso*-propanol. Enzyme digestion degrades the cell walls aiding penetration of antibody in thick sections, and may also help to remove some fluorescent material.

Film

For colour recording, I use Kodak EPP100 colour slides, commercially developed using E-6 processing. Although the film is used at its rating of 100 ASA, which can lead to long exposure times (up to 2 min is not uncommon), no serious photobleaching has been noted (although it is advisable to keep both photographic and image-viewing exposures as short as possible since prolonged/repeated illumination can eventually lead to photobleaching of the fluorescence).

For black and white images, I use Kodak T-max 400 film, processed with diafine 2-bath film developer (supplied by Agar, No. Y-2363). This film can be rated at 800 ASA, but its rating of 400 ASA gives exposures that are sufficiently short for photobleaching not to be a problem.

I have not used digital or CCD cameras for recording IIF images. However, the wider availability of these devices, which is part of a much wider increase in LM hardware of all sorts, makes such options increasingly affordable and attractive. That, plus the great variety of graphics packages and other software that are available, means that even more exciting possibilities exist for image capture, manipulation, and interpretation.

Humid box

I use a plastic 'sandwich box', 11.5 × 17.5 × 3 cm deep. The humid environment is created by placing a thoroughly wetted tissue in the base of the box and applying the lid. Place slides in box so that they do not touch each other! (see '**Controls**' section above). Also, ensure that the box is placed flat, otherwise solutions can run off the slide causing the sections to dry out during incubation. Where required, and unless specified otherwise, dark incubation conditions can be created by covering the humid box with a cardboard box, or similar, or placing in a cupboard.

3-maleimidobenzoic acid *N*-hydroxysuccinimide ester (MBS)

Used as a 'pre-fixative' to stabilise MFs (after Sonobe and Shibaoka, 1989).
Store aliquots of a 100 mM solution of MBS (Sigma, No. M-2786) in dimethyl sulphoxide (DMSO), at –20 °C. Used for pre-fixation at 100 μM in 25 mM **PIPES buffer**, pH 6.9.
Note – to prevent precipitation of the MBS when making the pre-fixation solution, keep the water/PIPES mix swirling during the addition of the MBS.

Mayer's egg albumen

The following recipe is taken from Clark (1981). Shake together 50 ml white of hen's egg, 50 ml glycerine and 1 crushed aspirin in a tall measuring cylinder, until there are loads of bubbles in suspension. Allow the mixture to stand, remove and discard the upper layer, filter and retain the lower layer. The mixture will keep for several years in the refrigerator at +4 °C.

To apply to **multi-well slides**, allow the mixture to be taken up into a glass Pasteur pipette by capillary action. With a finger covering the large diameter end of pipette, touch the tip of the pipette on to a well. This will dispense approx. 2 μl of the mixture that can then be spread over the well with the little finger. Repeat this procedure as many times as necessary, leave slides to dry for approx. 7 hours at RT before storing in the refrigerator.

To apply to ordinary microscope slides, smear a drop of the mixture over the surface of the slide with a finger. Leave to dry overnight at RT, and store as for multi-well slides. Such slides are best identified by suitable marks made with a diamond scribe.

Microscope objectives

Although cytoskeletal structures are individually small (approx. 25–7 nm in diameter), they are visible at the LM level because their apparent size is increased by the shells of antibodies with which they are coated in IIF (so much so that an individual MT has an apparent diameter of approx. 250 nm – Williamson, 1991). Even so, it is necessary to make observations with appropriate microscope objectives. Generally, it is necessary to use as a minimum a × 40 objective, possibly making use of an optovar to increase the magnification of the observed image. Ideally, it is desirable to use a × 100 oil-immersion objective, or (better!) a × 63 water-immersion objective, with the greatest numerical aperture available. Ultimately, however, the choice of objectives (and microscope and filters, etc.) will be dictated by the resources available to the investigator.

MTSB (microtubule-stabilising buffer)

12.5 mM **PIPES** buffer, pH 6.9
1.25 or 5 mM MgSO$_4$.7H$_2$O,
1.25 or 5 mM ethylene glycol-bis(p-aminoethyl ether)N,N,N',N'-tetraacetic acid (EGTA)

Multi-well slides

Several versions of these slides (sometimes called 'multitest' slides) are available. I use the 8-well version (ICN, No. 6040805), which uses approx. 100 μl solution per slide (cf. approx. 200 μl per ordinary glass slide with Steedman's wax sections). Depending on the volumes of antibodies, amount of tissue, and time you have available, you might like to consider 10, 12 or even 14-well slides. To identify the slides, it is best to write on the white area with a pencil; ink will dissolve in the acetone, and sticky labels are liable to come unstuck during the extensive washing steps. Multi-well slides are routinely used straight from the box as supplied. However, upon first use, some of the blue colouration from the multi-well slides dissolves in the acetone; this does not seem to harm the immunolocalisation, but can be a little disconcerting if you are not prepared for it. After use, **slide-washing** can be performed, and the slides reused.

Orientation

It is necessary to consider the ultimate plane of sectioning before embedding; however, I recommend that tissue blocks are allowed to settle so that their radial surface is at the bottom of the capsule. In that orientation RLSs are readily cut. Removal of a cylinder approx. 3 mm high and containing the tissue block allows either TLSs or TSs to be taken, with appropriate orientation of the cylinder in the microtome.

PBSA

A solution of 1% (w/v) bovine serum albumen (BSA) (fraction V, Sigma, No. A2153), 0.1% (v/v) fish skin gelatin (Sigma, No. G-7765) in **PBS**. Filter through 0.45 μm filter, store aliquots at –20 °C.

p-formaldehyde

Stock solution at 7.4% (w/v) in water, either made up fresh, or stored frozen at –20 °C (Doonan and Clayton, 1986).
To make: heat approx. 100 ml of water in a beaker in a microwave till barely hand-holdable, add approx. 20 ml to 1.85 g p-formaldehyde in 25 ml volumetric flask, shake to dissolve (goes in readily but not a clear solution), add a few (approx. 5)

drops of molar NaOH until solution clears, and make up to 25 ml. Check pH and adjust to approx. 7.0 using HCl (but note comments in the chapter by Regan and Sundberg in this volume).

Phosphate-buffered saline (PBS)

Made up as: NaCl 8 g, KCl 0.2 g, Na_2HPO_4 1.15 g, KH_2PO_4 0.2 g, NaN_3 0.2 g. l^{-1} of water, pH should be approx. 7.3–7.6. If the pH differs from this, check that you have used the correct phosphates in making up the solution.

PIPES buffer

Piperazine-N,N'-bis-[2-ethylsulphonic acid] (Sigma, No. P6757).
Stock at 0.2 M, pH 6.9 stored in refrigerator (lasts for many months). To make: add solid to approx. 750 ml of water, check pH (which will be very low), use molar KOH to increase pH. Take pH to approx. 6.5, leave to stir for approx. 30 min, then increase to 6.9, and make up to 1L.
Note – adjusting the pH will require a lot of KOH, so be patient.

Resin curing

For best results, stand the capsules on a piece of UV-transmitting plastic (I use the lid of the **humid box**). Rest this on a shelf in the freezer with the **UV lamp** approx. 10 cm *below* the capsules, and aluminium foil placed loosely *over* the capsules (to help ensure UV light reaches top and sides of capsules).
Note – the blocks generally develop a pink colour when removed from the freezer. This colour disappears after a few hours leaving the blocks colourless. I do not know why the pinkness develops, but the blocks – and the tissues – appear to be unharmed.

Slide-washing

1   Soak slides for 2 h-overnight in a solution of 2% (v/v) 'Micro' detergent in warm water.
2   Rinse slides in running tap water for 5 min.
3   Soak slides for 2 h–overnight in solution of 5% (v/v) acetic acid.
4   Rinse slides in running tap water for 5 min.
5   Rinse slides in deionised water.
6   Polish dry with a tissue, and store ready for reuse.

Steedman's wax

Melt 9 parts of polyethylene glycol 400 distearate at 60 °C, then add 1 part of hexadecanol. Stir until uniform, cool and solidify in an aluminium foil-lined dish (this makes it much easier to remove the solid wax from the dish). Melt freshly only as much wax as is needed for infiltration and embedding. The wax cuts well at 6–10 μm and will form ribbons, hence it is best suited to ordinary – not multi-well – microscope slides. See also Norenburg and Barrett (1987), Brown and Lemmon (1995), and Vitha *et al.* (2000b) for additional comments on this embedding medium for IIF.

Toluidine blue

Used as 0.01% (w/v) in **PBS** to reduce autofluorescence (after Smith and McCully, 1978), and to stain BMM-sections or Steedman's wax-sections in ribbon, to check for quality of fixation, orientation, etc.

UV lamp

2×6 W blacklite-blue UV lamps (TAAB, No. B7920)

'Wax fixative'

Solution of freshly prepared 3.7% (w/v) **p-formaldehyde** in **MTSB** with 10% (v/v) **DMSO**.

TROUBLE-SHOOTING SECTION

It is possible that several 'problems' may be encountered in attempting to reproduce the procedures described above in different species, etc. Only two that I have found most problematic are mentioned here. Where the unexpected is found, a great deal of common sense advice can be found in such publications as Wick (1993), Brown and Lemmon (1995) and Goding (1996).

Polymerisation of BMM

Polymerisation of BMM (and LR White – see the TEM chapter by Chaffey in this volume) is an exothermic reaction (Lulham, 1979; Baskin *et al.*, 1992), and, if too rapid, may result in the formation of gas bubbles in (Bennett *et al.*, 1976; Lulham, 1979; Fig. 10b in Glauert and Young, 1989), and/or incomplete polymerisation of, the resin. To minimise the danger of heat-damage to antigens, the BMM is polymerised at –20 °C. Most success in reducing formation of bubbles has been achieved by placing the UV lamps *beneath* the capsules. Although bubble formation is not entirely eliminated, it is substantially reduced. Incomplete polymerisation of BMM is sometimes found; usually only a few blocks are affected. When it is not so bad that sections can still be cut (do not risk diamond knives for such blocks!), there is a bonus that resin-removal by acetone is quicker.

Section adhesion

Bennett *et al.* (1976) recommend drying-down methacrylate sections in a drop of water on uncoated slides at 60–80 °C. However, when I tried this it resulted in almost total destruction of *α*-tubulin antigenicity. It is likely that antigenicity of other cytoskeletal proteins will be similarly impaired with such an aggressive treatment. However, subjecting the sections to 10 sec at approx. 60 °C (after Baskin *et al.*, 1992), then drying-down overnight at approx. 40 °C promotes section-adhesion, does not appear to compromise antigenicity, and has the advantage of enhancing stretching and flattening of sections (which can otherwise be a problem given the size of section routinely used: 2 × 2 mm).

Adhesive is necessary, however, to ensure section-retention on the slide throughout the numerous washing and staining steps in the IIF protocols. Mayer's egg albumen has proved to be a most satisfactory adhesive for BMM-embedded sections and exhibits negligible background staining. Although I have also used poly-L-lysine for BMM, section-retention was extremely poor. I have not tried gelatin-coated slides, as used by Kobayashi *et al.* (1994) (A. Hardham, personal communication) in their IIF study of flax, but it is likely that sections of woody tissue will be adequately stuck to the slide with this adhesive (as suggested by O'Brien and McCully, 1981). However, ultimately, the choice of adhesive will depend on personal preference and a consideration of any effect it may have on the particular immunolocalisation procedure, especially background staining.

APPLICATIONS

The IIF protocols described in this chapter were developed specifically to study the MT and MF cytoskeleton in the SVS of angiosperm trees. The results obtained by use of IIF of *α*-tubulin and actin in the SVS of *Aesculus hippocastanum* are summarised below (précised from Chaffey *et al.*, 1996, 1997a, b, c, 1998a, 1999, 2000a, b; Chaffey and Barlow, 2000), and illustrated in Figure 2.

MT arrays

Cambium

In active cambia, the MTs of both ray and fusiform cambial cells are randomly arranged, and frequently overlap. Endoplasmic MTs are present in transvacuolar strands of ray cells. During cambial dormancy, the MTs are arranged in a parallel-aligned helical orientation.

When cambial derivatives differentiate, their MTs adopt new arrangements, although at this stage of differentiation the cell walls are unlignified and appear still to be primary.

Axial elements of the secondary xylem

During early stages of *fibre* differentiation, the MTs become progressively more ordered (Figure 2B), and more numerous as wall thickening proceeds, and eventually exhibit a dense, parallel-aligned, helical array (see Figures 1F, H in the 'Wood microscopical techniques' chapter in this volume). The MTs continue to be located parietally in the cytoplasm, but as a single layer, throughout the length of the fibre. No distinct MT arrangements are seen in association with the developing simple pits of this cell type. Microtubules appear to be similarly oriented in *axial parenchyma cells* at early stages of differentiation. At maturity, the MTs are oriented axially in these long-lived cells.

Vessel elements possess MTs showing several distinct arrays:

(i)   rings associated with early stages of bordered pit formation, and the developing aperture (Figure 2F);

(ii) rings associated with development of contact pits;

(iii) linear arrays associated with tertiary-thickening, both parallel to the developing ridge of wall-thickening (Figures 2E, F), and bridging the region between adjacent wall ridges; and

(iv) circles at the periphery of the developing perforations (Figure 2F).

Axial elements of the secondary phloem

The predominant MT array in these cells (axial parenchyma, companion cells, sieve elements and fibre-sclereids) at all stages of development is the helix (or series of parallel helices) (Figure 2A). Variations on this theme occur, e.g. the MTs are dense in developing fibre-sclereids (as for xylem fibres), and the helix becomes more transversely oriented at late stages of sieve element development. The only other MT array so far observed in this tissue is MT-rings apparently at the periphery of pit regions in axial parenchyma cells. It is note-worthy that no MT arrays have been seen in associa-tion with development of sieve areas.

Radial elements of phloem and xylem

At early stages of ray cell differentiation, MTs are random; thereafter, the MT arrangement appears to depend on cell type. In phloem ray cells, where all cells appear alike, the MTs are helically arranged around the long axis of the cell (Figure 2A). In xylem rays, two cell types are recognised. During differentiation, isolation cells tend to have helical or near-transverse MTs, whereas contact cells have rings of MTs at the periphery of developing contact pits during the stage of random MTs. Subsequently the MTs become more transversely oriented. Later in development, both cell types have thick, axially (parallel to the cell's long axis) bundled MTs. This array persists for many months (possibly for years) in these long-lived cells.

MF arrays

Cambium

The MFs in interphase fusiform cambial cells are bundled, frequently branched, and axially arranged, apparently extending from one end of the cell to the other. In ray cambial cells, MFs are randomly arranged and less common than for fusiform cells. These MF arrangements are main-tained throughout the cambial seasonal cycle.

Axial elements of the secondary xylem

At *early stages* of differentiation of all axial sec-ondary xylem cells, MFs retain their net-axial ori-entation. Although the net-axial orientation of MFs is also retained in *later stages* of cell differentiation (Figures 2C, D), additional MF arrangements

occur. In *fibres*, ellipses of MFs are associated with sites of pits. Similar MF arrangements are seen in *axial parenchyma cells*. At maturity, and similar to the MTs, the MFs are oriented axially in these long-lived cells.

Vessel elements possess MFs showing several distinct arrays:

(i)   helically oriented bundles apparently extend-ing the length of the cell;
(ii) rings associated with early stages of bordered pit formation, and the developing aperture;
(iii) circular arrays associated with development of contact pits;
(iv) linear arrays associated with tertiary-thicken-ing, both parallel to the developing ridge of wall-thickening, and bridging the region between adjacent wall ridges;
(v)   a circle at the periphery of the developing perforations; and
(vi) a meshwork at the perforation plate itself.

Axial elements of the secondary phloem

The predominant MF array seen in these cells is the helical-axial arrangement that is common in secondary xylem fibres. Ellipses of MFs are seen in association with development of simple pits in phloem fibres (as for xylem fibres), but no specific arrays of MFs have been seen in con-nection with the developing sieve areas of the sieve elements.

Radial elements of the phloem and xylem

As soon as the ray cell derivatives commence elongation, MF bundles are more readily observed, appearing either axially aligned (paral-lel to the axis of cell elongation), in the xylem, or helically arranged, in the phloem. Additionally, in the xylem, MF rings are seen in association with development of contact pits of contact cells, and ellipses with developing simple pits in iso-lation cells. As for the MTs, axially-oriented MF cables exist in xylem ray cells at maturity, and persist for many months in these long-lived cells.

Unfortunately, space does not permit a detailed consideration of the significance of these observa-tions. For interpretation, readers are directed to the original research papers, or to review articles such

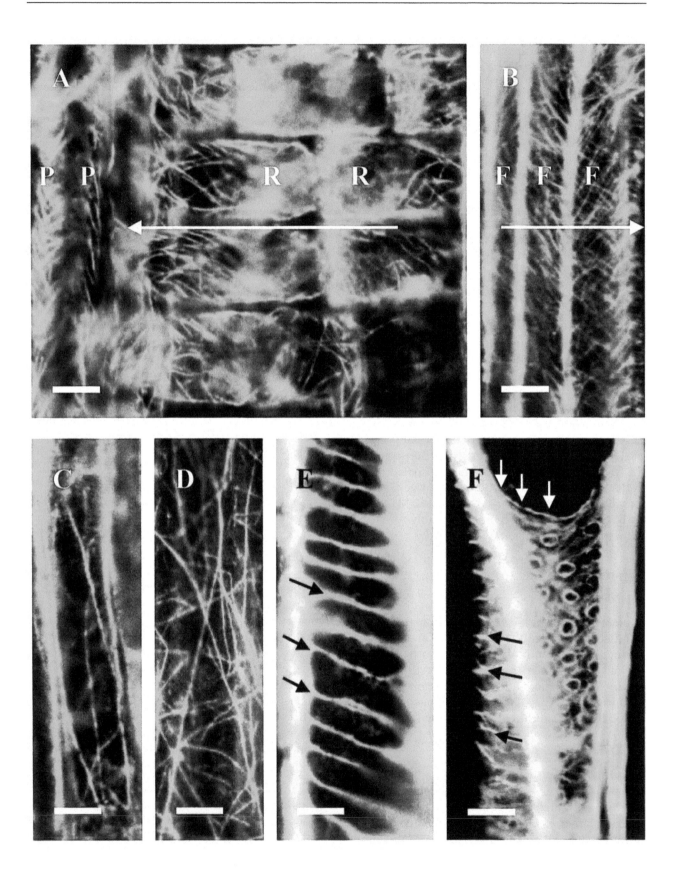

◄

Figure 2 Aspects of the cytoskeleton of the hardwood trees, *Aesculus hippocastanum* (A-C, E, F) and *Populus tremula* x *P. tremuloides* (D). All are epifluorescent images of radial longitudinal sections of stem showing localisation of α-tubulin (for microtubules) (A, B, E, F) or of actin (for microfilaments) (C, D) with fluorescein *iso*thiocyanate (FITC)-conjugated secondary antibody.
A Helical array of microtubules in differentiating secondary phloem ray cells. Note also the helical array of microtubules in the axial phloem elements. Arrow indicates direction of differentiation.
B Helical array of microtubules in differentiating secondary xylem fibres (see also Figure 3C in the transmission electron microscopy chapter by Chaffey in this volume, and cf. the orientation of microfilaments in Figure 2C). Arrow indicates direction of differentiation.
C, D Predominantly axial arrangement of microfilaments in developing secondary xylem fibre (C) (cf. the arrangement of microtubules in Figure 2B) and vessel element (D).
E Bundled microtubules (arrows) associated with tertiary wall thickenings in a vessel element (cf. Figure 2D in the Wood microscopical techniques chapter in this volume).
F Several microtubule arrays in vessel elements: bundles associated with tertiary-thickening (black arrows), ring of microtubules at periphery of developing perforation (white arrows), and rings of microtubules associated with developing bordered pits (cf. Figure 2D in the Wood microscopical techniques chapter in this volume).
Key: F, fibre, P, phloem cell, R, ray cell. Scale bars represent: 25 μm (A, B, D-F) and 30 μm (C).

Table 1 Some angiosperm tree species in which cytoskeletal features have been detected using indirect immunofluorescent localisation in butyl-methylmethacrylate-embedded material^.

Taxon	MT/F	Sec. xy	CZ*	Sec. ph.
Acer platanoides	MT	F[h]		S[h],P[h]
Salix viminalis	MT	F[h], V	f[ao],r[ao]	F[h],S[h],C[h],P[h],R[h]
S. viminalis	MF		f[l]	
S. burjatica x *S. viminalis*	MT	F[h], V	f[ao],r[ao]	F[h],S[h],C[h],P[h],R[h]
S. burjatica x *S. viminalis*	MF	F[l]	f[l]	
Liriodendron tulipifera	MT	F[h], V[hs]		F[h],S[h]
L. tulipifera	MF	F[l]	f[l]	
Hedera helix	MT	F[h], V[p]		A[h]
H. helix	MF	F[l]	f[l]	
Quercus ilex	MT	F[hp]	f	
Q. ilex	MF	F[l]	f[l]	
Platanus sp	MT	V	f[o],r[o]	F[h],A[h]
Platanus sp.	MF	F[l]	f[l],r	

Key: ^, omission of mention of a cell type merely indicates that the particular cytoskeletal component has not yet been visualised therein; MT/F, microtubule/microfilament; Sec. xy, secondary xylem; CZ, cambial zone; *, cortical cytoskeletal element unless specified otherwise; Sec. ph., secondary phloem; F, fibre; V, vessel element; P, axial parenchyma; R, ray parenchyma; f, fusiform cambial cell; r, ray cambial cell; S, sieve tube member; C, companion cell; A, unidentified axial element; [a], cortical and endoplasmic elements; [h], helically oriented cortical microtubules; [l], axially oriented cortical microfilament bundles; [o], randomly oriented cortical microtubules; [p], associated with developing pits; [s], associated with developing perforation plate.

as Chaffey (1999a, 2000) and Chaffey and Barlow (2000).

Other hardwood trees

The BMM-embedding and IIF technique described above has been used successfully on several tree species other than the *Aesculus* and *Populus tremula* × *tremuloides* for which it was developed. Cytoskeletal arrangements in *P. tremula* × *tremuloides*, with the exception of those relating to tertiary thickenings (which are not present in

this taxon), appear similar to those found in *Aesculus*. Table 1 summarises the use of IIF of MTs and MFs in the other angiosperm trees I have studied to date, and indicates the utility of this technique for study of the cytoskeleton in the hardwood SVS.

Comparison with softwood trees

Although I have concentrated on angiosperm trees here, gymnosperms also have a cytoskeleton (see the chapter by Funada in this volume).

Appreciation of the softwood cytoskeleton is useful to a better understanding of that in hardwood angiosperms, and vice versa. A detailed comparison of the softwood and hardwood cytoskeletons is beyond the scope of this chapter, but for completeness, the work of Funada and co-workers on gymnosperms (see the chapter by Funada in this volume), can be summarised thus:

- Both MTs and MFs have been immunolocalised in cambial and secondary xylem cells of a range of gymnosperm trees.
- Cortical MTs (CMTs) are randomly-arranged in both ray and fusiform cambial initials.
- CMTs change orientation to longitudinal, then to transverse as the tracheids expand.
- CMTs are oriented transversely in compression wood tracheids at the early stages of secondary wall formation, subsequently becoming obliquely oriented in a Z-helix.
- CMTs change orientation to transverse as ray cells expand, eventually becoming oblique/longitudinal at late stages of development.
- CMTs increase in density as secondary walls form in both tracheids and ray cells.
- During secondary wall formation in tracheids, the orientation of CMTs changes rotatively from a flat S-helix to a step Z-helix (and is paralleled by similar changes in orientation of wall microfibrils).
- CMT-free regions appear at sites of bordered pit formation in tracheids, a circle of CMTs delimits the periphery of the bordered pits, and decreases in diameter as the pit develops.
- Bands of CMTs are associated with spiral-thickening in gymnosperms that possess this feature.
- MFs are bundled and axially-oriented in fusiform cambial cells.
- During tracheid differentiation, the axial MF rientation either does not change, even though that of the CMTs does, or it may be transverse or oblique at the final stage of differentiation.

FUTURE PROSPECTS

As with all branches of tree cell biology, the limits to what is possible in the future are determined largely by the imagination of the investigator (although constrained by the available technology). However, without wishing to extend too far beyond the

horizon of the next few years, some lines of investigation that immediately suggest themselves, in the context of the role of the cytoskeleton in wood formation, are outlined below.

Immunolocalisation of other cytoskeletal proteins, and their inter-relationships

The procedures detailed above for BMM work well for MTs and MFs in the angiosperm – and gymnosperm – tree SVS. They also allow visualisation of β-tubulin, post-translationally modified α-tubulins, and myosin (Chaffey and Barlow, unpublished observations). It is likely that they will also work for other cytoskeletal proteins, such as IFs, MTOCs, MAPs and MF-associated proteins; they just haven't been tried yet. No cell component exists alone; very few – if any – act alone. Just as we are beginning to understand some of the interactions between MTs and MFs in wood formation (summarised above), so it is highly likely that other interactions will be found between these known members of the plant cytoskeleton and the others which are as yet unknown, and with other cell organelles (such as the plasmalemma – Lloyd et al., 1996).

Given the evidence for involvement of MTs and MFs in cell wall formation/modification during wood formation, there is a strong case for existence of a 'cell wall-plasmalemma-cytoskeleton continuum' (e.g. Wyatt and Carpita, 1993; Cyr 1994; Miller et al., 1997a, b) in these cells (Chaffey et al., 1998b). It will be instructive to examine the plasmalemma for further evidence of the domains which are postulated to exist in connection with development of the various types of pits and perforations (e.g. Chaffey et al., 1997b), particularly in view of the several different cytoskeletal arrays which can exist on the same cell face at the same time (e.g. Chaffey et al., 1997c). In that connection, cytoskeletal proteins that come to mind are MAPs, MF-associated proteins, and γ-tubulin associated with MTOCs.

In view of the way plant cytoskeletal work lags behind that in animals, there is an enormous range of 'animal antibodies' available commercially which can be used to probe the wood cells, to give clues to the variety of cytoskeletal components which may exist in the plant system. Of course, there is no substitute for having plant-derived antibodies directed against the cytoskeletal proteins of interest. However, until that time, and bearing in mind that practically all the published work on plant MTs and MFs uses animal-directed antibodies, there is still a lot that can usefully be done using such antibodies.

Correlation of IIF localisation with TEM localisation techniques

The IIF techniques described here for cytoskeletal proteins are good for providing a relatively low resolution picture of the distribution and arrangement of these proteins within and between cells of the tree SVS. However, it is essential to obtain information at the fine-structural level of resolution to obtain the fuller picture (see e.g. the TEM chapter by Chaffey in this volume). To that end it is necessary to develop immunolocalisation techniques for the cytoskeleton at the TEM level. Although techniques exist for this sort of study in cells and tissues of the primary plant body (e.g. Hawes, 1988), attempts to develop immunogold localisation of the cytoskeleton in the tree SVS have so far been largely unsuccessful. However, it is probably only a question of persistence, and I am optimistic that within the next few years such procedures will be 'routine' in trees and the requisite ultrastructural resolution achieved.

New model systems

To date, cytoskeletal work in the tree SVS has only been performed on fixed material (see also the chapter by Funada in this volume). This has largely been necessitated by the difficulties in working with trees (see the chapter 'An introduction to the problems of working with trees' in this volume). Although examination of static images from sequential cells along a radial file gives insight into the development of an individual cell type, it is no substitute for examining 'the real thing'. To that end, it will be extremely instructive to investigate the cytoskeleton – and other aspects of wood cell biology – *in vivo*. Techniques are being developed for studying the cytoskeleton in this way in other plant cells (e.g. Kost *et al.*, 1998; Himmelspach *et al.*, 1999). Unfortunately, those techniques are largely designed for easily accessible cells and tissues. To apply such procedures to the tree SVS will be a considerable challenge.

However, one relatively old approach to tree cell biology, which has recently made a welcome comeback, is the use of *in vitro* wood culture

(e.g. Leitch and Savidge, 2000). Whilst it must be borne in mind that such a system is not completely natural, it is a good compromise between the extremes of the intact tree SVS and cultured *Zinnia* mesophyll cells (e.g. Chaffey, 1999a). Exploitation of this system seems to be an obvious next step from the standpoint of wood cell biology and should be amenable to a wide range of non-invasive imaging techniques in a way that the tree is not. The opportunity to witness cytoskeletal dynamics in living cells is an extremely exciting prospect, and the possibility of manipulating the cultural conditions thus permitting hypothesis-testing, is an attractive proposition. With such an approach it should be possible to investigate whether 'designer wood' can be created by manipulation of the cytoskeleton, as recently proposed (Chaffey, 2000).

Another major step forward is the adoption of *Populus* as a model hardwood species for study of all aspects of tree biology (see Chaffey, 1999a, 2001; and the chapter entitled 'An introduction to the problems of working with trees' in this volume). My current cytoskeletal work is almost exclusively on *Populus*. Concentration on poplar is also evident in the work of Hawkins *et al.*, Itoh, Micheli *et al.*, and Regan and Sundberg (see their chapters in this volume). Such effort directed at a single 'model' is likely to advance our understanding of hardwood tree biology in a way similar to that achieved in the model herb *Arabidopsis* (e.g. Pyke, 1994; Somerville 2000). And even we 'tree people' must not be completely blind to the advantages of this ephemeral crucifer. Recent work on secondary growth in *Arabidopsis* suggests that it might be useful as a model for the same process in trees (e.g. Regan *et al.*, 1999b). Although it is extremely unlikely that such work will replace that on the tree SVS, it may well prove a useful system for *developing* hypotheses of wood formation, which can then be *tested* in the tree (e.g. Chaffey, 1999b, 2001).

Whilst acknowledging that there can be problems in devoting 'excess' effort to one or a few species, such a focused approach to tree cell biology should enable great advances in knowledge in a relatively short period of time. However, we must not forget that there are many different tree-types, such as storied vs non-storied; rayed vs rayless; diffuse-porous vs ring-porous, temperate vs tropical, softwood vs hardwood, all of which need to be understood.

Technological advances

Technological improvements continue to be made. One that is happening already is the reduction in price of the CLSM. As that technology becomes more widespread, we will probably see a marked decline in epifluorescent studies. It is worth mentioning that resin-free BMM-sections of the tree SVS processed as described above, are suitable for CLSM viewing (Chaffey *et al.*, 1997a, 1998a, 1999, 2000a, b; Chaffey, 2000; Chaffey and Barlow, 2000; see also Figures 1E, F, H in the 'Wood microscopical techniques' chapter in this volume). Innovations in fluorochrome technology continue to be announced, e.g. the Alexa dyes (Kumar *et al.*, 1999; Panchuk-Voloshina *et al.*, 1999), which permit better IIF studies. And it can be widely expected that plant-specific antibodies will continue to be developed, allowing greater specificity – and certainty – in IIF studies on plants.

ACKNOWLEDGEMENTS

I thank Peter Barlow for sparking my interest in the plant cytoskeleton (and for his comments on an earlier draft of this chapter) and John Barnett for introducing me to the cambium, Jill Parker for initiation into Steedman's Wax-embedding, and Sharon Regan for (finally!) persuading me to try freeze-fixation of woody samples.

REFERENCES

Publications of interest with reference to immunocytochemistry:
Newman and Hobot (1993); Goding (1996) [particularly recommended, especially chapters 12 and 13]; Polak and van Noorden (1997); Javois (1999).

Publications of general interest with reference to plant immunocytochemistry:
Knox (1982); Knox and Singh (1985); Roberts (1986); Hawes (1988); Vigil and Hawes (1989) [this is the book version of the special issue of *Cell Biology International Reports* volume 13, issue 1, 1989, and includes the articles by Gubler (1989) and Wick *et al.* (1989)]; Singh *et al.* (1993); Harris (1994); Brown and Lemmon (1995).

Articles that provide some background to the use of antibodies in plant science:
Knox and Butcher (1988); Dewey *et al.* (1991).

Andersland, J.M., Fisher, D.D., Wymer, C.L., Cyr, R.J. and Parthasarathy, M.V. (1994) Characterization of a monoclonal antibody prepared against plant actin. *Cell Motility and Cytoskeleton*, **29**, 339–344.

Baluška, F., Parker, J.S. and Barlow, P.W. (1992) Specific patterns of cortical and endoplasmic microtubules associated with cell growth and tissue differentiation in roots of maize (*Zea mays* L.). *Journal of Cell Science*, **103**, 191–200.

Baluška, F., Volkmann, D. and Barlow, P.W. (1999) Hormone-cytoskeleton interactions in plant cells. In *Biochemistry and Molecular Biology of Plant Hormones*, edited by P.J.J. Hooykaas, M.A. Hall and K.R. Libbenga, pp. 363–390. Amsterdam, New York, Oxford: Elsevier.

Barlow, P.W. and Baluška, F. (2000) Cytoskeletal perspectives on root growth and morphogenesis. *Annual Review of Plant Physiology and Plant Molecular Biology* **51**, 289–322.

Baskin, T.I., Busby, C.H., Fowke, L.C., Sammut, M. and Gubler, F. (1992) Improvements in immunostaining samples embedded in methacrylate: localization of microtubules and other antigens throughout developing organs in plants of diverse taxa. *Planta*, **187**, 405–413.

Baskin, T.I., Miller, D.D., Vos, J.W., Wilson, J.E. and Hepler, P.K. (1996). Cryofixing single cells and multicellular specimens enhances structure and immunocytochemistry for light microscopy. *Journal of Microscopy*, **182**, 149–161.

Bennett, H.S., Wyrick, A.D., Lee, S.W. and McNeil, J.H. (1976) Science and art in preparing tissues embedded in plastic for light microscopy, with special reference to glycol methacrylate, glass knives and simple stains. *Stain Technology*, **51**, 71–97.

Berlyn, G.P. (1963) Methacrylate as an embedding medium for woody tissues. *Stain Technology*, **38**, 23–28.

Bibikova, T.N., Blancaflour, E.B. and Gilroy, S. (1999) Microtubules regulate tip growth and orientation of root hairs of *Arabidopsis thaliana*. *Plant Journal*, **17**, 657–665.

Boevink, P., Oparka, K., Cruz, S.S., Martin, B., Betteridge, A. and Hawes, C. (1998) Stacks on tracks: the plant Golgi apparatus traffics on an actin/ER network. *Plant Journal*, **15**, 441–447.

Bourrett, T.M., Czymmek, K.J. and Howard, R.J. (1999) Ultrastructure of chloroplast protuberances in rice leaves preserved by high-pressure freezing. *Planta*, **208**, 472–479.

Brown, R.C. and Lemmon, B.E. (1995) Methods in plant immunolight microscopy. *Methods in Cell Biology*, **49**, 85–107.

Brown, R.C., Lemmon, B.E. and Mullinax, J.B. (1989) Immunofluorescent staining of microtubules in plant tissues: improved embedding and sectioning techniques using polyethylene glycol (PEG) and Steedman's wax. *Botanica Acta*, **102**, 54–61.

Burns, R.G. and Surridge, C.D. (1994) Tubulin: conservation and structure. In *Microtubules*, edited by J.S. Hyams and C.W. Lloyd, pp. 3–31. New York, Toronto: Wiley-Liss, Inc.

Burry, R.W. (2000) Specificity controls for immunocytochemical methods. *Journal of Histochemistry and Cytochemistry*, **48**, 163–165.

Chaffey, N.J. (1994) Structure and function of the membranous grass ligule: a comparative study. *Botanical Journal of the Linnean Society*, **116**, 53–69.

Chaffey, N.J. (1996) Structure and function of the root cap of *Lolium temulentum* L. (Poaceae): parallels with the ligule. *Annals of Botany*, **78**, 3–13.

Chaffey, N.J. (1999a) Cambium: old challenges – new opportunities. *Trees*, **13**, 138–151.

Chaffey, N.J. (1999b) Wood formation in forest trees: from *Arabidopsis* to *Zinnia*. *Trends in Plant Science*, **4**, 203–204.

Chaffey, N.J. (2000) Cytoskeleton, cell walls and cambium: new insights into secondary xylem differentiation. In *Cell and Molecular Biology of Wood Formation*, edited by R. Savidge, J. Barnett and R. Napier, pp. 31–42. Oxford: Bios Scientific Publishers.

Chaffey, N.J. (2001) Cambial cell biology comes of age. In *Trends in European Forest Tree Physiology Research*, edited by S. Huttunen, J. Bucher, P. Jarvis, R. Matyssek and B. Sundberg. Kluwer Academic Press (in press).

Chaffey, N.J. and Barlow, P.W. (2000) Actin in the secondary vascular system of woody plants. In *Actin: a Dynamic Framework for Multiple Plant Cell Functions*, edited by F. Baluška, P.W. Barlow, C. Staiger and D. Volkmann, pp. 587–600. Dordrecht, The Netherlands: Kluwer Academic Press.

Chaffey, N.J., Barlow, P.W. and Barnett, J.R. (1996) Microtubular cytoskeleton of vascular cambium and its derivatives in roots of *Aesculus hippocastanum* L. (Hippocastanaceae). In *Recent Advances in Wood Anatomy*, edited by L.A. Donaldson, B.G. Butterfield, P.A. Singh and L.J. Whitehouse, pp. 171–183. Rotorua, NZ: New Zealand Forest Research Institute.

Chaffey, N.J., Barlow, P.W. and Barnett, J.R. (1997a) Microtubules rearrange during differentiation of vascular cambial derivatives, microfilaments do not. *Trees*, **11**, 333–341.

Chaffey, N.J., Barlow, P.W. and Barnett, J.R. (1997b) Formation of bordered pits in secondary xylem vessel elements of *Aesculus hippocastanum* L.: an electron and immunofluorescent microscope study. *Protoplasma*, **197**, 64–75.

Chaffey, N.J., Barlow, P.W. and Barnett, J.R. (1998a) A seasonal cycle of cell wall structure is accompanied by a cyclical rearrangement of cortical microtubules in fusiform cambial cells within taproots of *Aesculus hippocastanum* L. (Hippocastanaceae). *New Phytologist*, **139**, 623–635.

Chaffey, N.J., Barlow, P.W. and Barnett, J.R. (2000a) Structure-function relationships during secondary phloem development in *Aesculus hippocastanum*: microtubules and cell walls. *Tree Physiology* **20**, 777–786.

Chaffey, N.J., Barlow, P.W. and Barnett, J.R. (2000b) A cytoskeletal basis for wood formation in angiosperm trees: the involvement of microfilaments. *Planta* **210**, 890–896.

Chaffey, N.J., Barlow, P.W., Barnett, J.R. and Sundberg, B. (1998b) The cell wall-plasmalemma-cytoskeleton continuum: its role in secondary vascular development in trees. *Abstracts of the 8th International Cell Wall Meeting at the John Innes Centre, 1–5 September, 1998*.

Chaffey, N.J., Barnett, J.R. and Barlow, P.W. (1999) A cytoskeletal basis for wood formation in angiosperm trees: the involvement of cortical microtubules. *Planta*, **208**, 19–30.

Chaffey, N.J., Barnett, J.R. and Barlow, P.W. (1997c) Visualization of the cytoskeleton within the secondary vascular system of hardwood species. *Journal of Microscopy*, **187**, 77–84.

Chu, B., Wilson, T.J., McCune-Zierath, C., Snustad, P.D. and Carter, J.V. (1998) Two β-tubulin genes, TUB1 and TUB8, of *Arabidopsis* exhibit largely nonoverlapping patterns of expression. *Plant Molecular Biology*, **37**, 785–790.

Clark, G. (ed.) (1981) *Staining procedures*, 4th edn. Baltimore, London: Williams and Wilkins.

Clayton, L. (1985) The cytoskeleton and the plant cell cycle. In *The Cell Division Cycle in Plants*, edited by J.A. Bryant and D. Francis, pp. 113–131. Cambridge, London, New York: Cambridge University Press.

Conn, P.M. (ed.) (1999) *Methods in Enzymology, Volume 302, Green Fluorescent Protein*. New York, London: Academic Press.

Cross, A.R. and Williams, R.C., Jr (1991) Kinky microtubles: bending and breaking induced by fixation *in vitro* with glutaraldehyde and formaldehyde. *Cell Motility and the Cytoskeleton*, **20**, 272–278.

Cyr, R.J. (1991) Microtubule-associated proteins in higher plants. In *The Cytoskeletal Basis of Plant Growth and Form*, edited by C.W. Lloyd, pp. 57–67. London: Academic Press.

Cyr, R.J. (1994) Microtubules in plant morphogenesis: role of the cortical array. *Annual Review of Cell Biology*, **10**, 153–180.

Davies, E., Fillingham, B.D. and Abe, S. (1996) The plant cytoskeleton. In *Cytoskeleton in Specialized Tissues and in Pathological States, Volume 3*, edited by J.E. Hesketh and I.F. Pryme, pp. 405–449. Greenwich, CT, London: JAI Press Inc.

Derksen, J., Wilms, F.H.A. and Pierson, E.S. (1990) The plant cytoskeleton: its significance in plant development. *Acta Botanica Neerlandica*, **39**, 1–18.

de Ruijter, N.C.A. and Emons, A.M.C. (1999) Actin-bundling proteins in plant cells. *Plant Biology*, **1**, 26–35.

Dewey, M., Evans, D., Coleman, J., Priestley, R., Hull, R., Horsley, D. and Hawes, C. (1991) Antibodies in plant science. *Acta Botanica Neerlandica*, **40**, 1–27.

Doonan, J.H. and Clayton, L. (1986) Immunofluorescent studies on the plant cytoskeleton. In *Immunology in Plant Science*, edited by T.L. Wang, pp. 111–136. Cambridge, London, New York: Cambridge University Press.

Emons, A.M.C., Derksen, J. and Sassen, M.M.A. (1992) Do microtubules orient plant cell wall microfibrils? *Physiologia Plantarum*, **84**, 486–493.

Emons, A.M.C. and Mulder, B.M. (2000) How the deposition of cellulose microfibrils builds cell wall architecture. *Trends in Plant Science*, **5**, 35–40.

Emons, A.M.C., Pierson, E. and Derksen, J. (1991) Cytoskeleton and intracellular movements in plant cells. In *Biotechnology – Current Progress, Volume 1*, edited by P.N. Cheremisinoff and L.M. Ferrante, pp. 311–335. Lancaster, Basel: Technomic Publishing Co. Inc.

Eun, S.-O. and Wick, S.M. (1998) Tubulin isoform usage in maize microtubules. *Protoplasma*, **204**, 235–244.

Fischer, K. and Schopfer, P. (1997) Interaction of auxin, light, and mechanical stress in orienting microtubules in relation to tropic curvature in the epidermis of maize coleoptiles. *Protoplasma*, **196**, 108–116.

Fischer, K. and Schopfer, P. (1998) Physical strain-mediated microtubule reorientation in the epidermis of gravitropically or phototropically stimulated maize coleoptiles. *Plant Journal*, **15**, 119–123.

Fisher, D.D. and Cyr, R.J. (1998) Extending the microtubule/microfibril paradigm: cellulose synthesis is required for normal cortical microtubule alignment in elongating cells. *Plant Physiology*, **116**, 1043–1051.

Fosket, D.E. (1989) Cytoskeletal proteins and their genes in higher plants. In *The Biochemistry of Plants, Volume 15, Molecular Biology*, edited by A. Marcus, pp. 393–454. San Diego, London: Academic Press.

Fosket, D.E. and Morejohn, L.C. (1992) Structural and functional organization of tubulin. *Annual Review of Plant Physiology and Plant Molecular Biology*, **43**, 201–240.

Fosket, D.E., Tonoike, H., Han, I.-S. and Colón, A. (1993) What is the significance of the relatively large tubulin multigene families for plant morphogenesis? In *Morphogenesis in Plants*, edited by K.A. Roubelakis-Angelakis and K. Tran Thanh Van, pp. 55–87. New York: Plenum Press.

Franklin, A.E. and Cande, W.Z. (1999) Nuclear organization and chromosome segregation. *The Plant Cell*, **11**, 523–534.

Funada, R., Furusawa, O., Shibagaki, M., Miura, H., Miura, T., Abe, H. and Ohtani, J. (2000) The role of cytoskeleton in secondary xylem differentiation in conifers. In *Cell and Molecular Biology of Wood Formation*, edited by R. Savidge, J. Barnett and R. Napier, pp. 254–264. Oxford: Bios Scientific Publishers.

Giddings, Jr, T.H. and Staehelin, L.A. (1991) Microtubule-mediated control of microfibril deposition: a re-examination of the hypothesis. In *The Cytoskeletal Basis of Plant Growth and Form*, edited by C.W. Lloyd, pp. 85–99. London: Academic Press.

Gilmer, S., Clay, P., MacRae, T.H. and Fowke, L.C. (1999) Acetylated tubulin is found in all microtubule arrays of two species of pine. *Protoplasma*, **207**, 174–185.

Gilroy, S. (1997) Fluorescence microscopy of living plant cells. *Annual Review of Plant Physiology and Plant Molecular Biology*, **48**, 165–190.

Glauert, A. and Young, R.D. (1989) The control of temperature during polymerization of Lowicryl K4M: there *is* a low-temperature embedding method. *Journal of Microscopy*, **154**, 101–113.

Goddard, R.H., Villemur, R., Silflow, C.D. and Wick, S.M. (1998) Generation of chicken polyclonal antibodies against distinct maize tubulins. *Protoplasma*, **204**, 226–234.

Goddard, R.H., Wick, S.M., Silflow, C.D. and Snustad, D.P. (1994) Microtubule components of the plant cell cytoskeleton. *Plant Physiology*, **104**, 1–6.

Goding, J.W. (1996) *Monoclonal Antibodies: Principles and Practice*, 3rd edn. London, New York, Tokyo: Academic Press.

Goodbody, K.C. and Lloyd, C.W. (1994) Immunofluorescence techniques for analysis of the cytoskeleton. In *Plant Cell Biology: a Practical Approach*, edited by N. Harris and K.J. Oparka, pp. 221–243. Oxford: Oxford University Press.

Granger, C.L. and Cyr, R.J. (2000) Microtubule reorganization in tobacco BY-2 cells stably expressing GFP-MBD. *Planta*, **210**, 502–509.

Green, P.B. (1969) Cell morphogenesis. *Annual Review of Plant Physiology*, **20**, 365–394.

Gubler, F. (1989) Immunofluorescence localisation of microtubules in plant root tips embedded in butyl-methyl methacrylate. *Cell Biology International Reports*, **13**, 137–145.

Harris, N. (1994) Immunocytochemistry for light and electron microscopy. In *Plant Cell Biology: a Practical Approach*, edited by N. Harris and K.J. Oparka, pp. 157–176. Oxford: Oxford University Press.

Hawes, C. (1988) Subcellular localization of macromolecules by microscopy. In *Plant Molecular Biology: a Practical Approach*, edited by C.H. Shaw, pp. 103–130. Oxford: Oxford University Press.

Hepler, P.K. (1985) The plant cytoskeleton. In *Botanical Microscopy*, edited by A.W. Robards, pp. 233–262. Oxford: Oxford University Press.

Hepler, P.K., Cleary, A.L., Gunning, B.E.S., Wadsworth, P., Wasteneys, G.O. and Zhang, D.H. (1993) Cytoskeletal dynamics in living plant cells. *Cell Biology International*, **17**, 127–142.

Hepler, P.K. and Gunning, B.E.S. (1998) Confocal fluorescence microscopy of plant cells. *Protoplasma*, **201**, 121–157.

Hepler, P.K. and Hush, J.M. (1996) Behavior of microtubules in living plant cells. *Plant Physiology*, **112**, 455–461.

Herman, B. (1998) *Fluorescence Microscopy*, 2nd edn. Oxford: Bios Scientific Publishers.

Himmelspach, R., Wymer, C.L., Lloyd, C.W. and Nick, P. (1999) Gravity-induced reorientation of cortical microtubules observed *in vivo*. *Plant Journal*, **18**, 449–453.

Hush, J.M., Hawes, C.R. and Overall, R.L. (1994a) A novel method for the visualization of microtubules in plant tissues. *Proceedings of the Linnean Society of New South Wales* **114**, 11–19.

Hush, J.M. and Overall, R.L. (1996) Cortical microtubule reorientation in higher plants: dynamics and regulation. *Journal of Microscopy*, **181**, 129–139.

Hush, J.M., Wadsworth, P., Callaham, D.A. and Hepler, P.K. (1994b) Quantification of microtubule dynamics in living plant cells using fluorescence redistribution after photobleaching. *Journal of Cell Science*, **107**, 775–784.

Hussey, P.J., Lloyd, C.W. and Gull, K. (1988) Differential and developmental expression of β-tubulins in a higher plant. *Journal of Biological Chemistry*, **263**, 5474–5479.

Hussey, P.J., Snustad, D.P. and Silflow, C.D. (1991) Tubulin gene expression in higher plants. In *The Cytoskeleton in Plant Growth and Development*, edited by C.W. Lloyd, pp. 15–27. London: Academic Press.

Ishiko, A., Shimizu, H., Masunaga, T., Kurihara, Y. and Nishikawa, T. (1998) Detection of antigens by immunofluorescence on ultrathin cryosections of skin. *Journal of Histochemistry & Cytochemistry*, **46**, 1455–1460.

Janßen, M., Hunte, C., Schulz, M. and Schnabl, H. (1996) Tissue specification and intracellular distribution of actin isoforms in *Vicia faba* L. *Protoplasma*, **191**, 158–163.

Javois, L. (ed.) (1999) *Immunocytochemical Methods and Protocols*. Totowa, New Jersey: Humana Press.

Johnson, G.D. and Araujo, G.M. deC.N. (1981) A simple method of reducing the fading of immunofluorescence during microscopy. *Journal of Immunological Methods*, **43**, 349–350.

Johnson, G.D., Davidson, R.S., McNamee, K.C., Russell, G., Goodwin, D. and Holborrow, E.J. (1982) Fading of immunofluorescence during microscopy: a study of the phenomenon and its remedy. *Journal of Immunological Methods*, **55**, 231–242.

Kandasamy, M., McKinney, E. and Meagher, R.B. (1999) The late pollen-specific actins in angiosperms. *Plant Journal*, **18**, 681–691.

Kiss, J.Z., Giddings, T.H. Jr, Staehelin, L.A. and Sack, F.D. (1990) Comparison of the ultrastructure of conventionally fixed and high pressure frozen/freeze substituted root tips of *Nicotiana* and *Arabidopsis*. *Protoplasma*, **157**, 64–74.

Knox, P. and Butcher, G. (1988) Using antibodies in plant science. *Plants Today*, **Nov–Dec**, 200–205.

Knox, R.B. (1982) Methods for locating and identifying antigens in plant tissues. In *Techniques in Immunocytochemistry*, edited by G.R. Bullock and P. Petrusz, pp. 205–238. London, New York, Paris: Academic Press.

Knox, R.B. and Singh, M.B. (1985) Immunofluorescence applications in plant cells. In *Botanical Microscopy*, edited by A.W. Robards, pp. 205–232. Oxford: Oxford University Press.

Kobayashi, I., Kobayashi, Y. and Hardham, A.R. (1994) Dynamic reorganization of microtubules and microfilaments in flax cells during the resistance response to flax rust infection. *Planta*, **195**, 237–247.

Koropp, K. and Volkmann, D. (1994) Monoclonal antibody CRA against a fraction of actin from cress roots recognizes its antigen in different plant species. *European Journal of Cell Biology*, **64**, 153–162.

Kost, B., Mathur, J. and Chua, N.-H. (1999) Cytoskeleton in plant development. *Current Opinion in Plant Biology*, **2**, 462–470.

Kost, B., Spielhofer, P. and Chua, N.-H. (1998) A GFP-mouse talin fusion protein labels plant actin filaments *in vivo* and visualizes the actin cytoskeleton in growing pollen tubes. *Plant Journal*, **16**, 393–401.

Kristen, U., Lockhausen, J., Menhardt, W. and Dallas, W.J. (1989) Computer-generated three-dimensional representation of dictyosomes reconstructed from serial ultrathin sections. *Journal of Cell Science*, **93**, 385–389.

Kumar, R.K., Chapple, C.C. and Hunter, N. (1999) Improved double immunofluorescence for confocal laser scanning microscopy. *Journal of Histochemistry & Cytochemistry*, **47**, 1213–1217.

Lambert, A.-M. (1993) Microtubule-organizing centers in higher plants. *Current Opinion in Cell Biology*, **5**, 116–122.

Lambert, A.-M. (1995). Microtubule-organizing centers in higher plants: evolving concepts. *Botanica Acta*, **108**, 535–537.

Lambert, A.M. and Lloyd, C.W. (1994) The higher plant microtubule cycle. In *Microtubules*, edited by J.S. Hyams and C.W. Lloyd, pp. 325–341. New York, Toronto: Wiley-Liss, Inc.

Ledbetter, M.C. and Porter, K.R. (1963) A "microtubule" in plant cell fine structure. *Journal of Cell Biology*, **19**, 239–250.

Leitch, M.A. and Savidge, R.A. (2000) Tissue culture for the study of cambial activity and wood formation – a resurgence of interest in an old technique. In *Cell and Molecular Biology of Wood Formation*, edited by R. Savidge, J. Barnatt and R. Napier, pp. 493–512. Oxford: Bios Scientific Publishers.

Li, Y.-Q., Moscatelli, A., Cai, G. and Cresti, M. (1997) Functional interactions among cytoskeleton, membranes, and cell wall in the pollen tube of flowering plants. *International Review of Cytology*, **176**, 133–199.

Liu, B., Cyr, R.J. and Palevitz, B.A. (1996) A kinesin-like protein, KatAp, in the cells of *Arabidopsis* and other plants. *Plant Cell*, **8**, 119–132.

Lloyd, C.W. (ed.) (1982) *The Cytoskeleton in Plant Growth and Development*. London: Academic Press.

Lloyd, C.W. (1986) Microtubules and the cellular morphogenesis of plants. In *Developmental Biology: a Comprehensive Synthesis, Volume 2, the Cellular Basis of Morphogenesis*, edited by L.W. Browder, pp. 31–57. New York, London: Plenum Press.

Lloyd, C.W. (1987) The plant cytoskeleton: the impact of fluorescence microscopy. *Annual Review of Plant Physiology*, **38**, 119–139.

Lloyd, C.W. (ed.) (1991) *The Cytoskeletal Basis of Plant Growth and Form*. London: Academic Press.

Lloyd, C.W. (1994) Why should stationary plant cells have such dynamic microtubules? *Molecular Biology of the Cell*, **5**, 1277–1280.

Lloyd, C.W., Drøbak, B.K., Dove, S.K. and Staiger, C.J. (1996) Interactions between the plasma membrane and the cytoskeleton in plants. In *Membranes: Specialized Functions in Plants*, edited by M. Smallwood, J.P. Knox and D.J. Bowles, pp. 1–20. Oxford: Bios Scientific Publishers.

Lloyd, C.W., Slabas, A.R., Powell, A.J., MacDonald, G. and Badley, R.A. (1979) Cytoplasmic microtubules of higher plant cells visualized with anti-tubulin antibodies. *Nature (London)*, **279**, 239–241.

Longin, A., Souchier, C., Ffrench, M. and Bryon, P. (1993) Comparison of anti-fading agents used in fluorescence microscopy: image analysis and laser confocal microscopy study. *Journal of Histochemistry & Cytochemistry*, **41**, 1833–1840.

Ludin, B. and Matus, A. (1998) GFP illuminates the cytoskeleton. *Trends in Cell Biology*, **8**, 72–77.

Lulham, C.N. (1979) Glycol methacrylate embedding for light microscopy. *Journal of Histotechnology*, **2**, 68–71.

Marc, J. (1997) Microtubule-organizing centres in plants. *Trends in Plant Science*, **2**, 223–230.

Marc, J., Granger, C.L., Brincat, J., Fisher, D.D., Kao, T.-h., McCubbin, A.G. and Cyr, R.J. (1998) A GFP-MAP4 reporter gene for visualizing cortical microtubule rearrangements in living epidermal cells. *Plant Cell*, **10**, 1927–1939.

McCurdy, D.W. and Williamson, R.E. (1991) Actin and actin-associated proteins. *The Cytoskeletal Basis of Plant Growth and Form*, edited by C.W. Lloyd, pp. 3–14. London: Academic Press.

Meagher, R.B. (1991) Divergence and differential expression of actin gene families in higher plants. *International Review of Cytology*, **125**, 139–163.

Meagher, R.B. and McLean, B.G. (1990) Diversity of plant actins. *Cell Motility and the Cytoskeleton*, **16**, 164–166.

Menzel, D. (1993) Chasing coiled coils: intermediate filaments in plants. *Botanica Acta*, **106**, 294–300.

Miller, D.D., de Ruijter, N.C.A., Bisseling, T. and Emons, A.M.C. (1999) The role of actin in root hair morphogenesis: studies with lipichito-oligosaccharide as a growth stimulator and cytochalasin as an actin perturbing drug. *Plant Journal*, **17**, 141–154.

Miller, D.D., de Ruijter, N.C.A and Emons, A.M.C. (1997a) From signal to form: aspects of the cytoskeleton-plasma membrane-cell wall continuum in root hair tips. *Journal of Experimental Botany*, **48**, 1881–1896.

Miller, D., Hable, W., Gottwald, J., Ellard-Ivey, M., Demura, T., Lomax, T. and Carpita, N. (1997b) Connections: the hard wiring of the plant cell for perception, signaling, and response. *Plant Cell*, **9**, 2105–2117.

Moscatelli, A., Del Castino, C., Lozzi, L., Cai, G., Scali, M., Tiezzi, A. and Cresti, M. (1995) High molecular weight polypeptides related to dynein heavy chains in *Nicotiana tabacum* pollen tubes. *Journal of Cell Science*, **108**, 1117–1125.

Nebenführ, A., Gallagher, L.A., Dunahay, T.G., Frohlick, J.A., Mazurkiewicz, A.M., Meehl, J.B. and Staehelin, L.A. (1999) Stop-and-go movements of plant Golgi stacks are mediated by the acto-myosin system. *Plant Physiology*, **121**, 1127–1141.

Newman, G.R. and Hobot, J.A. (1993) *Resin Microscopy and On-section Immunocytochemistry*. Berlin, Heidelberg, New York, London: Springer.

Nick, P. (1998) Signaling to the microtubular cytoskeleton in plants. *International Review of Cytology*, **184**, 33–80.

Nick, P. (1999) Signals, motors, morphogenesis – the cytoskeleton in plant development. *Plant Biology*, **1**, 169–179.

Nick, P., Wang, Q.-Y. and Freudenreich, A. (1997) Approaches to understand signal-triggered responses of the plant cytoskeleton. *Recent Research Developments in Plant Physiology*, **1**, 153–180.

Norenburg, J.L. and Barrett, J.M. (1987) Steedman's polyester wax embedment and deembedment for combined light and scanning electron microscopy. *Journal of Electron Microscopy Techniques*, **6**, 35–41.

O'Brien, T.P. and McCully, M.E. (1981) *The Study of Plant Structure: Principles and Selected Methods*. Melbourne: Termacarphi Pty Ltd.

Pain, N.A., O'Connell, R.J., Bailey, J.A. and Green, J.R. (1992) Monoclonal antibodies which show restricted binding to four *Colletotrichum* species: *C. lindemuthianum*, *C. malvarum*, *C. orbiculare* and *C. trifolii*. *Physiological and Molecular Pathology*, **40**, 111–126.

Panchuk-Voloshina, N., Haugland, R.R., Bishop-Stewart, J., Bhalgat, M.K., Millard, P.J., Leung, W.-Y. and Haugland, R.P. (1999) Alexa dyes, a series of new fluorescent dyes that yield exceptionally bright, photostable conjugates. *Journal of Histochemistry & Cytochemistry*, **47**, 1179–1188.

Pierson, E.S., Derksen, J. and Traas, J.A. (1986) Organization of microfilaments and microtubules in pollen tubes grown *in vitro* and *in vivo* in various angiosperms. *European Journal of Cell Biology*, **41**, 14–18.

Polak, J.M. and van Noorden, S. (1997) *Introduction to Immunocytochemistry*, 2nd edn. Oxford: Bios Scientific Publishers.

Preston, R.D. (1988) Cellulose-microfibril-orienting mechanisms in plant cell walls. *Planta*, **174**, 67–74.

Pyke, K. (1994) *Arabidopsis* – its use in the genetic and molecular analysis of plant morphogenesis. *New Phytologist*, **128**, 19–37.

Regan, S., Bourquin, V., Tuominen, H. and Sundberg, B. (1999a) Accurate and high resolution *in situ* hybridization analysis of gene expression in secondary stem tissues. *Plant Journal*, **19**, 363–369.

Regan, S., Chaffey, N.J. and Sundberg, B. (1999b) Exploring cambial growth with *Arabidopsis* and *Populus*. *Journal of Experimental Botany*, **50** (Supplement), 33.

Reichelt, S., Knight, A.E., Hodge, T.P., Baluška, F., Šamaj, J., Volkmann, D. and Kendrick-Jones, J. (1999) Characterization of the unconventional myosin VIII in plant cells and its localization at the post-cytokinetic cell wall. *Plant Journal*, **19**, 555–567.

Robards, A.W. (1991) Rapid-freezing methods and their applications. In *Electron Microscopy of Plant Cells*, edited by J.L. Hall and C. Hawes, pp. 257–312. London, New York, Toronto: Academic Press.

Robards, A.W. and Humpherson, P.G. (1967) Microtubules and angiosperm bordered pit formation. *Planta*, **77**, 233–238.

Roberts, K. (1986) Antibodies and the plant cell surface: practical approaches. In *Immunology in Plant Science*, edited by T.L. Wang, pp. 89–110. Cambridge, London, New York: Cambridge University Press.

Roberts, K., Burgess, J., Roberts, I. and Linstead, P. (1985) Microtubule rearrangement during plant cell growth and development: an immunofluorescence study. In *Botanical Microscopy*, edited by A.W. Robards, pp. 263–283. Oxford: Oxford University Press.

Samaj, J., Baluška, F. and Volkmann, D. (1998) Cell-specific expression of two arabinogalactan protein epitopes recognized by monoclonal antibodies JIM8 and JIM13 in maize roots. *Protoplasma*, **204**, 1–12.

Schellenbaum, P., Vantard, M. and Lambert, A.-M. (1992) Higher plant microtubule-associated proteins (MAPs): a survey. *Biology of the Cell*, **76**, 359–364.

Schroeder, M., Wehland, J. and Weber, K. (1985) Immunofluorescence microscopy of microtubules in plant cells, stabilization by dimethylsulfoxide. *European Journal of Cell Biology*, **38**, 211–218.

Seagull, R.W. (1989) The plant cytoskeleton. *Critical Reviews in Plant Sciences*, **8**, 131–167.

Seagull, R.W. (1994) Plant cytoskeleton. *Encyclopedia of Agricultural Science*, **3**, 241–257.

Shaw, P.J., Fairbairn, D.J. and Lloyd, CW. (1991) Cytoplasmic and nuclear intermediate filament antigens in higher plant cells. In *The Cytoskeletal Basis of Plant Growth and Form*, edited by C.W. Lloyd, pp. 69–81. London: Academic Press.

Shibaoka, H. (1991) Microtubules and the regulation of cell morphogenesis by plant hormones. In: *The Cytoskeletal Basis of Plant Growth and Form*, edited by C.W. Lloyd, pp. 159–168. London: Academic Press.

Shibaoka, H., Asada, T., Yamamoto, S. and Sonobe, S. (1996) The use of model systems prepared from tobacco BY-2 cells for studies of the plant cytoskeleton. *Journal of Microscopy*, **181**, 145–152.

Shimmen, T. and Yokota, E. (1994) Physiological and biochemical aspects of cytoplasmic streaming. *International Review of Cytology*, **155**, 97–139.

Silflow, C.D., Oppenheimer, D.G., Kopczak, S.D., Ploense, S.E., Ludwig, S.R., Haas, N. and Snustad, D.P. (1987) Plant tubulin genes: structure and differential expression during development. *Developmental Genetics*, **8**, 435–460.

Simmonds, D.H., Seagull, R.W. and Setterfield, G. (1985) Evaluation of techniques for immunofluorescent staining of microtubules in cultured plant cells. *Journal of Histochemistry and Cytochemistry*, **33**, 345–352.

Singh, M.B., Taylor, P.E. and Knox, R.B. (1993) Special preparation methods for immunocytochemistry of plant cells. In *Immunocytochemistry: a Practical Approach*, edited by J.E. Beesley, pp. 77–102. Oxford, New York, Toronto: Oxford University Press.

Smith, M.M. and McCully, M.E. (1978) Enhancing aniline blue fluorescent staining of cell wall structures. *Stain Technology*, **53**, 79–85.

Smertenko, A., Blume, Y., Viklický, V., Opartný, Z. and Dráber, P. (1997) Post-translational modifications and multiple tubulin isoforms in *Nicotiana tabacum* L. cells. *Planta*, **201**, 349–358.

Somerville, C. (2000) The twentieth century trajectory of plant biology. *Cell*, **100**, 13–25.

Sonobe, S. (1997) Cell model systems in plant cytoskeleton studies. *International Review of Cytology*, **175**, 1–27.

Sonobe, S. and Shibaoka, H. (1989) Cortical fine actin filaments in higher plant cells visualized by rhodamine-phalloidin after pretreatment with m-maleimidobenzoyl N-hydroxysuccinimide ester. *Protoplasma*, **148**, 80–86.

Staiger, C.J. (2000) Signaling to the actin cytoskeleton of plants. *Annual Review of Plant Physiology and Plant Molecular Biology* **51**, 257–288.

Staiger, C., Baluška, F., Volkmann, D. and Barlow, P.W. (2000) *Actin: A Dynamic Framework for Multiple Plant ell Functions*. Dordrecht, The Netherlands: Kluwer Academic Press.

Staiger, C.J. and Lloyd, C.W. (1991) The plant cytoskeleton. *Current Opinion in Cell Biology*, **3**, 33–42.

Sullivan, K.F. and Kay, S.A. (eds) (1999) Green fluorescent proteins. *Methods in Cell Biology Volume 58*. London, New York: Academic Press.

Swartz, C.J. and Santi, P.A. (1996) Immunofluorescent artifacts due to the pH of anti-fading mounting media. *BioTechniques*, **20**, 398–400.

Tewinkel, M., Kruse, S., Quader, H., Volkmann, D. and Sievers, A. (1989) Visualization of actin filament patterns in plant cells without pre-fixation: a comparison of differently modified phallatoxins. *Protoplasma*, **149**, 178–182.

Tiwari, S.C., Wick, S.M., Williamson, R.E. and Gunning, B.E.S. (1984) Cytoskeleton and integration of cellular function in cells of higher plants. *Journal of Cell Biology*, **99**, 63s–69s.

Traas, J.A. (1990) The plasma membrane-associated cytoskeleton. In *The Plant Plasma Membrane*, edited by C. Larsson and I.M. Møller, pp. 269–292. Berlin: Springer.

Trewavas, A.J. and Malhó, R. (1997) Signal perception and transduction: the origin of the phenotype. *Plant Cell*, **9**, 1181–1195.

Tsien, R.Y. (1998) The green fluorescent protein. *Annual Review of Biochemistry*, **67**, 509–544.

Tsien, R.Y. and Waggoner, A. (1990). Fluorophores for confocal microscopy: photophysics and photochemistry. In *Handbook of Biological Confocal Microscopy*, edited by J.B. Pawley, revised 1st edn, pp. 169–178. New York, London: Plenum Press.

Uehara, K. and Hogetsu, T. (1993) Arrangement of cortical microtubules during formation of bordered pits in the tracheids of *Taxus*. *Protoplasma*, **172**, 145–153.

Uribe, X., Torres, M.A., Capellades, M., Puigdomènech, P. and Rigau, J. (1998) Maize α-tubulin genes are expressed according to specific patterns of cell differentiation. *Plant Molecular Biology*, **37**, 1069–1078.

Vigil, E.L. and Hawes, C.R. (eds) (1989) *Cytochemical and Immunological Approaches to Plant Cell Biology*. London, San Diego, Tokyo: Academic Press.

Vitha, S., Baluška, F., Braun, M., Samaj, J., Volkmann, D. and Barlow, P.W. (2000a) Comparison of cryofixation and aldehyde fixation for plant actin immunocytochemistry: aldehydes do not destroy F-actin. *Histochemical Journal* **32**, 457–466.

Vitha, S., Baluška, F., Jasik, J., Volkmann, D. and Barlow, P.W. (2000b) Steedman's wax for F-actin visualization. In *Actin: a Dynamic Framework for Multiple Plant Cell Functions*, edited by C. Staiger, F. Baluška, D. volkmann and P.W. Barlow, pp. 619–636. Dordrecht, The Netherlands: Kluwer Academic Press.

Vitha, S., Baluška, F., Mews, M. and Volkmann, D. (1997) Immunofluorescence detection of F-actin on low melting point wax sections from plant tissues. *Journal of Histochemistry and Cytochemistry*, **45**, 89–95.

Volkmann, D. and Baluška, F. (1999) Actin cytoskeleton in plants: from transport networks to signaling networks. *Microscopy Research and Technique*, **47**, 135–154.

Wasteneys, G.O., Gunning, B.E.S. and Hepler, P.K. (1993) Microinjection of fluorescent brain tubulin reveals dynamic properties of cortical microtubules in living plant cells. *Cell Motility and the Cytoskeleton*, **24**, 205–213.

Wasteneys, G.O., Sugimoto, K. and Baskin, T. (1998) Comparison of fixation methods for analysis of a plant microtubule mutant. *Cell Biology International*, **21**, 905–907.

Wasteneys, G.O., Willingale-Theune, J. and Menzel, D. (1997) Freeze shattering: a simple and effective method for permeabilizing higher plant cell walls. *Journal of Microscopy*, **188**, 51–61.

Wick, S.M. (1991) Spatial aspects of cytokinesis in plant cells. *Current Opinion in Cell Biology*, **3**, 253–260.

Wick, S.M. (1993) Immunolabeling of antigens in plant cells. *Methods in Cell Biology*, **37**, 171–200.

Wick, S.M., Cho, S.-O. and Mundelius, A.R. (1989) Microtubule deployment within plant tissues: fluorescence studies of sheets of intact mesophyll and epidermal cells. *Cell Biology International Reports*, **13**, 95–106.

Wick, S.M. and Duniec, J. (1986) Effects of various fixatives on the reactivity of plant cell tubulin and calmodulin in immunofluorescence microscopy. *Protoplasma*, **133**, 1–18.

Wick, S.M., Seagull, R.W., Osborn, M., Weber, K. and Gunning, B.E.S. (1981) Immunofluorescence micro-

scopy of organized microtubule arrays in structurally stabilized meristematic cells. *Journal of Cell Biology*, **89**, 685–690.

Williamson, R.E. (1991) Orientation of cortical microtubules in interphase plant cells. *International Review of Cytology*, **129**, 135–206.

Williamson, R.E., Grolig, F., Hurley, U.A., Jablonski, P.P., McCurdy, D.W. and Wasteneys, G.O. (1989) Methods for studying the plant cytoskeleton. *Modern Methods of Plant Analysis*, **10**, 203–218.

Wyatt, S.E. and Carpita, N.C. (1993) The plant cytoskeleton-cell-wall continuum. *Trends in Cell Biology*, **3**, 413–417.

Wymer, C.L., Beven, A.F., Boudonck, K. and Lloyd, C.W. (1999) Confocal microscopy of plant cells. In *Methods in Molecular Biology, Volume 122, Confocal Microscopy Methods and Protocols*, edited by S. Paddock, pp. 103–130. Totowa, NJ: Humana Press Inc.

Wymer, C.L., Shaw, P.J., Warn, R.M. and Lloyd, C.W. (1997) Microinjection of fluorescent tubulin into plant cells provides a representative picture of the cortical microtubule array. *Plant Journal*, **12**, 229–234.

Yokota, E., Mimura, T. and Shimmen, T. (1995) Biochemical, immunochemical and immunohistochemical identification of myosin heavy chains in cultured cells of *Catharanthus roseus*. *Plant & Cell Physiology*, **36**, 1541–1547.

Yuan, M., Warn, R.M., Shaw, P.J. and Lloyd, C.W. (1995) Dynamic microtubules under the radial and outer tangential walls of microinjected pea epidermal cells observed by computer reconstruction. *Plant Journal*, **7**, 17–23.

Yuan, M., Shaw, P.J., Warn, R.M. and Lloyd, C.W. (1994) Dynamic reorientation of cortical microtubule, from transverse to longitudinal, in living plant cells. *Proceedings of the National Academy of Sciences of the United States of America*, **91**, 6050–6053.

Zhang, D., Wadsworth, P. and Hepler, P.K. (1990) Microtubule dynamics in living dividing plant cells: confocal imaging of microinjected fluorescent brain tubulin. *Proceedings of the National Academy of Sciences of the United States of America*, **87**, 8820–8824.

Wood Formation in Trees ed. N. Chaffey
© 2002 Taylor & Francis
Taylor & Francis is an imprint of the
Taylor & Francis Group
Printed in Singapore.

Immunolocalisation and Visualisation of the Cytoskeleton in Gymnosperms Using Confocal Laser Scanning Microscopy

Ryo Funada

Department of Forest Science, Faculty of Agriculture, Hokkaido University, Sapporo 060-8589, Japan

ABSTRACT

This chapter provides protocols for visualisation of the cytoskeleton, namely, cortical microtubules and actin filaments, by chemical fixation, cryo-sectioning, immunostaining, and confocal laser scanning microscopy (CLSM). This new technique allows construction of three-dimensional images of the cytoskeleton in differentiating secondary xylem cells. CLSM is a powerful tool for observations of dynamic changes in the orientation and localisation of components of the cytoskeleton in relatively large areas. The roles of the cytoskeleton during differentiation of the secondary xylem (wood formation) in gymnosperm trees are briefly summarised.

Key words: actin filaments, confocal laser scanning microscopy, cortical microtubules, gymnosperms, immunostaining, secondary xylem cells

INTRODUCTION

Observations of the cytoskeleton in gymnosperm trees prior to the introduction of confocal laser scanning microscopy

The activity of the vascular cambium (cambium) leads to increases in the stem diameter of trees (see Catesson, 1994; Larson, 1994). The periclinal division of cambial cells produces secondary phloem (phloem) on the outside, relative to the cambium, and secondary xylem (xylem) on the inside. In gymnosperm trees (softwoods), cambial cells differentiate into xylem cells, namely, axial tracheids (tracheids), axial parenchyma cells, ray parenchyma cells and ray tracheids. However, more than 90% of cambial cells differentiate into tracheids. During such differentiation, cambial cells increase in length and/or diameter and their cell walls thicken and develop characteristic profiles. The process of differentiation can be followed by examining cells in a radial file. Thus, cambial cells and their derivatives appear to provide a suitable model system for observations of cytodifferentiation in secondary tissues *in situ* (see Catesson *et al.*, 1994; Murakami *et al.*, 1999; Chaffey, 2000; Funada, 2000; Funada *et al.*, 2000).

Cell walls are reinforced by cellulose microfibrils, which resist cell expansion in response to turgor pressure. The orientation of cellulose microfibrils in the primary walls of cambial derivatives determines the direction of expansion of cells, thereby controlling the shape and size of xylem cells. In addition, the orientation of cellulose microfibrils in the secondary walls, in particular in the middle (S_2) layer of the secondary wall, is closely related to the physical properties of xylem cells. Thus, the orientation of these microfibrils determines the properties of wood.

It is generally accepted that cortical microtubules regulate the arrangement of cellulose microfibrils by guiding the movement of cellulose synthase complexes in the plasma membrane (see Giddings and Staehelin, 1991). The average diameters of cytoskeletal components are small, about 24 nm for microtubules and 7 nm for actin filaments. Thus, details of cytoskeletal components have been examined by transmission electron microscopy (TEM) since the

143

original observations of cortical microtubules in plant cells by Ledbetter and Porter (1963). TEM reveals the fine structure of the cytoskeleton (see e.g. the TEM chapter by Chaffey in this volume) and the interactions between its components. However, it is difficult to examine the cytoskeleton in relatively large areas of any plant material by TEM because this technique requires small ultrathin sections, which are generally slightly oblique. Thus, successive changes in the orientation of the cytoskeleton during differentiation cannot easily be followed.

With the successful introduction of indirect immunofluorescence microscopy for examination of sections thicker than those suitable for TEM, it has become possible to visualise the cytoskeleton over a relatively large area within plant tissues (see Lloyd, 1987). Cytoskeletal components, in particular cortical microtubules, in the cambial cells and their derivatives in woody plants have been observed by a similar technique (Uehara and Hogetsu, 1993; Abe et al., 1995b; Prodhan et al., 1995a; Chaffey et al., 1997a, b, c, 1998, 1999; see also the cytoskeletal chapter by Chaffey in this volume). These observations have allowed an increase in our understanding of the role of the cytoskeleton in the differentiation of the secondary xylem and phloem. In addition, high-resolution scanning electron microscopy (SEM), such as field emission-SEM (FE-SEM), also allows investigation of cytoskeletal components in a relatively large area of plant tissues (Abe et al., 1994; Prodhan et al., 1995a; Vesk et al., 1996).

Advantages of confocal laser scanning microscopy (CLSM)

Examination of immunofluorescently-stained specimens by conventional fluorescence microscopes suffers from many drawbacks, although it has contributed significantly to study of the role of the cytoskeleton during secondary vascular differentiation in angiosperm trees (see the cytoskeletal chapter by Chaffey in this volume). The relatively widespread availability of the CLSM, however, offers significant advantages over traditional epifluorescent observations. For instance, the CLSM can be used to great advantage when thick biological specimens are analysed after fluorescence staining or when reflecting specimens with surface structure are analysed (for detailed theory of CLSM, e.g., see Minsky, 1988; Brakenhoff et al., 1989; Wilson, 1990; Matsumoto, 1993; Pawley, 1995;

Sheppard and Shotton, 1997; Hepler and Gunning, 1998).

Lasers have a number of unique properties compared with other light sources. They have a high degree of monochromaticity, high brightness, low noise, small divergence, high degree of spatial and temporal coherence, a Gaussian beam profile, and can be focused onto very small regions. Unlike conventional microscopes, the CLSM only detects information that originates from the focal plane. All other information, namely, unfocused images from other planes, is removed by the small pinhole. Thus, the single image obtained consists of an optical section in the x-y plane (perpendicular to the optical axis) with a very low depth of focus. The information is passed through a beam splitter (dichroic mirror), a space filter (small pinhole), and then to a photo-detector (e.g., photo-multiplier tube). The photo-detector converts light energy into electrical energy. A two-dimensional image is obtained by scanning the specimen with a laser beam, which is usually servo-controlled by a galvano mirror scanner (x and y).

High-speed rotation of a Nipkow disc, which has thousands of apertures in a spiral pattern, can also be used to scan spots of light over the specimen. This scanning system offers the advantage of real-time imaging, but it has been infrequently employed for the CLSM due to its associated technical problems, such as the low intensity of the illumination due to reduced light transmission through the apertures of the disk.

As an example, an image obtained by CLSM in the confocal mode is compared with an image of the same section obtained in the non-confocal mode in Figure 1. Information that is out of focus in the non-confocal mode creates a contrast-reducing background. By contrast, the confocal mode with the pinhole yields a sharp image with high-contrast due to elimination of information that is out of focus and improvement of resolution.

CLSM does not require mechanical sectioning of specimens. Therefore, it allows repeated measurements of both fixed and living samples, and damage to the surface of specimens caused during sectioning is avoided. The position of the focal plane is changed with a computer-controlled motor and, therefore, optical sections from the interior of a thick section can be recorded in series along the z-axis at user-defined intervals (Figure 2A). This suggests that the extended-focus image can be obtained by adding together each of

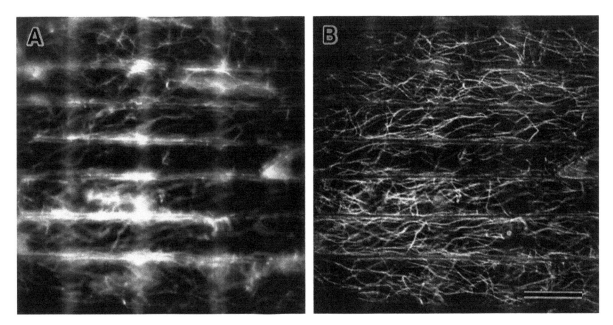

Figure 1 Immunofluorescence images of cortical microtubules in a single section of ray parenchyma cells of *Taxus cuspidata* observed under two sets of conditions. A, Non-confocal mode. B, Confocal mode (a single optical image). Averaging of images is eight times. The confocal mode yields a sharp high-contrast image that results from elimination of signals that are not in the focal plane. The orientation of cortical microtubules is oblique or longitudinal with respect to the cell axis. The left side of this, and other micrographs, corresponds to the outer side of the tree. All images are obtained from radial longitudinal sections approx. 50 μm thick. Bar = 25 μm.

the optical sections. Each image obtained via the photomultiplier can be stored automatically via a computer. Since the data obtained are digital, processing and enhancement of images is easy using the software with which the microscope is equipped (Figure 2B). Therefore, the CLSM allows easy three-dimensional (3-D) reconstruction of a specimen, revealing, for example, the 3-D structure of xylem cells, and permitting precise measurements of the dimensions of xylem cells in woody plants (Knebel and Schnepf, 1991; Donaldson and Lausberg, 1998; Matsumura *et al.*, 1998; Kitin *et al.*, 1999). See also the 'Wood microscopical techniques' chapter in this volume.

One of the best techniques for determining the shapes and dimensions of whole samples is reconstruction from standard serial thin sections. This method is, however, time-consuming since it requires the preparation of numerous sections in series and appropriate orientation of images. CLSM allows the rapid 3-D analysis of large and relatively thick sections, providing great advantage over conventional anatomical techniques (e.g. Kitin *et al.*, 2000). One limitation of CLSM is that the intensity of fluorescent signals from a specimen

decreases with increasing depth along the z-axis, in particular, when thick samples are used. Such attenuation leads to errors in quantitative measurements of cell dimensions. However, it may be possible to minimise errors by appropriate correction for attenuation of signals (e.g. Gray *et al.*, 1999).

CLSM, in combination with indirect immunostaining, has been used to construct 3-D images of the cytoskeleton in differentiating xylem cells of several woody plants (Abe *et al.*, 1995a; Funada *et al.*, 1997; Furusawa *et al.*, 1998; Chaffey *et al.*, 1997a; 1998), as well as to monitor dynamic changes in the orientation and localisation of components of the cytoskeleton over large areas (see Funada 2000; Funada *et al.*, 2000). CLSM has revealed that the orientation of cortical microtubules in tracheids and ray parenchyma cells of gymnosperm trees changes dynamically during differentiation of the xylem and that such changes are closely related to the process of differentiation. In addition, since living tissues can be observed without fixation, CLSM can record series of images at specific time-intervals, providing what is essentially four-dimensional information, namely, spatial information in three dimensions together

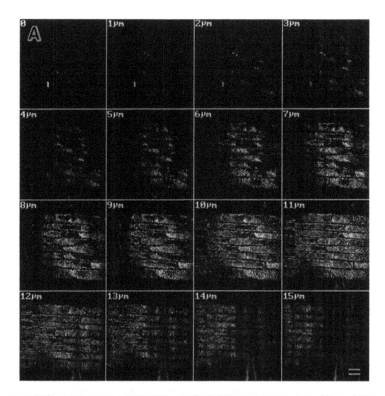

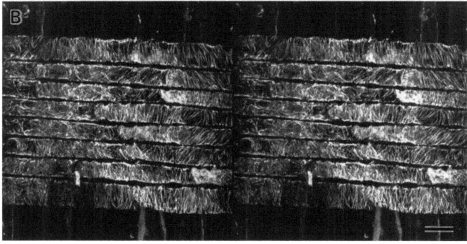

Figure 2   Confocal immunofluorescence images of cortical microtubules in ray parenchyma cells of *Taxus cuspidata*. Averaging of images is eight times. A, Consecutive 16 optical sections recorded at 1-μm intervals along the z-axis (gallery image). B, A pair of stereo images observed from the projection of all the consecutive optical sections in A. The orientation of cortical microtubules is transverse or oblique with respect to the cell axis during radial elongation of cells. Bar = 25 μm.

with information in the temporal dimension. After microinjection of a fluorescent dye, CLSM allows us to observe dynamic reorientation of cortical microtubules in living plant cells with the passage of time (Yuan *et al.*, 1994; 1995; Lloyd *et al.*, 1996; Himmelspach *et al.*, 1999). The applications of

CLSM in plant cell biology have been reviewed by Hepler and Gunning (1998).

This chapter describes the application of CLSM to visualisation of 3-D arrangements of components of the cytoskeleton in cambial cells and cambial derivatives of gymnosperm trees. The cytoskeletal components, namely, cortical micro-tubules and actin filaments, are immunostained with specific monoclonal antibodies in relatively thick cryo-sections.

TECHNIQUES

Sampling procedure

Equipment needed:	Chemicals needed:
Sharp knife	Piperazine-N',N'-bis [2-ethanesulfonic acid] (PIPES) buffer (Dojindo, No. 347-02224)
Scalpel	
Razor blades	

Duration of procedure: approx. 15–20 min.

1 Take blocks of plant material from an area of 30 mm × 30 mm, containing outer bark, mature and differentiating phloem, cambial zone cells and mature and differentiating xylem, from the stem of a gymnosperm tree using a sharp knife and a scalpel.
2 Cut longitudinal slivers, with dimensions approx. 30 mm (longitudinal) × 5–10 mm (radial) × 2 mm (tangential), from the block with a razor blade under 50 mM PIPES buffer. The pH of the buffer is adjusted to approx. 7.0 with 1 M NaOH.
3 Remove the outer bark.

Fixation for immunostaining

Equipment needed:	Chemicals needed:
Glass vials with tops (approx. 20 ml capacity)	p-formaldehyde (TAAB, No. F-003)
Razor blades	Glutaraldehyde (WAKO, No. 072-02262)
Freezer at −80 °C	Dimethyl sulphoxide (DMSO) Nonidet P-40 (Sigma, I-3021)

	PIPES buffer (see previous section) Ethylene glycol-bis(β-aminoethyl ether)-N,N,N',N'-tetraacetic acid (EGTA) MgSO$_4$ NaOH Distilled water Liquid nitrogen m-maleimidobenzoic acid N-hydroxysuccin-imide ester (MBS) (PIERCE, No. 22311)

Duration of procedure: approx. 18 h including overnight fixation.

Microtubules

1 Fix longitudinal slivers for about 20 min under a vacuum for removal of air from inter-cellular spaces in vials in a mixture of 3.6% p-formaldehyde and 0.2% glutaraldehyde that contains 10% DMSO and 0.1% Nonidet P-40 in 50 mM PIPES buffer, pH 7.0, supplemented with 5 mM EGTA and 5 mM MgSO$_4$.
Note – p-formaldehyde, which has monoalde-hyde groups, forms bridges via amino and peptide linkages between neighbouring peptide chains (Roland and Vian, 1991). Glutaraldehyde, which is a dialdehyde, increases the potential for the cross-linking of peptide chains. Therefore, glutaraldehyde provides better preservation of cell structure than p-formaldehyde, but it may eliminate the recognition of proteins by antibodies. The addition of small amounts of glutaraldehyde, for example 0.1–0.2%, to p-formaldehyde can improve the quality of fixation for immunos-taining (see also the cytoskeletal chapter by Chaffey in this volume). Detergents, such as Nonidet P-40, facilitate penetration of sections by antibodies, in particular, when sections are relatively thick (e.g. 50 μm).
2 Fix again in fresh fixative overnight at room temperature (RT). Since microtubules tend to depolymerise at low temperatures, it is neces-sary to avoid low temperatures during the fixation of samples.

3   Remove the terminal 5-mm ends of the slivers with a razor blade to eliminate damaged regions. Since fusiform cambial cells and differentiating tracheids of gymnosperm trees generally reach up to 4 mm in length, cells at the terminal ends might be expected to have lost their cytoplasm during sampling.

4   Trim the slivers into small blocks of 5–10 mm (longitudinal) × 5–10 mm (radial) × 2 mm (tangential).

5   Wash briefly in distilled water.

6   Freeze in liquid nitrogen. The frozen samples are stored in a freezer at –80 °C, where they can remain for at least six months.

Microfilaments

Since actin filaments are sensitive to aldehyde fixatives, treatment with a protein cross-linking reagent prior to fixation with a mixture of *p*-formaldehyde and glutaraldehyde is useful for stabilising these filaments.

1   Pre-fix longitudinal slivers in a solution of 0.2 mM MBS (a protein cross-linking reagent) in 2% DMSO and 0.05% Nonidet P-40 in distilled water for 60 min at RT (Sonobe and Shibaoka, 1989).
    Note – MBS is first dissolved in 2% DMSO and then the solution is diluted in distilled water.

2   Fix in the same fixative as that used to fix cortical microtubules overnight at RT.

Cryo-sectioning

Equipment needed:	Chemicals needed:
Freezing stage at –20 °C	Distilled water
Sliding microtome (P-401; Pika Seiko Ltd.)	NaCl
Disposable stainless-steel blades	KCl
Polystyrene microplates	KH_2PO_4 $Na_2HPO_4.12H_2O$

Duration of procedure: approx. 1–2 h.

1   Put each small block in a drop of distilled water on a freezing stage (MA-101; Komatsu Electronics Inc.) at –20 °C.

2   Cut longitudinal radial sections at a thickness of approx. 50 μm with a sliding microtome equipped with a disposable stainless-steel blade. See the chapter by Uggla and Sundberg in this volume for further details of the cryo-sectioning technique.

3   Wash three times for 3 min each with phosphate-buffered saline (PBS; 100 ml containing 0.8 g NaCl, 0.02 g KCl, 0.02 g KH_2PO_4, 0.287 g $Na_2HPO_4.12H_2O$ in water; pH 7.3–7.4) in a polystyrene microplate.

Immunostaining of cortical microtubules and actin filaments

Equipment needed:	Chemicals needed:
Parafilm™ (American Can Co.)	PBS (see above section for formulation)
Petri dish	Glycerol
Pipettes	*p*-phenylene diamine (WAKO, No. 164-01532)
Drying oven at 30 °C (DX300; Yamato Inc).	NaN_3
Freezer at –80 °C	Bovine serum albumen (BSA, Sigma, A-7906)
Refrigerated centrifuge (1700; Kubota Inc).	Anti-α-tubulin anti body (Amersham, No. N356)
Glass slides	Anti-actin antibody (Amersham, No. N350)
Coverslips	Fluorescein *isothio*-cyanate (FITC)-conjugated anti-mouse immunoglobulin G (IgG) antibody (Amersham, No. N1031)
Filter papers Glass vials with tops (approx. 30 ml capacity) Aluminium foil Freezer at –20 °C	

Duration of procedure: approx. 2.5 h (ensure any frozen reagents are allowed sufficient time to thaw!).

Immunolocalisation of a single cytoskeletal protein

1   Put sections on a sheet of Parafilm™ in a Petri dish, and replace lid.

2 Apply the primary mouse-raised monoclonal antibodies (approx. 20 μl per section) either against chicken brain α-tubulin, or against chicken gizzard actin, with a pipette and incubate for 60 min at 30 °C in darkness in the drying oven.

Note – the antibodies (aliquots) should be stored in a freezer at –80 °C. Each antibody is diluted 1:500 in PBS that contains 0.1% NaN_3 and 1 mg ml^{-1} BSA (PBSB). After dilution, the antibody solution is clarified at 12,000 × g for 5 min at 0 °C with a refrigerated centrifuge. The resultant solution should be stored at 2–8 °C and not frozen. Repeated freezing and thawing can denature antibodies.

3 Wash three times for 3 min each with PBS.

4 Apply the FITC-conjugated secondary antibodies (approx. 20 μl per section) with a pipette and incubate for 60 min at 30 °C in darkness. These antibodies are diluted 1:10 in PBSB and the solution of diluted antibodies should be stored at 2–8 °C in darkness.

5 Wash three times for 3 min each with PBS in darkness. The sections can be stored in PBS in darkness for 2–3 days at RT.

6 Mount sections stored in PBS on glass slides in 50% glycerol in PBS that has been supplemented with an antioxidant, such as 0.1% p-phenylene diamine with coverslips (see the cytoskeletal chapter by Chaffey in this volume for formulation of this mountant) to reduce photo-bleaching. In particular, antifade mounting agents should be used for FITC-stained sections to protect against photo-bleaching because this dye is much more stable at or above pH 8 than at pH 7 (Hale and Matsumoto 1993). The diluted mounting agent should be stored in darkness (e.g. in vials wrapped in aluminium foil) in a freezer at –20 °C.

7 Wipe off excess mounting medium with a filter paper.

Immunolocalisation of two cytoskeletal proteins

Duration of procedure: approx. 2.5 h (ensure any frozen reagents are allowed sufficient time to thaw!).

For double immunostaining, primary antibodies that have been raised in different species are needed to avoid significant cross-reactions with the secondary antibodies. For example, a rat monoclonal antibody against chicken brain α-tubulin (Harlam Sera-Lab, No. MAS-078b) and a mouse monoclonal antibody against chicken gizzard actin (Amersham, No. N350) can be used on a single section.

1 Apply a mixture of the two primary antibodies (approx. 20 μl per section, diluted as detailed above) and incubate for 60 min at 30 °C in darkness.

2 Wash three times for 3 min each with PBS.

3 Apply a mixture of FITC-conjugated antibodies (approx. 20 μl per section, diluted as detailed above) raised in goat against rat IgG (no cross-reaction with mouse IgG) (Cappel, No. 55755) and tetramethylrhodamine-5-(and-6)-isothiocyanate (TRITC)-conjugated antibodies raised in goat against mouse IgG (Cappel, No. 55539) and incubate for 60 min at 30 °C in darkness.

A fluorescent phallatoxin, such as rhodamine-labelled phalloidin (Molecular Probes, No. R-415), has frequently been used to stain actin filaments (e.g., Sonobe and Shibaoka, 1989; Cleary *et al.*, 1993). However, no staining of actin filaments in differentiating xylem cells of gymnosperm trees, such as *Abies*, *Pinus*, and *Taxus*, has been detected with rhodamine-labelled phalloidin. Similar absence of staining has been observed with cambial cells and cambial derivatives of angiosperm trees (see the cytoskeletal chapter by Chaffey in this volume).

Optical sectioning by confocal laser scanning microscopy

Equipment needed:

Confocal laser scanning microscope
Magnetic Optical (MO) disks

Duration of procedure: approx. 3 days.

Confocal laser scanning microscopes have been produced by several companies (e.g., Carl Zeiss, Olympus, Nikon, Leica, Bio-Rad, and Noran). In this chapter, observations were made with CLSMs from Carl Zeiss (LSM-310 and LSM-410 invert; Carl

Zeiss Co.). An outline of the operating instructions is given below.

1   Place a section on the stage of the microscope. An objective lens at low magnification, such as 10×, with conventional transmitted light, with reflected light or in the fluorescence mode, is used to select the area to be examined.

2   Select a wavelength of lasers and an emission filter appropriate to the investigation. Since the peak wavelengths for excitation and emission of conjugated-FITC are 496 and 518 nm, respectively (Brelje et al., 1993), an argon ion laser (488 nm, blue) or a krypton-argon ion laser (488 nm) is used to excite FITC. A bandpass filter (510–525, 515–540 or 515–565 nm) should be used as the emission filter for detection of excited FITC. In the case of TRITC-stained sections, a combination of a helium neon laser (543 nm, green) and a long-pass filter (570 or 590 nm) should be used since the peak excitation and emission wavelengths of conjugated TRITC are 554 and 576 nm, respectively. A krypton-argon ion laser (568 nm, green-yellow) also excites TRITC-stained sections. Lignified cell walls of secondary xylem cells have red-orange auto-fluorescence (Chaffey et al., 1997b). An adequate combination of a wavelength of lasers and an emission filter reduces effectively the background due to auto-fluorescence.

For double immunostaining with FITC-conjugated and TRITC-conjugated antibodies, a combination of a bandpass filter (510–525 nm) and a long-pass filter (590 nm) should be used to allow efficient separation of the two emissions. Instead of TRITC, other fluorochromes such as Lissamine rhodamine (the peak excitation and emission wavelengths are 570 and 590 nm, respectively), Texas red (the peak excitation and emission wavelengths are 592 and 610 nm, respectively) and cyanine (CY) 5.18 (the peak excitation and emission wavelengths are 649 and 667 nm, respectively), which have long peak wavelengths of emission, are also available when the confocal laser scanning microscopes equipped with a krypton-argon ion laser (568 nm and 647 nm, green-yellow and red) can be used.

3   Select the objective lens (e.g. 40 or 63×), the wavelength of laser beam (e.g. 364, 488, 543 nm or 568 nm), the scan field size

(e.g. 512 × 512 or 1024 × 1024 pixels), the intensity of laser light (choice of optical attenuating filters), electron bandwidth (choice of low-pass filters), scan time, zoom factor, pinhole size, and region of interest on the sample, on the control panel. The pixel size (μm/pixel), which is adjusted by zoom factor, should be determined by width of field-views, optimum objective resolution and photo-bleaching. A high zoom factor (small pixel size) scans over small areas of the specimen. Thus, it produces a high magnification in the final image on the display but increases the photo-bleaching. By contrast, a low zoom factor (large pixel size) increases the field of view and reduces the photo-bleaching.

4   Start the laser beam scanning.

5   Adjust the position of pinhole. This is one of the most important steps and it is a prerequisite for enhancement of the image from only a single focal plane.

6   Adjust alternately contrast and brightness. The contrast function controls the high voltage at the photo-multiplier and the electronic pre-amplification factor. The oversaturation of contrast should be avoided. The brightness function controls the electron offset, which determines the overall brightness of the image. However, the setting of high brightness increases background noise. Thus, a compromise must be found.

7   Produce a single optical image. Numbers for averaging images in a line or a frame are selected to eliminate effectively the noise. Four- or eight-times averaging is usually used, but at the expense of photo-bleaching.

8   Obtain a series of optical sections along the z-axis (referred to as z-sectioning or z-series). The starting position for optical sectioning, the distance between planes (e.g. 0.5 or 1 μm along the z-axis), the number of sections, the refractive index of the objective, and the refractive index of the immersion fluid must all be selected appropriately. 15–25 optical sections along the z-axis at 1 μm intervals are usually obtained for the cytoskeleton in differentiating tracheids and ray parenchyma cells of gymnosperm trees (Figure 2). The confocal depth (resolution on the z-axis) is determined by the numerical aperture (NA) of lens, the wavelength of the exciting laser and the size of the

pinhole. For example, the confocal depth is approx. 0.5 μm when an oil immersion lens with an NA of 1.4 is used with an exciting laser at 488 nm. The use of longer wavelengths of the exciting laser such as a krypton-argon ion laser (647 nm) provides lower confocal depth than the use of shorter wavelengths such as an argon ion laser (364 nm or 488 nm). Therefore, if we need the same confocal depth at a different wavelength, we must adjust the size of the pinhole. An objective lens with a high NA is needed for high resolution. However, the working distance of the lens becomes the factor that limits the depth of optical sectioning. The confocal depth is also determined by the refractive index of the immersion fluid. The refractive index of water and oil is 1.33 and 1.52, respectively (Hell and Stelzer, 1995). Therefore, a water immersion lens has higher resolution than an oil immersion lens when the NA is the same. An objective dry lens 40× (NA 0.75 or 0.95) is often used for the wide field views of cortical microtubules, although the confocal depth becomes smaller than the objective lens with a higher NA.

9   Store automatically the sequence of images in the host computer's memory, on a hard disk or on a Magnetic Optical (MO) disk (e.g. 230 Mbytes). A single optical grey-level image (8 bits) is 0.257 Mbytes when the scan field size of 512 × 512 pixels is used. Therefore, 20 optical sections along the z-axis require approx. 5 Mbytes to store. The pseudocolour image (R, G, and B) needs three times storage space compared with the grey-level image. In general, the Tag Image File Format (TIFF) is used.

10   Obtain images with transmitted light for observations of the structure of cells. Other optical imaging techniques, such as differential interference contrast (DIC), phase contrast, polarisation contrast and dark-field contrast can also be used.

Construction of three-dimensional images

Duration of procedure: approx. 10–60 sec per each construction.

Each series of confocal images can be processed for 3-D reconstruction of the sample with the software of the CLSM. The following methods can be used for depiction of 3-D information, the choice depends on the particular CLSM:

1   gallery, the simultaneous display of several images (e.g. Figure 2A);

2   projection, a single projection image or a series of projections after rotation for an animation (e.g. Figure 3A);

3   animation, scrolling of a series of images, resembling a film;

4   depth-coding, a coloured projection for visualisation of depth information;

5   stereo-imaging, a stereo image that can be observed with red/green spectacles (an anaglyph) or a pair of stereo images (split stereo images) (e.g. Figure 2B);

6   orthogonal sections, sections from a three-dimensional packet with borders that parallel the principal planes, namely, x-y, x-z and y-z; and

7   overlay, superimposed images of two or three images of the same area, for example, a fluorescent image and a transmitted image. Pseudocolours are used to mimic the original fluorochromes (e.g. green for FITC and red for TRITC) (e.g. Figure 5C).

Viewing of images

Equipment needed:
Film recorder
Black-and-white film
Colour slide film
Digital colour printer

Photographs of images can be taken with a 35-mm film recorder on black-and-white film (e.g. Ilford Pan-F film) or on colour slide film (e.g. Fujichrome Provia or Kodak Ektachrome). Images can also be transferred to a computer using software such as Adobe Photoshop and printed with a high-resolution digital colour printer (e.g. Sony UP-D8800).

APPLICATIONS

Examination of the cytoskeleton in gymnosperm trees

A summary is given below of results related to the orientation and localisation of components of the cytoskeleton, as visualised by CLSM, in differentiating tracheids and ray parenchyma cells of gymnosperm trees. Full details of the results can be found in the original papers or in review articles (Abe *et al.*, 1995a; Funada *et al.*, 1997, 2000; Furusawa *et al.*, 1998; Funada, 2000). Representative results are shown in Figures 3–5.

The arrangement of the cytoskeleton in secondary xylem cells changes dynamically during differentiation. There is considerable evidence that the cytoskeleton, in particular the cortical microtubules, is closely related to the orientation and localisation of newly deposited cellulose microfibrils in the differentiating secondary xylem cells of woody plants (see Chaffey, 2000; Funada, 2000; Funada *et al.*, 2000).

Cortical microtubules

In fusiform and ray cambial cells of gymnosperm trees, cortical microtubules are arranged at random. The predominant orientation of cortical microtubules on the radial walls of expanding tracheids is longitudinal at the early stage of cell expansion. However, the orientation changes progressively from longitudinal to transverse as the tracheids expand. Finally, ordered and transversely oriented cortical microtubules are visible when the radial expansion of the tracheids is complete. By contrast, the cortical microtubules in differentiating ray parenchyma cells are oriented transversely to the cell axis at the early stage of cell elongation (Figure 2). As cells elongate radially, the orientation of cortical microtubules changes from transverse to either oblique or longitudinal with respect to the cell axis. After the completion of cell elongation, the density of distribution of cortical microtubules in ray parenchyma cells is lower than that in elongating cells. The cortical microtubules are oriented obliquely or longitudinally with respect to the cell axis.

After cessation of radial cell expansion, the cortical microtubules of differentiating tracheids are aligned in well-ordered arrays during formation of the secondary walls, in which cellulose microfibrils are aligned close to and in parallel with one another. The number of microtubules increases as the secondary walls form in differentiating tracheids and ray parenchyma cells. Increases in the densities of distribution of cortical microtubules in differentiating xylem cells might be synchronised with the start of formation of the secondary wall.

Such increases in the number of cortical microtubules might depend on the synthesis of tubulin *de novo*, which is regulated at the transcriptional level (see Fukuda, 1996).

During formation of the secondary wall, the orientation of cortical microtubules in differentiating tracheids changes rotatively from a flat S-helix to a steep Z-helix with clockwise rotation (as viewed from the lumen side). This shift in the orientation of cortical microtubules is completed within three or four tracheids in a radial file when cambial activity is high (Figure 3A). Then approx. 10 tracheids, with cortical microtubules oriented in a steep Z-helix at almost the same angle, are aligned in a radial file. After further differentiation, the orientation of cortical microtubules changes from a steep Z-helix to a flat S-helix in the tracheids (Figure 3B). This shift is completed within one or two tracheids in a radial file and it is more abrupt than the shift from a flat S-helix to a steep Z-helix.

Not only the orientation of cortical microtubules but also the orientation of the newly deposited cellulose microfibrils changes progressively in the differentiating tracheids (Harada and Côté 1985; Abe *et al.*, 1991; 1995a, b, 1997; Kataoka *et al.*, 1992) and wood fibres as well (Roland and Mosiniak, 1983; Prodhan *et al.*, 1995a, b). The cellulose microfibrils in the primary walls of tracheids are not well-ordered. The predominant orientation of cellulose microfibrils in the differentiating tracheids is longitudinal with respect to the cell axis at the early stage of formation of the primary wall. The longitudinally oriented cellulose microfibrils, with their considerable tensile strength, impede the longitudinal expansion of cells and, thus, they facilitate the lateral expansion of fusiform cambial derivatives. As the tracheids expand, the orientation of the cellulose microfibrils changes from longitudinal to transverse. Finally, when cell expansion ceases, the cellulose microfibrils are well-ordered and aligned parallel to one another. The cellulose microfibrils change their orientation rotatively from a flat S-helix to a steep Z-helix during formation of the secondary wall. With cessation of this rotation, a thick middle (S_2) layer develops as a result of the repeated deposition of cellulose microfibrils with a consistent texture. Then the orientation of the cellulose microfibrils changes from a steep Z-helix to a flat S-helix.

The parallelism between the orientation of cortical microtubules and that of cellulose microfibrils during differentiation suggests that the orientation

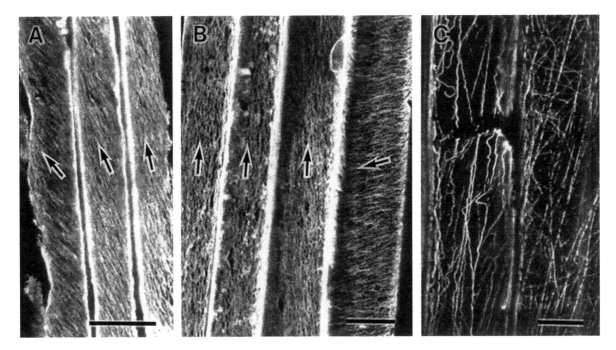

Figure 3 Confocal immunofluorescence images of the arrangement of components of the cytoskeleton (viewed from the lumen side). 15–20 optical sections along the z-axis at 1 μm intervals. Averaging of images is eight times. A and B, Cortical microtubules in differentiating tracheids of *Abies sachalinensis* (projection images). Note the successive changes in the orientation of cortical microtubules (arrows) from a flat helix to a steep Z-helix and back to a flat S-helix during formation of the secondary wall. C, Actin filaments in differentiating tracheids of *Pinus densiflora* (projection image). The axial orientation of actin filaments is retained during formation of the secondary wall. Bars = 25 μm.

of the cortical microtubules might control that of the cellulose microfibrils in the differentiating secondary xylem cells of woody plants, although exceptions to the co-alignment of cellulose microfibrils and cortical microtubules in some plants have been reported (see Giddings and Staehelin, 1991; Emons *et al.*, 1992). The orientation of cortical microtubules does indeed control the orientation of newly deposited cellulose microfibrils on the inner surface of the primary wall, thereby determining the direction of expansion of cambial cells. In addition, the behaviour of cortical microtubules, in terms of their angles to the cell axis and the speed of their rotation during formation of the secondary wall, controls the structure of the secondary wall, thereby determining, in turn, the mechanical properties of the wood. Thus, manipulation of cortical microtubules might allow control of properties of the wood (see Barnett *et al.*, 1998; Chaffey, 2000; Funada, 2000).

The patterns of orientation of cortical microtubules during formation of the secondary walls are affected by a variety of stimuli, such as gravity and plant hormones. Cortical microtubules in dif-

ferentiating tracheids of compression wood, which is formed on the lower sides of the inclined stems of gymnosperm trees, are oriented transversely at the early stage of formation of the secondary wall. Their orientation then changes progressively during formation of the secondary wall until the cortical microtubules are oriented obliquely, at an angle of about 45°, to the tracheid axis in a Z-helix. In such tracheids, the orientation is retained until differentiation is completed with little change in the oblique orientation (Furusawa *et al.*, 1998). By contrast, the orientation of cortical microtubules in normal wood tracheids changes rotatively from a flat S-helix to a steep Z-helix and then back to a flat S-helix during formation of the secondary wall, as mentioned above (Figures 3A and B). Similar effects of inclination on changes in the alignment of cortical microtubules in the secondary wall have been observed in the differentiating tension wood fibres in angiosperm trees (Nobuchi and Fujita, 1972; Robards and Kidwai, 1972; Fujita *et al.*, 1974; Prodhan *et al.*, 1995a).

During formation of the primary wall, the cortical microtubules in differentiating tracheids disappear locally from areas of prospective inter-tracheal bordered pits. Such localised disappearance of cortical microtubules might determine the site of each future pit region (Funada *et al.*, 1997). At a later stage, circular bands of cortical microtubules are localised around the edges of developing bordered pits (Figures 4B and C). Similar circular bands of cortical microtubules around the pit apertures of bordered pits have been observed in tracheids of *Taxus cuspidata* (Uehara and Hogetsu, 1993) and *Quercus ilex* (Chaffey *et al.*, 1997c) and in vessel elements of *Salix fragilis* (Robards and Humpherson, 1967) and *Aesculus hippocastanum* (Chaffey *et al.*, 1997b, c, 1999). As tracheids differentiate, the circular bands of cortical microtubules around the edges of inter-tracheal bordered pits become small centripetally. These circular bands of cortical microtubules might be involved in the localised deposition of cellulose microfibrils at pit borders.

During the final stages of formation of the secondary wall, obliquely oriented bands of cortical microtubules can be seen in the tracheids of some conifers, in which spiral thickenings are formed, such as *Taxus* (Uehara and Hogetsu, 1993; Figure 4A). These bands of cortical microtubules are first approximately 3 μm in width and then become narrow and rope-like structures (Furusawa *et al.*, 1998). Transverse bands of cortical microtubules have also been observed in the vessel elements of angiosperm trees, in which spiral thickenings are formed, at the final stage of differentiation (Chaffey *et al.*, 1997c, 1999; Chaffey, 2000). Spiral thickenings are formed by the deposition of localised ridges of parallel bundles of cellulose microfibrils on the innermost surfaces of tracheids. Bands of cortical microtubules are superimposed on these helical thickenings. The involvement of cortical microtubules in the formation of bordered pits and spiral thickenings suggests that the localised appearance or disappearance of cortical microtubules might control the localised deposition of cellulose microfibrils, thereby controlling modifications of wood structure.

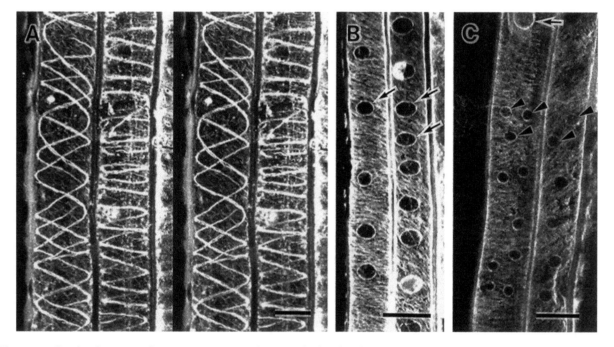

Figure 4 Confocal immunofluorescence images showing the localised arrangements of cortical microtubules that can be seen during formation of the helical wall thickenings and pits in differentiating tracheids of *Taxus cuspidata*. Single projections of 15–25 optical sections along the z-axis at 1 μm intervals. Averaging of images is eight times. A, Bands of helically oriented cortical microtubules during formation of the helical wall thickenings (a pair of stereo images). B and C, Circular bands of cortical microtubules are visible around the edges of developing bordered pits (projection images). Cortical microtubules in intertracheal pits (arrows) and cross field pits between tracheids and ray parenchyma cells (arrow heads). Bars = 25 μm.

Actin filaments

Bundles of actin filaments are oriented axially in fusiform cambial cells (Funada *et al.*, 2000). Their orientation in differentiating tracheids does not change during formation of the primary and secondary walls (Figure 3C) even though the orientation of cortical microtubules changes, as described above (Figures 3A and B). In some tracheids, transversely or obliquely oriented actin filaments have been observed at the final stage of xylem differentiation; in other tracheids the axial orientation is maintained. Thus, there appears to be no clear relationship, in terms of orientation, between cortical microtubules and actin filaments in differentiating tracheids (Figure 5). In addition, bundles of axially oriented microfilaments (actin filaments) are present in fusiform cambial cells and their axial orientation is retained in cambial derivatives during early stages of xylem differentiation in angiosperm trees (Chaffey *et al.*, 1997a, 2000; Chaffey, 2000; Chaffey and Barlow, 2000). Actin filaments might not play a major role in the changes in orientation of cortical microtubules in the differentiating secondary xylem cells that are derived from cambial cells.

FUTURE PROSPECTS

The recent advances in CLSM have provided powerful tools for the study of plant cell biology. The CLSM provides a significant improvement in lateral and axial resolution compared with the conventional microscope. In addition, CLSM with the software of image analysis allows the relatively rapid and easy 3-D visualisation of large and thick samples. In particular, CLSM might be very useful for 3-D analysis of soft and fragile developing cells, such as cambial cells and their derivatives. Thus, CLSM will be a standard tool for wood biologists in the near future (see also 'the Wood microscopical techniques' chapter in this volume).

As mentioned above, cortical microtubules in secondary xylem cells in gymnosperm trees change their arrangement during differentiation. However, mechanisms for the changes in arrangement of cortical microtubules are not fully understood. The detailed 3-D analysis of distribution of cortical microtubules, other components of cytoskeleton, such as actin filaments and intermediate filaments, and microtubule-associated proteins in fixed xylem cells might be important to understand these

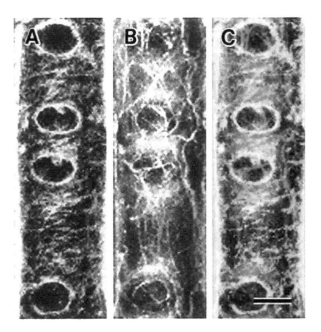

Figure 5 Confocal immunofluorescence images showing the arrangement of double immunostained components of the cytoskeleton in a differentiating tracheid of *Taxus cuspidata*. Single projection of 20 optical sections along the z-axis at 1 μm intervals. Averaging of images is eight times. A, The orientation of the cortical microtubules, stained with FITC-conjugated antibodies, is transverse with respect to the cell axis. B, The orientation of actin filaments, stained with TRITC-conjugated antibodies, is axial or oblique with respect to the cell axis. C, Superimposed image of cortical microtubules (green) and actin filaments (red). The overlap region is yellow. Bar = 10 μm.

mechanisms because it is likely that they interact during xylem differentiation. In addition, four-dimensional analysis (3-D analysis over time) with a microinjection of fluorescent analogues in living xylem cells allow us to follow the dynamics of cytoskeleton. The use of the CLSM will continue to provide valuable new information about the roles of the cytoskeleton in wood formation.

ACKNOWLEDGEMENTS

The author thanks Prof. K. Fukazawa, Prof. J. Ohtani, Dr Y. Sano, Mr M. Shibagaki, Mr O. Furusawa, Dr H. Abe, Dr A.K.M.A. Prodhan, Dr P. Kitin, Mr H. Miura, Mr T. Miura, Miss H. Imaizumi, Dr Y. Utsumi and Mr K. Yazaki for their co-operation during the preparation of this chapter. The author's research has been supported by Grants-in-Aid for Scientific Research from the Ministry of Education,

Science and Culture, Japan (Nos. 06404013, 07760156, 08760157, 09760152 and 11460076) and the Future program of Japan Society for the Promotion of Science (No. JSPS-RFTF 96L00605).

REFERENCES

Abe, H., Funada, R., Imaizumi, H., Ohtani, J. and Fukazawa, K. (1995a) Dynamic changes in the arrangement of cortical microtubules in conifer tracheids during differentiation. *Planta*, **197**, 418–421.

Abe, H., Funada, R., Ohtani, J. and Fukazawa, K. (1995b) Changes in the arrangement of microtubules and microfibrils in differentiating conifer tracheids during the expansion of cells. *Ann. Bot.*, **75**, 305–310.

Abe, H., Funada, R., Ohtani, J. and Fukazawa, K. (1997) Changes in the arrangement of cellulose microfibrils associated with the cessation of cell expansion in tracheids. *Trees*, **11**, 328–332.

Abe, H., Ohtani, J. and Fukazawa, K. (1991) FE-SEM observation on the microfibrillar orientation in the secondary wall of tracheids. *IAWA Bull. n.s.*, **12**, 431–438.

Abe, H., Ohtani, J. and Fukazawa, K. (1994) A scanning electron microscopic study of changes in microtubule distributions during secondary wall formation in tracheids. *IAWA J.*, **15**, 185–189.

Barnett, J.R., Chaffey, N.J. and Barlow, P.W. (1998) Cortical microtubules and microfibril angle. In *Microfibril Angle in Wood*, edited by B.G. Butterfield, pp. 253–271. Christchurch: IAWA/IUFRO.

Brelje, T.C., Wessendorf, M.W. and Sorensen, R.L. (1993) Multicolor laser scanning confocal immunofluorescence microscopy: practical application and limitations. In *Cell Biological Applications of Confocal Microscopy*, edited by B. Matsumoto, pp. 97–181. San Diego: Academic Press.

Catesson, A.M. (1994) Cambial ultrastructure and biochemistry: changes in relation to vascular tissue differentiation and the seasonal cycle. *Int. J. Plant. Sci.*, **155**, 251–261.

Catesson, A.M., Funada, R., Robert-Baby, D., Quinet-Szély, M., Chu-Bâ, J. and Goldberg, R. (1994) Biochemical and cytochemical cell wall changes across the cambial zone. *IAWA J.*, **15**, 91–101.

Chaffey, N.J. (2000) Cytoskeleton, cell walls and cambium: new insights into secondary xylem differentiation. In *Cell and Molecular Biology of Wood Formation*, edited by R. Savidge, J. Barnett and R. Napier, pp. 31–42. Oxford: BIOS Scientific Publishers.

Chaffey, N.J. and Barlow, P.W. (2000) Actin in the secondary vascular system of woody plants. In *Actin: A Dynamic Framework for Multiple Plant Cell Functions*, edited by F. Baluska, P.W. Barlow, C. Staiger and D. Volkmann, pp. 587–600. Dordrecht, The Netherlands Kluwer.

Chaffey, N.J., Barlow, P.W. and Barnett, J.R. (1997a) Cortical microtubules rearrange during differentiation of vascular cambial derivatives, microfilaments do not. *Trees*, **11**, 333–341.

Chaffey, N.J., Barlow, P.W. and Barnett, J.R. (1998) A seasonal cycle of cell wall structure is accompanied by a cyclical rearrangement of cortical microtubules in fusiform cambial cells within taproots of *Aesculus hippocastanum* L. (Hippocastanaceae). *New Phytol.*, **139**, 623–635.

Chaffey, N.J., Barlow, P.W. and Barnett, J.R., (2000) A cytoskeletal basis for wood formation in angiosperm trees: the involvement of microfilaments. *Planta*, **210**, 890–896.

Chaffey, N.J., Barnett, J.R. and Barlow, P.W. (1997b) Cortical microtubule involvement in bordered pit formation in secondary xylem vessel elements of *Aesculus hippocastanum* L. (Hippocastanaceae): a correlative study using electron microscopy and indirect immunofluorescence microscopy. *Protoplasma*, **197**, 64–75.

Chaffey, N.J., Barnett, J.R. and Barlow, P.W. (1997c) Visualization of the cytoskeleton within the secondary vascular system of hardwood species. *J. Microsc.*, **187**, 77–84.

Chaffey, N., Barnett, J. and Barlow, P. (1999) A cytoskeletal basis for wood formation in angiosperm trees: the involvement of cortical microtubules. *Planta*, **208**, 19–30.

Cleary, A.L., Brown, R.C. and Lemmon, B.E. (1993) Organisation of microtubules and actin filaments in the cortex of differentiating *Selaginella* guard cells. *Protoplasma*, **177**, 37–44.

Donaldson, L.A. and Lausberg, M.J.F. (1998) Comparison of conventional transmitted light and confocal microscopy for measuring wood dimensions by image analysis. *IAWA J.*, **19**, 321–336.

Emons, A.M.C., Derksen, J. and Sassen, M.M.A. (1992) Do microtubules orient plant cell wall microfibrils? *Physiol. Plant.*, **84**, 486–493.

Fujita, M., Saiki, H. and Harada, H. (1974) Electron microscopy of microtubules and cellulose microfibrils in secondary wall formation of poplar tension wood fibers. *Mokuzai Gakkaishi*, **20**, 147–156.

Fukuda, H. (1996) Xylogenesis: initiation, progression, and cell death. *Annu. Rev. Plant Physiol. Plant Mol. Biol.*, **47**, 299–325.

Funada, R. (2000) Control of wood structure. In *Plant Microtubules: Potential for Biotechnology*, edited by P. Nick, pp. 51–81. Springer-Verlag, Heidelberg.

Funada, R., Abe, H., Furusawa, O., Imaizumi, H., Fukazawa, K. and Ohtani, J. (1997) The orientation and localization of cortical microtubules in differentiating conifer tracheids during cell expansion. *Plant Cell Physiol.*, **38**, 210–212.

Funada, R., Furusawa, O., Shibagaki, M., Miura, H., Miura, T., Abe, H. and Ohtani, J. (2000) The role of cytoskeleton in secondary xylem differentiation in conifers. In *Cell and Molecular Biology of Wood Formation*, edited by R. Savidge, J. Barnett and R. Napier, pp. 255–264. Oxford: BIOS Scientific Publishers.

Furusawa, O., Funada, R., Murakami, Y. and Ohtani, J. (1998) Arrangement of cortical microtubules in compression wood tracheids of *Taxus cuspidata* visualized by confocal laser microscopy. *J. Wood Sci.*, **44**, 230–233.

Giddings, T.H. Jr. and Staehelin, L.A. (1991) Microtubule-mediated control of microfibril deposition: a re-examination of the hypothesis. In *The Cytoskeletal Basis of Plant Growth and Form,* edited by C.W. Lloyd, pp. 85–99. London: Academic Press.

Gray, J.D., Kolesik, P., Høj, P.B. and Coombe, B.G. (1999) Confocal measurement of the three-dimensional size and shape of plant parenchyma cells in a developing fruit tissue. *Plant J.,* **19**, 229–236.

Hale, I.L. and Matsumoto, B. (1993) Resolution of subcellular detail in thick tissue sections: immunohistochemical preparation and fluorescence confocal microscopy. In *Cell Biological Applications of Confocal Microscopy,* edited by B. Matsumoto, pp. 289–324. San Diego: Academic Press.

Harada, H. and Côté, W.A. Jr. (1985) Structure of wood. In: *Biosynthesis and Biodegradation of Wood Components,* edited by T. Higuchi, pp. 1–42. Orlando: Academic Press.

Hell, S.W. and Stelzer, E.H.K. (1995) Lens aberrations in confocal fluorescence microscopy. In *Handbook of Biological Confocal Microscopy,* 2nd edn, edited by J.B. Pawley, pp. 347–354. New York: Plenum Press.

Hepler, P.K. and Gunning, B.E.S. (1998) Confocal fluorescence microscopy of plant cells. *Protoplasma,* **201**, 121–157.

Himmelspach, R., Wymer, C.L., Lloyd, C.W. and Nick, P. (1999) Gravity-induced reorientation of cortical microtubules observed *in vivo. Plant J.,* **18**, 449–453.

Kataoka, Y., Saiki, H. and Fujita, M. (1992) Arrangement and superimposition of cellulose microfibrils in the secondary walls of coniferous tracheids. *Mokuzai Gakkaishi,* **38**, 327–335.

Knebel, W. and Schnepf, E. (1991) Confocal laser scanning microscopy of fluorescently stained wood cells: a new method for three-dimensional imaging of xylem elements. *Trees,* **5**, 1–4.

Kitin, P., Funada, R., Sano, Y., Beeckman, H. and Ohtani, J. (1999) Variations in the length of fusiform cambial cells and vessel elements in *Kalopanax pictus. Ann. Bot.,* **84**, 621–632.

Kitin, P., Funada, R., Sano, Y. and Ohtani, J. (2000) Analysis by confocal microscopy of the structure of cambium in the hardwood *Kalopanax pictus. Ann. Bot.* **86**, 1109–1117.

Larson, P.R. (1994) *The Vascular Cambium: Development and Structure.* Heidelberg: Springer-Verlag.

Ledbetter, M.C. and Porter, K.R. (1963) A 'microtubule' in plant cell fine structure. *J. Cell Biol.,* **19**, 239–250.

Lloyd, C.W. (1987) The plant cytoskeleton: the impact of fluorescence microscopy. *Annu. Rev. Plant Physiol.,* **38**, 119–139.

Lloyd, C.W., Shaw, P.J., Warn, R.M. and Yuan, M. (1996) Gibberellic-acid-induced reorientation of cortical microtubules in living plant cells. *J. Microsc.,* **181**, 140–144.

Matsumoto, B. (1993) *Cell Biological Applications of Confocal Microscopy (Methods in Cell Biology, vol. 38).* San Diego: Academic Press.

Matsumura, J., Booker, R.E., Donaldson, L.A. and Ridoutt, B.G. (1998) Impregnation of radiata pine wood by vacuum treatment: identification of flow paths using fluorescent dye and confocal microscopy. *IAWA J.,* **19**, 25–33.

Minsky, M. (1988) Memoir on inventing the confocal scanning microscope. *Scanning,* **10**, 128–138.

Murakami, Y., Funada, R., Sano, Y. and Ohtani, J. (1999) The differentiation of contact cells and isolation cells in the xylem ray parenchyma of *Populus maximowiczii. Ann. Bot.,* **84**, 429–435.

Nobuchi, T. and Fujita, M. (1972) Cytological structure of differentiating tension wood fibers of *Populus euroamericana. Mokuzai Gakkaishi,* **18**, 137–144.

Pawley, J.B. (1995) *Handbook of Biological Confocal Microscopy,* 2nd edn. New York: Plenum Press.

Prodhan, A.K.M.A., Funada, R., Ohtani, J., Abe, H. and Fukazawa, K. (1995a) Orientation of microfibrils and microtubules in developing tension-wood fibers of Japanese ash (*Fraxinus mandshurica* var. *japonica*). *Planta,* **196**, 577–585.

Prodhan, A.K.M.A., Ohtani, J., Funada, R., Abe, H. and Fukazawa, K. (1995b) Ultrastructural investigation of tension wood fibre in *Fraxinus mandshurica* Rupr. var. *japonica* Maxim. *Ann. Bot.,* **75**, 311–317.

Robards, A.W. and Humpherson, P.G. (1967) Microtubules and angiosperm bordered pit formation. *Planta,* **77**, 233–238.

Robards, A.W. and Kidwai, P.A. (1972) Microtubules and microfibrils in xylem fibres during secondary wall formation. *Cytobiologie,* **6**, 1–21.

Roland, J.C. and Mosiniak, M. (1983) On the twisting pattern texture and layering of the secondary cell walls of lime wood. proposal of an unifying model. *IAWA Bull. n.s.,* 4, 15–26.

Roland, J.C. and Vian, B. (1991) General preparation and staining of thin sections In *Electron Microscopy of Plant Cells,* edited by J.L. Hall and C. Hawes, pp. 1–66. London: Academic Press.

Sheppard, C.J.R. and Shotton, D.M. (1997) *Confocal Laser Scanning Microscopy.* Oxford: BIOS Scientific Publishers.

Sonobe, S. and Shibaoka, H. (1989) Cortical fine actin filaments in higher plant cells visualized by rhodamine-phalloidin after pretreatment with *m*-maleimidobenzoyl N-hydroxysuccinimide ester. *Protoplasma,* **148**, 80–86.

Uehara, K. and Hogetsu, T. (1993) Arrangement of cortical microtubules during formation of bordered pit in the tracheids of *Taxus. Protoplasma,* **172**, 145–153.

Vesk, P.A., Vesk, M. and Gunning, B.E.S. (1996) Field emission scanning electron microscopy of microtubule arrays in higher plant cells. *Protoplasma,* **195**, 168–182.

Wilson, T. (1990) *Confocal Microscopy.* London: Academic Press.

Yuan, M., Shaw, P.J., Warn, R.M. and Lloyd, C.W. (1994) Dynamic reorientation of cortical microtubules, from transverse to longitudinal, in living plant cells. *Proc. Natl. Acad. Sci. USA,* **91**, 6050–6053.

Yuan, M., Warn, R.M., Shaw, P.J. and Lloyd, C.W. (1995) Dynamic microtubules under the radial and outer tangential walls of microinjected pea epidermal cells observed by computer reconstruction. *Plant J.,* **7**, 17–23.

0 Cell Walls of Woody Plants: Autoradiography and Ultraviolet Microscopy

Wood Formation in Trees ed. N. Chaffey
© 2002 Taylor & Francis
Taylor & Francis is an imprint of the
Taylor & Francis Group
Printed in Singapore.

Keiji Takabe

Graduate School of Agriculture, Kyoto University, Kyoto 606–8502, Japan

ABSTRACT

Autoradiography can be used to visualise the biochemical events that occur along the metabolic pathway in the cells. Careful selection of radiolabelled precursor gives us more reliable results. This chapter includes detailed protocols for the administration of radiolabelled precursors to woody plants to study cell wall formation, and the visualisation of the radioactive products under both light and transmission electron microscopes. Lignin is the most characteristic component of the cell wall in woody plants. It absorbs ultraviolet (UV) light, whereas, the other main cell wall components, that is, cellulose, pectins and hemicelluloses, do not. Therefore, localisation of lignin within the cell wall is observable in unstained material under a UV-microscope. This chapter also includes details of the preparation of woody specimens for UV-microscopy and the determination of softwood lignin by UV-microspectrometry.

Key words: autoradiography, cell wall formation, light microscope, lignin, transmission electron microscope, UV-microspectrometry

INTRODUCTION

Brief summary of use of autoradiography in the study of cell wall formation

Autoradiography is a useful technique for visualising physiological and biochemical processes within cells and cell walls. The incorporation site of radiolabelled precursor is easily identified as black deposits after developing the nuclear emulsion coated on the specimen. Choosing good radiolabelled precursors is most important for good results. Radiolabelled precursor of lignin, for example, enables us to show which cells are undergoing lignification and which part of the cell wall accumulates monolignols (Figure 1).

Use of autoradiography in cell wall formation has revealed both the distribution of cell wall components within the cell walls and the role of cell organelles involved in synthesis of cell wall components. Setterfield and Bayley (1959) and Ray (1967) investigated the manner of deposition of cellulose and hemicelluloses, and Saleh *et al.* (1967), Fujita and Harada (1979), and Takabe *et al.* (1981) demonstrated the lignification process of cell wall in woody plants. Thereafter, Terashima and his co-workers showed the outstanding results of lignification and heterogeneous distribution of lignin in softwoods and hardwoods by autoradiography coupled with specific labelling of lignin precursors (summarised and reviewed by Terashima, 1993). Northcote and Wooding (1965), Northcote and Pickett-Heaps (1966), and Wooding (1968) showed the role of the Golgi apparatus in synthesis and transport of polysaccharides. Pickett-Heaps (1968), Fujita and Harada (1979) and Takabe *et al.* (1985) investigated the cell organelles involved in lignification by using lignin precursors.

Brief summary of UV-microscopy

A UV-microscope is a powerful tool for the detection of lignin, because lignin absorbs UV light, whereas other cell wall components, that is,

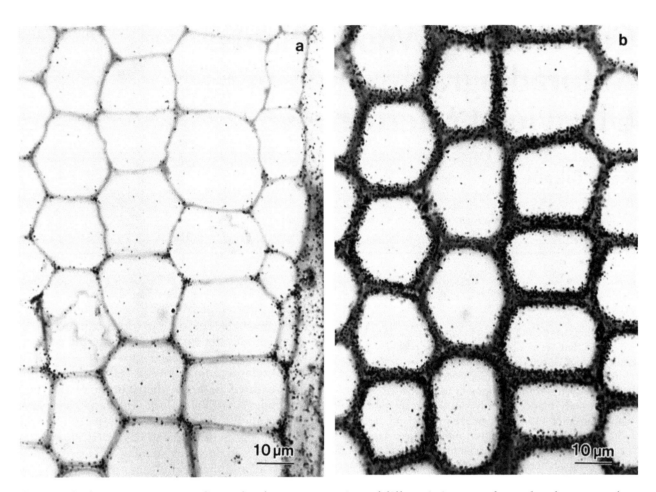

Figure 1   Light microscopic autoradiographs of transverse sections of differentiating secondary xylem from stem of *Cryptomeria japonica* fed ^{3}H-phenylalanine. a. Silver grains are observed at cell corner regions in the tracheids of S_1 formation stage. b. Numerous silver grains are distributed on the secondary walls in the tracheids just after S_3 formation.

cellulose, pectins and hemicelluloses do not (Figure 2). The first attempt to observe lignification of the cell wall by UV-microscopy was achieved by Wardrop (1957), who detailed the process of lignification of cell wall in *Pinus radiata*. Despite their age, it is emphasised that most of his results are still valid today.

In the 1960s and 70s, Goring and his co-workers further developed the technique of UV-microscopy for determination of lignin within the cell wall. Fergus *et al.* (1969) established the technique for measuring lignin contents of cell wall in softwoods by UV-microscopy coupled with densitometry of UV-photographic negatives. Scott *et al.* (1969) determined the lignin contents in wood cell wall by a more advanced method of UV-microscopy. They

further showed that their method is ideally suited to the study of the removal of lignin during cooking. Musha and Goring (1975) demonstrated the distribution of syringyl and guaiacyl lignins in hardwoods. On the basis of the difference of UV absorption spectra between syringyl and guaiacyl lignins, they measured the spectra from vessel, fibre and ray cell walls, and showed heterogeneous distribution of lignins in hardwoods. Afterwards, many workers demonstrated the lignin distribution in softwoods and hardwoods according to their methods.

In the 1990s, a newly developed UV-microspectrometer enabled us easily to obtain UV-absorption spectra from a very limited area of the cell wall. Takabe *et al.* (1992) and Yoshinaga *et al.* (1997) revealed the distribution of syringyl and guaiacyl

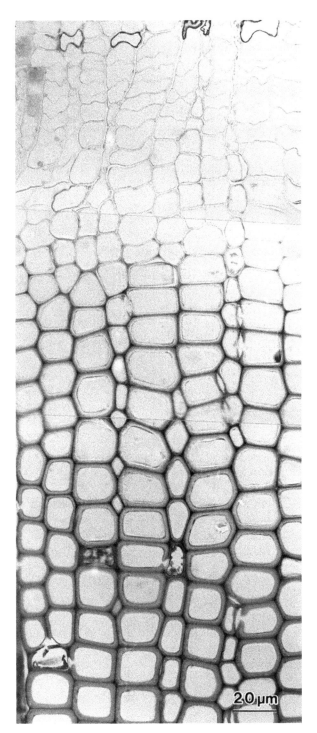

Figure 2 Transverse section of differentiating secondary xylem from stem of *Cryptomeria japonica* photographed in UV light at 280 nm.

lignins in fibre and vessel secondary walls within an annual ring in beech and oak woods respectively. Thereafter, Okuyama *et al.* (1998) established the method for determining lignin contents within the cell wall in softwoods by UV-microspectrometry.

TECHNIQUE

Note – the techniques described below use radioactive materials. Appropriate safe-working procedures should be employed in all aspects of this work, and all local regulations relating to handling and disposal of these materials observed.

Autoradiography

Administration of radiolabelled precursor

Equipment needed:	Chemicals needed:
Absorbent cotton, or glass wool	Radiolabelled precursor
Knife	
Parafilm	
Pipettes	
Razor blade	
Saw	

Duration of procedure: approx. one day.

1 Cut a trunk into 20–30 cm lengths with a saw.
2 Seal both cut surfaces of the excised pieces with parafilm.
3 Scrape away a small part of the bark with a knife or a razor blade.
4 Make a V-shaped groove, 2 mm wide and 10 mm long, into the differentiating xylem with a razor blade.
5 Pack absorbent cotton or glass wool into the groove.
6 Add a drop of radiolabelled precursor continuously to absorbent cotton or glass wool for scheduled hours (Figure 3).
7 Cut into small blocks (2 × 3 × 5 mm, tangential, radial and longitudinal directions) with a knife and razor blade, and fix (see next section).

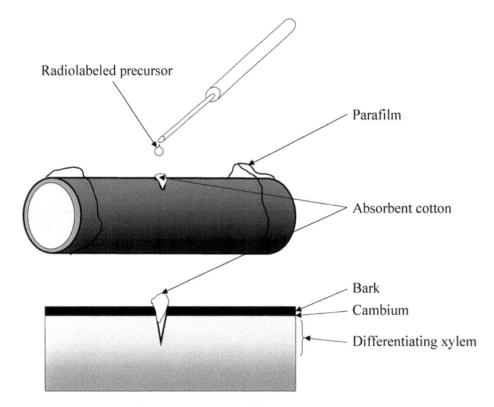

Figure 3   Administration of radiolabelled precursor to the differentiating secondary xylem.

Fixation and embedding

Equipment needed:	Chemicals needed:
Embedding capsules	Epoxy resin (as supplied by TAAB, Polyscience, etc) I recommend using the hardest grade resin
Pipettes	Ethanol
Refrigerator	Glutaraldehyde
Scintillation counter	Osmium tetroxide
Glass vials (10–20 ml) with tops	0.067 M (1/15) phosphate buffer (pH 7.1–7.4); mix 0.067M potassium dihydrogen phosphate aqueous solution with 0.067M disodium hydrogen phosphate aqueous solution to adjust pH
Rotator	Propylene oxide
Oven	Distilled water
	Ice

Duration of procedure: approx. 6 days.

1   Fix for 2–4 h in 3% glutaraldehyde in phosphate buffer in ice-water or in a refrigerator.

2   Wash in phosphate buffer 5 × 15 min in ice water.

3   Post-fix for 2 h in 1% osmium tetroxide in phosphate buffer in ice-water or in a refrigerator.

4   Wash in phosphate buffer 5 × 15 min at room temperature (RT).

5   Dehydrate in a graded water:ethanol series: 30%, 50%, 70%, 80%, 90% for 15 min each, followed by 3 × 15 min in 95%, 100% ethanol.
   Note – during the above steps 1–5, radioactivity in the fixatives, buffer and graded ethanol series must be monitored with a scintillation counter.

6   Infiltrate with propylene oxide 3 × 30 min.

7   Infiltrate with propylene oxide:epoxy resin (1:1) overnight using a rotator.

8   Embed specimen in fresh epoxy resin.

9   Transfer specimens to an oven at 35 °C, leave overnight, increase to 45 °C for 1 day, and finally to 60 °C for 1 day to cure resin.

Detection of radiolabel by light microscope autoradiography

Equipment needed:	Chemicals needed:
Coplin jar	Canada balsam
Cover slips	Developer (D-19, Kodak)
Light-tight box	Distilled water
Filter paper	95% ethanol
Glass knives	Fixer (Fuji Fix, Fuji Film or F-5, Kodak)
Glass microscope slides	Nuclear emulsion (Sakura NR-H2 emulsion, Kodak NTB-2 or 3)
Knife	1% aqueous safranin solution
Light microscope fitted with bright field optics	Stripping films (London Kodak AR-10)
Microtome	Silica gel
Platinum loop	
Razor blade	
Refrigerator	
Slide box	
Staining dish with removable slide rack	
Tape	
Trays (approx. 20 × 15 × 3 cm)	
Water bath	
Vinyl tape	
Kodak safelight filter No. 1 illuminator, or equivalent	

Duration of procedure: approx. one to several weeks.

1   Trim a specimen with a knife and razor blade.
2   Cut 1 μm-thick sections using a microtome equipped with a glass knife and water-filled boat.
3   Pick up sections with a platinum loop.

4   Mount several sections on an uncoated glass slide and allow to air-dry.
     Note – I recommend preparing several sets of glass slides for different exposure periods. Exposure period depends on the specific activity of radiolabelled precursor, amount of incorporation into the specimen, thickness of section and so on.
     Note – The slides should be coated simultaneously with emulsion, stored in dark condition for exposure, and developed at one- or two-week intervals. Although the slides can be stored in dust-free condition for several months before they are coated with emulsion, ideally they should be used for autoradiography as soon as possible.
     Note – the following steps 5–13 and 16–18 must be carried out in a darkroom, under an appropriate safelight (Kodak safelight filter No. 1 illuminator, or equivalent, is recommended). Relative humidity of darkroom is not important.

If using stripping film (Figure 4), proceed as follows:

5   Cut stripping film into an appropriate size (approx. 5 × 2 cm) with a razor blade.
     Note – stripping film is supplied on a glass plate.
6   Soak the film in a tray filled with 95% ethanol for 5 min.
7   Strip off a piece of film from a glass plate.
8   Float a piece of film, emulsion face down, on distilled water in a tray at RT, and leave for several min.
9   Cover the section with film.
10   Remove distilled water from the glass slide with filter paper.
11   Air-dry the glass slide.

If using nuclear emulsion (Figure 5), proceed as follows:

5′   Melt the nuclear emulsion in a water bath at 40 °C. I recommend using a Coplin jar for melting emulsion.
     Note – nuclear emulsion arrives as a gel in a container. Before using the emulsion, I recommend that you melt it in a water bath, subdivide into smaller volumes and store individually in 10–20 ml glass-vials in order

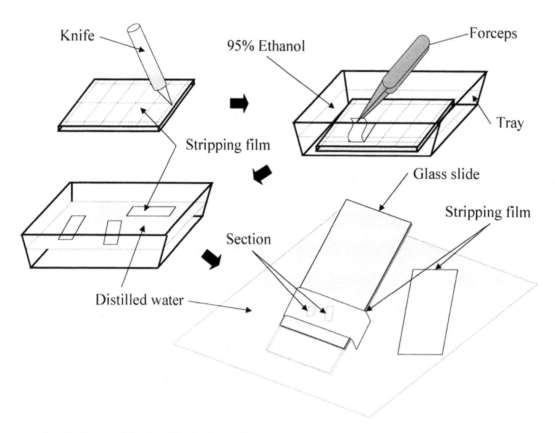

Figure 4   Application of stripping-film to the sections

to avoid accidentally exposing the whole batch to light. The vials must be wrapped with aluminium foil, kept in a light-tight box, and stored in a refrigerator.

6′ Dilute the emulsion with an equal volume of distilled water warmed at 40 °C.
7′ Dip a glass slide, on which sections are mounted, perpendicularly into emulsion,

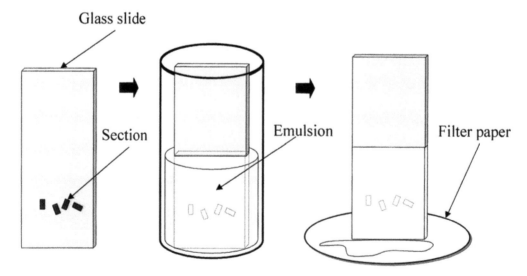

Figure 5   Coating of sections with nuclear emulsion

leave for several seconds, remove and get rid of excess emulsion from slide with filter paper.

8'   Air-dry the glass slide at RT.

The following steps apply irrespective of whether nuclear emulsion or stripping film was used:

12   Store the coated glass slides in a slide box containing silica gel and seal the box with vinyl tape.

13   Put the slide box in a light-tight box.

14   Store box in a refrigerator for 1 to several weeks.

15   To develop image, prepare developer and keep at 20 °C. I recommend using a staining dish with a removable glass slide rack for developing.

16   Develop film or emulsion by dipping glass slides into developer for 7 min at 20 °C.

17   Wash developed glass slides in distilled water for 1 min.

18   Fix film or emulsion by dipping glass slides into fixer at 20 °C for 10 min.

19   Wash glass slides in running tap water for 10 min.

20   Air-dry the glass slides.

21   Stain the sections with 1% aqueous safranin solution for 10–30 min.

22   Wash sections several times with distilled water to remove excess staining solution, followed by washing with 50% ethanol till section is clearly observable.

23   Air-dry slides.

24   Mount sections beneath a cover slip with Canada balsam.

25   Observe any radiolabel under a bright field light microscope.
     Note – radiolabel appears as black dots under a light microscope. Refer to Figure 1.

Detection of radiolabel by TEM autoradiography

Equipment needed:	*Chemicals needed:*
Coplin jar	Carbon
Copper grids from 75–300 mesh	Colloidon
Light-tight box	Developer (D-19, Kodak)
Diamond knife	Distilled water
Transmission electron microscope (TEM)	Fixer (Fuji Fix, Fuji Film or F-5, Kodak)
Filter paper	Lead nitrate
Forceps	Nuclear emulsion (Sakura NR-H2 emulsion, Kodak NTB-2 or 3)
Glass rods (diameter: 5–7 mm, length: 5 cm) fixed on stand (Figure 6)	Silica gel
Grid holder (Figure 6)	
Grid storage box	Uranyl acetate
Utility knife	Sodium citrate
Glass Petri dish	Sodium hydroxide
Platinum loop (diameter of loop should be slightly greater than that of glass rod)	
Razor blades	
Refrigerator	
Ultramicrotome	
Vacuum evaporator	
Vinyl tape	
Water bath	
Double-sided sticky tape	
Volumetric flask (50 ml capacity)	

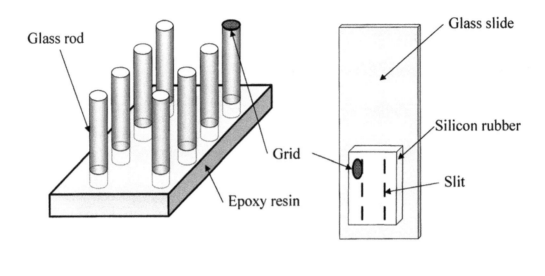

Glass rods set in epoxy resin   Grid holder

Figure 6 Glass rods and grid holder used in the 'wire-loop method' (see Figure 7). Glass rods are fixed onto the stand using epoxy resin or araldite adhesive. Silicon rubber sticks to a glass slide with araldite adhesive. Grid is fixed on silicon rubber by being inserted into slit.

Duration of procedure: approx. one to several months.

1 Trim specimen with a knife and razor blade.
2 Cut ultra-thin sections with a ulramicrotome equipped with a diamond knife.
3 Mount sections on a colloidon-coated copper grid.
4 Stain ultra-thin sections with 2% aqueous uranyl acetate solution for 30 min at 45 °C.
5 Wash sections with distilled water.
6 Stain sections with Reynolds' lead citrate stain (Reynolds, 1963) for several min at RT.

Make Reynolds' stain as follows: add 1.33 g of lead nitrate, 1.76 g of sodium citrate, and 30 ml of distilled water to a 50 ml volumetric flask, and shake well for 1 min, and intermittently for 30 min. Add 8.0 ml of 1M NaOH, and dilute to 50 ml with distilled water.

7 Wash sections with distilled water.
8 Carbon-coat the sections with a vacuum evaporator.

If following the 'wire-loop method' (Figure 7), proceed as follows:

9 Fix the grids on top of glass rods using double-sided sticky tape
   Note – the following steps 10–14 must be carried out in a darkroom.
10 Melt nuclear emulsion using water bath at 40 °C. I recommend using a Coplin jar for melting the emulsion.
11 Dilute the emulsion 10 times with distilled water at 40 °C.
   Note – excess diluted emulsion can be kept in usable condition for several months, if kept in the dark and stored in a refrigerator.
12 Dip a wire loop into the emulsion, hold it there for several seconds, then lift it out of the emulsion. Check that a thin layer of emulsion has formed in the loop.
13 Pass the loop over the glass rod to cover the ultrathin sections with emulsion.
14 Leave the glass rods in a dust free place for several hours.

If using the 'touching method' (Figure 8), proceed as follows:

   Note – the following steps 9'–13' must be carried out in a darkroom.
9' Dissolve nuclear emulsion using a water bath at 40 °C. I recommend using a Coplin jar for dissolving the emulsion.

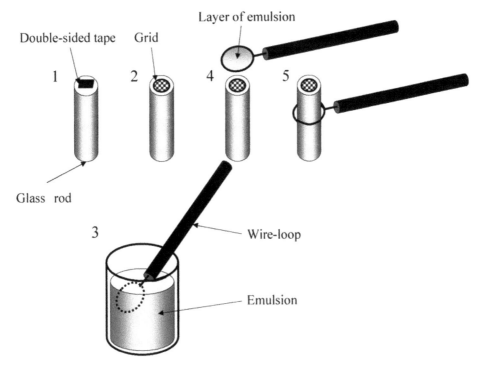

Figure 7   Coating of nuclear emulsion on the ultrathin sections by the 'wire-loop method'.

10′   Dilute emulsion 10 times with distilled water at 40 °C.

11′   Using forceps, touch the ultra-thin sections on the grid on to the surface of the emulsion.

12′   Remove excess emulsion from the grid with filter paper.

13′   Air-dry grids in a Petri dish.
Note – the following steps 15–17 and 21–23 must be carried out in a darkroom.

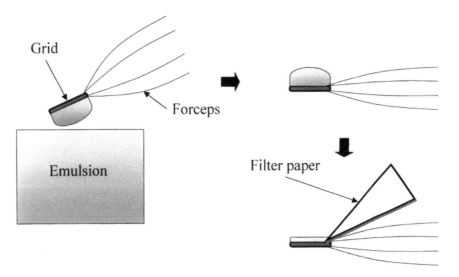

Figure 8   Coating of nuclear emulsion on the ultrathin sections by the 'touching method'.

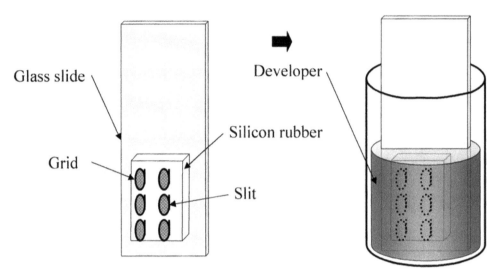

Figure 9   Developing nuclear emulsion.

15   Store the coated grids in a grid storage box. I recommend using a grid storage box with many rhombic holes. The grids should be inserted perpendicularly into the holes.
16   Store grid case in a light-tight box containing silica-gel.
17   Seal the box with vinyl tape.
18   Store in a refrigerator for several weeks.
     Note – exposure period depends on specific activity of radiolabelled precursor, incorporation of the precursor into specimen, thickness of section and so on. I recommend to develop the emulsion coated on sections at intervals of a few weeks.
19   Make up developer and keep at 20 °C.
20   Affix grids on a grid holder (Figure 9), or stick grids on a glass slide with a double-sided tape.
21   Develop the emulsion by dipping grid holder or glass slide into developer for 3.5 min at 19 °C (Figure 9).
22   Wash the grid holder or glass slide in distilled water for 1 min.
23   Fix the emulsion by dipping grid holder or glass slide into fixer for 10 min.
24   Wash grid holder or glass slide in distilled water for 10 min.
25   Air-dry grids in a dust-free place.
26   Observe ultra-thin sections under a TEM.

UV-microscopy

Specimen preparation

Equipment needed:	Chemicals needed:
Diamond knife	Epoxy resin (as supplied from TAAB, Polyscience, etc.). I recommend using the hardest grade resin.
Flat embedding moulds	Ethanol:benzene (1:2) mixture
Glass vials (10–20 ml) with tops	37% formaldehyde solution
Knife	Distilled water
Oven	Glacial acetic acid
Platinum loop	Ethyl alcohol
Quartz microscope slide	Glycerine
Quartz cover slips	Propylene oxide
Razor blade	
Rotator	
Ultramicrotome	

Duration of procedure: approx. one week.

For differentiating xylem:

1 Excise differentiating xylem with a knife, cut into small blocks (2 × 3 × 4 mm, tangential, radial and longitudinal directions) in water with a razor blade and put them in glass vials.
2 Fix blocks in formalin: acetic acid: (ethyl) alcohol (FAA) cooled in ice-water for 1 day.
Note – FAA is better than glutaraldehyde for UV microscopy because the latter has UV absorption maximum at 280 nm (as for lignin), and made up as follows:

37% formaldehyde solution	10 ml
Distilled water	35 ml
Glacial acetic acid	5 ml
Absolute ethyl alcohol	50 ml

3 Wash the blocks in distilled water 6 × 15 min.

For mature xylem:

1′ Excise mature xylem and cut into small blocks with a razor blade.
2′ Extract polyphenols from specimen with ethanol:benzene (1:2) mixture for several hours.
Note – FAA fixation is not necessary for mature wood.
Note – polyphenols affect the UV-absorption spectra, because they absorb UV-light.
3′ Wash the blocks in distilled water 6 × 15 min.
4 Dehydrate in a graded water:ethanol series: 30%, 50%, 70%, 80%, 90% for 15 min each, followed by 3 × 15 min in 95%, 100% ethanol.
5 Infiltrate with propylene oxide 3 × 30 min.
6 Infiltrate with propylene oxide:epoxy resin (1:1) overnight using a rotator.
7 Embed the specimen in fresh epoxy resin using a flat embedding mould.
8 Transfer moulds to oven at 35 °C and leave overnight, then at 45 °C for 1 day, and finally at 60 °C for 1 day for curing resin.
9 Trim specimen with a knife and razor blade.
10 Cut 0.5–1 μm-thick sections with a microtome equipped with a diamond knife or a glass knife.
Note – the former is much better for preparing the required smooth, flat surfaces of sections than the latter.

11 Pick up the sections with a platinum loop.
12 Mount the sections on an uncoated quartz slide.
13 Add 1–3 drops of immersion glycerine on top of the sections.
14 Cover the sections with a quartz cover slip.

Observation under a UV-microscope

Equipment needed:	Chemicals needed:
Camera (Carl Zeiss)	Acetic acid (for stop-bath)
CCD camera suitable for UV light (Carl Zeiss)	Developer
Film	Fixer
Microspectrometer (Carl Zeiss) (MPM 800)	Immersion glycerine
UV monitor (Carl Zeiss)	

Duration of procedure: approx. one day.

1 Mount a quartz microscope slide onto the specimen stage of microspectrometer.
2 Focus onto a section using a low magnification objective lens under visible light.
3 Place a drop of immersion glycerine on a quartz cover slip.
4 Refocus onto a section using a 32× quartz objective.
5 Set a monochromator of illumination light at 280 nm.
6 Change illumination from visible light to UV light.
7 Check illumination of section using a monitor. If illumination is not uniform, it will be necessary to centre, and adjust the height of the condenser lens.
8 Observe a section on a monitor connected with CCD camera suitable for UV light.
9 Take several UV-photomicrographs at slightly different levels of focus. I usually use black and white film with ISO 100. Exposure times are 10 and 60 seconds for 32× and 100× lenses, respectively.
10 Develop film according to manufacturer's instructions.

Measurement of guaiacyl lignin content by
UV-microspectrometry

Equipment needed:	Chemicals needed:
Microspectrometer (MPM 800) (Carl Zeiss) controlled by software, Lambda Scan (Carl Zeiss) Quartz microscope slide (Carl Zeiss) Quartz cover slip (Carl Zeiss) Universal surface shape profiler (SE-3E) (Kosaka, Japan) Atomic force microscope (optional!)	Immersion glycerine

Duration of procedure: approx. 1–2 days.

1   Cut and mount 0.5–1 μm-thick sections on a quartz microscope slide (as above).
2   Measure the thickness of sections with a surface shape profiler or an atomic force microscope.
3   Add 1–3 drops of immersion glycerine to the sections.
4   Cover sections with a quartz cover slip.
5   Mount a quartz slide onto the specimen stage of the microspectrometer.
6   Focus onto a section using a low magnification objective under visible light.
7   Add a drop of immersion glycerine to a quartz cover slip.
8   Refocus on a section using a 32× quartz lens.
9   Set a monochromator of illumination light at 280 nm.
10 Change illumination from visible to UV light.
11 Check the illumination of UV light on a monitor connected to a CCD camera that is suitable for UV light. If the illumination is not uniform, centre and adjust height of condenser lens.
12 Observe a section on a monitor and select carefully the portion from which the UV absorption spectrum is to be measured.
      Note – it is necessary to choose a smooth surface of cell wall, because cracking and/or wrinkles of the sections strongly affect the

absorption spectrum and consequently lead to the miscalculation of lignin content.
13 Select the diameter of the measuring circle.
      Note – I recommend using the 0.5 μm diameter measuring circle for cell corner middle lamella, and the 1.5 μm diameter measuring circle for secondary wall.
14 Set the bandwidth of the monochromator at 5 nm.
      Note – the bandwidth of the monochromator is changeable from 1 to 30 nm. I recommend adjusting it to 5 nm for reliable spectra.
15 Measure stray light 50 times from 250 to 350 nm with 1 nm step increment.
16 Focus onto the section.
17 Measure the light intensity that passes through the lumen 50 times, from 250 to 350 nm, with 1 nm step increment.
      Note – it is much better to measure the light intensity passing through the lumen of the cell from whose cell wall the spectrum is to be measured.
18 Adjust the measuring circle onto the cell wall from which the spectrum is to be measured, and focus carefully onto the section again.
19 Measure the light intensity passing through the cell wall 50 times from 250 to 350 nm with 1 nm step increment.
20 Calculate the UV absorption at each wavelength. In practice, the Lambda Scan software calculates this automatically according to the following equation:

$$A = 2 - \log_{10}((O - P)/(S - P) \times 100)$$

Where O, S and P are intensity of UV light passing through the cell wall, intensity of UV light passing through the cell lumen and stray UV light, respectively (see Figure 10).
21 Calculate lignin content according to Lambert-Beer's law:

$$A = \varepsilon \times C \times d$$

Where A, ε, C and d are UV absorption, an absorption coefficient ($cm^{-1}.l.g^{-1}$), lignin concentration ($g.l^{-1}$) and the distance of light pass (cm), respectively.
      Note – in practice, slight modification of this relationship is necessary for calculating lignin content from UV absorption at 280 nm obtained by microspectrometry. The MPM 800 illuminates the section with non-parallel light, which causes slightly higher absorption than

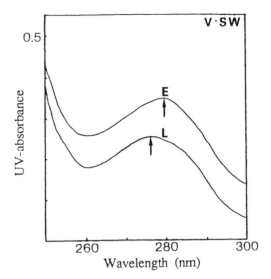

Figure 10 UV absorption spectra of the secondary wall of a secondary xylem vessel from Japanese beech. E: earlywood, L: latewood. Arrows show the position of absorption maxima.

with parallel light. According to Okuyama *et al.* (1998), the increase of UV absorption is 6%. Therefore, the absorption at 280 nm must be divided by 1.06. The absorption coefficient varies from 12.8 to 18.7 cm^{-1}.l.g^{-1}. Fergus *et al.* (1969) recommend using a value of ε = 15.6 cm^{-1}.l.g^{-1} for the calculation of lignin content. Therefore, the equation is modified to:

$$A/1.06 = 15.6 \times C \times d' \times 10^{-4}$$

Where d' is section thickness (in μm). Lignin content (g.l^{-1}) is then calculated as follows:

$$C = (A \times 10^4) / (1.06 \times 15.6 \times d)$$

The value of lignin content (g.g^{-1}) is calculated by converting the unit gram per litre to gram per cubic centimetre followed by multiplying by specific volume. According to Stone and Scallan (1967), specific volume of the water-swollen cell wall of black spruce is 1.07 cm^3.g^{-1}. Therefore, lignin content (g.g^{-1}) is calculated according to the following equation:

$$C' = (A \times 1.07 \times 10^4) / (1.06 \times 15.6 \times 10^3 \times d)$$

Where C' is lignin content (g.g^{-1}).

Note – hardwood lignins are co-polymers of guaiacyl and syringyl moieties. Since the ratio of guaiacyl- to syringyl-units differs between species, types of cells and cell wall

layers, it is impossible to determine lignin content in hardwood solely by UV-microspectrometry.

Visible light (VIS)-microspectrometry for detection of syringyl lignin

Equipment needed:	Chemicals needed:
Cover slip	Ammonium hydroxide
Forceps	Dental wax
Glass microscope slide	Distilled water
Microspectrometer	3% hydrochloric acid
Sliding microtome	Immersion glycerine
Spatula	1% aqueous potassium permanganate solution

Duration of procedure: approx. one day.

1 Prepare 10–20 μm-thick sections using a sliding microtome.
2 Mount a section on an uncoated glass slide with a spatula and forceps.
3 Add a drop of distilled water onto a section.
4 Cover the section with a cover slip.
5 Mount a glass slide onto the specimen stage of the microspectrometer.
6 Focus onto a section using a low magnification objective under visible light.
7 Place a drop of immersion glycerine on a cover slip.
8 Refocus onto a section using 32× quartz objective lens.
9 Set a monochromator of illumination light at 550 nm.
10 Change illumination from visible to monochromatic light.
11 Check the illumination of monochromatic light. If illumination onto section is not uniform, centre and adjust the height of the condenser lens.
12 Select the diameter of the measuring circle.
13 Set the bandwidth of monochromator at 5 nm.
14 Measure stray light 5 times from 400 to 700 nm with 5 nm step increment.
15 Focus onto the section.
16 Measure the light intensity passing through the lumen 5 times from 400 to 700 nm with 5 nm step increment.

Note – the following steps 17–22 are the Mäule (1901) reaction.

17   Treat another section with 1% aqueous potassium permanganate solution for 5 min in a vial.

18   Wash sections 3 × 2 min with distilled water.

19   Treat the sections with 3% hydrochloric acid till the colour of sections changes from black or dark-brown to pale-brown.

20   Wash the sections 3 × 1 min with distilled water.

21   Mount the sections on a glass slide.

22   Add 1–3 drops of saturated ammonium hydroxide to the sections.

23   Cover the sections with a cover slip.

24   Seal the cover slip with dental wax to avoid damage to objective lens from ammonium hydroxide.

25   Mount a glass microscope slide onto the specimen stage of the microspectrometer.

26   Adjust the measuring circle on the cell wall from which spectrum is measured, and focus carefully onto the section again.

27   Measure the light intensity passing through the cell wall 5 times from 400 to 700 nm with 5 nm step increment.

28   Calculate the VIS absorption at each wavelength. In practice, this is calculated automatically by the Lambda Scan software, according to the following equation:

$$A = 2 - \log_{10}((O - P)/(S - P) \times 100)$$

Where O, S and P are intensity of VIS light passing through the cell wall, intensity of VIS light passing through the cell lumen, and stray light, respectively (Figure 11).

APPLICATIONS

Summary of results obtained by autoradiography

The first attempt to investigate the deposition of cell wall by autoradiography was accomplished by Setterfield and Bayley (1959). They administered tritiated sucrose to oat coleoptiles, and observed its incorporation into the cell wall by autoradiography. Their results indicated that cell wall materials are deposited throughout the depth of the thickened cell wall. In addition to this, cell wall formation occurs evenly over the entire wall rather than at restricted centres of synthesis.

Northcote and Wooding (1965) used autoradiography to reveal the function of the Golgi apparatus in the development of sieve tubes in *Acer pseudoplatanus*. They showed that tritiated glucose was incorporated into both the developing cell wall and the callose around the sieve pores. Radioactivity was observed in the Golgi apparatus. These results suggest that labelled glucose is incorporated into cell wall materials via the Golgi apparatus.

Northcote and Pickett-Heaps (1966) investigated the role of the Golgi apparatus in polysaccharide synthesis and transport in root-cap cells of wheat by autoradiography coupled with chemical analysis. After short-pulse incorporation of tritiated glucose, the radioactivity was located on the Golgi apparatus, but little was present on the cell wall. However, incubation with unlabelled glucose after pulse-incorporation of labelled glucose caused a progressive decrease of label on the Golgi apparatus and progressive increase of label on the cell wall. Chemical analysis revealed that labelled glucose was incorporated mainly into arabinose and galactose. Therefore Northcote and Pickett-Heaps deduced that the material transported in the Golgi apparatus may be pectic substances.

Pickett-Heaps (1966, 1967) further investigated the function of the Golgi apparatus in cell wall formation by autoradiography. His observations suggested that the Golgi apparatus and the endoplasmic

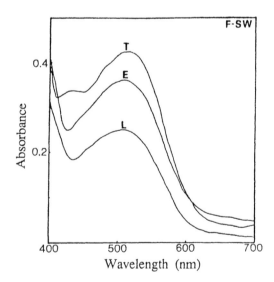

Figure 11   VIS-light absorption spectra of fibre secondary wall in Japanese beech after Mäule colour reaction.
E: earlywood, L: latewood, T: terminal zone (i.e. latewood formed at the close of a season's growth).

reticulum are involved in synthesis of cell wall precursors. Interestingly, he observed the label derived from lignin precursors under a transmission electron microscope and suggested that both the endoplasmic reticulum and the Golgi apparatus are involved in lignin synthesis (Pickett-Heaps, 1968).

Saleh *et al.* (1967) studied the lignification process of developing cells in cottonwood, Douglas fir and wheat by autoradiography. In cottonwood, radiolabelled lignin was first detected in the primary wall at cell corners and then the middle lamella. Lignification of vessels in vascular bundles in wheat was completed first, followed by the cell between vessels. The phloem and the parenchyma cells were lignified last. These results agreed with earlier results obtained by staining methods and by UV microscopy.

Wooding (1968) observed the incorporation of tritiated glucose and phenylalanine into walls of sycamore vascular tissue by autoradiography coupled with chemical analysis. He suggested that cellulose synthesis occurs at, or very close to, the plasma membrane, and hemicellulose is synthesised in the Golgi apparatus. However, he could not distinguish the site of synthesis of lignin precursor in the cell, because there was no labelling on the cytoplasm in the specimen fed phenylalanine.

Ray (1967) studied cell wall deposition in *Avena sativa* var. Victory by autoradiography coupled with extraction of cell wall materials. He demonstrated that cellulose is deposited appositionally at the cell wall surface, whereas hemicelluloses are deposited into the interior of the cell wall.

Fujita and Harada (1979) fed tritiated lignin precursors to differentiating compression wood of Japanese cedar and investigated the lignification process of cell wall by light- and electron-microscopic autoradiography. Radioactivity was observed at the intercellular layer and cell corner regions in the cells of depositing S_1 and transitional cells of S_1 to S_2 formation. It was again observed abundantly at the outer region of S_2 after rapid thickening of secondary wall. These results support the lignification process of compression wood tracheid revealed by UV-microscopy. They also observed the radioactivity on the Golgi apparatus, suggesting its involvement in lignification.

Takabe *et al.* (1981) studied closely the lignification process of differentiating tracheid of Japanese black pine by autoradiography, UV-microscopy and fluorescence microscopy. Lignification starts at cell corner in the cell just before S_1 formation. Lignification of the compound middle lamella proceeds continuously during the stages of S_1 and S_2 thickening, and is completed by the time of S_3 thickening. Lignification of secondary wall starts at the outer portion of cell wall in the tracheid beginning S_2 formation, extends centripetally to the inner portion. Then, rapid lignification occurs over all of the secondary wall in the tracheids after S_3 formation is completed.

Subsequently, Takabe *et al.* (1985) investigated more detailed lignification process in the cell wall of Japanese cedar by autoradiography. They showed that labelled phenylalanine was incorporated rapidly into the lignin of the compound middle lamella, whereas it was converted slowly to secondary wall lignin. They further observed radioactivity on the Golgi apparatus, and rough- and smooth-endoplasmic reticula, suggesting their involvement in synthesis of monolignols (Figure 12).

Terashima and his co-workers (1993) revealed the relation between lignification and heterogeneous distribution of lignin in softwoods and hardwoods by autoradiography coupled with specific labelling of lignin precursors. They administered the (aromatic ring-2-^{3}H) *p*-glucocoumaryl alcohol, (aromatic ring-2-^{3}H) coniferin, and (aromatic ring-2-^{3}H) syringin to the differentiating xylem as precursors of *p*-hydroxyphenyl-, guaiacyl-, and syringyl lignins, respectively (see depiction of lignin biosynthetic pathway in Figure 6 in the chapter by Hawkins *et al.* in this volume). Then, they observed the distribution of radioactivity on the developing cell wall by autoradiography. In addition, they administered the appropriate lignin precursors (tritiated at the C-2 and the C-5 position of the aromatic ring) to differentiating xylem, and showed the distribution of condensed and non-condensed lignins, respectively, within the cell wall by comparing the radioactivity on the cell wall. Their excellent results are summarised in Figure 13.

Summary of results obtained by UV-microscopy

Lignification process in softwood and hardwood trees

Wardrop (1959) observed transverse sections of differentiating xylem in radiata pine, and showed that lignification starts in the primary wall adjacent to cell corner region, and then extends to the intercellular layer and primary wall. Lignification of secondary wall starts from outer portion, and proceeds toward lumen lagging behind cell wall thickening. He also observed the longitudinal

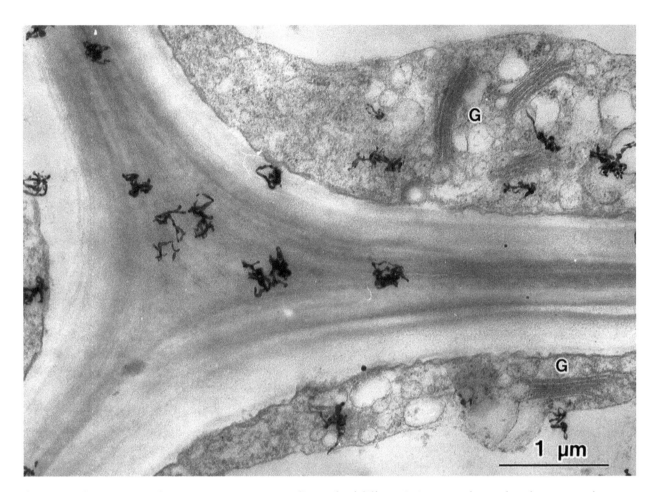

Figure 12   Transmission electron microscopic autoradiograph of differentiating secondary xylem from stem of *Cryptomeria japonica*. Radiolabelled products are observed on the cell wall and near the Golgi apparatus (G).

section of the same specimen and indicated that lignification starts near the middle of the cells and proceeds towards their tips.

Imagawa (1975) observed the lignification process of tracheids in Japanese larch under a UV-microscope, and showed that lignification starts at the intercellular layer in the cell corner and the pit border, and extends to the radial and the tangential walls. Interestingly, the intercellular layer in the cell corners adjacent to ray cell lignifies earlier than other corners. Then lignification proceeds towards the lumen lagging behind cell wall formation.

Afterward, Fujita *et al.* (1978) applied UV microscopy to investigate the lignification process of compression wood in Japanese cedar, and showed that lignification of compression wood also starts at cell corner middle lamella, extends to middle lamella, and then proceeds towards lumen, lagging

behind secondary wall thickening. Interestingly, lignin content in the outer region of S_2 layer is greater than that in the middle lamella.

Takabe *et al.* (1981) investigated the lignification process in Japanese black pine under a UV microscope, and showed that lignification starts at cell corner middle lamella when S_1 formation starts, then extends to compound middle lamella during S_1 and S_2 formation stages, and ceases at S_3 formation stage. Lignification of secondary wall starts at the outermost part in the tracheid of S_2 formation, and proceeds towards the lumen centripetally, lagging behind secondary wall thickening.

Recently, Yoshinaga *et al.* (1997a) revealed the lignification process in Japanese oak by UV-microspectrometry. Cell walls of vessel elements and vasicentric tracheids are quickly thickened and lignified. During their lignification, guaiacyl lignin

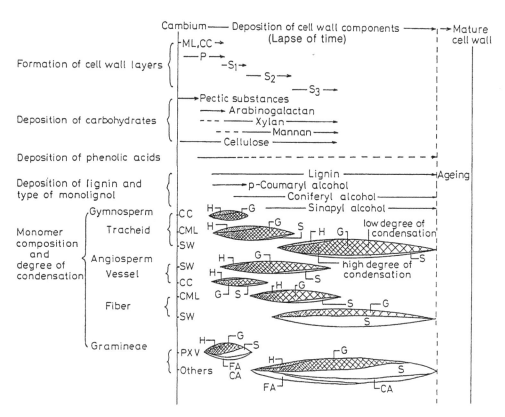

Figure 13 A conceptual illustration of the relationship between the deposition of cell wall polymers and formation of the heterogeneous structure of lignin. Abscissa shows the lapse of time or sequence of deposition of cell wall components during the differentiation of cell walls from the early stage (cambium) to the final stage (mature cell wall). Cell wall layers are formed successively in the order of middle lamella (ML), primary wall (P), outer, middle, and inner layers of secondary wall (S_1, S_2 and S_3). Deposition of carbohydrates occurs in the order of pectin substances, hemicelluloses associated to primary wall such as arabinogalactan or xylan. Monolignols are provided in the order of p-coumaryl, coniferyl and synapyl alchohol. Lignification starts and ends in the order of cell corner (CC), P, compound middle lamella (CML), secondary wall (SW). Combination of these differences in deposition of cell wall components causes heterogeneity in the structure of macromolecular lignin with respect to morphological regions. The lignins formed at an early stage are always more condensed than those formed at a late stage. p-hydroxyphenylpropane (H) and guaiacylpropane (G) units are rich in the former, while syringylpropane (S) units in the latter. In the case of gramineae, protoxylem vessel (PXV) wall lignifies early. Ferulic acid (FA) is incorporated earlier than p-coumaric acid (CA). (reproduced with permission from reference Terashima et al. (1993) Copyright: American Society of Agronomy, Inc., Crop Science Society of America, Inc. and Soil Science Society of America, Inc.).

is mainly accumulated in the cell walls. In contrast, cell wall thickening of fibres proceeds quickly in the early stage, and then slowly in the later stage. Accumulation of lignin in the fibre cell wall is rich in guaiacyl type in the early stage, but the syringyl type dominates in the later stage.

Distribution of guaiacyl and syringyl lignins within the cell wall

Goring and his co-workers accomplished their pioneering work on lignin distribution within the cell wall by UV-microscopy. Scott et al. (1969) devel-

oped the method for determining lignin content in softwood tracheids. They prepared thin sections, took photographs in monochromatic light and subjected the photographic negatives to densitometric analysis. Then they calculated the lignin content of the cell wall according to the Lambert-Beer law with some modifications. Soon after that, Fergus et al. (1969) determined the lignin concentration in spruce wood by UV-microscopy. They showed that the lignin concentration of the middle lamella is about twice that in the secondary wall, and lignin concentration of cell corner middle lamella is nearly four times that in secondary wall.

Musha and Goring (1975) developed the technique of UV-microspectrometry for investigating the distribution of guaiacyl and syringyl lignins in hardwood cell walls. They measured UV absorbances from 240 to 320 nm directly with a photomultiplier by recording the intensity of light passing through the particular region of cell wall and comparing it to that passing through the lumen. Their results indicated that fibre and ray parenchyma secondary walls are predominant in syringyl lignin, though vessel secondary wall and cell corner middle lamella contains mostly guaiacyl lignin.

Subsequently, Fukazawa and Imagawa (1981), and Takano et al. (1983) analysed quantitatively the lignin distribution within an annual ring in softwoods by using a UV microscope coupled with an image analyser. They showed that lignin content of earlywood is higher than that of latewood, though it is high in the terminal zone of latewood.

Takabe et al. (1992) investigated the lignin distribution of Japanese beech within an annual ring by UV-microspectrometry and VIS-microspectrometry coupled with the Mäule colour reaction. The proportion of guaiacyl lignin in vessel secondary wall is high in earlywood, but becomes low in latewood. In contrast, syringyl lignin in vessel secondary wall is quite low in earlywood, but becomes appreciable in latewood. The content of syringyl lignin in fibre secondary wall is high in earlywood, becomes low in latewood, and again increases in the terminal zone of an annual ring. Interestingly, appreciable amount of guaiacyl lignin is present in fibre secondary wall in the terminal zone.

Recently, Yoshinaga et al. (1997b) reported the cellular distribution of guaiacyl and syringyl lignins within an annual ring in Japanese oak. The cells located far from vessel elements tend to increase the proportion of syringyl lignin in the secondary wall, suggesting that the proportion of syringyl lignin tends to increase as cell function change from water conduction to mechanical support.

FUTURE PROSPECTS

Autoradiography has been used not only to visualise physiological and biochemical events in cells, but also to investigate the chemical nature of substances in the cell and cell walls. This technique has revealed the involvement of cell organelles in synthesis of cell wall components. However, autoradiography visualises only the radiolabelled materials insoluble with water and/or organic solvents, because the specimen is washed with buffer and passed through organic solvents. More improved techniques will be necessary to study the localisation of radiolabelled low-molecular-weight substances in the cell. The use of an appropriate precursor, radiolabelled at specific positions, offers exciting potential to clarify the chemical nature of targeting substances in the cell wall. Terashima et al. (1993), for example, revealed the heterogeneous distribution of lignin in softwoods and hardwoods by using lignin precursors labelled at specific positions. The precise design of radiolabelled precursor is necessary to reveal more detailed chemical nature of cell wall substances.

Tremendous development of microscopic instruments enables us easily to obtain reliable data even from a very small area with high sensitivity. UV absorption spectra were, previously, obtained by complicated calculation based on the densitometer traces of UV photographic negatives taken from 240 to 320 nm with 2.5 nm increment. Because this procedure was a very time-consuming process, tedious efforts were needed to obtain the spectra. In recent years, a newly developed microspectrometer permits measurement of the UV absorption spectrum from a specified area within a few min. Okuyama et al. (1998) established the method for measuring lignin content within the cell wall in softwoods by microspectrometry. However, their method is applicable only to softwoods, because they contain predominantly guaiacyl lignin in the cell wall. Hardwood lignins are well known to be copolymers of syringyl and guaiacyl lignins. Proportion of syringyl to guaiacyl lignins varies among tissues and species, and therefore it is quite difficult to determine the lignin content in hardwoods by UV microspectrometry. Specific markers for syringyl and/or guaiacyl lignins are necessary to determine the lignin contents in hardwoods.

REFERENCES

Fergus, B.J., Procter, A.R., Scott, J.A.N. and Goring, D.A. I. (1969) The distribution of lignin in sprucewood as determined by ultraviolet microscopy. *Wood Science and Technology*, **3**, 117–138.

Fujita, M. and Harada, H. (1979) Autoradiographic investigations of cell wall development. Tritiated phenylalanine and ferulic acid assimilation in relation to lignification. *Mokuzai Gakkaishi*, **25**, 89–94.

Fujita, M., Saiki, H. and Harada, H. (1978) The secondary wall formation of compression wood tracheids. Cell wall thickening and lignification. *Mokuzai Gakkaishi*, **24**, 158–163.

Fukazawa, K. and Imagawa, H. (1981) Quantitative analysis of lignin using an UV microscopic image analyzer. Variation within one growth increment. *Wood Science and Technology*, **15**, 45–55.

Imagawa, H., Fukazawa, K. and Ishida, S. (1975) Study on the lignification in tracheids of Japanese larch, *Larix leptolepis* Gord. *Research Bulletins of the College Experimental Forests, Hokkaido University, Japan*, **33**, 127–138.

Musha, Y. and Goring, D.A.I. (1975) Distribution of syringyl and guaiacyl moieties in hardwoods as indicated by ultraviolet microscopy. *Wood Science and Technology*, **9**, 45–58.

Mäule, C. (1901) Das Verhalten verholzter Membranen gegen Kaliumpermanganat, eine Holzreaktion neuer Art. *Beiträge zur Wissenschaftlichen Botanik.*, **4**, 166–185.

Northcote, D. H. and Pickett-Heaps, J. D. (1966) A function of the Golgi apparatus in polysaccharide synthesis and transport in the root-cap cells of wheat. *Biochemistry Journal*, **98**, 159–167.

Northcote, D.H. and Wooding, F.B.P. (1965) Development of sieve tubes in *Acer pseudoplatanus*. *Proceedings of the Royal Society of London. Series B, Biological Sciences*, **163**, 524–536.

Okuyama, T., Takeda, H. Yamamoto, H. and Yoshida, M. (1998) Relation between growth stress and lignin concentration in the cell wall: Ultraviolet microscopic spectral analysis. *Journal of Wood Science*, **44**, 83–89.

Pickett-Heaps, J.D. (1966) Incorporation of radioactivity into wheat xylem walls. *Planta*, **71**, 1–14.

Pickett-Heaps, J.D. (1967) Further observations on the Golgi apparatus and its functions in cells of the wheat seedling. *Journal of Ultrastructure Research*, **18**, 287–303.

Pickett-Heaps, J.D. (1968) Xylem wall deposition. Radioautographic investigations using lignin precursors. *Protoplasma*, **65**, 181–205.

Ray, P.M. (1967) Radioautographic study of cell wall deposition in growing plant cells. *The Journal of Cell Biology*, **35**, 659–674.

Reynolds, ES. (1963) The use of lead citrate at high pH as an electron-opaque stain in electron microscopy. *The Journal of Cell Biology*, **17**, 208–212.

Saleh, T.M. Leney, L. and Sarkanen, K.V. (1967) Radioautographic studies of cottonwood, Douglas fir and wheat plants. *Holzforschung*, **21**, 116–120.

Scott, J.A.N., Procter, A.R., Fergus, B.J. and Goring, D.A. I. (1969) The application of ultraviolet microscopy to the distribution of lignin in Wood. Description and validity of the technique. *Wood Science and Technology*, **3**, 73–92.

Setterfield, G. and Bayley, S.T. (1959) Deposition of cell walls in oat coleoptiles. *Canadian Journal of Botany*, **37**, 861–870.

Takabe, K., Fujita, M., Harada, H. and Saiki, H. (1981) Lignification process of Japanese black pine (*Pinus thunbergii* Parl.) tracheids. *Mokuzai Gakkaishi*, **27**, 813–820.

Takabe, K., Fujita, M., Harada, H. and Saiki, H. (1985) Autoradiographic investigation of lignification in the cell walls of cryptomeria (*Cryptomeria japonica* D. Don). *Mokuzai Gakkaishi*, **31**, 613–619.

Takabe, K., Miyauchi, S., Tsunoda, R. and Fukazawa, K. (1992) Distribution of guaiacyl and syringyl lignins in Japanese beech (*Fagus crenata*). Variation within an annual ring. *IAWA Bulletin n. s.*, **13**, 105–112.

Takano, T., Fukazawa, K. and Ishida, S. (1983) Within-a-ring variation of lignin in *Picea glehnii* by UV microscopic image analysis. *Research Bulletins of the College Experimental Forests*, **40**, 709–722.

Terashima, N., Fukushima, K., He, L-F. and Takabe, K. (1993) Comprehensive model of the lignified cel wall. In *Forage Cell Wall Structure and Digestibility*. edited by H.G. Jung, D.R. Buxton, R.D. Hatfield and J. Ralph, pp. 247–270. Madison, WI 53711, USA, ASA-CSSA-SSSA.

Wooding, F.B.P. (1968) Radioautographic and chemical studies of incorporation into sycamore vascular tissue walls. *Journal of Cell Science*, **3**, 71–80.

Yoshinaga, A., Fujita, M. and Saiki, H. (1997a) Secondary wall thickening and lignification of oak xylem during latewood formation. *Mokuzai Gakkaishi*, **43**, 377–383.

Yoshinaga, A., Fujita, M. and Saiki, H. (1997b) Cellular distribution of guaiacyl and syringyl lignins within an annual ring in oak wood. *Mokuzai Gakkaishi*, **43**, 384–390.

Wardrop, A.B. (1957) The phase of lignification in the differentiation of wood fibres. *Tappi*, **40**, 225–243.

1 Cell Walls of Woody Tissues: Cytochemical, Biochemical and Molecular Analysis of Pectins and Pectin Methylesterases

Wood Formation in Trees ed. N. Chaffey
© 2002 Taylor & Francis
Taylor & Francis is an imprint of the
Taylor & Francis Group
Printed in Singapore.

Fabienne Micheli*, Fabienne F. Ermel, Marianne Bordenave,
Luc Richard and Renée Goldberg

Institut Jacques Monod and Université Pierre et Marie Curie, F-75252 Paris Cedex 05, France
* *Author for correspondence.*

ABSTRACT

Techniques detailed in this chapter present converging approaches for the study of pectins, a major cell wall constituent, and pectin methylesterases which are key enzymes of pectin metabolism. A large number of methods is presented, ranging from well proven standard techniques, such as light microscope staining, and cytochemical dissection, to newly perfected techniques such as immunolocalisation, cryo-sectioning, and micro-fraction analysis. Where necessary, modifications of existing methods are provided which overcome specific problems arising from working with tree material.

Key words: cryo-sections, histochemistry, immunolocalisation, pectinases, pectins, protein extraction.

INTRODUCTION

Changes to the cell wall are amongst the most obvious and fundamental aspects of the differentiation of cambial derivatives (see Table 1 in the Introduction by Chaffey to this volume), and aspects of cell wall biology have been studied for many years (e.g. see introductions to chapters in this volume by Chaffey (TEM), Takabe, Samaj and Boudet, and Itoh). Among the various cell wall constituents, pectins are of special interest since these polymers are known to modulate cell wall porosity and electrostatic potential (Carpita and Gibeaut, 1993). Moreover, they are deposited very early in the apoplasm, which suggests that they may also intervene during the first steps of cell differentiation (McCann and Roberts, 1996). Indeed, when cells divide, the new cell walls of the daughter cells are known to be rich in pectic material.

Three kinds of pectic polymers have been isolated and further characterised from cell suspensions, and subsequently recognised in almost all plant cells. The general structure of these pectic polymers consists of homogalacturonan linear chains (HG or smooth regions) interspersed with highly branched galacturonic chains whose backbone is composed of repeats of the disaccharide 2-α-L-rhamnosyl-1,4-α-D-galactosyluronic acid, arabinans, galactans and arabinogalactans being attached to O-4 of the rhamnosyl residues (RGI or hairy regions). The third pectic molecule (RGII or rhamnogalacturonan II) is a small complex polysaccharide containing twelve different glycosyl residues including rare 'diagnostic' monosaccharides such as apiose, 2-O-methyl-L-fucose, and aceric acid. Homogalacturonans are deposited in the apoplasm as highly methylated polymers, demethylesterification occurring later on inside the cell wall via pectin methylesterases (PMEs).

For many years, very little information has been available concerning the evolution of pectic polymers during the formation of cambial derivatives in trees. Recently, however, immunological tools have been used, providing interesting data on the respective localisation of RGI and variously methylated homogalacturonan. Moreover, cryo-dissection methods (see chapter by Uggla and

Sundberg in this volume), which allow sampling of well-defined cell regions or layers, now permit detailed biochemical as well as molecular investigations, such as our own work on the tissue-specific distribution of the PME isoforms and the expression of their genes.

The methods described below cover a wide range of techniques for analysis of pectins and PMEs at the light and ultrastructural levels.

THE TECHNIQUES

Note – equipment and chemicals printed in bold type are considered in more detail in the relevant 'CHEMICALS, EQUIPMENT, etc' section.

IN SITU LOCALISATION

Standard techniques for light (LM) and transmission electron (TEM) microscopy are given in the chapter on 'Wood microscopical techniques', and those of Rensing and Chaffey, respectively. Although those chapters should be referred to where indicated in the protocols described in this chapter, most of the equipment and chemicals required are listed in the appropriate section below for the reader's convenience.

Choice of sample size

A priori, the smaller the sample, the better the fixation. However, large samples may be needed to obtain more complete images of the material under study (e.g. Ermel *et al.*, 1999). Even then, to allow a good penetration of both the fixative and of the embedding medium, the thickness of samples should not exceed 2 mm, whatever their length and width (up to 1.5 cm).

Fixation

Equipment needed:	Chemicals needed:
Thin, artist's paint brushes (1–2)	Appropriate fixative solutions
Glass microscope slides	Dehydrating chemicals
Petri dishes (1–2)	
Black paper	
Vacuum apparatus	
Razor blades (several)	
Glass vials and tops	

Fixation is not necessary when staining pectins for the LM. In other cases, the choice of fixative mixture depends on whether polysaccharides or cell wall enzymes are to be studied. Cytochemical detection of pectins can be carried out on glutaraldehyde-fixed material, even after osmium tetroxide post-fixation. A *p*-formaldehyde-glutaraldehyde mixture, without osmication, is much better for enzyme localisation. Fixation processes used for immunolocalisation of the cytoskeleton described in the chapter by Chaffey – without the MBS-pre-fixation step, and without DMSO in the fixative – are suitable for the pectin localisation techniques described below. That chapter should be consulted for detailed fixation protocols; we will only add some complementary comments here.

General considerations

Very careful excision, and infiltration of fixative under vacuum are often necessary for very fragile cells or tissues (e.g. sieve tubes and very active cambial cells).

Excision of material

When fixing for TEM, excise tissue slivers from tree under a film of fixative; i.e. using a small, thin, clean paint brush (or a pipette), place some drops of fixative on the first sectioned surface, and continue covering each successive sectioned surface with a film of fixative. Put the slivers directly into fixative. For longitudinal sections/observations, cut slivers approx. 0.5 mm thick. Slices approx. 1 mm thick (maybe more for softwoods) are better for transverse sections since cell length (e.g. cambial cell length) can reach 0.3–0.5 mm (up to 1 mm or more in softwoods).

Vacuum-infiltration

Degas the vial containing fixative and samples under vacuum for 5–15 min. Tissue samples should fall to the bottom of the vial when the vacuum is removed. Replace the fixative in the vial with fresh fixative.

Trimming slivers

When possible, trim the slivers and cut them into small blocks (approx. 0.5–1 × 1 × 2 mm would be the best) whilst they are still in the fixative solution (work under a laboratory hood!). Hold the sliver firmly on a glass microscope slide, pressing with one finger on one end of the sliver (which will be

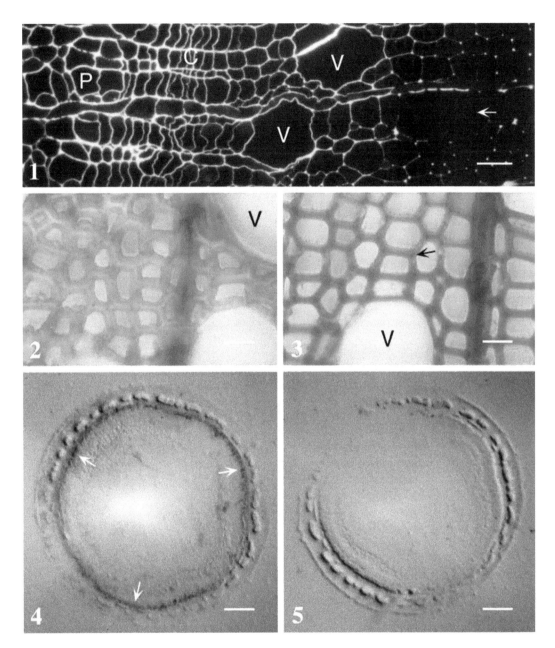

Figure 1 Semi-thin, transverse section of young *Populus tremula* x *tremuloides* stem. Fixation: glutaraldehyde 4%, JIM5-staining of the cell walls. The staining decreases with xylem differentiation and disappears when the walls are fully lignified (arrow). C: cambial zone; P: phloem; V: vessels. Bar: 50 μm.

Figure 2 Thick, freezing microtome cut, transverse section of one-year-old xylem from *Populus* x *euramericana*. Fixation: ethanol; Ruthenium red-staining. Primary and secondary walls are only faintly stained. V: vessel. Bar: 25 μm.

Figure 3 Same material as Figure 2, but extracted for 48 h with NH$_4$OH prior to ruthenium red staining. Stronger wall staining than in Figure 2, especially in the compound middle lamella and cell junctions (arrow). V: vessel. Bar: 25 μm.

Figure 4 Bright-field image of tissue-print of a small branch of *Populus* x *euramericana* treated with an antibody raised against PME from *Glycine max*. Binding reaction visualised with alkaline phosphatase. A brown-pink staining is present in the cambial zone and its immediate derivatives (arrows). Variations in staining intensity are due to irregular pressure applied to the sample during tissue-printing. Bar: 0.5 mm.

Figure 5 Same material as Figure 4, but the antibody was omitted from the incubation medium. No brown-pink staining. The dark lines are the shadows of lignified cell imprints (phloem fibres and xylem) where the pressure applied was the strongest. Bar: 0.5 mm.

discarded later). Discard also the other end. Cut away and dispose of all unnecessary tissues, especially if they are impermeable to water (e.g. cork) or full of air (e.g. pith parenchyma), since this will otherwise prevent rapid entry of the fixative. To assist in seeing the sliver, it may be useful to place the glass slide over a Petri dish and to put some black paper under the dish. Change the razor blade every 2 or 3 sections.

Polyphenol-containing material

Fixation of tissues with a high polyphenol content is problematic since this substance diffuses easily imparting a dark colour to the cytoplasm. Adding 0.1% caffeine to the fixative prevents this artefact (Mueller and Greenwood, 1978).

Embedding

Equipment needed:	Chemicals needed:
Embedding moulds	Historesin, e.g. Technovit 7100 kit (Heraus Kulzer) which contains chemicals for pre-infiltration, infiltration and embedding Acrylic resin (e.g. **LR White)** Ethanol

For light microscopy (using historesin)
Duration of procedure: 1–2 weeks.

1   Fix material (see e.g. chapter by Chaffey in this volume).
2   Dehydrate with ethanol-water series: 25% (30 min), 50% (30 min), 70% (1 h, then overnight), 95% (3 × 1 h).
3   Treat with pre-infiltration solution: 15 min under vacuum, then 5 h.
4   Treat again with fresh pre-infiltration solution: 1–7 days (according to sample size).
5   First treatment with infiltration solution: 15 min under vacuum, followed by 5 h.
6   Treat again with fresh infiltration solution 1–7 days (according to sample size).
7   Put the embedding solution into moulds at room temperature (RT), in a fume cupboard; place and carefully orient samples in moulds.
8   Polymerise for 4 h at RT.

9   Cover the block surfaces with a film of water to isolate them from air and polymerise for 1 h at RT.
10 Remove water.
11 Polymerise for a further 12 h at RT.

For immunolocalisation
Duration of procedure: approx. 1 week.

Use a hydrophilic resin such as **LR White** (see also TEM chapter by Chaffey).

Pectin localisation

Equipment needed:	Chemicals needed:
Small Petri dishes (5.5 cm diameter)	Ruthenium red
Equipment for fixation/embedding	**Glutaraldehyde**
Light microscope	**Mounting medium**
Transmission electron microscope	**Cacodylate buffer**
Razor blades	**Osmium tetroxide** Propylene oxide Hydroxylamine hydrochloride $FeCl_3$ Ethanol HCl NaOH **Spurr** or other epoxy resin

Standard staining of pectins for light microscopy
This is generally performed on fresh or ethanol-stored, unembedded material, sectioned by hand or with a freezing microtome, but de-paraffined sections can also be used.

Ruthenium red-staining (Figures 2, 3)
This reagent stains acidic pectins red (Jensen, 1962).
Duration of procedure: 10 min when sections are mounted in water.

1   Stain for 3–5 min in 0.02% (w/v) aqueous ruthenium red.
    Note – hand-cut sections can be stained in a small Petri dish containing water plus a very

few granules of ruthenium red. Wait for complete dissolution of reagent before staining.

2 Wash thoroughly in water.

3 Mount in water or Canada balsam or other standard mounting medium.

Note – sections mounted in balsam can be kept for several years.

Hydroxylamine-ferric chloride reaction

Methylesterified pectins are stained red (Jensen, 1962) with this procedure. The reaction is thought to be specific for esterified pectins, which are transformed into iron hydroxamate (Gee et al., 1959 in Albersheim, 1965).

Duration of procedure: approx. 15 min.

1 Treat for 5 min in a 60% ethanol-water solution containing 7% hydroxylamine hydrochloride and 7% NaOH (w/v).

2 Treat for 5 min in 6 M HCl in 95% ethanol.

3 Mount in a mixture of 10% $FeCl_3$ in 1:1 (v/v) mixture of 60% ethanol: 0.1 M HCl.

Staining small blocks of tissues for electron microscopy

Ruthenium red (Luft, 1971)

Duration of procedure: approx. 3–4 h for fixation and washes.

Note – same buffer as used in step 1 for all buffer-steps.

1 Fix with 4% glutaraldehyde in **cacodylate buffer**, pH 7.2 containing 0.1% ruthenium red, for 1 h 30 min (degas samples).

2 Wash thoroughly in buffer containing 0.1% ruthenium red.

3 Post-fix for 1 h with 1% osmium tetroxide in buffer containing 0.1% ruthenium red, pH 7.2.

4 Wash thoroughly in buffer.

5 Dehydrate and embed in Spurr or other epoxy resin (see e.g. TEM chapter by Chaffey in this volume).

6 View sections in TEM without further contrasting. Sites of pectins are revealed as electron-dense deposits.

Hydroxylamine-ferric chloride reaction (modified from Albersheim, 1965)

Duration of procedure: approx. 5–6 h for fixation, hydroxylamine ferric staining and dehydration.

1 Fix with glutaraldehyde, post-fix with osmium tetroxide (see also TEM chapter by Chaffey in this volume).

2 Wash thoroughly in the same buffer used for fixation.

3 15 min in 25% ethanol-water.

4 15 min in 60% ethanol-water.

5 20 min in a fresh 60% ethanol-water solution containing 14% hydroxylamine hydrochloride and 14% NaOH (w/v).

6 Wash for 5 min in 0.1 M HCl in 60% ethanol-water solution.

7 Stain for 20 min in a fresh solution of $FeCl_3$ (20% (w/v) in a 60% ethanol-water solution).

8 Wash in 60% ethanol-water.

9 Dehydrate and embed in Spurr or other epoxy resin (see e.g. TEM chapter by Chaffey in this volume).

10 View in TEM without further contrasting. Sites of pectins are revealed as electron-dense deposits.

Indirect localisation by subtractive methods

Subtractive methods are based on enzymatic or mild chemical extraction of pectins from the cell wall before, during or after fixation. Sections from extracted and unextracted samples can be compared, after specific staining for pectins, or after a general staining for polysaccharides, e.g. using periodic-acid-Schiff's reagent (PA-S) (see 'Wood microscopical techniques' chapter in this volume) or periodic acid-thiocarbohydrazide-silver proteinate (PATAg) (see TEM chapter by Chaffey in this volume). The sites from which wall material have been extracted appear clear compared with controls. Enzymatic extraction is specific if performed with highly purified enzymes attacking a single type of glycosidic linkage (Roland and Vian, 1991; Jauneau et al., 1998). Otherwise, the extent of cell wall solubilisation can be surprising! But, to our knowledge, purified enzymes have not been used on trees.

In contrast, mild chemical extractions of wall components have been repeatedly performed on woody plants (e.g. Catesson and Roland, 1981; Baïer et al., 1994; Catesson et al., 1994). The lack of specificity of this approach is counterbalanced by allowing direct comparison between biochemical and cytochemical data.

Equipment needed:	Chemicals needed:
Nickel or gold grids, 200 mesh	Periodic acid (for PATAg staining)
Petri dish with lid (for PATAg staining)	Thiocarbohydrazide (for PATAg staining)
Parafilm (for PATAg staining)	Silver proteinate (for PATAg staining)
Equipment for fixation/ embedding	Acetic acid (for PATAg staining)
Water bath at 100 °C	Ethanol
Orbital shaker	Propylene oxide
Transmission electron microscope	**Glutaraldehyde**
	Osmium tetroxide
	Ethylene diamine tetracetic acid (EDTA), or, Diaminocyclohexane tetracetic acid (CDTA)
	Cacodylate buffer
	Spurr or other epoxy resin

Sequential extraction of pectins (either on fresh samples or on samples fixed in glutaraldehyde and washed in buffer) (Figures 6, 7, 9)

Duration of procedure: approx. 24 h for extraction and fixation processes.

1   Incubate samples for 2 h in distilled water (approx. pH 5.6), at 100 °C (put the open vials in a bath of boiling water).
2   Wash in cold water.
3   Stand overnight in 1% aqueous EDTA, with constant shaking.
4   Wash in water and then in cacodylate buffer.
5   If specimens are unfixed, treat now with 4% glutaraldehyde in cacodylate buffer, pH 7.2 for 1.5 h, and wash in buffer; otherwise, go directly to the following step.
6   Post-osmicate, dehydrate, and embed in Spurr or other epoxy resin (see e.g. TEM chapter by Chaffey in this volume).
7   Stain ultrathin sections with PATAg (see TEM chapter by Chaffey in this volume).

For separate extractions, omit either steps 1 and 2, or steps 3 and 4. Boiling water at pH 5.6 solubilises mostly RGI and highly methylesterified pectins. The calcium-chelating agents, EDTA and CDTA, extract low-methylesterified HG interconnected via calcium bridges.

Note – pectin-extraction increases bark slippage in trees with an active cambium. Although glutaraldehyde-pre-fixation may reduce this tearing, comparison between cytochemical and biochemical data will no longer be fully accurate. And, even then, most samples will tear. The only remedy is to cut a great number of not-too-small samples (may be 50) and hope that some will stay whole (it happens!).

Wall components other than pectins can also be chemically extracted (Roland and Vian, 1991). Dimethylsulphoxide (DMSO), NH_4OH, and NaOH are known to remove hemicelluloses from the cell wall matrix. Methylamine, and NH_4OH to a lesser degree, have been used to extract lignin whose presence often prevents both the extraction and cytochemical detection of other wall components (Czaninski and Monties, 1982; Harche and Catesson, 1985).

Immunolabelling

Among the various affinity techniques available for specific labelling of pectins, only immunolabelling has been used on a large scale on woody plants. Although various polyclonal antibodies were raised against pectins and other cell wall polysaccharides in the 70's and 80's (Hoson, 1991; van den Bosch, 1991, 1992), the technique has only exceptionally been applied to lignified organs or tissues (see Northcote et al., 1989 on bean xylem). However, the advent of hybridoma technology and the isolation of monoclonal antibodies (MAbs) (Table 1) has stimulated immunocytochemical studies of plant cell walls, woody plants included (Chaffey et al., 1997; Guglielmino et al., 1997; Hafrén, 1999).

Equipment needed:	Chemicals needed:
Nickel or gold grids, 200 mesh	Chemicals for PATAg-staining (see box above)
Poly-L-lysine-coated multi-well slides	Normal goat serum (Amersham, No. RPN 410)
Petri dish with lid	**Uranyl acetate**
Heating plate at 50–80 °C	Bovine serum albumen (BSA) (BioCell, No. BSA 10)
Paper tissues	**Primary antibodies**
Fluorescence microscope	**FITC-conjugated secondary antibodies**

Light microscope	**Gold-conjugated secondary antibodies**
Transmission electron microscope	**Silver enhancement kit**
Microtome	**Eukitt**
	Vectashield®
	TBS buffer
	TBS-T buffer

Immunofluorescence techniques (Figure 1)

Labelling of semithin sections with FITC-conjugated secondary antibodies

Duration of procedure: approx. 8–20 h.

Note – allow time for any frozen solutions to thaw prior to their use.

Note – all steps performed at RT unless specified otherwise.

1   Cut transverse, semi-thin sections (1 μm) from samples fixed with **glutaraldehyde** or **p-formaldehyde** and embedded in **LR White** (see also TEM chapter by Chaffey in this volume).

2   Collect sections on multi-well glass slides which have been coated with poly-L-lysine to promote section-adhesion. Let sections dry overnight at 60 °C or for 30 min–1 h at 80 °C.

3   Rinse the sections for 10 min with 0.05 M TBS buffer (± 0.1% BSA).

4   Treat for 15–30 min with normal goat serum diluted 1:30 in TBS-T (± 0.1% BSA).

5   Incubate for 4 h at RT, or overnight at 4 °C, with MAb JIM5 or 7, LM5 or 6, diluted, respectively, 1:45, 1:45, 1:30, 1:10 in TBS (± 0.1% BSA). Slides are put in a covered Petri dish beside a wet paper tissue to keep a humid atmosphere.

Note – the first time the MAbs are used, a range of decreasing concentrations should be examined, and the concentration chosen which optimises the signal:background ratio.

6   Wash 4 × 10 min in TBS (± 0.1% BSA).

7   Incubate in goat anti-rat IgG coupled to fluorescein *iso*thiocyanate (FITC), diluted 1:300 in TBS (± 0.1% BSA) for 1 h in the dark in humid atmosphere (as for step 5).

8   Rinse 6 times in TBS (± 0.1% BSA), in the dark.

9   Rinse twice in water, in the dark.

10 Mount in Vectashield® and store in refrigerator in the dark until ready to view.

Table 1   Monoclonal antibodies raised against pectin epitopes

Antibody	Antigen/epitope	Use on trees	References
JIM5	HG with DE<50%	Yes[1]	VandenBosch *et al.* (1989); Knox *et al.* (1990)
JIM7	HG with 35%<DE<90%	Yes[1]	Knox *et al.* (1990)
2F4	Ca^{2+}-requiring configuration of HG (DE<40%)	No	Liners *et al.* (1992); Liners and Van Cutsem (1992)
CCRC-M5	Sycamore RGI	No	Puhlman *et al.* (1994)
CCRC-M7	Arabinogalactan epitopes of maize and sycamore RGI	?	Puhlman *et al.* (1994); Steffan *et al.* (1995)
CCRC-RI	RG II	No?	Williams *et al.* (1996)
LM5	1,4-β-galactan	Yes[2]	Jones *et al.* (1997); Vicré *et al.* (1998)
LM6	1,5-α-arabinan	Yes[2]	Willats *et al.* (1998)
PAM1 and PAM2	Unesterified and unsubstituted homogalacturonan	No	Willats *et al.* (1999)

DE = degree of esterification

1 Chaffey *et al.* (1997); Guglielmino *et al.* (1997); Hafrén (1999)

2 Ermel *et al.* (2000)

Controls
Sections are treated either with

i)   pre-immune serum at the same dilution in buffer, or
ii)   with buffer alone, instead of the primary antibody.

FITC-labelled sections can be observed either in a fluorescence microscope (excitation filter 450–490 nm, barrier filter 529 nm) or with a confocal microscope (see also the chapter by Funada in this volume). The latter technique has several advantages over 'conventional' fluorescence microscopy:

i)   it allows the detection of a very low signal;
ii)   autofluorescence noise can be subtracted, which is specially useful for lignified tissues with a high content of autofluorescent phenolics;
iii) background noise can be reduced by 'averaging' and collecting the signals over, e.g., 8 frames.

Immunogold techniques (Figures 10, 11)
Observation of semi-thin sections with the light microscope
Duration of procedure: approx. 8–20 h.
Note – allow time to thaw any frozen solutions prior to their use.
Note – all steps performed at RT unless specified otherwise.

Steps 1–4 as for immunofluorescence above

5   1 h-treatment with a solution of goat anti-rat IgG conjugated with 10 nm colloidal gold probes, diluted 1:30 in TBS.
6   Rinse 6 times in TBS.
7   Rinse twice in water.
8   Intensify gold labelling with the silver enhancement kit for 1–4 min, depending on the sample.
9   Rinse sections thoroughly, dry on a heating plate (approx. 50 °C) and mount in Eukitt medium.
10   Observe slides with the light microscope either directly or with an epipolarisation filter (A pol filter 487960, Zeiss).

Observation with the electron microscope
Ultrathin sections are mounted on coated or uncoated nickel or gold (in case of PATAg counterstaining) grids.
Duration of procedure: approx. 9–21 h.
Note – allow time to thaw any frozen solutions prior to their use.
Note – all steps performed at RT.

Steps 1–7 as above for semi-thin sections
8   Stain for 25 min with 1% uranyl acetate; alternatively stain with PATAg (gold grids only) (as described in TEM chapter by Chaffey in this volume).
9   Rinse 6 × 5 min in water.
10   Let grids dry in storage box.

Double-labelling technique (Figure 8)

It is possible to label ultrathin sections with two different antibodies using one of the following techniques:

a   Collect ultrathin sections on uncoated nickel grids. Follow the above protocol with the first antibody, using a 5 nm colloidal gold probe. Then, turn the grid over in order to label the other side of the section with the second antibody, this time using a 15 nm gold probe.
b   Label with the first antibody and a 5 nm colloidal gold probe by floating the unmounted sections on the successive incubation media (see Roland and Vian, 1991). Sections are then collected on coated or uncoated nickel grids, the labelled side against the grid. Process the unlabelled side of the section with the second antibody and a 15 nm gold probe.

Whatever the technique used, two kinds of labelling should be done in order to check the significance of the result. For instance, in the case of JIM5/JIM7 double-labelling:

●   half the sections are treated first with JIM5, followed by a 5 nm colloidal gold probe, and secondly with JIM7 followed by a 15 nm gold probe;

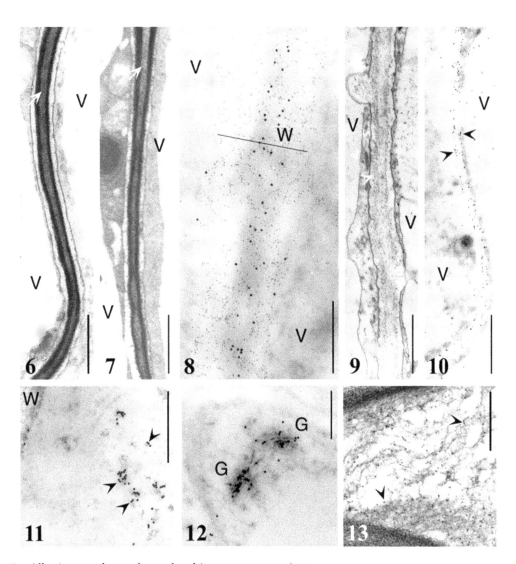

Figures 6–13 All micrographs are from ultrathin transverse sections.
Figure 6 *Acer pseudoplatanus*. Unextracted material, PATAg-staining. The middle lamella (arrow) is only just a little less dense than the primary walls. Bar: 1 μm.
Figure 7 *Fraxinus excelsior*. EDTA-treated material, PATAg-staining. The middle lamella (arrow) is less dense than the primary walls. Bar: 1 μm.
Figure 8 *Populus x euramericana*. Double immunogold localisation of low-esterified and highly-esterified galacturonan with JIM5 (large gold particles) and JIM7 (small gold particles), respectively. Contrary to JIM7, JIM5 labelling is restricted to the middle lamella. The thickness of the cell wall (W) is indicated by a short line. Bar: 0.5 μm.
Figure 9 *Acer pseudoplatanus*. Material treated with boiling water, PATAg-staining. Most of the PATAg-reactive material has been solubilised (cf. Figure 6). Note the discrete microfibrillar skeleton (white arrowhead). Bar: 1μm.
Figure 10 *Populus tremula* x *tremuloides*. Immunogold labelling of RGI galactan residues with LM5 in the wall (arrowheads). Bar: 1 μm.
Figure 11 *Populus tremula* x *tremuloides*. Cambial cells. Immunogold labelling of RGI galactan residues with LM5. PATAg-staining. Gold particles are present on the PATAg-stained Golgi vesicles (arrowheads). Bar: 0.5 μm
12–13: Immunogold localisation of PME in cambial zone of *Populus* x *euramericana*. Fixative: formaldehyde-glutaraldehyde.
Figure 12 Cambial cells. Antibody raised against flax PME. Labelling of Golgi cisternae and vesicles (G). Bar: 0.5 μm.
Figure 13 Tricellular cell junction between cambial derivatives towards phloem. Antibody raised against soya bean PME. PATAg-staining. Labelling of the loose, central, polysaccharidic network (arrowheads). Bar: 0.5 μm.
Figures 6–10 Cell wall common to two very young vessel elements (V), at the beginning of the stage of diametral enlargement. 6, 7, 9: fixation: glutaraldehyde-osmium; 8, 10: fixation: glutaraldehyde.

- the other half are treated first with JIM7, followed by a 5 nm colloidal gold probe, and secondly with JIM5 followed by a 15 nm gold probe.

Pectin methylesterase localisation

Antibodies raised against PME

In our experiments on poplar and aspen, we used two different polyclonal antibodies:

- an antiserum raised against PME extracted from the cell walls of *Glycine max* cell suspensions (Goldberg *et al.*, 1992);
- an antibody raised against the neutral PME from *Linum usitatissimum* (Mareck *et al.*, 1995)

PME immunolocalisation

Equipment needed:	Chemicals needed:
Nickel or gold grids, 200 mesh	**Anti-PME primary antibody**
Petri dish with lid	BSA
Grid storage box	**TBS-T buffer**
Poly-L-lysine-coated **multi-well slides**	**FITC-conjugated secondary antibody**
Fluorescence microscope	**Gold-conjugated secondary antibody**
Transmission electron microscope	**Osmium tetroxide**
Paper tissues	**Uranyl acetate**
Microtome	Chemicals for PATAg-staining (see box above)

On ultrathin sections (Figures 12, 13)

Duration of procedure: approx. 20 h–3 d, depending on grid staining.
Note – allow time for frozen solutions to thaw prior to use.

1 Collect sections on nickel (or gold) grids.
2 Wash for 15 min with 0.1 M TBS-T, pH 7.6.
3 Treat grids for 15 min with normal goat serum diluted 1:30 in 0.1 M TBS-T, pH 7.6 containing 0.1% BSA (in some instances 1% BSA was used).

4 Incubate overnight at 4 °C with anti-PME serum (primary antibody) diluted in the same TBS-T buffer (1:5000 re *Linum* PME, 1:500 for *Glycine* PME). Use a covered Petri dish with a damp paper tissue beside the grids to provide a humid atmosphere.
5 Several 5 min washes in TBS-T buffer.
6 5 min bath in TBS-T plus 0.1% BSA.
7 Incubate 1 h at RT in secondary antibody (anti-rabbit IgG coupled to colloidal gold) diluted 1:30 in TBS-T buffer.
8 Several 5 min washes in TBS-T buffer.
9 Several 5 min washes in water.
10 Optional: overnight post-fixation with osmium tetroxide vapours followed by several 5 min washes with water.
11 25 min staining with 1% uranyl acetate; alternatively, stain with PATAg (gold grids only) (as described in TEM chapter by Chaffey in this volume).
12 Let grids dry in storage box.

Controls

Treat grids either with

i) pre-immune serum instead of primary antibody, or
ii) with TBS-T and BSA alone.

Note – the procedure needs to be adjusted in relation to the material and the antibody used. You can vary the concentration of BSA (0.1–2%) or Tween (0–1%), and, slightly, the buffer pH.

On semi-thin sections
Duration of procedure: approx. 24–48 h.

1 Semi-thin sections (1 μm) are collected on poly-L-lysine-coated multi-well glass slides
2–6 As for steps 2–6 above re ultrathin sections.
7 Incubate for 1 h in secondary antibody (FITC-conjugated goat anti-rabbit IgG diluted 1:200).
8 Several 5 min washes in TBS-T buffer.
9 Similar to steps 9–10 described above for pectin immunolocalisation with FITC-labelled probe

On tissue-prints (Figures 4, 5)

Tissue-printing uses the ability of some supports (e.g. nitro-cellulose membrane) to absorb and retain proteins. It offers a simple, rapid method for protein localisation at low magnification (see Cassab, 1992).

Equipment needed:	Chemicals needed:
Nitro-cellulose membrane (Cellulosenitrate, Schleicher and Schnell, No. 401196)	5-bromo-4-chloro-3-indolylphosphate (BCIP)/Nitroblue tetrazolium (NBT) (Sigma, No. B5655)
Razor blades	$CaCl_2. 2H_2O$
Forceps	Powdered skimmed milk or BSA
Gloves	**Primary antibodies**
Glass microscope slides	**Secondary antibodies**
Filter or Whatman paper	**TBS buffer**
Stereomicroscope Pencil	**TBS-T buffer**

Duration of procedure: approx. 1.5 days.
Note – allow sufficient time for frozen solutions to thaw prior to use.

1   Draw a grid pattern on nitro-cellulose membranes with a pencil. It will be useful to number the prints.
2   Soak the membrane in 0.2 M $CaCl_2. 2H_2O$ for 30 min. Air-dry for 30 min–1 h on filter paper.
3   Place a small branch, or another tree piece, no larger than 1 cm in diameter on a glass slide. Cut 2–3 mm long segments. Do not hesitate to change the razor blade every two or three sections.
4   Put segments on grid squares with forceps, section surfaces against the nitro-cellulose. Press each sample very firmly on nitro-cellulose with a gloved finger for 30–60 sec before removing it carefully with forceps.
5   Air-dry the membrane for 2–3 min.
6   Incubate the membranes 1–2 h in TBS-T buffer plus 5% powdered skimmed milk (or BSA).
7   Incubate overnight at 4 °C with constant shaking in TBS-T plus 3% milk (or BSA) containing anti-*Glycine* PME MAb, diluted 1:1000.
8   Wash thoroughly in distilled water.

9   Rinse for 2 × 10 min in TBS-T.
10 Incubate for 1 h in TBS-T plus 3% milk containing the second antibody under constant shaking. For poplar prints, we used an anti-rabbit antiserum conjugated to alkaline phosphatase (AP) diluted 1:10 000, but the suitability of the conjugated probe must be checked carefully; for instance, untreated mung bean prints stain strongly with AP substrate.
11 Wash for 3 × 10 min in TBS-T.
12 Incubate in AP substrate solution (one pellet BCIP/NBT in 10 ml TBS buffer) shake with hand; colour appears in 5–10 min.
13 Stop staining reaction with water.
14 Dry the membrane and keep it in the dark at RT.

Controls
Tissue-printed membranes are incubated in:

i)   TBS-T without primary antibody, or
ii) TBS-T containing pre-immune serum instead of primary antibody

CHEMICALS, EQUIPMENT, ETC FOR *IN SITU* LOCALISATION:

Fixatives

Glutaraldehyde

4% glutaraldehyde (grade I, 25%, Sigma, No. G 5882) in pH 7.2 **cacodylate** or **phosphate buffer**

PAF mixture

4% *p*-formaldehyde (Sigma, No. P 6148) plus 0.2% glutaraldehyde in 0.1 M **phosphate buffer** at pH 7.2 for 2 h

Osmium tetroxide

Osmium tetroxide (Sigma, No. O 5500) 1% in **cacodylate** or **phosphate buffer**, 1 h

Buffer

Cacodylate buffer

0.1 M sodium cacodylate (trihydrate) (also known as sodium dimethylarseniate) (Fisher Scientifics,

No. S 0595, or Electron Microscopy Sciences, No. 12300), pH 7.2

Phosphate buffer

0.1 M NaK phosphate buffer, pH 7.2

> solution A (0.1 M Na_2HPO_4. $2H_2O$, 35.6 g/l): 75 ml
> solution B (0.1 M KH_2PO_4, 27.2 g/l): 25 ml
> adjust pH before use

Tris-Buffered Saline (TBS)

Tris-buffered saline, pH 7.6

> Tris ultra pure (ICN Biomedicals, No. 819623): 2.42 g
> 1 M HCl: 3.8 ml
> NaCl: 8 g
> Dilute to 1000 ml with water
> Check pH before use
> Store at 4 °C if frequently used. Otherwise, store as frozen aliquots

TBS-T

> TBS plus 0.1% Tween 20 (polyoxyethylene sorbitol monolaurate), (Sigma, No. P 1379)
> Store at 4 °C

Embedding Media

Spurr epoxy resin

Low viscosity kit (Electron Microscopy Sciences, No. 14300)
Choose hard or standard mixture (see instructions given in the kit) according to the material (hard embedding mixture is better for wood). Kit stored at RT, complete mixture stored at –80 °C

LR White acrylic resin

LR White, medium grade, uncatalysed (Polysciences, No. 17411)
Stored at 4 °C

Multi-well slides (Polylabo, Superteflon)

Since each worker must choose the number and size of wells according to his or her own experiments: e.g., number of immunoreaction and controls, section size, etc. the appropriate catalogue number must be checked with the supplier.

Poly-L-lysine (Sigma, No. P8920)-coating of multi-well slides:

1   Prepare 0.01% poly-L-lysine solution from commercial 0.1% solution; keep at 4 °C.
2   Wash slides for 12 h in a solution of 0.5% HCl (35%) in 95% ethanol.
3   Rinse for 3 h under running tap water.
4   Dry for 1 h at 60 °C.
5   Put one or two drop(s) of poly-L-lysine in each well, making a convex dome; wait for 5 min.
6   Stand the slide on its long side to drain off excess poly-L-lysine.
7   Dry overnight at RT, or for 1 h at 60 °C.

Stains

Uranyl acetate

1% uranyl acetate in water, filter solution twice and keep in the dark, in refrigerator.
Filter on millipore in a syringe before use

Antibodies

Primary

JIM5, JIM7, LM5, and LM6 were kindly given by Dr JP Knox (University of Leeds, UK)
Stored at –20 °C as frozen aliquots

Secondary

goat anti-rat IgG conjugated with FITC (Sigma, No. F-6258)
goat anti-rat IgG conjugated with colloidal gold probes (5, 10, 15 nm: BioCell, EMGAT 5 10, 15)
Stored at 4 °C

Silver enhancement kit: (BioCell, SEKL 15)

Used according to manufacturer's instructions

Mounting media

Canada balsam (Prolabo, No. 20 246 298)

Vectashield® (Vector Laboratories, No. H 1000)

Eukitt (Kindler, from Polylabo, No. 826021)

BIOCHEMICAL AND MOLECULAR ANALYSIS

Sampling

Preparation of macro-fractions by scraping

Equipment needed:	Chemicals needed:
Scalpel	Liquid nitrogen
Razor blades (several)	
Falcon tubes (50 ml), 1 per tissue fraction	
Freezer at –80 °C	

Scrape individual tissue macro-fractions as illustrated in Figure 14 using scalpel and razor blades, collect in a 50 ml Falcon tube, freeze in liquid nitrogen, and store at –80 °C.

Collection and identification of tissue fractions

During the active stage of the cambial seasonal cycle, the sampling method uses the unlignified enlarging xylem vessels as a breaking zone to separate the centre of the stem (comprising enlarging, lignifying and mature xylem) from the external part of the stem (containing cambial zone, phloem, cortical parenchyma and bark). Scraping the centre of the stem gives the 'a$_3$ fraction' containing the enlarging xylem. Scraping of the inside of the external part of the stem provides the 'a$_2$ fraction' which comprises the cambial zone and the phloem derivatives formed after the last fibres. The 'a$_1$ fraction' consists of the rest of the external part of the stem and comprised the old phloem with fibres, the cortical parenchyma and the bark.

In the dormant stage, due to the absence of an appropriate 'breaking zone', samples are taken from the outside (bark) to the inside (mature xylem) of the stem. The fraction 'r$_1$' includes tissue from the bark to the last formed phloem fibres and is based on the colour and the texture of the tissue (the phloem contains a lot of fibres and is green). The fraction 'r$_2$' is predominantly cambial tissue which is soft and has a very light yellow colour. The limit of the 'r$_2$' fraction is reached when the scalpel contacts the white and very hard mature xylem.

Transverse sections of the stem are made after scraping of each fraction, and observed under microscope to verify the composition of each macro-fraction.

Note – the scraping method is generally poor for analytical studies because tissues are only crudely isolated. However, this method can provide large amounts of tissues making it suitable for preparative purposes prior to protein purification.

Preparation of micro-fractions by cryo-sectioning

Cryo-sectioning at 25 μm was performed according to the method of Uggla *et al.* (1996) (see description of procedure in the chapter by Uggla

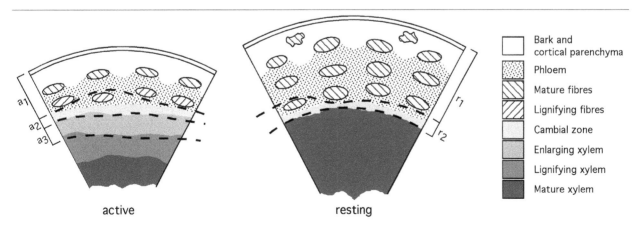

Figure 14 Schematic transverse sections of active and resting hybrid aspen shoots showing the tissues collected for biochemical analysis on macro-fractions. Three fractions (a$_1$, a$_2$, a$_3$) were scraped from active shoots and only two (r$_1$ and r$_2$) from resting shoots.

and Sundberg in this volume). By contrast to the scraping technique, the cryo-sectioning permits much improved protein analysis at a tissue level, but cannot be used for experiments that require large amounts of plant material (Figure 15).

Macro-fraction analysis

Biochemical analysis

Protein extraction

The following methods for extraction of ionically wall-bound and soluble PME from macro-fractions are based on those of Goldberg *et al.* (1986). In these protocols, the grinding step is crucial for a good protein extraction: scraped fractions must be reduced to a very fine powder free from pieces of tissues, particularly of lignified cells (e.g. phloem fibres, and xylem vessels).

Equipment needed:	Chemicals needed:
Enamelled sandstone mortar and pestle	Liquid nitrogen
Falcon tubes (25 or 50 ml)	Ice-cold **extraction buffer**
Cheese cloth	Triton X-100 (Sigma, No. T-8787)
Magnetic stirrer	Ice-cold bidistilled water
Refrigerator at 4 °C Cold room at 4 °C Freezer at –80 °C	Ice-cold **elution buffer**

Cell wall-bound proteins

Duration of procedure: approx. 1 day.
Note – except where indicated otherwise, all steps must be performed at 4 °C.

1 Weigh the freshly scraped fractions.
2 Grind the scraped material in liquid nitrogen using the pestle and mortar until a very fine powder is obtained, collect the powder in Falcon tubes and store at –80 °C.
3 Resuspend the ground tissue in an adequate volume of ice-cold extraction buffer (around 25 ml per gram of fresh weight). Mix the extract for 15 min on a magnetic stirrer.
4 Filter the extract through cheese cloth, retain the cell wall debris in the cloth. Discard the filtrate.

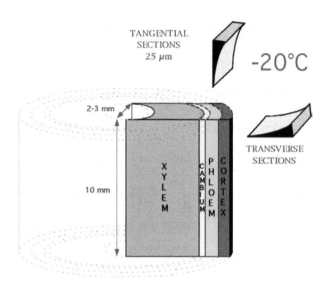

Figure 15 Schematic representation of the cryo-sectioning method. Axillary stem (1 to 2 cm in diameter) was cut transversely into 1 cm-long segments. These segments were further radially cut to produce blocks (2–3 mm × 10 mm) made of extraxylary tissues and the last annual ring. Blocks were cut in serial tangential sections of 25 μm from the bark to the mature xylem at –20 °C using a cryostat equipped with a steel knife. For radial localisation of the tangential sections, transverse sections of the specimen were hand-cut with a razor blade after each third tangential section, then examined under the microscope to determine tissue-type according to anatomical criteria.

5 Repeat steps 2–4 again.
6 Resuspend the cell walls in 25 ml per gram of fresh weight in 0.1% (v/v) Triton X-100. Mix for 10 min on a magnetic stirrer.
7 Filter the extract through cheese cloth, retain the cell walls.
8 Wash the cell walls with at least 1 l of ice-cold bidistilled water, until the filtrate is devoid of bubbles.
9 Resuspend the cell walls in 25 ml per gram of fresh weight of ice-cold bidistilled water.
10 Add 25 ml of ice-cold elution buffer, and mix for 30 min on a magnetic stirrer.
11 Filter the extract through a cheese cloth, collect the filtrate, and keep at 4 °C.
12 Repeat steps 9 to 11. Add the second filtrate to the first and keep at 4 °C.
13 Keep the cell walls at 4 °C for subsequent expression of PME activity on cell wall weight basis.

Soluble proteins

An alternative method of protein extraction is used for the biochemical analysis of some PME iso-enzymes that may exist free in the apoplast or more-weakly bound to the cell wall (Bordenave and Goldberg, 1994), and which would be expected to be washed out under the procedure described above.

Duration of procedure: approx. 6 hours.

Note – except where indicated otherwise, all steps must be conducted at 4 °C.

1   Grind each sample of scraped material in liquid nitrogen using a pestle and mortar until a very fine powder is obtained, collect the powder in Falcon tubes and store at –80 °C.
2   Resuspend the ground tissues in an adequate volume of ice-cold extraction buffer (around 25 ml per gram of fresh weight), and mix for 15 min on a magnetic stirrer.
3   Filter the extract through cheese cloth, and retain the cell walls.
4   Keep the filtrate containing soluble proteins at 4 °C.
5   Keep the cell walls at 4 °C in order to calibrate the PME activity on cell wall amount.

It is possible to combine these two protocols to permit analysis of both the cell wall-bound and the soluble proteins from the same sample. To do this, at step 4 of the cell wall bound protein extraction protocol, *keep the filtrate* for soluble PMEs analysis, then continue with the cell wall bound protein extraction procedure from step 5.

Quantification of cell wall amount

After elution of proteins was completed, cell walls were rinsed three times with 100 ml distilled water, then frozen in liquid nitrogen, and lyophilised. Weighing of lyophilised cell walls was performed on a high precision balance.

Assay of PME activity

Using an automatic titrator

PME activity was measured titrimetrically by following the increase of carboxyl groups as described by Bordenave and Goldberg (1993). The carboxyl groups liberated by PME from 0.25% *Citrus* pectin (Sigma, No. P-9131) were titrated with 10 mM NaOH (Prolabo, No. 31627.290) under nitrogen, the pH being maintained at chosen values ranging from 5.2 to 8.4 with an automatic titrator (Radiometer, No. TTT 80).

Spectrophotometric assay

Spectrophotometric assay of PME activity from macro-fractions was performed according to Richard *et al.* (1994). Thus, 10 μl of protein extract was added to 1 ml of **pectin methyl-red substrate solution**. Absorbance was measured at 525 nm. A calibration curve was obtained by adding 1–100 nEq H$^+$ to 1 ml of **pectin methyl-red substrate solution**. This curve showed a relationship between the A$_{525\,nm}$ and the amount of nEq H$^+$ added, and was used to convert the activity A$_{525\,nm}$ h^{-1} into nEq H$^+$ h^{-1}.

In contrast to assay of PME activity with an automatic titrator, the spectrophotometric assay of PME activity has the advantage of using smaller amounts of protein extract. However, spectrophotometric assay does not permit maintenance of the pH of the reaction mixture during measurement.

Iso-electric focusing

Cell wall proteins were fractionated by iso-electric focusing (IEF) on ultrathin (0.5 mm) polyacrylamide slab gel (10% acrylamide, 0.26% bis-acrylamide) containing 10% ampholines (Pharmacia, No. 17-0456-01) in the pH range 3–10, as described by Bordenave and Goldberg (1993). The IEF was run on an LKB Multiphor 2117 (Pharmacia) using an LKB 2103 power supply of 15 W at 15 mA for the first 30 minutes, and 20 W at 50 mA for one hour. 0.1 M NaOH and 0.1 M H$_2$SO$_4$ were used, respectively, as catholyte and anolyte. The PME active bands were revealed with the agar-pectin sandwich technique (Bertheau *et al.*, 1984), using ruthenium red (Sigma, No. M-7267).

Depending on the study, each sample can be calibrated either according to the amount of cell wall or with respect to the amount of total PME activity before loading on the IEF gel. Active PMEs were revealed on the gel according to the conditions described by Bertheau *et al.* (1984) in which agar-pectin was replaced by acrylamide-pectin to enhance resolution. The apparent isoelectric point (pI) was determined by reference to pI markers (Bio-Rad, No. 161–0310). The IEF gels were scanned, and densitometric measurements were made using NIH Image 1.54 (Wayne Rasband, NIH, USA). These values were used to determine the relative percentage of the PME isoform activity within each fraction.

Molecular analysis

All molecular biology techniques carried out on tree tissue macro-fractions were performed as described for herbaceous plants such as *Arabidopsis*. DNA extraction, RNA extraction and expression analysis by reverse Northern were performed as described by Dellaporta *et al.* (1983), Ausubel *et al.* (1987) and Micheli *et al.* (1998), respectively. Other usual methods of molecular biology described by Sambrook *et al.* (1989) can be applied to tree material. However, particular attention must be paid to the preliminary tissue-grinding step in order to improve extractability (see *'Biochemical analysis'* above).

Biochemical analysis of micro-fractions

Protein extraction

According to the experiment, cell wall proteins could be extracted from one or more consecutive cryo-tangential sections of 25 μm each (Micheli *et al.*, 2000).

Equipment needed:	*Chemicals needed:*
0.1 ml micropotter (Bioblock Scientific, No. C14261)	Triton X-100 (Sigma, No. T-8787)
Motor (Ika-Werk, Typ RW 18)	NaCl (Sds, No. 1380517)
Pasteur pipette	Ice
Vortex mixer	Bidistilled water
1.5 ml Eppendorf tubes	
Centrifuge	
Cold room at 4 °C	

In considering the choice of micropotter, the following criteria should be borne in mind:

● its capacity must be lower than 0.5 ml;
● glass is preferred to Teflon (which is not sufficiently strong to permit grinding of tough tissues like phloem fibres or xylem vessels);
● however, the glass must be resistant to erosion.

After testing several micropotters from different manufacturers, that manufactured by Bioblock Scientific was chosen as being the best suited to this purpose.

Cell wall-bound proteins
The following method allows elution of proteins that are ionically bound to the cell walls. The small amount of starting plant material (as cryo-sections) has required improvements to the extraction methods conventionally used for cell wall protein analysis from macro-fractions.
Duration of procedure: approx. 1 day.
All steps must be performed at 4 °C.

1   Transfer the sections into a 0.1 ml micropotter connected to a motor.
2   Add 100 μl 0.1% (v/v) Triton X-100 for each section.
3   Grind the sections very finely using a medium speed of the motor (150 rpm).
4   Transfer the homogenate into an Eppendorf tube using a Pasteur pipette.
5   Centrifugate the homogenate for 5 min at 3000 rpm.
6   Discard the supernatant.
7   Resuspend the pellet in 1 ml of bidistilled water to remove cellular compounds and traces of detergent.
8   Centrifuge for 5 min at 3000 rpm.
9   Discard the supernatant.
10 Repeat steps 7 & 8 twice.
11 Adjust the volume of the pellet to 25 μl with bidistilled water.
12 Add 25 μl of 2 M NaCl.
13 Vortex for 30 sec.
14 Incubate for 1 h on ice to elute the ionically wall-bound proteins.
15 Centrifuge for 15 min at 3000 rpm to separate the pelleted cell walls from the eluted ionically wall-bound protein of the supernatant.
16 Keep the supernatant for PME analysis.
17 Keep the pelleted cell walls in order to calibrate the PME activity on basis of cell wall amount (see *'Quantification of cell wall amount'* below).

Soluble proteins

Duration of procedure: approx. 6 h.
All steps must be performed at 4 °C.

1   Transfer the sections into a 0.1 ml micropotter connected to a motor.
2   Add 100 μl, 0.1% (v/v) Triton X-100 for each section.
3   Grind sections very finely using a medium speed of the motor (150 rpm).

4   Transfer the homogenate into an Eppendorf tube using a Pasteur pipette.

5   Centrifugate the homogenate for 5 min at 3000 rpm.

6   Keep the supernatant (which contains the soluble proteins for PME analysis).

7   Keep the pelleted cell walls in order to calibrate the PME activity on the cell wall amount (see 'Quantification of cell wall amount' below).

As described for macro-fractions, it is possible to combine both protocols so as to analyse the cell wall-bound as well as the soluble proteins from the same sample. To do this, at step 6 of the first protocol, keep the supernatant for analysis of soluble PMEs, then follow the cell wall bound protein extraction procedure from step 7.

Quantification of cell wall amount

Equipment needed:	Chemicals needed:
Spectrophotometer (e.g. Shimadzu)	Molecular biology grade agarose (Appligene, No. 130021)

Determination of cell wall amount

Because of the low amount of tissues, the dry weight of the cell wall within the samples was determined using a method based on turbidimetry.

1   Resuspend the pelleted cell walls in 0.1% melted-agarose (molecular biology grade) in order to prevent sedimentation of the cell wall particles.

2   Measure the turbidimetry of the homogenate at 600 nm using a spectrophotometer.

Calibration

The absorbance at 600 nm is converted to an equivalent dry weight value by reference to a standard curve made from a large-scale preparation of cell walls isolated from plant material. Increasing amounts of dried cell walls are weighed, resuspended in 0.1% melted-agarose BM grade, then assayed spectrophotometrically at 600 nm. The standard curve obtained from hybrid aspen stem indicates that 1 unit of $A_{600\,nm}$ corresponds to 1.15 mg of dried cell walls.

Microassay of PME activity

Equipment needed:	Chemicals needed:
Microplate	**Pectin methyl-red solution**
Microplate reader (Bio-Rad, No. 170–6621)	

In accordance with the low amounts of proteins obtained from cryo-sections, PME activity was measured in a small reaction volume. Thus, 5 μl of proteins were add to 150 μl of pectin methyl-red substrate solution in a microplate cupule, then spectrophotometrically assayed at 525 nm on a microplate reader.

The PME activity expressed in $A_{525\,nm}$ h^{-1} was converted into nEq H$^+$ h^{-1} using a calibration curve as described above (see *'Macro-fraction analysis'*).

Iso-electric focusing

The qualitative analysis of PMEs by IEF was performed as described above in the section entitled *'Macro-fraction analysis'*.

CHEMICALS, EQUIPMENT, ETC. FOR BIOCHEMICAL AND MOLECULAR ANALYSIS

Pectin methyl-red substrate solution

- 0.5% (w/v) *Citrus* pectin (Sigma, No. P-9131)
- 0.2 M NaCl (Sds, No. 1380517)
- 1.6% (v/v) of a saturated ethanolic solution of methyl-red (Sigma, No. M-7267)
- Adjust pH to 6.1 with 1 M NaOH (Prolabo, No. 31627.290)

Extraction buffer

- 5 mM MES (4-morpholinoethanesulfonic acid) (Boehringer, No. 223794)
- 4 mM L-cysteine hydrochloride monohydrate (Sigma, No. C-7880)
- Adjust pH to 6.0 with 1 M NaOH (Prolabo, No. 31627.290)

Elution buffer

- 50 mM succinic acid (Sigma, No. S-7501)
- 2 M NaCl (Sds, No. 1380517)
- Adjust pH to 5.0 with 1 M NaOH (Prolabo, No. 31627.290)

APPLICATIONS

Pectin localisation

The techniques described above have been used, rather successfully, to localise the main constituents of the walls, and especially pectins, in the tree SVS. For instance, they have demonstrated the differences in structure and composition between radial and tangential walls in the cambial zone of trees (Catesson and Roland, 1981), and suggested early differences in pectin biosynthesis between the phloem and the xylem differentiation pathways (Baïer et al., 1994). These findings have been confirmed by more recent observations, using MAbs raised against pectins (Chaffey et al., 1997; Guglielmino et al., 1997; Ermel et al., 2000). The figures included in this chapter illustrate various approaches to the in situ localisation of pectins and pectinases in the cell walls of the SVS of trees, as well as the benefit of combining the results obtained from different techniques, whether standard or specialised.

Figure 1 shows the localisation of homogalacturonan with a low degree of esterification in the cambial zone and young xylem using JIM5 antibodies. Staining decreases with differentiation and lignification. Similarly, staining of acidic pectins with ruthenium red gives a negative result in lignified and lignifying walls (Figure 2). In fact, this is a false negative staining due to the presence of lignin which prevents the reagent from reaching the reactive sites, as demonstrated in Figure 3. This micrograph shows a section partly delignified following a 48 h treatment with 28% ammonia solution prior to staining with ruthenium red. The treatment allows the red colour to develop in the compound middle lamella of xylem cells. However, use of delignification cannot be recommended prior to JIM5-staining, since pectin epitopes may not remain intact. Ruthenium red-staining is a quick and easy method to use for LM observations. It gives a fine red colour, and it is not as expensive as it may appear since only a very small amount is needed for each experiment. The hydroxylamine reaction requires more time but also gives good results. Both techniques can be used on fresh material and hand-made sections, which can be most advantageous. On the other hand, the more precise affinity techniques (immunolocalisation, lectin-gold or enzyme-gold) should be preferred for TEM investigations.

Results obtained at the ultrastructural level by parallel cyto-dissection and immunolocalisation are illustrated in Figures 6–10. They concern the composition and structure of cell walls common to two very young vessel elements at the beginning of their diametral enlargement. Comparison of Figures 6 and 7 shows that EDTA action is limited to a slight extraction of middle lamella polysaccharides, which suggests a very low amount of calcium-bonded galactan chains at these sites. Labelling of acidic pectins with JIM5 confirms their nearly exclusive location in the middle lamella (Figure 8). In contrast, most of the wall material is solubilised with boiling water (Figure 9). This agrees with the presence in the whole wall of highly esterified galacturonan labelled with JIM7 (Figure 8), and with the presence of RGI whose galactan side chains are labelled with LM5 (Figure 10), since both kind of pectins are soluble in hot water. Note, in Figure 9, the extreme reduction of the wall microfibrillar skeleton in cells whose diameter will enlarge 8–10 times or more. Figure 11 illustrates the secretion of LM5-labelled RGI from the cambial cells. Gold particles are present on Golgi vesicles whose polysaccharidic content was lightly stained with PATAg (3 h in thiocarbohydrazide). Cell wall labelling with LM5 is faint in cambial cells.

Figures 4 and 5 demonstrate, on tissue-prints of poplar twig, the presence of PME in the cambial zone and its immediate derivatives. This localisation closely agrees with the data obtained from biochemical analysis of branch microfractions (Figure 16; Micheli et al., 2000a,b).

At the ultrastructural level, the enzyme can be visualised in the Golgi apparatus, but not in the walls, of cambial cells (Figure 12). The enzyme appears to be localised in cell junctions, and more precisely in the central polysaccharidic network (Figure 13) in differentiating phloem and xylem cells (see also Guglielmino et al., 1997).

Biochemical analysis of pectin methylesterases

Figure 16 shows IEF of soluble PME isoforms extracted by the method described above from hybrid aspen stem at the active cambial stage. Calibration was done in order to have the same activity between samples (80 nmoles H^+/h).

At the active stage, the pattern of the PMEs across the stem comprises a number of isoforms distinguishable by their acidic (A1–A2), neutral (N3) and basic (B1–4) apparent pIs. The N3 isoform (apparent pI of 7.6) occurred in all tissues of the stem. The B1, B2 and B3 isoforms,

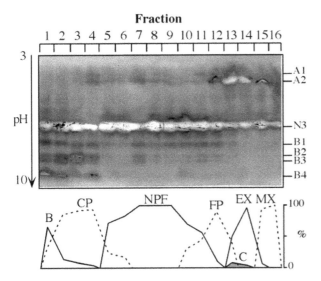

Figure 16 Iso-electrofocusing of soluble proteins eluted from aspen stem tissues during the active period. PME activities, calibrated on 80 nmoles H⁺/h, were revealed by the acrylamide-pectin sandwich technique. Names of isoforms are indicated on the right of the gel. Percentage of tissues contained in each fraction is represented at the bottom of the figure. Cambium is shaded in grey. B, bark; CP, cortical parenchyma; NFP, non-functional phloem; FP, functional phloem; C, cambium; EX, expanding xylem; MX, mature xylem.

pI of 7.7, 8.1 and 8.5, respectively, are present in most tissues of the stem. The cortical parenchyma comprised an additional isoform showing pI about 9.3 and named B4.

The most striking observation made on stems at the active stage relates to the occurrence of two acidic PME isoforms (A1 and A2), with apparent pIs of 5.2 and 5.6, respectively. Interestingly, optimum signal intensity of the A2 isoform covered the cambium and the immediate adjacent tissues, corresponding to the young phloem and xylem derivatives. Some traces of A2 extended from the young phloem and xylem derivatives to mature phloem and xylem, respectively. The A1 isoform was continuously, but slightly, distributed from the cambial region to the mature xylem at the active stage.

This qualitative analysis permitted the determination of the precise localisation of the PME isoforms across the hybrid aspen stem. This analysis could be further refined by a similar study on individual 25 μm tissue fractions.

FUTURE PROSPECTS

Few of the techniques available for pectin localisation have been used in the past on woody material. Other approaches, briefly discussed below, might give complementary information and should not be neglected.

i) An easy and rapid method, using Ni^{2+} or Co^{2+}, was propounded by Varner and Taylor (1989) to localise polygalacturonates on free-hand sections.

ii) The lectin-gold and pectinase-gold complex techniques could be adapted to woody plants. The pectinase-gold method (see Roland and Vian, 1991) can be used with PMEs or pectin lyases to detect methylesterified pectins, and with pectate lyases or polygalacturonases to label low-esterified homogalacturonan. It results in a high yield of specific labelling. However, the technique is little used, even on simple cell systems, probably because conditions of absorption and labelling have to be determined for each system (Jauneau *et al.*, 1998). In contrast, lectin-gold complexes (Benhamou, 1989, 1996) have been widely used, if exceptionally on trees, to investigate the distribution of various sugar residues able to bind specifically to a given lectin. However, the efficiency of labelling is still under discussion (Jauneau *et al.*, 1998).

Besides these techniques, which still need to be adapted to tree material, it is most probable that new monoclonal antibodies against specific sequences will become available in the near future. Confocal microscopy and image analysis will probably ensure a better exploitation of data obtained through *in situ* localisation techniques (Ermel *et al.*, 2000).

iii) Whatever the method used to localise pectins *in situ* (chemical staining, immunolocalisation or other affinity techniques), a negative result does not mean an absence of pectin molecules since the reactive sites can be masked by other wall components, especially by the presence of lignin (Czaninski and Monties, 1982), which is highly abundant in trees. Substractive methods, as we have seen, can be of help to discriminate between false and true negative reactions.

Cytochemical dissection of cell walls on resin-embedded sections can also usefully complement pectin-staining or -labelling. This kind of study was recently illustrated in a paper on apple cell walls (Roy *et al.*, 1997) in which pectin de-esterification and calcium chelation were carried out on grids. The development of microdissection and microanalysis methods is well under way. This will allow a more precise understanding of data obtained *in situ*.

iv) The advent of new approaches is illustrated by the very recent development of Fourier transform infra-red (FTIR) microspectroscopy. The technique offers the possibility to determine the structural and regulatory functions of pectins during plant growth and differentiation (McCann and Roberts, 1996).

It is to be hoped that the further development of the techniques presented here will give more precise images of the topological relations between pectin chains and between pectins and other wall components. It will also allow a better understanding of the origin and fate of polysaccharides present in the extracellular matrix. However, most important is the synthesis of data obtained from various technical approaches, ensuring interpretation to be as close as possible to the real thing.

The cryo-sectioning method (see Uggla *et al.*, 1996; chapter by Uggla and Sundberg in this volume) permits the collection of very small fractions from bark to mature xylem, whose tissue composition is precisely known. In this chapter, we report biochemical methods of protein analysis of extracted enzyme activity which have been adapted for small amounts of material and we demonstrate the possibility to detect and analyse PME activity in specific tissue fractions consisting of only a few cell layers. The cryo-sectioning technique could easily be extended to study PME gene expression at the molecular level, as recently described by Hertzberg and Olsson (1998) for homeobox genes concerned with xylem differentiation.

ACKNOWLEDGEMENTS

When working on the present chapter, FM and FFE have been financially supported by grants from the Commission of the European Communities, Agriculture and Fisheries (FAIR) specific RTD programme, CT 98-3972, 'Wood Formation Processes: the key to improvement of the raw material'. The authors are grateful to the following colleagues for past and present technical discussions on the pecularities of tree material and for allowing the use of some of their documents: Marie-Laure Follet-Gueye, Nadia Guglielmino, Michèle Liberman, Anne-Marie Catesson, Brigitte Vian, Yvette Czaninski and Michèle Grosbois.

REFERENCES

Albersheim, P. (1965) A cytoplasmic component stained by hydroxylamine and iron. *Protoplasma*, **60**, 131–135.

Ausubel, F.M., Brent, R., Kingston, R.E., Moore, D.D., Seidman, J.G., Smith, J.A. and Struhl, K. (1987) *Current protocols in molecular biology*, edited by Greene Publishing Associates and Wiley-Interscience, Vol I, pp. 4.1.2 John Wiley & Sons, New York.

Baïer, M., Goldberg, R., Catesson, A.M., Liberman, M., Bouchemal, N., Michon, V. and Hervé du Penhoat, C. (1994) Pectin changes in samples containing poplar cambium and inner bark in relation to the seasonal cycle. *Planta*, **193**, 446–454.

Benhamou, N. (1989) Preparation and application of lectin-gold complexes. In *Colloidal Gold: Principles, Methods and Applications*, edited by M.A. Hayat, Vol I, pp. 93–143. London: Academic Press.

Benhamou, N. (1996) Gold cytochemistry applied to the study of plant defense reactions. In *Histology, Ultrastructure and Molecular Cytology of Plant-Microorganism Interactions*, edited by N. Nicole and V. Gianinazzi-Pearson, pp. 55–77. Dordrecht, The Netherlands: Kluwer Academic Publishers.

Bertheau, Y., Madgidi-Hervan, E., Kotoujanski, A., Nguyen The, C. and Coleno, A. (1984) Detection of depolymerase isoenzymes after electrophoresis. *Analyt. Biochem.*, **139**, 383–389.

Bordenave, M. and Goldberg, R. (1993) Purification and characterization of pectin methylesterases from mung bean hypocotyl cell walls. *Phytochem.*, **33**, 999–1003.

Bordenave, M. and Goldberg, R. (1994) Immobilized and free apoplastic pectinmethylesterases in mung bean hypocotyl. *Plant Physiol.*, **106**, 1151–1156.

Carpita, N.C. and Gibeaut, D.M. (1993) Structural models of primary cell walls in flowering plants: consistency of molecular structure with the physical properties of the wall during growth. *Plant J.*, **3**, 1–30.

Cassab, G.I. (1992) Localization of Cell Wall Proteins in *Tissue Printing*, edited by P.D. Reid and R.F. Pont-Lezica, pp. 23–39. San Diegofd, New York, Boston, USA: Academic Press.

Catesson, A.M. and Roland, J.C. (1981) Sequential changes associated with cell wall formation and fusion in the vascular cambium. *IAWA Bull.*, **2**, 151–162.

Catesson, A.M., Funada, R., Robert-Baby, D., Quinet-Szelly, M., Chu-Bâ, J. and Goldberg, R. (1994) Biochemical and cytochemical cell wall changes across the cambial zone. *IAWA J.*, **15**, 91–101.

Chaffey, N.J., Barlow, P.W. and Barnett, J.R. (1997) Cortical microtubules involvement in bordered pit formation in secondary xylem vessel elements of *Aesculus hippocastanum* L. (Hippocastanaceae): a correlative study using microscopy and indirect immunofluorescence microscopy. *Protoplasma*, **197**, 64–75.

Chaffey, N.J., Barlow, P.W. and Barnett J.R. (1998) A seasonal cycle of cell wall structure is accompanied by a cyclical rearrangement of cortical microtubules in fusiform cambial cells within taproots of *Aesculus hippocastanum* L. (Hippocastanaceae). *New Phytol.*, **139**, 623–635.

Chaffey, N.J., Barnett, J.R. and Barlow P.W. (1997) Endomembranes, cytoskeleton and cell walls: aspects of the ultrastructure of the vascular cambium of tap roots of *Aesculus hippocastanum* L. (Hippocastaneaceae). *Int. J. Plant Sc.*, **158**, 97–109.

Czaninski, Y. and Monties, B. (1982) Etude cytochimique ultrastructurale des parois du bois de Peuplier après extraction ménagée. *C.R. Acad. Sc. sér. III*, **295**, 551–556.

Dellaporta, S.L., Wood, J. and Hicks, J.B. (1983) A plant DNA minipreparation: version II. *Plant Mol. Biol. Reporter*, **1**, 19–21.

Ermel, F.F., Kervella, J., Catesson, A.M. and Poëssel, J.L. (1999) Localized graft incompatibility in pear/quince (*Pyrus communis/Cydonia oblonga*) combinations: multivariate analysis of histological data from 5-month-old grafts. *Tree Physiol.*, **19**, 645–654.

Ermel, F.F., Follet-Gueye, M.L., Cibert, C., Vian, B., Morvan, C., Catesson, A.M. and Goldberg, R. (2000) Differential localization of RG I arabinan and galactan side chains in cambial derivatives. *Planta*, **210**, 732–740.

Goldberg, R., Bordenave, M., Pierron, M., Prat, R. and Mutaftschiev, S. (1992) Enzymatic processes in growing walls: possible control by pectin methylesterases. In *Plant Cell Walls as Biopolymers with Physiological Functions*, edited by Y. Masuda, pp. 269–274. Osaka, Japan: Yamada Science Foundation.

Goldberg, R., Morvan, C. and Roland, J.C. (1986) Composition, properties and localization of pectins in young and mature cells of the mung bean hypocotyl. *Plant Cell Physiol.*, **27**, 419–427.

Guglielmino, N., Liberman, M., Catesson, A.M., Mareck, A., Prat, R., Mutaftschiev, S. and Goldberg, R. (1997) Pectin methylesterases from poplar cambium and inner bark: localization, properties and seasonal changes. *Planta*, **202**, 70–75.

Guglielmino, N., Liberman, M., Jauneau, A., Vian, B., Catesson, A.M., Goldberg, R., (1997) Pectin immuno-localization and calcium visualization in differentiating derivatives from poplar cambium. *Protoplasma*, **199**, 151–160.

Harche, M. and Catesson, A.M. (1985) Cell wall architecture in alfa (*Stipa tenacissima* L.) fibres. *IAWA Bull.*, **6**, 61–69.

Hafrén, J. (1999) Ultrastructure of the wood cell wall. Doctoral Thesis, Royal Institute of Technology, Stockholm, Sweden.

Hertzberg, M. and Olsson, O. (1998) Molecular characterisation of a novel plant homeobox gene expressed in the maturing zone of *Populus tremula × tremuloides*. *Plant J.*, **16**, 285–295.

Hoson, T. (1991) Structure and functions of plant cell walls: immunological approaches. *Int. Rev. Cytol.*, **130**, 233–268.

Jauneau, A., Roy, S., Reis, D. and Vian, B. (1998) Probes and microscopical methods for the localisation of pectins in plant cells. *Int. J. Plant Sci.*, **159**, 1–13.

Jensen, W.A. (1962) *Botanical Histochemistry*. San Francisco and London: WH Freeman & Co.

Jones, L., Seymour, G. and Knox, J.P. (1997) Localization of pectic galactan in tomato cell wall using a monoclonal antibody specific to (1–4)-β-D-galactan. *Plant Physiol.*, **113**, 1405–1412.

Knox, J.P. (1997) The use of antibodies to study the architecture and developmental regulation of plant cell walls. *Int. Rev. Cytol.*, **171**, 79–120.

Knox, J.P., Linstead, P.J., King, J.J., Cooper, C. and Roberts, K. (1990) Pectin esterification is spatially regulated both within cell wall and between developing tissues of root apices. *Planta*, **181**, 512–521.

Liners, F. and Van Cutsem, P. (1992) Distribution of pectic polysaccharides throughout walls of suspension-cultured carrot cells. *Protoplasma*, **170**, 10–21.

Liners, F., Thibault, J.F. and Van Cutsem, P. (1992) Influence of the degree of polymerisation of oligalacturonates and of esterification pattern of pectin on their recognition by monoclonal antibodies. *Plant Physiol.*, **99**, 1099–1104.

Luft, J.H. (1971) Ruthenium red and violet. I Chemistry, purification, methods of use for electron microscopy and mechanism of action. *Anat. Rec.*, **171**, 347–368.

Mareck, A., Gaffé, J., Morvan, O., Alexandre, C. and Morvan, C. (1995) Characterization of isoforms of pectin methylesterase of *Linum usitatissimum* using polyclonal antibodies. *Plant Cell Physiol.*, **36**, 409–417.

McCann, M.C. and Roberts, K. (1996) Plant cell wall architecture: the role of pectins. In *Pectins and Pectinases*, edited by J. Visser, A.G.J. Voragen, pp. 91–107. Amsterdam: Elsevier.

Micheli, F., Bordenave, M. and Richard, L. (2000a) Pectin methylesterases: possible markers for cambial derivatives differentiation? In *Cell and molecular biology of wood formation*, edited by R. Savidge, J. Barnett, R. Napier, pp. 295–304. Oxford: Bios Scientific Publ. Ltd.

Micheli, F., Holliger, C., Goldberg, R. and Richard, L. (1998) Characterization of the pectin methylesterase-like gene *AtPME3*: a new member of a gene family comprising at least 12 genes in *Arabidopsis thaliana*. *Gene*, **220**, 13–20.

Micheli, F., Sundberg, B., Goldberg, R. and Richard. L. (2000b) Radial distribution pattern of pectin methylesterases across the cambial region of hybrid aspen at activity and dormancy. *Plant Physiol.*, **124**, 191–199.

Mueller, W.C. and Greenwood, A.D. (1978) The ultrastructure of phenolic cells fixed with caffein. *J. Exp. Bot.*, **29**, 757–764.

Northcote, D.H., Davey, R. and Lay, J. (1989) Use of antisera to localize callose, xylan and arabinogalactan in the cell plate, primary and secondary walls of plant cells. *Planta*, **178**, 353–366.

Puhlman, J., Bucheli, E., Swain, M.J., Dunning, N., Albersheim, P., Darvill, A.G. and Hahn, M.G. (1994) Generation of monoclonal antibodies against plant cell wall polysaccharides. I. Characterization of a monoclonal antibody to a terminal α-(1–2)-linked fucosyl-containing epitope. *Plant Physiol.*, **104**, 699–710.

Richard, L., Qin, L.X., Gadal, P. and Goldberg, R. (1994) Molecular characterization of a putative pectin methylesterase cDNA and its expression in *Arabidopsis thaliana* (L.). *FEBS Lett.*, **355**, 135–139.

Roland, J.C. and Vian, B. (1991) General preparation and staining of thin sections. In *Electron Microscopy of Plant Cells*, edited by J.L. Hall and C. Hawes, pp. 1–66. London: Academic Press.

Roy, S., Watada, A.E. and Wergin, W.P. (1997) Characterization of the cell wall microdomain surrounding plasmodesmata in apple fruit. *Plant Physiol.*, **114**, 539–547.

Sambrook, J., Fritsh, E.F. and Maniatis, T. (1989) *Molecular cloning: a laboratory manual*. Cold Spring Harbor, NY: Cold Spring Harbor Laboratory Press.

Steffan, W., Kovàc, P., Albersheim, P., Darvill, A.G. and Hahn, M.G. (1995) Characterization of a monoclonal antibody that recognizes an arabinosylated 1-6-D-galactan epitope in plant complex carbohydrates. *Carbohydr. Res.*, **275**, 295–307.

Uggla, C., Moritz, T., Sandberg, G. and Sundberg, B. (1996) Auxin as a positional signal in pattern formation in plants. *Proc. Natl. Acad. Sci. USA*, **93**, 9282–9286.

VandenBosch, K.A. (1991) Immunogold labelling. In *Electron Microscopy of Plant Cells*, edited by J.L. Hall and C. Hawes, pp. 181–218. London: Academic Press.

VandenBosch, K.A. (1992) Localization of proteins and carbohydrates using immunogold labelling in light and electron microscopy. In *Molecular Plant Pathology*. Vol.II. A Practical Approach, edited by S. Gurr, M.J. McPherson and D.J. Bowles, pp. 31–43. Oxford, UK: Oxford University Press.

VandenBosch, K.A., Bradley, D.J., Knox, J.P., Perotto, S., Butcher, G.W. and Brewin, N.J. (1989) Common components of the infection thread matrix and intercellular space identified by immunocytochemical analysis of pea nodules and uninfected roots. *EMBO J.*, **8**, 335–342.

Varner, J.E. and Taylor, R. (1989) New ways to look at the architecture of plant cell walls. *Plant Physiol.*, **91**, 31–33.

Vicré, M., Jauneau, A., Knox, J.P. and Driouich, A. (1998) Immunolocalization of β-(1–4) and β-(1–6)-D-galactan epitopes in the cell wall and Golgi stacks of developing flax root tissues. *Protoplasma*, **203**, 26–34.

Willats, W.G.T., Gilmartin, P.M., Mikkelsen, J.D. and Knox, J. (1999) Cell wall antibodies without immunization: generation and use of de-esterified homogalacturonan block-specific antibodies from a native phage display library. *The Plant J.*, **18**, 57–65.

Willats, W.G.T., Marcus, S.E. and Knox, J.P. (1998) Generation of a monoclonal antibody specific to (1–5)-α-1—L-arabinan. *Carbohydr. Res.*, **308**, 149–152.

Williams, M.N.V., Freshour, G., Darvill, A.G., Albersheim, P. and Hahn, M.G.M. (1996) An antibody PAb selected from a recombinant phage display library detects deesterified pectic polysaccharide rhamnogalacturonan II in plant cells. *Plant Cell*, **8**, 673–685.

2 Immunolocalisation of Enzymes of Lignification

Wood Formation in Trees ed. N. Chaffey
© 2002 Taylor & Francis
Taylor & Francis is an imprint of the
Taylor & Francis Group
Printed in Singapore.

Jozef Šamaj[1]* and Alain Michel Boudet[2]

[1] *Institute of Plant Genetics and Biotechnology, Slovak Academy of Sciences, Akademická 2, P.O. Box 39A, SK-950 07 Nitra, Slovak Republic*
[2] *UMR 5546 CNRS – UPS, Pôle de Biotechnologie Végétales, 24 chemin de Borde Rouge, BP17 Auzeville, F-31326 Castanet Tolosan, France*
* Author for correspondence

ABSTRACT

Lignin represents one of the most abundant cell wall biopolymers of terrestrial plants and has crucial functions in plant defence and mechanical support of the plant body. However, the fine details of lignin biosynthesis and the biological significance of lignin structural and functional heterogeneity are still far from fully understood. Lignification is a very complex process dependent on the activity of two key enzymes, cinnamoyl CoA reductase (CCR) and cinnamyl alcohol dehydrogenase (CAD), which are specific for the lignin-branch biosynthetic pathway. However, our knowledge of the spatio-temporal localisation and activity of these crucial lignin-specific enzymes *in situ* is very limited, particularly in woody plants. Therefore we aimed to develop a simple and reliable protocol for immuno-localisation of CAD 1, CAD 2 and CCR for tree species applicable at both the light and electron microscopy levels. These techniques are suitable for the study of lignification during plant development, pathogen attack and/or wounding, and seasonal changes in cambial activity. Moreover, immunolocalisation together with other *in situ* localisation techniques, like promoter-GUS histochemistry, *in situ* hybridisation and enzyme cytochemistry, will help to improve substantially our understanding of lignification process in trees.

Key words: cinnamoyl-CoA-reductase, cinnamyl alcohol dehydrogenase, immunogold localisation, lignin

INTRODUCTION

Lignification is considered to be a typical feature of terrestrial plants which has developed during their evolution. This important biological process concerns the deposition of lignins, complex phenolic polymers, in cell walls and cell wall domains of growing and differentiating plant cells. After cellulose, lignin is the second most abundant biopolymer on the Earth occurring as a major component of woody plants. This compound not only reinforces cell walls of conductive and non-conductive vascular cells, but recently it has also been found in primary cell walls of actively growing young cells, where it was suggested to be a limiting factor of wall extensibility and cell elongation (Müsel *et al.* 1997). Moreover, lignin in surface/peridermal layers of plant organs forms an efficient barrier against pathogens (Nicholson and Hammerschmidt, 1992), and also participates in wound healing (Hawkins and Boudet, 1996). All this indicates the importance of lignification, and highlights the necessity for its precise control, during normal plant development and plant-environment interactions.

Lignin is synthesised via the common phenyl-propanoid pathway and the lignin-specific branch pathway. This latter pathway is dependent on two key enzymes: cinnamoyl-CoA-reductase and cinnamyl alcohol dehydrogenase, which are responsible for monolignol biosynthesis.

Cinnamoyl-CoA-reductase (CCR, EC 1.2.1.44) is the first committed enzyme of the lignin branch pathway which is responsible for conversion of cinnamoyl CoA esters to their corresponding cinnamaldehydes (Boudet *et al.*, 1995). Recently, the cDNA encoding CCR was cloned from *Eucalyptus gunii* and the enzyme was shown to be expressed in developing xylem of poplar stems using *in situ* hybridisation (Lacombe *et al.*, 1997).

Cinnamyl alcohol dehydrogenase (CAD, EC 1.1.1.195), the second reductive enzyme of the lignin-specific branch, catalyses the reversible conversion of cinnamylaldehydes to the corresponding hydroxycinnamyl alcohols (monolignols). The term 'lignification enzyme' for CAD might be an oversimplification because monolignols produced by different CAD enzymes (CAD 1 and CAD 2) can both serve as precursor molecules of the lignin polymer, as well as for production of dimeric and oligomeric derivatives such as lignans and neolignans (reviewed by Boudet *et al.*, 1995; Whetten and Sederoff, 1995) which are not directly involved in lignification. This indicates that CAD is a polyfunctional enzyme involved in the synthesis of phenolic compounds with broad biological functions. While lignin reinforces cell walls it also has a role in defence, as for the lignans and neolignans, which are well known antiviral, antifungal, bactericidal and insecticidal agents (Lewis and Davin, 1994). Moreover, lignans and neolignans also have some properties which are similar to plant growth regulators, namely cytokinins (Binns *et al.*, 1987; Lyn *et al.*, 1987). Nevertheless, the main function of CAD is considered to be associated with lignification.

Two distinct CADs – CAD 1 and CAD 2 – have been characterised and purified from *Eucalyptus gunii* (Goffner *et al.*, 1992; Hawkins and Boudet, 1994). These two enzymes differ greatly in their molecular mass, the inability of CAD 1 to use sinapyl aldehydes and sinapyl alcohols as substrates, and immunoreactivity to specific antibodies. For CAD 2, the corresponding cDNA has been cloned and characterised from gymnosperms and angiosperms: loblolly pine (O'Malley *et al.*, 1992), spruce (Galliano *et al.*, 1993), eucalyptus (Grima-Pettenati *et al.*, 1993) and *Aralia cordata* (Hibino *et al.*, 1993). Very recently, the cDNA encoding CAD 1 was cloned from eucalyptus (Goffner *et al.*, 1998). Its product was characterised as a structurally and biochemically novel aromatic alcohol dehydrogenase in higher plants which was shown to be expressed in periderm and lignified tissues of poplar. Goffner *et al.* (1998) proposed that, besides CAD 2, which is most likely responsible for constitutive lignification, CAD 1 might function as an alternative enzyme, especially during defence lignin biosynthesis. Interestingly, both CAD 1 and CCR have been found to be closely related genetically, with a high degree of sequence homology (Goffner *et al.*, 1998).

Both lignification enzymes, CCR and CAD, are considered to be good targets for genetic manipulation of lignin quality and/or quantity in woody

plants using recombinant DNA and anti-sense DNA technologies (e.g. Boudet and Grima-Pettenati, 1996; Boudet, 1998). Transgenic plants down-regulated in CCR show an orange-brown xylem phenotype, presence of unusual cell-wall-bound phenolics, and severe alteration in lignin quantity accompanied by structural defects, e.g. collapsed xylem vessels and vessels with reduced size (Piquemal *et al.*, 1998). This indicates that CCR has an important role in the *quantitative* control of lignin synthesis (Boudet, 1998). On the other hand, transgenic trees with down-regulated CAD show a red-xylem phenotype and altered monolignol composition, with lignins enriched in cinnamaldehydes (Halpin *et al.*, 1994). A similar phenotype was found in a loblolly pine null mutant defective in CAD (Ralph *et al.*, 1997). These data indicate that CAD plays a role in control of lignin *quality*, namely in incorporation of cinnamylalcohol residues into lignin. In the absence of CAD these residues are replaced by corresponding aldehydes (Boudet, 1998).

Both CCR and CAD have been biochemically characterised in woody species (for review, see Boudet *et al.*, 1995; Whetten and Sederoff, 1995). The expression of the corresponding CCR and CAD genes was studied using Northern analysis (Grima-Pettenati *et al.*, 1993), promoter-GUS histochemistry (Feuillet *et al.*, 1995; Hawkins *et al.*, 1997; Šamaj *et al.*, 1998) and *in situ* hybridisation (Hawkins *et al.*, 1997, Lacombe *et al.*, 1997). All these techniques demonstrated the close spatial relationship that exists between lignified tissues and the gene expression.

However, we should keep in mind that the level of gene transcription may not accurately reflect the occurrence of the protein within a given cell or tissue due to possible post-transcriptional and post-translational modifications. Therefore, it is important to show localisation and function of enzymes *in situ* applying immunolocalisation and enzyme cytochemistry techniques. For example, the activity of the CAD enzyme has been localised by enzyme cytochemistry in four herbaceous and woody species (Baudracco *et al.*, 1993) and by tissue-printing in developing stems of poplar (Roth *et al.*, 1997). Although the resolution of the latter technique is rather limited, it provided the first information about CAD activity at the tissue level in a model tree species.

As for tissue-specific enzyme activity, very little was known about their precise localisation within plant tissues and particular cell types. Therefore, in order to understand more about the detailed expression pattern of CCR and CAD genes, we decided to study the distribution of the corresponding proteins using immunogold localisation. Three prerequisites are important to perform successful immunolocalisation: production of specific antibody, good fixation which preserves antigenic sites, and optimisation of the immunolabelling protocol. We have developed simple and reliable protocols for immunolocalisation of lignification enzymes at the light and electron microscopy levels based on immunogold labelling. These protocols are particularly suitable for woody species, but they can be very easily adapted also for herbaceous species. In this chapter we describe these immunocytochemical techniques and also cover their applications. We also discuss the correlation of immunocytochemical data with the cell-specific, developmental expression pattern of the CAD 2 promoter as revealed by GUS histochemistry (Hawkins *et al.*, 1997; Šamaj *et al.*, 1998; see also the chapter by Hawkins *et al.* in this volume) and localisation of CAD 2 and CCR transcripts using *in situ* hybridisation (Hawkins *et al.*, 1997; Lacombe *et al.*, 1997; see also the chapter by Regan and Sundberg in this volume) in poplar trees in order to make such a study more comprehensive. Using immunogold localisation techniques we show that CCR, CAD 1 and CAD 2 are present in the same cells in which their corresponding genes are expressed.

TECHNIQUES

Note — chemicals in bold type are considered in more detail in the SOLUTIONS section.

Plant material

Young trees of hybrid aspen, *Populus tremula* L. × *P. alba* L. (3–4 years old) grown outside (Nitra, Slovakia) and poplar clones P1, 2, 13, 15 and 21 (4 months old, corresponding to INRA clone 7/7/B4 *Populus tremula* × *P. alba*) grown in greenhouse, which have been individually transformed with the *Eucalyptus gunii* Hook CAD 2 promoter fused to the GUS gene (Feuillet *et al.*, 1995), were used to perform CAD 1, CAD 2 and CCR immunogold localisation. Small pieces of stem apices and stems (5 × 3 × 3 mm) were excised in different positions (0, 2, 5, 10, 20, 30, 40 cm) from the shoot apex, and processed as described below.

Processing for immunogold labelling

Equipment needed:	*Chemicals needed:*
Glass vials (10 ml capacity) with caps	Acetone
Glass microscope slides and cover slips	**Biobond adhesive solution**
Glass or diamond knife	Ethanol
Light microscope (equipped with bright field (BF), differential interference contrast (DIC), dark field (DF))	LR White resin (London Resin Company Limited, medium grade)
Freezer at –20 °C	**Phosphate-buffered saline (PBS)**
Refrigerator at +4 °C	***p*-formaldehyde solution**
	Washing solution
Embedding oven at 50 °C	
Forceps	**Formvar solution**
Slide storage box	**Blocking solution**
Transmission electron microscope	Glutaraldehyde (kept in frozen aliquots)
Gelatine embedding capsules (small size, 5 mm diameter)	Silver enhancement kit (BioCell, No. SEKL 15)
Ultramicrotome	**Incubation solution**
Single-edge razor blades	Toluidine blue
Nickel grids for EM (200 mesh or slot)	
Humid box	**Antibodies**
	DPX neutral mounting medium (Aldrich, No. 31,761-6)
Pasteur pipettes	NaN₃
	NH₄Cl
	Uranyl acetate
	Lead citrate (EMS, No. 17 800)
	NaCl

Sample preparation

Duration of procedure: approx. 150–155 h.
Steps 6–8 performed at room temperature (RT).

1 Excise pieces of tissue (5 × 3 × 3mm) in cold (–20 °C) 90% acetone.
2 Pre-fix with cold (–20 °C) 90% acetone (in order to prevent wound-induction of 'lignification' genes) for 30 min.

3   Rinse in cold (4 °C) PBS for 5 min.
4   Fix in freshly prepared mixture of 4% (v/v) *p*-formaldehyde and 0.25% (v/v) glutaraldehyde in PBS at 4 °C for 48 h.
5   Wash 3 × 5 min in PBS and 1 × 5 min in 100 mM NaCl at 4 °C.
6   Dehydrate in graded water/ethanol series: 15%, 30%, 50%, 75%, 90%, 96%, 30 min each followed by 3 × 30 min in dry 100% ethanol at RT.
7   Infiltrate with 33% and 66% LR White resin diluted with 100% ethanol, 12 h each followed by 3 × 12 h in fresh (100%) resin at RT.
8   Replace all resin with fresh resin, leave for another 2 h at RT.
9   Transfer samples to gelatine capsules, embed with fresh resin and polymerise in an embedding oven at 50 °C for 24 h. Orient samples in capsules bearing in mind the plane of section to be cut.

Immunolabelling for light microscopy

Duration of procedure: approx. 14–16 h
All steps performed at RT in a humid environment (humid box with wet filter paper).

1   Cut sections at 1 *μ*m, collect on Biobond-covered glass slides.
2   Block aldehyde groups with 50 mM NH$_4$Cl for 10 min.
3   Wash with deionised distilled water for 10 min using Pasteur pipette.
4   Incubate with blocking solution for 30 min.
5   Rinse with washing solution for 5 min.
6   Incubate with primary antibodies: rabbit anti-CAD 2 serum or IgGs at serial dilutions ranging from 1:20 to 1:500, rabbit anti-CCR serum and rabbit anti-CAD 1 serum, both at dilutions 1:100 in incubation solution for 3 h at RT.
7   Rinse 6 × 5 min with washing solution.
8   Incubate with secondary antibody: goat F(ab′)$_2$ anti-rabbit fragments conjugated to 5 nm gold diluted 1:100 in incubation solution for 1 h at RT.
9   Rinse 6 × 5 min with washing solution.
10   Post-fix with 3% glutaraldehyde in PBS for 10 min.
11   Wash extensively with PBS and water (5 × 5 min each).

12   Use silver enhancement kit (follow manufacturer's instructions) for 8–10 min (monitor intensity of labelling using microscope).
13   Rinse with deionised distilled water.
14   Counterstain with 0.02% aqueous toluidine blue for 1 min.
15   Rinse with deionised distilled water.
16   Air-dry and mount in DPX mounting medium.

Immunolabelling for electron microscopy:

Duration of procedure: approx. 16–18 h.
Ultrathin sections on nickel grids (either uncoated or formvar-coated) floated on 20 ml drops of appropriate solutions at RT.
Follow steps 2–11 as for the LM protocol above, except that primary antibodies are more diluted (in range 1:200 to 1:2000), and the secondary antibodies are either conjugated to 5 nm or to 10 nm gold. Smaller gold particles are subsequently silver-enhanced using a silver enhancement kit (BioCell) for 2–3 min, then washed.

12   Post-contrast with 1% aqueous solutions of uranyl acetate and lead citrate.
13   Air-dry.

Immunoblotting

Polyclonal anti-CAD 2 serum was prepared from 100 mg of pure recombinant *Eucalyptus gunii* CAD 2 protein by Eurogenetec (Seraing) according to standard procedures. Anti-CAD 2 immunoglobulin G (IgG) fraction was purified after Harlow and Lane (1988) on a Protein A-Sepharose CL-4B (Sigma) column. Both anti-CAD 2 serum and IgG were tested against purified CAD 2 protein and crude extracts of poplar in Western blots and dot blots. Western blots were performed as described in Goffner *et al.* (1992) and Grima-Pettenati *et al.* (1993). All reactions were specific for antibody concentrations used for immunolocalisations.

Controls

Negative control samples for CAD 1, CAD 2 and CCR were treated as above except that the primary antibodies were omitted from the incubation solution, or pre-immune rabbit sera were used instead of primary antibodies.

For positive controls, a polyclonal serum raised against heat shock protein 70 was used as a primary antibody diluted 1:1000 in incubation solution.

SOLUTIONS

Phosphate-buffered saline (PBS):

10 mM Na_2HPO_4, 10 mM NaH_2PO_4, 150 mM NaCl, pH 7.4

p-formaldehyde:

Add 8 g of p-formaldehyde powder to 100 ml water, heat to 60 °C in fume hood, add a few drops of 1M NaOH to dissolve all p-formaldehyde (you may need to adjust back to 100 ml to allow for evaporation), cool at RT, add 100 ml of double-strength **PBS**.

Biobond adhesive solution (BioCell, No. BB20):

2% in acetone (100 ml will cover approx. 100 microscopic slides)

Formvar solution:

1 g formvar (Sigma, No. F-6146) in 70 ml of chloroform

Blocking solution:

0.8% bovine serum albumen (BSA) (Sigma, fraction V), 0.1% fish gelatin (FG) (BioCell, No. GEL10), 2mM NaN_3, 5% normal goat serum (NGS) (Sigma, No. G9023), in **PBS**

Washing solution:

0.8% BSA, 0.1% FG, 2mM NaN_3 in **PBS**

Incubation solution:

0.8% BSA, 0.1% FG, 2mM NaN_3, 1% NGS in **PBS**

Antibodies

Primary:
polyclonal rabbit anti-CAD 2, anti-CCR and anti-CAD 1 sera (prepared by Eurogenetec, Seraing, according to standard procedures)
polyclonal anti-CAD 2 IgGs prepared from anti-CAD 2 serum according to Harlow and Lane (1988) on a Protein A-Sepharose CL-4B column

Secondary:
Goat F(ab')$_2$ anti-Rabbit IgGs conjugated to 5 nm gold particles (BioCell, No. LM.GFAR5)
Goat anti-Rabbit IgGs conjugated to 10 nm gold particles (BioCell, No. EM.GAR10)

APPLICATIONS

Immunolocalisation is a powerful technique for localising gene products – enzymes and proteins – to specific tissues, cells and/or subcellular compartments. In plants, this technique has been successfully used to localise some phenylpropanoid enzymes (Table 1) such as phenylalanine ammonia-lyase (PAL) and cinnamate-4-hydroxylase (C4H) in French bean hypocotyls (Smith et al., 1994); PAL in hybrid aspen (Osakabe et al., 1996); and PAL and CAD in differentiating tracheary elements of Zinnia (Nakashima et al., 1997).

Antibody characterisation

The specificity of polyclonal anti-CAD 2 and anti-CAD 1 sera as well as purified IgG fractions was verified on dot blots and Western blots of crude protein extracts from poplar and eucalyptus stems. Both anti-sera and IgG fractions specifically recognised a 43 kD band corresponding to the CAD 2 polypeptide (Šamaj et al., 1998) and a 35.8 kD band corresponding to CAD 1 polypeptide (Goffner et al., 1992, 1998) at the dilutions used for immunomicroscopy. No cross-reactivity was observed with other protein bands. Some preliminary biochemical tests have been done also with the anti-CCR antibody which shows good reactivity with the 37 kD band corresponding to CCR (data not published).

Immunolocalisation of lignification enzymes

Immunolocalisation experiments were performed in stems of two woody species: eucalyptus and poplar. Poplar proved to be more amenable to localisation techniques. The limited success achieved in eucalyptus (e.g. high nonspecific background levels) is likely to be related to the high content of endogenous phenolics and secondary metabolites in this species. This was also found to be a problem for in situ hybridisation using mRNA probes specific for CAD 2 and CCR (see Hawkins et al., 1997; Lacombe et al., 1997). Therefore, we performed our study on poplar stems, in which

Table 1 Use of immunolocalisation techniques for detection of phenylpropanoid and lignification enzymes in plant cells.
CAD = cinnamyl alcohol dehydrogenase; C4H = cinnamate-4-hydroxylase; CW = cell wall; EM = electron microscopy;
ER = endoplasmic reticulum; GA = Golgi apparatus; LM = light microscopy; PAL = phenylalanine ammonia-lyase

Enzyme	Plant species	Type of technique	Resolution	Location	Reference
PAL	*Phaseolus vulgaris*	Immunogold/silver Immunogold/silver	LM EM	Vascular tissue, xylem parenchyma, epidermis after wounding, cytosol	Smith *et al.* (1994)
PAL	*Populus kitkamiensis*	Immunogold/silver	LM	Xylem, phloem fibres	Osakabe *et al.* (1996)
PAL	*Zinnia elegans*	Immunofluorescence Immunogold	LM EM	Secondary thickenings of CW, cytosol, GA	Nakashima *et al.* (1997)
C4H	*Phaseolus vulgaris*	Immunogold/silver Immunogold/silver	LM EM	Vascular tissue, epidermis after wounding, ER	Smith *et al.* (1994)
CAD	*Zinnia elegans*	Immunofluorescence Immunogold	LM EM	Secondary thickenings of CW, cytosol, GA	Nakashima *et al.* (1997)
CAD 2	*Populus tremula × P. alba*	Immunogold/silver Immunogold	LM EM	Leaf axils, cambium, vascular tissue and surrounding parenchyma, cytosol, ER, GA	Samaj *et al.* (1998)
CAD 1	*Populus tremula × P. alba*	Immunogold/silver	LM	Periderm, cambium, vascular tissue and surrounding parenchyma	Goffner *et al.* (1998)

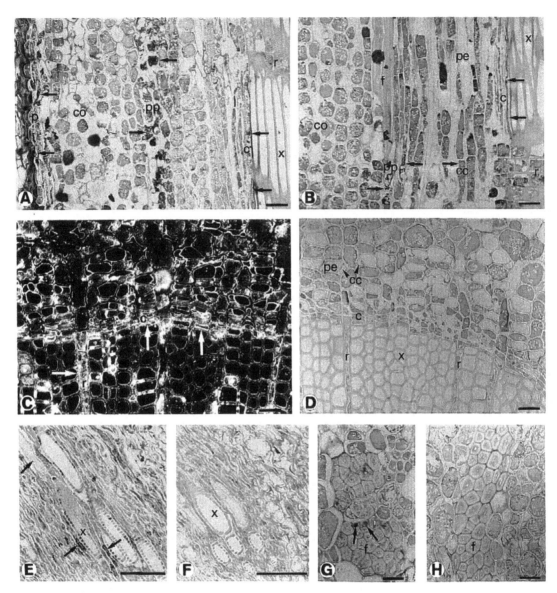

Figure 1 Immunogold localisation of CCR, CAD 1 and CAD 2 in poplar shoots as revealed with light microscopy.
For immunostaining, the LR White sections were treated with polyclonal antisera against CCR (A), CAD 1 (B, C) and
CAD 2 (E, G) and subsequently with secondary antibody conjugated to colloidal gold. Immunogold-labelled enzymes
appear as brown-black spots and areas after silver-enhancement of gold particles in bright field (A, B, E, G) or as a
white spots and label in dark field microscopy (C). **A** and **B** Radial longitudinal sections through mature stems labelled
with CCR antibody (**A**) and CAD 1 antibody (**B**). Most intense labelling is indicated by arrows depicting periderm (p),
phloem parenchyma cells (pp) close to the phloem fibres (f), cambium (c), young xylem (x) and ray cells (r). Note that
CAD 1 is also present in phloem parenchyma cells and companion cells (cc) located close to the phloem elements (pe).
C Transverse section through mature stem labelled with CAD 1 antibody viewed with dark field optics. Arrows
indicate most intense silver labelling of cambium (c), very young xylem (x) and ray cells (r). **D** Control transverse
section through mature stem (negative control to B and C) labelled with pre-immune CAD 1 serum showing no
immunolabelling of any cell type. **E** CAD 2 immunostaining in procambial cells and developing tracheids (arrows) on
radial longitudinal section through the shoot apex. **F** Corresponding negative control to E with CAD 2 pre-immune
serum showing no labelling. **G** Immunostaining of CAD 2 in the phloem fibres as revealed on transverse section
through the mature stem. Arrows indicate label within the lumen of phloem fibres. **H** Negative control to G with CAD
2 pre-immune serum.
Bars are equal to 50 μm for A, B, C, D, G and H; and 25 μm for E and F
Figures 1E–G are reproduced with kind permission of Springer–Verlag GmbH & Co. KG from Şamaj *et al.* (1998), *Planta*
204: 437–443.

heavily lignified cells are confined to the xylem, phloem fibres and the periderm. Within the xylem, CCR, CAD 1 and CAD 2 were detected in ray parenchyma cells and young vessel elements (Figures 1A, B, C). Overall, there were no dramatic differences in the expression pattern of these enzymes, except for differences in label intensity between the same cell types. All three enzymes were found in cambium, but only when an improved protocol for woody species sample preparation (after Hawkins *et al.*, 1997) was used (Figures 1A, B, C).

In the phloem, all three enzymes were abundantly localised to parenchymatic cells immediately surrounding phloem fibres (Figures 1A, B, C), but different expression patterns were found for other cell types in this tissue. Whereas CCR and CAD 2 were almost absent in cells surrounding phloem elements, e.g. in companion cells and phloem parenchyma, CAD 1 was moderately expressed in these cell types. On the other hand, we were only able to detect CAD 2 unambiguously directly within the lumen of heavily lignified phloem fibres (Figure 1G). This does not necessarily mean that CCR and CAD 1 are not present in phloem fibres, but they might be there in amounts below the detection limit of our method.

An intense signal was detected in the periderm for CCR (Figure 1A), and for CAD 1 (see Goffner *et al.*, 1998), however, it was not so strong for CAD 2 (not shown). This might indicate that both CCR and CAD 1 participate in concert in the synthesis of defence lignin which is deposited in surface cell layers of periderm. Immunostaining was negligible for all three enzymes in both stem cortex and pith (Figures 1 A, B). No immunolabelling was observed in negative controls performed with pre-immune sera instead of primary antibodies (Figures 1D, F, H). For better comparison and overview, localisation patterns for all three enzymes, as revealed by the immunogold method, are summarised in Table 2.

Additionally, in order to perform developmental studies we have localised CAD 2 in poplar shoot apices. No CAD 2 was localised within the primary apical meristem, but it was localised to the shell zones of axillary buds and leaf axils (Šamaj *et al.*, 1998). CAD 2 was also associated with procambial strands and developing protoxylem (Figure 1 E), which were not labelled in controls where pre-immune serum was used instead of primary antibody (Figure 1F).

Table 2 Overview of immunolabelling pattern for CCR, CAD 1 and CAD 2 in different tissues and cell types of mature poplar stems. Scale of labelling intensity: (+++) strong, (++) moderate, (+) weak, (–) none

	CCR	CAD 1	CAD 2
Periderm	+++	+++	+
Cortex	–	–	–
Parenchyma around phloem fibres	+++	+++	++
Phloem fibres	–	-	++
Phloem parenchyma	–	++	–
Companion cells	–	++	–
Phloem elements	–	–	–
Cambium	+++	+++	+++
Young xylem	+++	+++	+++
Ray parenchyma cells	+++	+++	+++
Old xylem	–	–	–
Xylem axial parenchyma	–	–	++
Pith	–	–	–

Subcellular localisation of lignification enzymes in stem cambial and ray cells

Immunogold electron microscopy localisation, either alone or in combination with silver enhancement, confirmed at the subcellular level the results obtained with light microscopy. It revealed that CAD 2 and CCR were localised in the cambium/developing xylem region and xylem ray cells. The CAD 2 enzyme was unambiguously localised in the cytoplasm of cambial cells. General observation showed that part of the immunostaining appeared to be associated with the membranes of the endoplasmic reticulum (ER) (Figure 2A inset) while the remainder was cytosolic (Figure 2A). Such a distribution pattern was regularly observed in poplar plants transformed with the promoter CAD 2–GUS construct (Figure 2A) as well as in non-transformed controls (Figure 2C). These results indicate that transformation events as such did not affect the protein distribution. More detailed observations revealed that membrane-associated gold particles were mostly attached to short elements of ER and Golgi-derived vesicles in cambial (Figure 2C, arrows) and ray (Figure 2D, arrowheads) cells. In control sections treated with pre-immune serum, no immunolabelling was observed in cambial (Figure 2B), ray or xylem cells (data not shown).

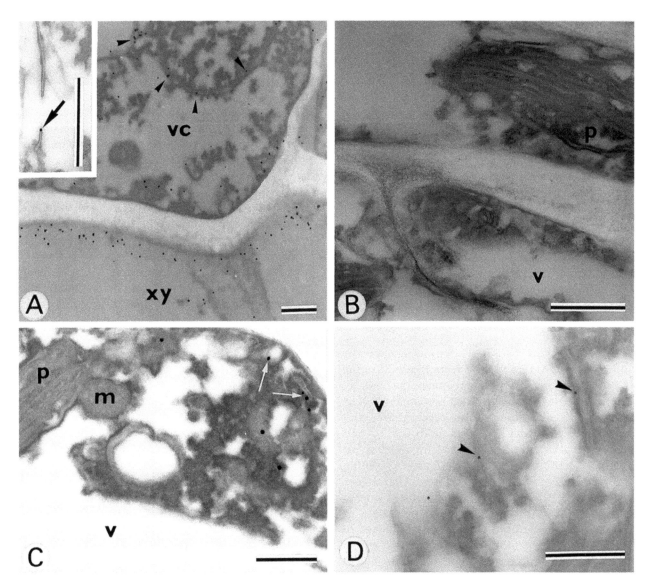

Figure 2 Subcellular localisation of CAD 2 in poplar stems as revealed by immunogold cytochemistry. Transverse ultrathin sections through cambium/young xylem zone were labelled with polyclonal rabbit anti-CAD 2 IgGs as primary antibody and goat anti-rabbit F(ab)$_2$ fragments conjugated to 5 or 10 nm colloidal gold as secondary antibody (to reduce non-specific background staining). Smaller gold particles were subsequently silver-enhanced for 5 to 8 min. Control sections were treated with pre-immune serum instead of primary antibody. **A.** Vascular cambium (vc) and young xylem (xy) of transformed poplar. Silver-enhanced gold particles are localised in the cytoplasm and associated with ER membranes of cambial cells, and within the remaining cytoplasm close to the developing secondary cell wall of the young xylem cell. Particles attached to the ER (other endomembranes can be excluded, e.g. by size) are indicated by arrowheads. *Insert*, detail of gold particle (arrow) attached to an element of ER. **B.** Control section to **A** through the vascular cambium showing no immunostaining. **C.** Cambial cell of wild type poplar. Membranes appear in negative contrast. Silver-enhanced gold particles attached to a short element of ER and a Golgi-derived vesicle are indicated by white arrows. **D.** Immunolabelling of xylem ray cell. Membranes appear in negative contrast. Gold particles associated with short elements of ER are indicated by arrowheads. m, mitochondrion; p, plastid; v, vacuole
Bars are equal to 0.5 μm for all images
Figures 2A–D are reproduced with kind permission of Springer–Verlag GmbH & Co. KG from Samaj *et al.* (1998), *Planta* **204**: 437–443.

Correlative studies using complementary *in situ* localisation techniques

As an integrated approach, three different techniques – promoter-GUS histochemistry, *in situ* hybridisation, and immunolocalisation – were used and compared in order to study CAD 2 gene expression and enzyme localisation in poplar trees (Hawkins *et al.*, 1997; Šamaj *et al.*, 1998). Taking into consideration that every *in situ* technique is prone to possible artefacts (e.g. high background and nonspecific signal for *in situ* hybridisation and immunolabelling, diffusion of enzyme product in GUS histochemistry) we aimed to correlate different techniques in order to achieve more reliable and comprehensive data about the spatial localisation of lignification enzymes.

We have shown that immunolocalisation of CAD 2 correlates with cell-specific expression of promoter CAD 2-GUS fusions (Šamaj *et al.*, 1998) as well as with *in situ* hybridisation of CAD 2 mRNAs (Hawkins *et al.*, 1997) in transgenic poplar shoots. Both CAD 2 and GUS were localised in the same types of cells in the shoot apices; particularly prominent were the determined meristematic cells in leaf axils and shell zones, procambium and developing tracheary elements. Both CAD 2 and GUS were also identified in cambium and fully or partially lignified cambial derivatives (young xylem, developing phloem fibres, chambered parenchyma cells around phloem fibres) within mature stems. Additionally, some new sites of GUS activity have been found in scale leaves of apical shoot buds and in transformed roots, e.g. in procambium, cambium, phellogen, young xylem and pericycle (Šamaj *et al.*, 1998). Employing immunogold cytochemistry at the subcellular level, CAD 2 was localised in the cytoplasm within cambial, ray and young xylem cells in stems, with gold particles randomly distributed along the ER and Golgi membranes. These results support the crucial role of CAD 2 in lignification.

We have previously demonstrated the importance of using a modified promoter-GUS histochemistry protocol for the study of gene expression in the secondary vascular tissues of trees (Hawkins *et al.*, 1997). This protocol for sample preparation is also suitable for immunolocalisation, after slight modification.

In subsequent work (Šamaj *et al.*, 1998), we showed that the cellular and subcellular fate of CAD 2 gene translation product is in close correlation to the gene activity during histogenetic events in the shoot apex and in the mature stems. In summary, our results with immunolocalisation of CAD 2, supported by GUS histochemistry, indicate that CAD 2 was not expressed in the apical meristems of shoots, but it was expressed in the leaf axillary zones, procambium, and secondary meristems, such as the vascular cambium in differentiated stems. This indicates that expression of the CAD 2 gene is spatio-temporally regulated and it is induced in those cells derived from primary meristem which are determined to be procambium or are involved in branching and development of lateral shoot organs (e.g. leaf primordia, leaf axils, shell zones as a basis for axillary bud development). As the promoter of the gene was also found to be active in the root pericycle during lateral root formation, it seems likely that the product of the gene could also be involved in branching events in this organ. In this respect it is interesting to mention that Walter *et al.* (1994) reported CAD activity in trichomes around axillary buds of transformed tobacco. CAD 2 was also detected in nonlignified megagametophytes of loblolly pine (O'Malley *et al.*, 1992) and its hypothetical role in lignan biosynthesis has been discussed (e.g. Goffner *et al.*, 1998; Šamaj *et al.*, 1998).

Within mature poplar shoots, CAD 2, CAD 1 and CCR have been found regularly in secondary meristems and in young xylem. This is a particularly important finding because it means that the gene is not developmentally 'switched off' in the nonlignified cambium and phellogen meristems, both of which play a crucial role in the development of lignified tissues within the secondary growing organs. These findings are in general consistent with the suggested crucial role of CCR and CAD in lignification. They further support the view that cells like young xylem, phloem fibres, and tracheary elements in the scale leaves can be lignified themselves, while parenchyma cells (e.g. chambered cells surrounding phloem fibres and tracheary elements in the scale leaves) located in their proximity, may provide lignin precursors for their own lignification and/or that of adjacent conductive elements.

From a developmental point of view such an observation is interesting since it suggests that the cells of the secondary meristem are, in contrast to those in the apical meristem, already partially differentiated. Convincing evidence supporting the hypothesis that lignification enzymes are active in the vascular cambium is provided by the work of Savidge and co-workers (reviewed by Savidge, 1996) who showed that coniferin, the glucoside of coniferyl alcohol, is present in cambial cells of *Pinus*. Interestingly, coniferin only starts to accumulate in cambial cells at the commencement of seasonal cambial activity and disappears with the onset of cambial dormancy. Similarly, the activity of the enzyme catalysing coniferin biosynthesis is only detectable when coniferin is present in the cambium. It would be of interest to see whether the seasonal variation in CAD 2, CAD 1 and CCR occurrence and activity is reflected in the seasonal, spatial expression pattern of the corresponding genes. Immunolocalisation techniques can be used to study this. The fact that not all of the cell derivatives on the phloem side of the vascular cambium are lignified suggests two possibilities. Either the monolignols resulting from the activity of lignification genes are not deposited in these cells, or, as suggested by the cell-specific expression pattern in the phloem, some of these genes are developmentally 'switched off' in certain cells, such as phloem elements which were not labelled with any antibody used in our experiments.

In conclusion, our results demonstrate the close spatial link that exists between lignified tissues and the localisation of CAD 2 and CCR in different cell types within poplar stems. This particular finding further supports the crucial role of these enzymes in lignification. However, the presence of CAD 1, CAD 2 and CCR in apparently non-lignified cells also indicates other roles for these enzymes, e.g. in plant defence and synthesis of defence compounds, such as lignans and neolignans.

CAD 2 was first localised *in situ* through enzyme cytochemistry in xylem and epidermis and/or periderm in stems and roots of five different plant species (*Phaseolus vulgaris*, *Vicia faba*, *Pisum sativum*, *Tilia campestris* and *Populus niger*) by Baudracco *et al.* (1993). Subsequently, CAD 2 promoter activity, investigated by GUS histochemistry, was reported to correlate with lignification areas in transformed poplar plants (Feuillet *et al.* 1995). Further investigations revealed new important sites of CAD 2 gene activity in mature poplar stems, especially within the vascular cambium and periderm, which was confirmed by *in situ* hybridisation (Hawkins *et al.*, 1997) and immunolocalisation (Šamaj *et al.*, 1998). Similarly, the CCR transcripts were localised to the

vascular cambium of poplar stems using *in situ* hybridisation (Lacombe *et al.*, 1997), the localisation of CCR in cambium was confirmed by immunolocalisation (e.g. shown here). All these data indicate that different *in situ* localisation techniques can be successfully combined and correlated to each other in order to reduce possible artefacts of individual techniques.

The subcellular localisation of the CAD 2 protein to the cytosol and ER raises questions concerning the mechanism of transport of the monolignols to their site of polymerisation in the cell wall. In conifers, glucosides of monolignols are accumulated in the vacuole and are presumably transported to the plasma membrane via vesicular trafficking at the appropriate time during xylem differentiation (Whetten and Sederoff, 1995). The majority of angiosperms do not accumulate monolignols as glucosides in the vacuole and must therefore possess another mechanism for transporting lignin precursors to the site of assembly. Smith *et al.* (1994) investigating the subcellular distribution of other enzymes of the phenylpropanoid pathway, localised PAL to the cytosol, while C4H was found to be associated with the ER and Golgi stacks. Our localisation of CAD 2 to the endomembrane system correlates well with its putative role in the synthesis and transport of lignin precursors, while its presence in the cytosol suggests the existence of another transport mechanism and further investigation is necessary.

FUTURE PROSPECTS

Immunofluorescence and monoclonal antibodies

Our experiments performed with polyclonal antisera against CAD 1, CAD 2 and CCR and with IgG fraction of CAD 2 antiserum revealed some interesting differences in the intensity and precise location of immunolabelling between them. Unfortunately, it is not easy to quantify these differences since we exclusively used immunogold labelling technique for bright and dark field light microscopy. However, the sensitivity of the immunolabelling technique can be further improved using immunofluorescence. This technique provides enhanced signal levels which can be more easily quantified with the help of high-resolution, low-light sensitive cameras (e.g. Hamamatsu) and special image analysis software.

We used polyclonal antisera and IgGs in our experiments. In order to reduce the nonspecific background labelling to minimum, it will be desirable to develop highly specific monoclonal antibodies against each enzyme or its isoform. This will further improve impact resulting from immunolocalisation experiments, and enable us to perform detailed studies focused on local differences in the intensity and precise spatial (cellular and subcellular) location of the label within the cell or tissue of interest. Highly specific monoclonal antibodies raised in different animals and coupled to gold particles of different sizes or to different fluorochromes will be a very helpful tool in order to elucidate whether lignification enzymes co-localise (as an indication that these enzymes might form enzyme complexes) when double or triple labelling technique in combination with electron microscopy or high resolution fluorescent microscopy will be used. Correlative studies comparing light and electron microscope techniques can be performed more easily applying the above mentioned approach, and this will help to identify possible co-localisation of CAD and CCR.

Advances in microscopy – confocal laser scanning microscopy and two/three photon microscopy as new powerful localisation tools

Substantial improvement in resolution and optical sectioning of cells are the main advantages of confocal laser scanning microscopy (CLSM) (Hepler and Guning, 1998; see also chapter by Funada in this volume). In CLSM, the object is illuminated by a laser beam and, in its point-scanning-variety, only a single point of the object is illuminated and imaged instead of imaging the entire field of view as is the case in conventional fluorescence microscopy. This results in reduced out-of-focus flare and improved object imaging.

The CLSM operates as a single photon system using photons of short wavelength, but recently multi-photon (two or three) microscopy has been developed. Such multiphoton microscopes further extend resolution and imaging limits, especially when long-wavelength photons are used. Two or three photons can interact at the focus point and cut the excitatory wavelength by one half or one third respectively. These long-wavelength photons penetrate better into bulky specimens (particularly important for woody species) with less scattering.

Moreover, they can be directly used for excitation of some biologically active molecules like serotonin or pyridine nucleotides. CLSM and multiphoton microscopy are vital tools for molecular cytological studies which can be performed also on living plants cells.

Although the CLSM has yet to be as widely used in the study of trees as it is in herbaceous species (particularly *Arabidopsis*), it has been applied to study of the cytoskeleton in gymnosperms (e.g. Abe *et al.*, 1995; see also the chapter by Funada in this volume) and the angiosperm tree, *Aesculus hippocastanum* (e.g. Chaffey *et al.*, 1997; Chaffey, 2000). And is beginning to be used to study aspects of cell biology of the model angiosperm tree, poplar (e.g. Chaffey, 1999), such as cell wall polysaccharide chemistry (Ermel *et al.*, 2000) and the cytoskeleton (Chaffey, Barlow and Sundberg, manuscript in preparation; see also Figures 1E, F, H in chapter on Wood microscopical techniques by Chaffey in this volume). Recently, the CLSM was used also for tissue localisation of phenolic compounds including lignin in *Picea abies* L. and some herbaceous species (Hutzler *et al.*, 1998). Other tree-related uses of the CLSM are considered in the chapter on Wood microscopical techniques in this volume.

Advancement of molecular-biological tools based on GFP technology

The gene encoding green fluorescent protein (GFP) from the jellyfish *Aequorea victorea* can be fused to other genes encoding proteins of interest, and such chimeras can be used to generate transformed plant cells (e.g. Sullivan and Kay, 1999). The fluorescing product of GFP can be subsequently used as a molecular marker for expression of the fused gene. The GFP can be visualised by fluorescence microscopy in living plant cells (especially when CLSM is used), allowing the targeting and fate of the protein of interest to be monitored in real time (e.g. Boevink *et al.*, 1998). There are different types of modified GFPs available which (as a consequence of their structural modifications) fluoresce in different colours – green, blue, yellow or red. Since genes encoding CAD 1, CAD 2 and CCR are already cloned and characterised, they can be fused to different types of GFPs in the near future. Such gene chimeras can be introduced to easily transformable trees, e.g. poplar (Leplé *et al.*, 1992) and used for high resolution localisation studies *in vivo*. Monitoring of GFP-based reporter signals will help to answer questions concerning subcellular locations of CAD and CCR, particularly whether these enzymes are located free in the cytoplasm or are associated with ER and Golgi membranes.

Moreover, GFP technology can be used in combination with fluorescence resonance energy transfer (FRET) microscopy in co-localisation studies (Gadella *et al.*, 1999). This latter technique, in combination with GFP technology or double/triple immunolocalisation, is able to answer the question of whether two proteins which co-localise are able to interact with each other and change their conformation (e.g. during enzyme reaction). This powerful combination of GFP and FRET can be used to look at interactions of CAD and CCR organised in putative enzyme complexes in cells of woody and herbaceous species. Structural organisation of enzymes in complexes has long been a matter of debate for some phenylpropanoid enzymes (Hrazdina and Wagner, 1985; Hrazdina, 1994), and has also recently been discussed for lignification enzymes, such as CAD 1 and CCR (Goffner *et al.*, 1998).

ACKNOWLEDGEMENTS

Jozef Šamaj is grateful to Eurosilva for his research fellowship. We thank our colleagues Prof. S. Hawkins, Drs D. Goffner, J. Grima-Pettenati and V. Lauvergeat for their help, fruitful discussion and interest during the course of this work. Drs G. Truchet, F. de Billy and J. Vasse (Laboratoire de Biologie Moleculaire des Relations Plantes-Microorganismes, INRA-CNRS, Castanet–Tolosan, France) are gratefully acknowledged for allowing us to use their equipment, and for their help.

REFERENCES

Abe H, Funada R, Imaizumi H, Ohtani J, Fukuzawa K (1995) Dynamic changes in the arrangement of cortical microtubules in conifer tracheids during differentiation. *Planta* **197,** 418–421.

Baudracco, S., Grima-Pettenati, J., Boudet, A. M. and Gahan, P. (1993) Quantitative cytochemical localisation of cinnamyl alcohol dehydrogenase activity in plant tissues. *Phytochemical Analysis*, **4**, 205–209.

Binns, A.N., Chen, R.H., Wood, H.N. and Lyn, D.G. (1987) Cell division promoting activity of naturally occurring dehydrodiconiferyl glucosides: Do cell wall components control cell division? *Proceedings National Academy of Sciences USA*, **84**, 980–984.

Boevink, P., Oparka, K., Santa Cruz, S., Martin, B., Betteridge, A. and Hawes, C. (1998) Stacks on tracks: the plant Golgi apparatus traffics on an actin/ER network. *Plant Journal*, **15**, 441–447.

Boudet, A.M., Lapierre, C. and Grima-Pettenati, J. (1995) Biochemistry and molecular biology of lignification. *New Phytologist*, **129**, 203–236.

Boudet, A.M. and Grima-Pettenati, J. (1996) Lignin genetic engineering. *Molecular Breeding*, **2**, 25–39.

Boudet, A.M. (1998) A new view of lignification. *Trends in Plant Science*, **3**, 67–71.

Chaffey, N.J. (1999) Cambium: old challenges – new opportunities. *Trees*, **13**, 138–151.

Chaffey, N.J. (2000) Cytoskeleton, cell walls and cambium: new insights into secondary xylem differentiation. In: *Cell and molecular biology of wood formation*. edited by R. Savidge, J. Barnett, and R. Napier. pp. 31–42: Oxford Bios Scientific Publishers.

Chaffey, N.J., Barlow, P.W., Barnett, J.R. (1997) Microtubules rearrange during differentiation of vascular cambial derivatives, microfilaments do not. *Trees*, **11,** 333–341.

Ermel, F.F., Follet-Gueye, M.-L., Cibert, C., Vian, B., Morvan, C., Catesson, A.-M., Goldberg, R. (2000) Differential localization of rhamnogalacturonan 1 arabinan and galactan side chains in cambial derivatives. *Planta*; **210**, 732–740.

Feuillet, C., Lauvergeat, V., Deswarte, C., Pilate, G., Boudet, A. and Grima-Pettenati, J. (1995) Tissue- and cell-specific expression of a cinnamyl alcohol dehydrogenase promoter in transgenic poplar plants. *Plant Molecular Biology*, **27**, 651–667.

Gadella, T.W.J., van der Krogt, G.N.M and Bisseling, T. (1999) GFP-based FRET microscopy in living plant cells. *Trends in Plant Science*, **4**, 287–291.

Galliano, H., Cabané, M., Eckerskorn, C., Lottspeich, F., Sandermann, H. and Ernst, D. (1993) Molecular cloning, sequence analysis and elicitor-/ozone-induced accumulation of cinnamyl alcohol dehydrogenase in Norway spruce (*Picea abies*). *Plant Molecular Biology*, **23**, 145–156.

Goffner, D., Joffroy, I., Grima-Pettenati, J., Halpin, C., Knight, M.E., Schuch, W. and Boudet, A.M. (1992) Purification and characterization of isoforms of cinnamyl alcohol dehydrogenase from eucalyptus xylem. *Planta*, **188**, 48–53.

Goffner, D., Van Doorsselaere, J., Yahiaoui, N., Šamaj, J., Grima-Pettenati, J. and Boudet, A.M. (1998) A novel aromatic alcohol dehydrogenase in higher plants: molecular cloning and expression. *Plant Molecular Biology*, **36**, 755–765.

Grima-Pettenati, J., Feuillet, C., Goffner, D., Borderies, G. and Boudet, A.M. (1993) Molecular cloning and expression of a *Eucalyptus gunii* cDNA clone encoding cinnamyl alcohol dehydrogenase. *Plant Molecular Biolology*, **21**, 1085–1095.

Halpin, C., Knight, M.E., Foxon, G.A., Campbell, M.M., Boudet, A.M., Boon, J.J., Chabbert, B., Tollier, M.T. and Schuch, W. (1994) Manipulation of lignin quality by down-regulation of cinnamyl alcohol dehydrogenase. *Plant Journal*, **6**, 339–350.

Harlow, E. and Lane, D. (1988) *Antibodies. A laboratory manual*, pp. 726, CHS, New York.

Hawkins, S. W. and Boudet, A.M. (1994) Purification and characterization of cinnamyl alcohol dehydrogenase isoforms from the periderm of *Eucalyptus gunii*. Hook. *Plant Physiology*, **104**, 75–84.

Hawkins, S. W. and Boudet, A.M. (1996) Wound-induced lignin and suberin deposition in a woody angiosperm (*Eucalyptus gunii* Hook.): histochemistry of early changes in young plants. *Protoplasma*, **191**, 96–104.

Hawkins, S., Šamaj, J., Lauvergeat, V., Boudet, A. and Grima-Pettenati, J. (1997) Cinnamyl alcohol dehydrogenase (CAD): identification of important new sites of promoter activity in transgenic poplar. *Plant Physiology*, **113**, 321–325.

Hepler, P.K. and Gunning, B.E.S. (1998) Confocal fluorescence microscopy of plant cells. *Protoplasma*, **201**, 121–157.

Hibino, T., Shibata, D., Chen, J.Q. and Higuchi, T. (1993) Cinnamyl alcohol dehydrogenase from *Aralia cordata*: cloning of the cDNA and expression of the gene in lignified tissues. *Plant and Cell Physiology*, **34**, 659–665.

Hrazdina, G. (1994) Compartmentation in phenolic metabolism. *Acta Horticulturae*, **381**, 86– 103.

Hrazdina, G. and Wagner, G.J. (1985) Metabolic pathways as enzyme complexes: Evidence for the synthesis of phenylpropanoids and flavonoids on membrane associated enzyme complexes. *Archives of Biochemistry and Biophysics*, **237**, 88–100.

Hutzler, P., Fischbach, R., Heller, W., Jungblut, T.P., Reuber, S., Schmitz, R., Veit, M., Weisenböck, G. and Schnitzler, J.P. (1998) Tissue localisation of phenolic compounds in plants by confocal laser scanning microscopy. *Journal of Experimental Botany*, **49**, 953–965.

Lacombe, E., Hawkins, S., van Doorsselaere, J., Piquemal, J., Goffner, D., Poeydomenge, O., Boudet, A.M. and Grima-Pettenati, J. (1997) Cinnamoyl CoA reductase, the first commited enzyme of the lignin branch biosynthetic pathway: cloning, expression and phylogenetic relationships. *Plant Journal*, **11**, 429–441.

Leplé, J.C., Brasileiro, A.C.M., Michel, M.F., Delmotte, F. and Jouanin, L. (1992) Transgenic poplars: expression of chimeric genes using four different constructs. *Plant Cell Reports*, **11**, 137–141.

Lewis, N.G. and Davin, L.B. (1994) Evolution of lignan and neolignan biochemical pathways. In *Evolution of natural products*, edited by D. Nes, pp. 202–246. ACS Symposium Series. Washington DC 562.

Lynn, D.G., Chen, R.H., Manning, K.S. and Wood, H.N. (1987) The structural characterization of endogenous factors from *Vinca rosea* crown gall tumors that promote cell division of tobacco cells. *Proceedings National Academy of Sciences USA*, **84**, 615–619.

Müsel, G., Schindler, T., Bergfeld, R., Ruel, K., Jacquet, G., Lapierre, C., Speth, V. and Schopfer, P. (1997) Structure and distribution of lignin in primary and secondary cell walls of maize coleoptiles analyzed by chemical and immunological probes. *Planta*, **201**, 146–159.

O'Malley, D.M., Porter, S. and Sederoff, R.R. (1992) Purification, characterization, and cloning of cinnamyl alcohol dehydrogenase in loblolly pine (*Pinus taeda* L.). *Plant Physiology*, **98**, 1364–1371.

Nakashima, J., Awano, T., Takabe, K., Fujita, M and Saiki, H. (1997) Immunocytochemical localisation of phenylalanine ammonia-lyase and cinnamyl alcohol dehydrogenase in differentiating tracheary elements derived from *Zinnia* mesophyll cells. *Plant and Cell Physiology*, **38**, 113–123.

Nicholson, R.L. and Hammerschmidt, R. (1992) Phenolic compounds and their role in disease resistance. *Annual Review of Phytopathology*, **30**, 369–389.

Osakabe, Y., Nanto, K., Kitamura, H., Kawai, S., Kondo, Y., Fujii, T., Takabe, K., Katayama, Y. and Morohoshi, N. (1996) Immunocytochemical localisation of phenylalanine ammonia-lyase in tissues of *Populus kitkamiensis*. *Planta*, **200**, 13–19.

Piquemal, J., Lapiére, K., Myton, K., O'Connell, A., Schuch, W., Grima-Pettenati, J. and Boudet, A.M. (1998) Down-regulation of cinnamoyl-CoA reductase induces significant changes of lignin profiles in transgenic tobacco plants. *Plant Journal*, **13**, 71–84.

Ralph, J., MacKay, J.J., Hatfield, R.D., O'Malley, D.M., Whetten, R.W. and Sederoff, R.R. (1997) Abnormal lignin in a loblolly pine mutant. *Science*, **277**, 235–239.

Roth, R., Boudet, A.M. and Pont-Lezica, R. (1997) Lignification and cinnamyl alcohol dehydrogenase activity in developing stems of tomato and poplar: a spatial and kinetic study through tissue printing. *Journal of Experimental Botany*, **48**, 247–254.

Šamaj, J., Hawkins, S., Lauvergeat, V., Grima-Pettenati, J. and Boudet, A. (1998) Immunolocalisation of cinnamyl alcohol dehydrogenase 2 (CAD 2) indicates a good correlation with cell-specific activity of CAD 2 promoter in transgenic poplar shoots. *Planta*, **204**, 437–443.

Savidge, R.A. (1996) Xylogenesis, genetic and environmental regulation – a review. *IAWA Journal*, **17**, 269–315.

Smith, C.G., Rodgers, M.W., Zimmerlin, A., Ferdinando, D. and Bolwell, G.P. (1994) Tissue and subcellular immunolocalisation of enzymes of lignin synthesis in differentiating and wounded hypocotyl tissue of French bean (*Phaseolus vulgaris* L.) *Planta*, **192**, 155–164.

Sullivan, K.F. and Kay, S.A. (eds) (1999) Green fluorescent proteins. *Methods in Cell Biology*. Vol. 58. Academic Press, London, New York.

Walter, M.H., Schaaf, J. and Hess, D. (1994) Gene activation in lignin biosynthesis: Patterns of promoter activity of a tobacco cinnamyl–alcohol dehydrogenase gene. *Acta Horticulturae*, **381**, 162–168.

Whetten, R. and Sederoff, R. (1995) Lignin biosynthesis. *Plant Cell*, **7**, 1001–1013.

Wood Formation in Trees ed. N. Chaffey
© 2002 Taylor & Francis
Taylor & Francis is an imprint of the
Taylor & Francis Group
Printed in Singapore.

3 Sampling of Cambial Region Tissues for High Resolution Analysis

Claes Uggla[1]* and Björn Sundberg[2]

[1]BioAgri AB, Box 914, SE-751 09 Uppsala, Sweden.
[2]Department of Forest Genetics and Plant Physiology, Swedish University of Agricultural Sciences, SE-901 83 Umeå, Sweden.
* Author for correspondence

ABSTRACT

Biochemical analysis of cambial tissues, and young and developing vascular tissues, requires that well-defined and tissue-specific samples are obtained. In this chapter, we describe a general method, based on tangential cryo-sectioning, that meets these demands by providing samples of pure tissue from the different developmental layers of the cambial region, i.e. the cambial zone, the different developmental zones on the xylem side of the cambium, and the functional phloem. All steps, including the sampling from tree-stems grown in the forest, the preparation, handling, and cryo-sectioning of samples, and the anatomical characterisation by microscopy, are covered. The method was originally developed for cell- and tissue-specific sampling of the cambial region of the gymnosperm *Pinus sylvestris*, but its usefulness in several other systems has been proven during the last few years. At the end of the chapter, some of these applications are briefly reviewed.

Key words: cryo-microtomy, interference contrast microscopy, plant anatomy, wood.

INTRODUCTION

With the development of microanalytical techniques, quantitative chemical analysis of very small amounts of plant material, comprising defined tissue(s) and cell types, can be performed. Such analysis can reveal spatial distribution patterns of molecular compounds in plant organs and tissues, and the information obtained will contribute significantly to our understanding of the regulation of plant growth and development.

Often the limiting step for such analysis is not the analytical sensitivity but problems with obtaining well-defined and preserved samples through various microdissection techniques. Characterisation of cell and tissue content in microdissected samples is an important issue, and with microanalytical techniques there also follows the problem of finding an appropriate basis of expression, since accurate weighing of such small samples is often associated with problems.

In this chapter we describe techniques for the sampling and preparation of defined tissues from the cambial region and the phloem of trees. We start in the forest with the hammer and chisel, ready for sampling, and finish up with a well-defined 1-mg sample ready for analysis. Some of the applications of these techniques, and results coming from them, will also be briefly described.

The cambial region consists of layers, or zones, of cells in different developmental stages. These are the cambial zone with meristematic cells (defined in accordance with Wilson *et al.*, 1966), and the zones of cambial cell derivatives undergoing expansion, and secondary wall formation towards mature xylem and phloem (Figure 1). Each of these developmental stages may be only a few cells wide, or less than 100 μm in radial thickness. Large variations in biochemistry, metabolic activity and gene expression are therefore expected across the cambial region tissues within very short distances. The method described below will provide samples of the cambial region and the phloem that are well-defined in terms of tissue

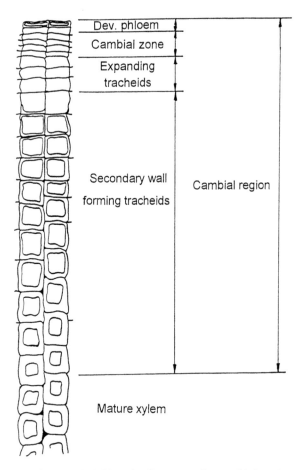

Figure 1 A diagrammatic drawing of two radial files of cells across the cambial region of *Pinus*. The different developmental zones, as defined in the text, are indicated. (From Uggla, 1998)

content and spatial localisation, and which can be used to elucidate distribution patterns of plant substances across tissues of the cambial region.

TECHNIQUES

The procedures that are covered in this section include the sampling of cambial region tissues from the stem of the tree, the preparation of samples for cryo-sectioning, the cryo-sectioning process itself, determination of the radial location of cryo-sections, and, finally, weighing of cryo-sections. All procedures are complementary and can be viewed as consecutive steps towards a final analysis. However, it is not possible to anticipate every application to which this technique may be used, and readers must feel free to modify each step to the requirements of their own particular analysis.

Sampling (Figure 2)

Equipment needed:	Chemicals needed:
Knife	Liquid nitrogen
Hammer	Dry ice (solid carbon dioxide)
Scalpel (e.g. No. 24) Chisel (approx. 2 cm wide) Containers (e.g. polystyrene boxes with a lid) for liquid nitrogen and for storage of samples Plastic bags for storage of samples	

Duration of procedure: approx. 20 min per sample (plus travelling time!).

Figure 2 A photograph of the sampling procedure from a mature Scots pine tree stem. After exposing the tissues of the upper and lower ends of the sample, the chisel is driven under the sample by the force of a hammer, and the sample can be smoothly removed from the stem.

To obtain a cambial region sample from a large tree, it is necessary to remove physically not only that part of the cambial region of immediate interest, but all of the tissues from the bark inwards, including at least 1 cm of the mature xylem. For analysis of the mature xylem, sampling may need to be even deeper, depending on annual ring width, etc. For most gymnosperms, and angiosperm trees of softer wood qualities, samples can be obtained as follows (e.g. Uggla *et al.*, 1996):

1   If necessary, make the bark surface even and remove some of the dead tissues using a knife or similar tool.

2   Mark on the surface of the stem a rectangular area, approx. 10 × 2 cm, by cutting through all the extraxylary tissues using a sharp scalpel.
    Note – it may be necessary to use a hammer to force the scalpel through the tissues. The rectangle should be oriented with its long dimension parallel to the stem axis, or, ideally, to the direction of the wood fibres, to facilitate the removal of the material.

3   Use the hammer and chisel to remove the tissues above and below the rectangle to a depth of about 1 cm into the mature xylem.
    Note – the channel so created should be deep enough to admit a chisel such that it can be driven almost vertically under the marked rectangle at the 1-cm-of-xylem-depth by the force of a hammer.

4   The rectangular block of tissue cut by the chisel, including the cambial region protected on both sides by bark and xylem, can then be smoothly removed from the stem, immediately frozen in liquid nitrogen or dry ice, and stored in a box containing dry ice.
    Note – sometimes the block of tissue may split when freezing in liquid nitrogen. In this case, freeze the block in dry ice instead.

5   If more material is needed, another rectangle can be likewise marked and chiselled out beside the first one, and so on.

This sampling method causes a minimum of damage to a large tree, and may therefore allow repeated sampling of the same individual tree during the season, before and after a treatment, or in different years. In these cases, however, sampling size should be minimized and sampling positions should be evenly spaced around the stem surface.

The most obvious position of sampling along the stem of a large tree is at breast height (approx. 1.3 m above ground level). Sampling positions at higher levels requires the use of a ladder, sky-lift or similar equipment, or that the tree is felled.

When trees are too small and slender to sample blocks of tissues using a hammer and chisel, it is more appropriate to sample and freeze complete, halved or quartered segments of stem.

Finally, it is recommended that you collect excess material, to allow for selection of areas that are homogeneous and undamaged. Also, freeze-drying will occur at sample surfaces and such areas should be avoided in sample preparation.

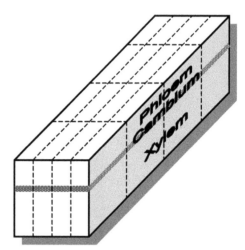

Figure 3   Drawing of sample prepared for band-sawing. The frozen block of tissue is trimmed and, in this case, cut to provide sixteen samples for cryo-microtomy, as indicated by the dotted lines.

Storage and Sample Preparation (Figure 3)

Equipment needed:	Chemicals needed:
Container (e.g. polystyrene box with a lid) for storage of samples Freezer at –80 °C Plastic bags/vials for storage of samples Band saw Measuring stick Pincers or large forceps for handling cold samples	Dry ice or liquid nitrogen for cooling of samples

Duration of procedure: approx. 1 h.

Collected material is most conveniently transported to the laboratory in polystyrene boxes containing dry ice, for long-term storage at –80 °C.

Storage of frozen tissues leads to freeze-drying and oxidation of the surface of the material. Therefore it is preferable to leave the collected material intact during long-term storage, and prepare just those specimens that are to be handled during the same day.

Specimens for sectioning in a cryostat microtome are prepared from the frozen sample (see section above; Figure 3) as follows:

Note – Thawing of the sample should be avoided during this procedure, and can be prevented by periodic immersion of the material in liquid nitrogen, or by leaving it on dry ice, at appropriate intervals during the sawing process.

1   Trim the frozen block a few millimetres on all sides using the band saw (to remove tissues which have been exposed to the air, and which may possibly be affected by freeze-drying and oxidation), and to make the block plane on all surfaces.
     Note – at least 5 mm of xylem and all of the phloem and living bark tissues should be left, to facilitate subsequent cryosectioning.
2   Cut specimens for cryo-sectioning by further use of the band saw.
     Note – depending on anatomical conditions, and requirements of the analysis, the size of the samples may vary considerably, but a tangential area of around 0.5 cm^2, or approx. 2–4 × 15–20 mm is often suitable.
3   Inspect the specimens for their suitability (tangential evenness and absence of damage) for subsequent tangential cryo-sectioning. A

specimen suitable for cryo-sectioning should be free from cracks or other inhomogeneities, such as resin pockets, large resin canals, fungal infections etc. Furthermore, it is of fundamental importance that the tangential layers of different tissues (i.e. the cambial zone, the phloem, and the annual ring) are plane and even in thickness within the complete specimen.

Cryo-Sectioning and Preparation of Sections for Microscopy

Equipment needed:	Chemicals needed:
Cryostat microtome equipped with device allowing reorientation of sectioning plane	Tissue-Tek® OCT™ Compound (Sakura Finetek, Europe)
Steel knife	Glycerol
Fine grindstone, polish leather and knife back, for knife polishing	Abrasive for knife polishing
Single-edge razor blades	
Forceps	
Glass microscope slides and coverslips	
Measuring stick	
Light microscope equipped with a micrometer eyepiece, and, preferably, differential interference contrast (DIC, Nomarski) optics	
Vials with lid and label (e.g. 1.5 ml Eppendorf tubes)	
Racks or boxes, for storage of vials	
Freezer at –80 °C	

Duration of procedure: approx. 0.5 days per series. However, this depends greatly on sectioning expertise, suitability of block, etc. In any event, do not anticipate cutting more than two series in a single day.

Requirements of the cryostat microtome

To be able to compare quantitative estimates within and between series of tangential sections obtained across the cambial region, it is essential that each section has a known and reproducible thickness. Therefore, it is important to control the accuracy of the microtome before commencing sectioning. In other words, if the display says 30 μm thickness, the sections should be 30 μm thick!

Wood and phloem are tough tissues, often containing considerable amounts of crystalline materials. Therefore, a plane wedge steel knife is recommended over disposable knives. The steel knife stays sharp much longer, and furthermore, it is not deformed by the heavy forces generated during sectioning. Good care of steel knives is crucial for acceptable sectioning quality. Before each session of sectioning, the knife should be polished. In our experience, the best results are obtained in a two-step procedure as follows:

1   First, polish the knife carefully on a *fine and plane* wetted grindstone, with the cutting edge leading (this is called honing).
2   Secondly, polish the knife on a stable, flat polish leather loaded with fine abrasive, this time with the cutting edge trailing (this is called stropping).

To obtain the right angle of the cutting edge to the polishing surface, a block (called knife back) is fixed at the back of the knife, thereby creating the correct distance to the grain plane. By examining the edge of the knife in a stereo microscope it is possible to identify, and mark with a pencil, damaged areas. These areas of the blade can then be avoided when sectioning in the microtome. After some use, however, the knife has to be thoroughly ground to restore its original sharpness.

The optimal knife angle in relation to cutting surface (angle of rake and clearance angle) depends on specimen properties as well as the knife. However, in our experience, a 13° clearance angle works well. For detailed information about knives, and their handling and setting for microtomy in general, see Gordon and Bradbury (1990).

Cryostat microtomes are usually equipped with an anti-roll plate to prevent sections from curling on the knife-edge. Adjustment of this anti-roll plate in relation to the knife partly depends on the sectioning thickness, and is of great importance for sectionability.

The temperature in the freezing chamber of the crytostat is also an important parameter for optimal sectioning. On the one hand, the temperature must be low enough to ensure freezing of all phloem and cambial region tissues (which often have high sugar solute contents). On the other hand, if the temperature is set too low the tissues become stiff which will increase the risk of breakage during sectioning. In our experience a temperature of –20 °C has been found to be most suitable.

Cryostat microtomes can be equipped with separate knife-cooling and specimen-cooling systems, making further adjustments of temperature conditions possible. If these opportunities exist, make use of them; it will save time and improve quality later in your project! Finally, make sure that both the knife and the specimen have reached the set temperature(s) in the chamber *before* beginning any sectioning work. See Bancroft (1990) for further information about the anti-roll plate and cryostat temperatures.

Installation of the specimen

A close and strong adherence of the specimen to the specimen holder is important for successful sectioning. This is why *at least* 5 mm, or, better, 8 mm, of mature xylem should remain on the specimen; this tough material serves to anchor the specimen to the specimen holder. Instead of fixing the specimen with a screw clamp chuck, we recommend the use of a flat and patterned surface to which the specimen is attached by means of the Tissue-Tek OCT Compound. This hardens at low temperatures and the specimen becomes tightly stuck to the holder, without induction of stress and tension in the tissues. By this method the xylem side of the specimen is relatively deeply surrounded by the freezing medium and placed on the specimen holder. Ensure that the specimen is oriented so that the fibres are parallel to the direction of sectioning.

The orientation of the specimen in relation to the plane of sectioning is of utmost importance in order to obtain tangential sections parallel to the plane of the tissues (see Figure 4). This is optimised during sectioning through the outer tissues of the specimen (i.e. cortex or old non-functional phloem). A section thickness of approx. 30 μm is often a good choice.

1   Prepare a plane surface on the specimen by cutting until complete cryo-sections are obtained.

2   Hand-cut transverse sections, as thin as possible, with a single-edge razor blade from *both* ends of the specimen.
Note – it is an advantage if the specimen holder can be rotated 180°, as it is more convenient to cut both of the transverse sections in the upper position.

3   Mount the unstained transverse sections in water and inspect under a light microscope using Nomarski optics (differential interference contrast, DIC), if available. By measuring the distance between the end of the section corresponding to specimen surface and the cambium in each corner of both the upper and lower transverse section with a micrometer eyepiece, deviations from the parallel between the cambium and specimen surface can be assessed.

4   If necessary, reorient the specimen in relation to the knife to obtain a better parallelism. It may be necessary to repeat this procedure to obtain an acceptable result.
Note – if satisfactory parallelism is not achieved until the functional phloem is reached, it is best to discard that specimen and start again with a new one.

Tangential sectioning and preparation of transverse sections for microscopy

When the specimen is correctly oriented, centripetal tangential sectioning can proceed, and collection of sections can begin. Thickness of the sections depends not only on the requirements of the amounts of tissue needed for analysis, but also on the sectionability of the tissues. While phloem and mature xylem hold together in fine sheets in sections 10–20 μm thick, the thin walled cambial, and particularly the expanding xylary, tissues often wrinkle and fall apart when sectioning at 30 μm. However, it is worth bearing in mind that it is usually not essential to obtain perfect sheet-like sections: for quantitative analysis, the principal requirement is only that the correct amount of the proper tissues is collected. The thickness of sections that is chosen will also depend on the radial width of the different developmental layers; e.g., when the zone of cells of interest is radially narrow, such as the cambial zone during dormancy, thinner sections will be needed in order to obtain a high tissue specificity in each section. The

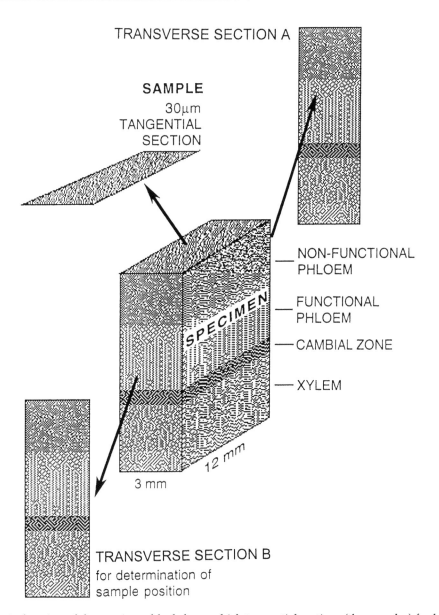

TRANSVERSE SECTION A

SAMPLE
30μm
TANGENTIAL
SECTION

NON-FUNCTIONAL
PHLOEM

FUNCTIONAL
PHLOEM

CAMBIAL ZONE

XYLEM

SPECIMEN

12 mm

3 mm

TRANSVERSE SECTION B
for determination of
sample position

Figure 4 Schematic drawing of the specimen block from which tangential sections (the samples) for high resolution analyses, and transverse sections for determination of sample position, are obtained. (From Uggla *et al.*, 1996; Copyright (1996) National Academy of Sciences, U.S.A.)

procedure of sectioning and collecting of sections is as follows:

For convenience, the individual sections are best stored directly in the same vials (e.g. 1.5 ml Eppendorf tubes) that will be used for subsequent sample preparation. Use pre-cooled forceps for the transfer of sections to the vials, and for all other section-handling.

Note – ensure that the vials are kept within the cooled chamber of the cryotome, and are placed there in advance of sectioning so that they have

time to cool before any cryo-sections are placed in them (otherwise the precious cryo-sections may thaw). When sectioning is finished, the vials are transferred to the freezer at –80 °C for storage.

1   After installation of the specimen in the cryo-microtome, commence sectioning, and store sections for analysis in pre-cooled vials.

2   For later determination of tissue composition in tangential sections, cut a pair of transverse sections with a razor blade from *both* ends of

the specimen after every second to fourth tangential sectioning, in the same way as for specimen orientation (see above section on Installation of the specimen).

3   Transfer the transverse sections to a glass microscope slide, saturate with water to avoid bubbles (remove excess water with a tissue) and mount under a cover glass in 100% glycerol.
Note – it is necessary to note the position and the orientation of the two transverse sections on the glass slide, so that the corresponding corners of the tangential sections can be identified. An immediate and preliminary inspection of these sections under the microscope will enable you to gauge the quality of the sections, allowing you to repeat the transverse sectioning if necessary.

4   After each transverse sectioning, measure the length of the specimen using a measuring stick.
Note – as a consequence of repeated transverse sectioning, the area of the tangential sections gradually decreases during the process of tangential sectioning. Measuring the length of the specimen after each transverse sectioning therefore allows the determination of area, hence volume, of the tangential cryo-sections. Similarly, it is important to bear in mind this gradual decrease in sample size for any sampling limits that might apply to the analytic procedures planned. From our experience, the last section obtained in a series of approx. 50 is less than half the size of the first section.

5   Store the glycerol-mounted samples horizontally at room temperature (RT) pending final microscopic inspection (see next section).

The complete procedure, including tangential and transverse sectioning, and mounting of transverse sections, should be done in one session, without breaks. This minimises the effect of shrinkage at the specimen surface due to the freeze-drying that occurs in the freezing chamber of the cryotome.

Determination of radial location and tissue composition of tangential sections by the microscopic analysis of transverse sections

For correct evaluation and interpretation of the analytical data it is of the utmost importance that the tissue composition of each sample is accurately determined. This is done using the series of hand-cut transverse sections prepared in parallel to the tangential sectioning (see above section on Tangential sectioning and preparation of transverse sections for microscopy).

Requirements of the microscope

Because the unstained, hand-cut transverse sections are thick, use of a microscope with the best properties is very important. Differential interference contrast (DIC; Nomarski optics), and use of polarised light have proved particularly useful to distinguish secondary cell walls from primary cell walls (e.g. Abe *et al.*, 1997), and to observe the S_3-layer of the secondary cell wall (Bailey, 1954).

Procedure

Microscopy should begin after at least one day of horizontal storage of the glycerol-mounted transverse sections, when the water and glycerol medium have mixed to homogeneity and disturbances from light refraction phenomena have disappeared. The goal is to determine the radial location of the outer short ends of the transverse sections. This will inform us about the radial position of the inner tangential border of the *tangential* section which was cut immediately prior to obtaining the transverse sections (see Figure 5; Uggla *et al.*, 1996, 1998).

1   Choose a pair of good quality transverse sections that was cut at the start of the sectioning series. Count the number of cells per radial cell file in each of the developmental zones (e.g. functional phloem, cambial zone, radially expanding xylem elements, and secondary-wall-forming xylem elements). If present, use the inner border of the developing annual ring as a base line! Do this for approx. *three* radial cell files at *three* positions (the middle and the two ends) in *each* of the two transverse sections.
Note – counting cells along radial cell files may be difficult if the quality of the transverse sections is poor, or if the inner border of the annual rings is not present. An alternative to cell counting in these cases is to measure the radial distance of the developmental zones using the micrometer eyepiece.

2   From the cell counts (alternatively the radial measures) of the developmental zones, a

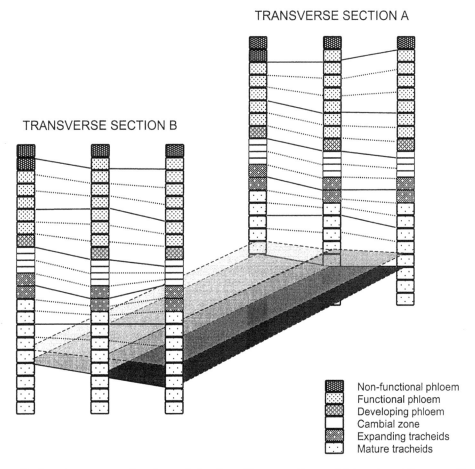

TRANSVERSE SECTION A

TRANSVERSE SECTION B

Non-functional phloem
Functional phloem
Developing phloem
Cambial zone
Expanding tracheids
Mature tracheids

Figure 5 Schematic drawing representing the radial cell profile across the cambial region at three positions (each of which is the mean of approx. three cell files). Radial cell profiles are determined by microscopic inspection of the two transverse sections, A and B, taken from either end of the block (as indicated in Figure 4). The developmental stage of the cells is indicated by the patterning. Bold lines between the cell files identify the same position within the transverse sections along the different cell files. The dotted lines indicate the interpolated position of the tangential sections taken for analysis. For illustration, the location of one of these tangential sections is shown by the dark (bottom surface) and shaded (top surface) areas; note that the section thickness, 30 μm, is approx. one xylem cell wide.

schematic drawing of number of cells (or radial distances) and cell types along the radial files can be constructed. When inspecting the transverse sections collected during tangential sectioning the position of the *tangential* sections cut prior to, and after, each pair of transverse sections can be estimated. From this information the proportion of tissues in each tangential section can be estimated (Figure 5).

Note – as the transverse sections were only taken after every second to fourth tangential cut, the intervening tangential sections have to be interpolated into the model.

Anatomical considerations

To separate the different developmental zones of the cambial region tissues it is important to have well-defined anatomical features for the different developmental stages. Although these may differ considerably between species, and are described in detail elsewhere (see Larson, 1994; Whitmore and Zahner, 1966; Wilson, 1964; Wodzicki, 1971), some general aspects on how to define the borders of the different developmental zones on the xylem side of the cambium are worth repeating.

The cambial zone consists of cells that divide frequently and expand. These cells gradually increase in their radial dimension, and may double their radial size from one mitosis to the next. At the border of the cambial zone, cell divisions cease but cell expansion proceeds, and the cells will eventually exceed the radial dimensions observed in the cambial zone; i.e. the transition from cambial zone to expansion zone is crossed. The next transition, from radial expansion to secondary-wall-formation of cambial derivatives, is readily recognised by the thickening of cell walls and the presence of birefringence under Nomarski optics. This birefringence is due to the crystalline nature of the cellulose microfibrils, and their orientation within secondary cell walls (Abe *et al.*, 1997). The transition from secondary-wall-forming xylem elements to mature autolysed tracheids can be localised by the presence (in living cells) or absence (in autolysed tracheids) of cytoplasm (Wilson *et al.*, 1966). Furthermore, by staining with acridine orange, the presence of RNA in the cytoplasm of living cells can be detected by UV-microscopy (Gahan, 1984; Uggla *et al.*, 1998). Another approach is to monitor the development of the S_3-layer of the secondary cell wall by polarised-light-microscopy (use a $\times$ 100 oil immersion objective). When the S_3-layer is present, the cell wall is completed and the cell ready to autolyse (Bailey, 1954; Uggla, 1998).

Basis of expression and weighing of tangential cryo-sections

Quantitative results obtained by tangential sectioning can be expressed on a cm^2 section area basis, or on the basis of the section volume. This gives a proper representation of the spatial distribution pattern of the analysed compound. However, when interpreting results and comparing with previously published data it may be of great value to express the quantification on a fresh weight (FW) or dry weight (DW) basis. In this case, it is not recommended to determine the FW of cryo-sections destined for analysis due to unavoidable thawing and possible degradation of components of interest. Collection of one or two series of tangential sections dedicated for FW and DW measurements should therefore be included in your experiment. The use of airtight vials for storage of sections is necessary to prevent evaporation of tissue-water before FW measurements are completed.

APPLICATIONS

Tangential cryo-sectioning to obtain tissue specific samples from cambial region tissues can be used for diverse purposes to study distribution patterns of e.g. gene transcripts, enzymes, metabolites, and growth regulators. Some applications where this technique has been used are briefly described below.

Mass spectrometry analysis of plant hormones in Scots pine and hybrid aspen

Tissue-specific sampling of cambial region tissues by tangential cryo-sectioning was initially developed to increase the spatial resolution of mass spectrometry (MS) analysis of the plant hormone, auxin (indole-3-acetic acid; IAA), in the cambial region of Scots pine (*Pinus sylvestris* L.) (Uggla *et al.*, 1996; Uggla, 1998). The technique revealed for the first time the IAA distribution across developing secondary vascular tissues. The results suggested that IAA serves as a positional signal in the control of wood formation, and exemplifies how the increased spatial resolution of sampling with this technique provides the potential for new insights into the physiological and molecular control of cambial growth.

The quantitative measurements of IAA in the individual cryo-sections were made by an isotope-dilution gas chromatography (GC)-selected reaction monitoring-MS technique, according to Edlund *et al.* (1995). Briefly, the frozen samples were homogenised in the 1.5 ml Eppendorf tube by a fast rotating piston, and the IAA extracted in phosphate-buffer, after the addition of [^{13}C-6]IAA as an internal standard. Samples were purified by the addition of an ion exchange resin, Amberlite XAD-7, followed by elution with dichloromethane. Samples were derivatised (by methylation and trimethylsilylation) before GC-MS analysis. The selected reaction monitoring technique used in MS analysis is extremely selective and very sensitive. Therefore, analysis of small sample volumes, such as cryo-sections, can be analysed with a minimum of purification and sample preparation. Both in Scots pine and hybrid aspen (*Populus tremula* L. $\times$ *P. tremuloides* Michx.), the highest amount of IAA (in the order of 10 ng.cm^{-2} of 30 μm section) was found in the cambial zone and its most recent derivatives (Figure 6). From this peak value, the level decreased steeply towards the phloem and the xylem, reaching low, and relatively stable, levels close to the transition between

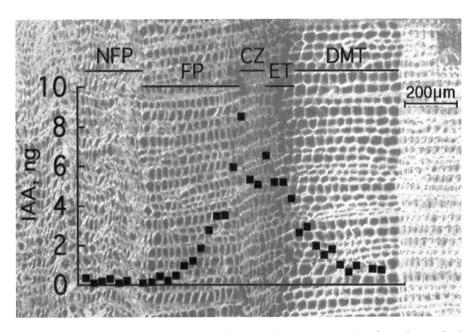

Figure 6 Differential interference contrast micrograph of a typical transverse section from the cambial region of *Pinus sylvestris*. The radial auxin gradient is overlain on the micrograph.
Key: NFP, zone of non-functional phloem from previous year's growth; FP, zone of differentiating and functional phloem; CZ, cambial zone; ET, zone of expanding differentiating tracheids; DMT, zone of differentiating tracheids forming secondary walls, as well as mature functional xylem. (From Uggla *et al.*, 1996; Copyright (1996) National Academy of Sciences, U.S.A.)

expanding and secondary-wall-forming cells. Within a radial distance of 250 μm, an approx. 50-fold difference in IAA level was evident (Figure 6, Uggla *et al.*, 1996; 1998; Tuominen *et al.*, 1997; Sundberg *et al.*, 2000).

Analysis of carbohydrates and enzymes of carbohydrate metabolism in Scots pine

Carbohydrate analysis by mass spectrometry

Formation of secondary xylem is a major sink for carbohydrates in growing trees. The main supply of carbohydrates is *via* the phloem transport system. As sucrose is the dominating sugar of the phloem sap in most tree species, its availability, and metabolism in the phloem/cambial regions, is of great interest for the understanding of cambial growth regulating mechanisms (Kozlowski and Pallardy, 1997). More-over, sugars are proposed to act as signalling molecules, possibly acting in concert with IAA (see section above; Sheen *et al.*, 1999).

Early steps in sucrose metabolism involve its cleavage either by sucrose-synthase to UDP-glucose and fructose, or by invertases to glucose and fructose

(see section below; Sturm and Tang, 1999). In addition to sucrose, two of these monosaccharides, fructose and glucose, were quantified across the entire phloem/cambial region of Scots pine, using the tangential cryo-sectioning technique, coupled to MS analysis (Uggla, 1998). In brief, the cryo-sections were freeze-dried and sugars extracted in water with phenyl-β-D-glucoside added as an internal standard. Subsequently, samples were trimethylsilylated and then analysed by GC-selected ion monitoring-MS.

Not surprisingly, sucrose content reached the highest levels, in the range of 100–300 mg.cm^{-2} of 30 μm section, in phloem tissues. From the phloem, a steep centripetal decrease in sucrose content was observed across the developing phloem and the cambial zone. This decrease in sucrose was opposed by an increase in fructose and glucose. In the cambial zone and radially expanding xylem derivatives, fructose level was higher than glucose, whereas glucose was generally more abundant during later stages of xylem development, in tissues undergoing secondary wall formation. These carbohydrate distribu-tions across cambial region tissues are most beneficially interpreted in relation to the differential activity of sucrose-metabolising enzymes as in the section below (Uggla, 1998; Sturm and Tang, 1999; see also the biochemistry chapter by Magel in this volume).

Enzymatic analysis of enzymes of carbohydrate metabolism

The activity of four enzymes involved in sucrose metabolism was assayed in freeze-dried cryosections obtained from the entire phloem/cambial region, using a microplate reader system (Uggla, 1998), according to the procedure of Egger and Hampp (1993). This technique is extensively outlined elsewhere, in the biochemistry chapter by Magel in this volume. Significant activity of the key enzyme in sucrose synthesis, sucrose phosphate synthase, was evident across the entire region. However, it was most conspicuous in phloem tissues, indicating a contribution to the pool of sucrose from starch reserves in phloem parenchyma, in addition to the direct contribution of sucrose *via* phloem transport (Egger *et al.*, 1996). The dominant sucrose-cleaving enzyme was sucrose synthase, which exhibited the highest activities in the zone of secondary wall formation, indicating that it may be specifically involved in polysaccharide (Quick and Schaffer, 1996) and/or lignin synthesis. One of the two forms of invertase, denoted soluble acid invertase, was most abundant in the zone of dividing and expanding cells, supporting the general view that this enzyme is specifically involved in cell expansion (e.g. Sturm and Tang, 1999). The other assayed invertase, denoted cell-wall bound acid invertase, showed only minor activity, mainly in mature xylem.

Analysis of differential gene expression in hybrid aspen

Tangential cryo-sectioning techniques are also suitable to provide information about gene expression patterns across cambial region tissues. Hertzberg and Olsson (1998) could demonstrate that two novel homeobox genes were found to be differentially expressed in pooled fractions from the cambial/expanding zone and from the secondary wall forming zone, respectively. Homeobox genes are considered to be key regulators of cell develop-

ment since they control the expression of other genes important for growth and differentiation (Gehring *et al.*, 1994). The function of these genes in cambial growth remains to be understood.

Amino acid analysis by high performance liquid chromatography in Scots pine

Although the cryo-sectioning technique has so far mostly been employed for investigations related to cambial growth and development, it is also suitable for analysis of compounds in the mature vascular systems. In contrast to xylem tissues, the functional phloem is usually very narrow, and rarely more than 10 or 20 cells wide in radial direction, making sampling of uncontaminated phloem sap by traditional means (e.g. centrifugation) difficult; collection of uncontaminated xylem sap is usually more easily performed. With tangential cryo-sectioning it is possible to produce sections containing 100% functional phloem or xylem, which solves the problem of contamination from surrounding tissues. However, in contrast to the sap obtained e.g. by centrifugation, the fact that extractives rather than pure sap are collected by cryo-sectioning allows the possibility of its contamination by the addition of compounds to the sample from the cytoplasm.

This was considered when analysing the content of amino acids in xylem and phloem tissues of Scots pine by Nordin *et al.* (2000). That study aimed to identify the major transport forms of nitrogen in xylem and phloem under different nutrient regimes in the field. The results showed that both the concentrations and the proportions of the amino acids glutamine and arginine in xylem and phloem were dependent on nitrogen supply to roots. Very briefly, high nitrogen regimes correlated to higher arginine/glutamine ratios in the xylem, but lower arginine/glutamine ratios in the phloem compared to low nitrogen regimes.

Analysis of isoforms of pectin methylesterase

Cryo-tangential sectioning has also been used to good effect for analysing the distribution of isoforms of pectin methylesterase (PME) in stems of hybrid aspen (Micheli *et al.*, 2000; see chapter by Micheli *et al.* in this volume). Using this technique they have identified, and examined the distribution of, seven isoforms of PME within secondary vascular tissues, during both the active and dormant periods of cambial growth. Although they obtained good resolution of different isoforms to tissues using three combined 25 μm sections, in future they will be attempting to analyse PME isoforms in *single* sections in order to refine the distribution in the cambial zone further (Micheli *et al.,* 2000).

FUTURE PROSPECTS

Although many of the details of the cell biology and physiology of the secondary vascular tissues are still unknown, we confidentially predict that more widespread use of the tangential cryo-sectioning technique described here will help in solving some of the remaining mysteries of the cambium and of cambial growth.

ACKNOWLEDGEMENTS

The work of developing the cryo-sectioning technique was financed by the Swedish Council for Forestry and Agricultural Sciences, the Swedish Natural Sciences Research Council, the Swedish Institute, and the Royal Academy of Forestry and Agriculture. Finally, we wish to thank the Editor, Dr Nigel Chaffey, for his ideas and inspiring running of this book project.

REFERENCES

Abe, H., Funada, R., Ohtani, J. and Fukazawa, K. (1997) Changes in the arrangement of cellulose microfibrils associated with the cessation of cell expansion in tracheids. *Trees,* **11,** 328–332.

Bancroft, J.D. (1990) Frozen and related sections. In *Theory and Practice of Histological Techniques,* edited by J.D. Bancroft and A. Stevens, pp. 81–92. Edinburgh: Churchill Livingstone.

Bailey, I.W. (1954) *Contributions to plant anatomy.* Waltham, MA: Chronica Botanica Company.

Edlund, A., Eklöf, S., Sundberg, B., Moritz, T. and Sandberg, G. (1995) A microscale technique for gas chromatography-mass spectrometry measurements of picogram amounts of indole-3-acetic acid in plant tissues. *Plant Physiology,* **108,** 1043–1047.

Egger, B., Einig, W., Schlereth, A., Wallenda, T., Magel, E., Loewe, A. and Hampp, R. (1996) Carbohydrate metabolism in one- and two-year-old spruce needles, and stem carbohydrates from three months before until three months after bud break. *Physiologia Plantarum,* **96,** 91–100.

Egger, B. and Hampp, R. (1993) Invertase, sucrose synthase and sucrose phosphate synthase in lyophilized spruce needles; microplate reader assays. *Trees,* **7,** 98–103.

Gahan, P.B. (1984) *Plant histochemistry and cytochemistry.* London: Academic Press.

Gehring, W., Affolter, M. and Bürglin, T. (1994) Homeo-domain proteins. *Annual Review of Biochemistry,* **63,** 487–526.

Gordon, K.C. and Bradbury, P. (1990) Microtomy and paraffin sections. In *Theory and Practice of Histological Techniques,* edited by J.D. Bancroft and A. Stevens, pp. 61–80. Edinburgh: Churchill Livingstone.

Hertzberg, M. and Olsson, O (1998) Molecular characterisation of novel plant homeobox gene expressed in the maturing xylem zone of *Populus tremula ¥ tremuloides. The Plant Journal,* **16,** 285–295.

Kozlowski, T.T. and Pallardy, S.G. (1997) *Physiology of Woody Plants.* San Diego: Academic Press.

Larson, P.R. (1994) *The Vascular Cambium.* Berlin: Springer-Verlag.

Micheli, F., Sundberg, B., Goldberg, R. and Richard, L. (2000) Radial distribution pattern of pectin methylesterases across the cambial region of hybrid aspen at activity and dormancy. Plant Physiology, **124,** 191–199.

Nordin, A., Uggla, C. and Näsholm, T. (2000) Nitrogen forms in bark, wood and foliage of nitrogen fertilized *Pinus sylvestris. Tree physiology,* **21,** 59–64.

Quick, W.P. and Schaffer, A.A. (1996) Sucrose metabolism in sinks and sources. In *Photoassimilate Distribution in Plants and Crops,* edited by E. Zamski and A.A. Schaffer, pp. 115–156. New York: Marcel Dekker.

Sheen, J., Zhou, L. and Jang, J.-C. (1999) Sugars as signaling molecules. *Current Opinion in Plant Biology,* **2,** 410–418.

Sturm, A. and Tang, G.-Q. (1999) The sucrose-cleaving enzymes of plants are crucial for development, growth and carbon partitioning. *Trends in Plant Science,* **4,** 401–407.

Sundberg, B., Uggla, C. and Tuominen, H. (2000) Cambial growth and auxin gradients. In *Cell and Molecular Biology of Wood Formation,* edited by R. Savidge, J. Barnett, and R. Napier. pp. 169–188. Oxford: Bios Scientific Publishers Ltd.

Tuominen, H., Puech, L., Fink, S. and Sundberg, B. (1997) A radial gradient of indole-3-acetic acid is related to secondary xylem development in hybrid aspen. *Plant Physiology,* **115,** 577–585.

Uggla, C. (1998) New perspectives on the role of auxin in wood formation. *PhD Thesis,* Silvestria 58, Swedish University of Agricultural Sciences, Umeå, Sweden.

Uggla, C., Mellerowicz, E.J. and Sundberg, B. (1998) Indole-3-acetic acid controls cambial growth in Scots pine by positional signalling. *Plant Physiology,* **117,** 113–121.

Uggla, C., Moritz, T., Sandberg, G. and Sundberg, B. (1996) Auxin as a positional signal in pattern formation in plants. *Proceedings of the National Academy of Science of the USA,* **93,** 9282–9286.

Whitmore, F.W. and Zahner, R. (1966) Development of the xylem ring in stems of young red pine trees. *Forest Science*, **12**, 199–210.

Wilson, B.F. (1964) A model for cell production by the cambium of conifers. In *The Formation of Wood in Forest Trees*, edited by M. Zimmermann, pp. 19–36. New York: Academic Press.

Wilson, B.F., Wodzicki, T.J. and Zahner, R. (1966) Differentiation of cambial derivatives: proposed terminology. *Forest Science*, **12**, 438–440.

Wodzicki, T.J. (1971) Mechanisms of xylem differentiation in *Pinus silvestris* L. *Journal of Experimental Botany*, **22** (72), 670–687.

Wood Formation in Trees ed. N. Chaffey
© 2002 Taylor & Francis
Taylor & Francis is an imprint of the
Taylor & Francis Group
Printed in Singapore.

4 Biochemistry and Quantitative Histochemistry of Wood

Elisabeth Magel

Physiological Ecology of Plants, University of Tübingen, Auf der Morgenstelle 1, D-72076 Tübingen, Germany

ABSTRACT

Knowledge of wood-specific metabolic processes and their regulation was, and still is, limited. This is due to the complexity, stiffness, and low content of living cells in the wood, together with the abundance of interfering phenolic extractives. This chapter describes the development and application of purification techniques, together with highly sensitive detection methods, which have allowed us to highlight metabolic pathways characteristic for secondary differentiation processes of woody tissues. For heartwood formation in the innermost parts of woody axes, a close interlinkage between sucrose cleavage and the biosynthesis of flavonoids was shown by quantification of the catalytic activities of sucrose synthase (SuSy) and chalcone synthase (CHS). By employing quantitative histochemistry and enzymic cycling techniques, the preparation and quantitative analysis of tissue fragments comprising only a few cells is possible. These methods were used to detect biochemical processes – catalytic activities of the key enzymes of the oxidative pentose phosphate pathway (glucose-6-phosphate dehydrogenase and 6-phosphogluconate dehydrogenase) and pool sizes of pyridine nucleotides – which were closely coupled to the formation of lignin and which were emphasised in the differentiation of xylem elements during cambial growth.

Key words: biochemical analyses, cambial differentiation, enzymic cycling, heartwood, quantitative histochemistry, *Robinia*

INTRODUCTION

For wood tissues, the successful application of biochemical and molecular methods is hampered by the very low percentage of living cells together with the rigidity and high phenolic content of the material. By developing sample preparation and extraction methods which are specifically adapted to the requirements of woody tissues, these difficulties can be overcome. Applying biochemical methods and quantitative histochemistry, we are able to investigate and quantify the distribution of enzyme activities in wood tissues of different age classes. Moreover, these methods enabled us to determine enzyme specific protein contents of wood tissues and the quantification of metabolite pool sizes. Most of the techniques were developed for stem tissues of the deciduous tree, *Robinia pseudoacacia* L. (black locust). However, all of the techniques and, after minor revision, all of the assay methods outlined in this chapter, can also be used for other angiosperm and even gymnosperm trees. Here we present data on the activities of enzymes of two closely interlinked metabolic pathways, for the turnover of sucrose, sucrose synthase (SuSy), and for the synthesis of flavonoids, chalcone synthase (CHS). These two pathways play a key role in secondary differentiation processes of woody axes, cambial growth and heartwood formation (for an overview of these processes see Hillis, 1987, and Magel, 2000). In the second part we report about possibilities to analyse metabolic pathways of spatially closely interlinked tissues of different structural properties. Tissue complexes comprising only a few structurally comparable cells are prepared by the help of a method called micro-dissecting. In our example, cell complexes of cambial layer which differentiate into either phloem or xylem were prepared and the pools of pyridine nucleotides as well as catalytic activities of the key enzymes of the oxidative pentose phosphate pathway, glucose-6-phosphate

dehydrogenase and 6-phosphogluconate dehydrogenase, were determined (Lowry and Passonneau 1972; Magel *et al.*, 1996).

TECHNIQUES AND PROTOCOLS

Biochemical studies

Sample preparation

Equipment needed:	Chemicals needed:
Dismembrator II® or Microdismembrator S®, agate balls (Braun)	Dry ice
Freeze-drier (composed of a vacuum pump (Edwards E2M5), freezing trap (KF -2-110, Saur), and a freezer (–35 °C) with the sample compartment,	Liquid nitrogen
Glass evacuation tubes	Vacuum grease
Glass vials with open tops (e.g. Whatman)	
Increment borer (5 mm diameter)	
Nylon mesh (size, e.g. 50 μm, Züricher Beuteltuchfabrik)	
Saw	
Chisels	
Freezers (–80 °C, –25 °C)	
Constant room (20 °C, 40% relative humidity (RH))	

Duration of procedure (approx.): Sample harvest: half day Freeze-drying: dependent on sample size, 2–8 weeks Homogenisation of samples: 30 min per sample.

Non-destructive collections are done by removing cores (5 mm in diameter, 10–20 cm in length) with an increment borer. Destructive sampling is by felling the tree and sawing off stem discs (2–5 cm width).

1 Immediately after withdrawal from the trunk, both increment cores as well as stem discs, were frozen in liquid nitrogen.
   Note – in order to preserve the metabolic state of the intact tissue during sample collection, sample preparation or handling, it is important that after 'freeze-stopping' of metabolic activity, the material is subsequently handled without interruption of the 'freezing chain' by thawing. Thus the material must be transported either under liquid nitrogen or on dry ice to the lab and stored at –80 °C.

2 Freeze-drying of the samples (increment cores or stem discs) is as outlined by Hampp *et al.* (1990).
   Note – in our simple, effective and inexpensive freeze-drier, the vacuum pump and sample compartment are separated by a freezing trap. The samples are kept at a constant –30 °C and are dried by reduction of the pressure to 0.001 mm of Hg or less. In order to avoid destruction of the samples by water contained in small ice crystals, the temperature is slowly increased to ambient toward the end of the drying procedure. Further handling of the material is performed under constant conditions (20 °C, 40% RH).

3 From the freeze-dried specimens, inner bark and individual growth ring tissues were chopped off separately using a chisel.

4 The tissue chips are then homogenised to a fine powder using the dismembrator II equipped with two agate balls.
   Note – normally, at least two intervals of 3.5 min each, with additional cooling in between, are needed to get homogeneous, powdered wood material. All specimens and tools must be pre-cooled with liquid nitrogen for at least 20 min.

5   Homogenised samples are collected in small glass vials (e.g. from Whatman) and redried in the freeze-drier for another two days.

Note – in order to allow vapour exchange, the tops of the caps are left open and the resulting hole is covered with nylon mesh (size, e.g. 50 μm, depending on the size of the samples).

6   Store the freeze-dried samples (compact or powdered) in glass evacuation tubes (= storage assemblies) under vacuum at –20 °C.

Note – before the samples are removed and used, the storage assembly is warmed to ambient room temperature (RT) in a constant atmosphere (20 °C, 40% RH); air should be admitted at a moderate rate.

Preparation of crude enzyme extracts from Robinia *stem tissues*

During extraction, phenolics released from the tissue and solubilised by extraction buffer media interfere with the native enzyme preparation. Inactivation of the proteins relies on alteration of the tertiary structure of the native protein, the oxidation of SH-groups and/or the formation of insoluble complexes. The latter is highly appreciated during the process of tanning of animal skin (leather), but greatly despised for biochemical investigations. In the preparation of native enzyme extracts, oxidation products of these plant phenolics deriving from airborne auto-oxidation or enzyme-catalysed (oxidases, phenol oxidases) reactions hinder access to unchanged, active enzyme proteins. Overcoming these difficulties requires special precautions, such as supplementing the extraction medium with phenol scavengers, e.g. polyethylenglycol (PEG), soluble (PVP) or insoluble (PVPP, e.g. Polyclar AT™) polyvinyl (poly) pyrrolidone, bovine serum albumen (BSA), and anti-oxidants such as ascorbic acid, dithiothreitol (DTT), or β-mercaptoethanol. In this report, we focus on the extraction and measurement of two enzymes from *Robinia* stem tissues. However, if wood tissues from sources other than *Robinia* are used, or enzymes other than sucrose synthase or chalcone synthase are analysed, each of the parameters will have to be adapted for the specific requirement of the individual tissue, both in the extraction procedure and the enzyme activity assay.

Measurement of the catalytic enzyme activities:

a) Measurement of the catalytic activity of sucrose synthase

Equipment needed:	Chemicals needed:
Adjustable microlitre pipettes	ATP (adenosine triphosphate) 90 mM
Centrifuge / microfuge equipped with cooling, at least 12000 g	BSA (bovine serum albumen)
Disposable pipette tips	Glucose-6-phosphate dehydrogenase (70 U/ml)
Orbital shaker	Hexokinase (750 U/ml)
Microplate reader with heating device (SLT)	Mercaptoethanol (2- or β-)
Microplates (flat bottom, Costar)	MgSO$_4$ 100 mM
Reaction (micro) tubes	NADP 120 mM
Whirlimix	Phosphoglucose-isomerase (350 U/ml)
	PVPP or Polyclar AT™ (= insoluble polyvinylpyrrolidone, MW 360000)
	Sucrose 1 M
	Triethanolamine buffer (0.625 M, pH 7.0)
	Tris/borate buffer (0.1 M / 0.3 M; 0.1 M Tris = 2-amino-2-(hydroxymethyl)-1,3-propane-diol) / 0.3 M boric acid, pH 7.3)
	UDP (uridine-5′ diphosphate) 50 mM
	Double-distilled water (ddH$_2$O)

Note – unless otherwise stated, all solutions are aqueous.

Note – all biochemicals including enzymes are purchased from Boehringer Mannheim.

Note – to avoid loss of enzyme activities during extraction procedure, temperature should not exceed 4 °C. Thus, all handling must be done on ice or in a cooling cabinet.

Extraction

Duration of procedure: approx. 30 min.

1   Mix 5 (bark) or 10 (wood) mg of lyophilised tissue powder with 20 mg PVPP (final conc. 5% w/v).
2   Add 400 μl of ice-cold Tris/borate (0.1 M / 0.3 M) buffer supplemented with 15 mM 2-mercaptoethanol and 1% (w/v) BSA, extract under occasional vortexing for 10 min at 4 °C.
3   Centrifuge at 10 000 x g for 5 min at 4 °C.
4   Use 20 μl aliquots of the cooled supernatants as crude extract in the enzyme activity assay.

Assay

Sucrose synthase (EC 2.4.1.13; SuSy) reversibly catalyses the conversion of sucrose into UDP-glucose and fructose. The measurement of the enzyme activity is based on published assays (e.g. Pozueta-Romero *et al.*, 1991) and was adapted to the specific requirements of black locust tissues. In our system, SuSy predominantly cleaves sucrose. Thus the catalytic activity of SuSy can be determined by the quantification of one of the reaction products, UDP-glucose or fructose, that are formed. For *Robinia*, the amount of fructose released in the specific step is simultaneously quantified by a multi-enzyme reaction and the sto-chiometrically generated amount of NADPH is followed *on-line* at 340 nm. At least three replicates per sample are performed. Blanks were run with ddH$_2$O instead of the specific substrate and with sample-free extraction buffer. The kinetics of the enzyme reactions are followed in a total volume of 100 μl in microtitre plates (important: the plates must be flat-bottomed).

Note – to avoid loss of enzyme activities, all handling must be done on ice.
Note – before routinely measuring enzyme activities, check the tissue specific enzyme charac-

teristics (e.g. temperature- and pH-optimum of the reaction, K$_m$-, K$_i$- and / or K$_a$-values of the enzyme for the substrates, inhibitors, and/or activator molecules, linearity of the assay, etc.). Duration of procedure: approx. 75 min.

1   Prepare a test mix of all reagents common to all the programmed reactions. Prepare only the quantity required for the total number of sample wells including controls; make sure that at least one extra aliquot is prepared to ensure adequate volume is available for the last well.
2   Combine the following amounts per well:
40 μl of 0.625 M triethanolamine buffer, pH 7.0
3 μl of 100 mM MgSO$_4$
3 μl of 90 mM ATP
1 μl of 120 mM NADP
1 μl of glucose-6-phophate dehydrogenase (70 U/ml)
1 μl of hexokinase (750 U/ml)
1 μl of phosphoglucose-isomerase (350 U/ml).
3   Dispense 50 μl of the test mix to each well of the microplate.
4   Add 30 μl of 1 M sucrose.
5   Add 10 μl of extraction medium to the sample free medium wells or 10 μl of the crude extract to the sample wells.
6   Add 10 μl of ddH$_2$O to the blanks of both the sample free medium and sample wells.
7   Start the reaction by adding 10 μl 50 mM UDP, gently mix and follow the enzyme activity by the formation of NADPH (recorded as an increase of the absorbance at 340 nm) for 60 min at 30 °C using a microplate reader system.
8   Calculate rates from the linear part of the recorded kinetics, usually between 5 and 60 min of incubation, by using the software package supplied with the instrument.

b) Measurement of the catalytic activity of chalcone synthase

Chalcone synthase (EC 2.3.1.74, CHS), the key enzyme of the various flavonoid pathways, catalyses the stepwise condensation of one molecule of 4-coumaroyl-CoA and three molecules of malonyl-CoA, deriving from the acetyl-CoA carboxylase reaction, to naringenin chalcone (Hahlbrock and Scheel, 1989; Heller and Forkmann, 1988).

Naringenin chalcone is transformed by the stereo-specific action of chalcone isomerase to naringenin. Naringenin chalcone and naringenin are the precursors of all other flavonoids. As enzyme activities are low, radio-tracer techniques are used for the quantification.

Equipment needed:

Adjustable microlitre pipettes
Centrifuge / microfuge equipped with cooling, at least $12000 \times g$
Disposable tips for pipettes
Fluorometer, e.g. SFM 25 (Kontron)
Reaction (micro) tubes
Scintillation counter (e.g. Beckman LS 1801)

Scintillation vials
Spectrophotometer
Vortex Whirlimix
Water bath or block heater for microtubes

Chemicals needed:

Ascorbic acid
BSA

$CaCl_2$
DOWEX 1×2–200
Ethylacetate (100%)
Potassium phosphate buffer (KPi), 0.2 M, pH 8.0, and 0.1 M, pH 8.0
Malonyl-CoA
[2-^{14}C]Malonyl-CoA
Naringenin
p-Coumaroyl-CoA
Polytheylene glycol (PEG) 20000
PVPP or Polyclar AT™ (MW 360000)
Scintillation cocktail
Sucrose

Note – unless otherwise stated, all solutions are aqueous.
Note – to avoid loss of enzyme activities during extraction procedure, temperature should not exceed 4 °C. Thus, all handling must be done on ice or in a cooling cabinet.

Extraction

Duration of procedure: approx. 75 min.

1   Pre-equilibrate 50 mg of DOWEX 1×2-200 with the extraction liquid (0.2 M KPi pH 8.0, 0.005% BSA, 0.15% PEG 20000, 15 mM sucrose, 0.1 mM $CaCl_2$ and 20 mM ascorbic acid).

2   Mix 50 mg of lyophilised tissue powder, 50 mg PVPP and 50 mg DOWEX 1×2-200 (pre-equilibrated as per step 1).

3   Add 750 μl of ice-cold extraction liquid (see step 1), extract under nitrogen with occasional vortexing/whirlimixing for 20 min on ice (4 °C).

4   Centrifuge at $10\,000 \times g$ for 10 min at 4 °C.

5   Use 100 μl aliquots of the cooled supernatants as crude extract in the enzyme activity assay.

Assay

For the CHS activity measurements, the methods of Britsch and Grisebach (1985) and Hinderer *et al.* (1983) were modified. At least two replicates were performed per sample. Blanks are done with ddH$_2$O instead of the specific, non-labelled substrate or with inactivated enzyme preparations.

Note – to avoid loss of enzyme activities, all handling must be done on ice.
Note – before routinely measuring enzyme activities, check the tissue specific enzyme characteristics (e.g. temperature- and pH-optimum of the reaction, K_m-values of the substrates, linearity of the assay, etc.). The enzymatically formed labelled product(s) have to be co-chromatographed with unlabelled substances by thin-layer chromatography (TLC) and identified by autoradiography.
Note – handling of radionucleotides is only allowed in restricted areas, requires permission, adequate laboratories, and security guidelines which must be strictly followed.
Duration of procedure: approx. 60 min.

1   Dispense 100 μl of 0.1 M KPi supplemented with 1.5% BSA into Eppendorf tubes.

2   Add 100 μl of the crude sample extract.
3   Pipette 5 μl malonyl-CoA (3.2 nmol [2-^{14}C] malonyl-CoA; 1.4 x 10^5 dpm) to each of the tubes.
4   Add 25 μl of ddH$_2$O to the blanks.
5   Start the reaction by adding 25 μl of 46 μM p-coumaroyl-CoA to the samples, carefully seal the tubes, and incubate at 30 °C for 30 min.
6   Stop the reaction by addition of 10 μl 0.1% (w/v) naringenin.
7   Extract the labelled reaction products and separate them from the substrates by adding 250 μl ethylacetate.
8   Mix for 2.5 min, and centrifuge at 12 000 x g for 1 min.
9   Count 50 μl aliquots of the ethylacetate phase in a scintillation counter and calculate the enzyme activity.

Protein determination

In order to express the catalytic enzyme activities, both on a dry weight and a protein basis, the protein content of the tissues and the extracts has to be determined. Two different approaches can be used.

For determination of the protein content according to Bradford (1976), the Bio-Rad Bradford microassay protein assay is used and the manufacture's protocol is followed. However, this protein quantification method is increasingly becoming replaced by a dot-blot assay which is not affected by detergents or other substances such as plant phenolics which are known to interfere with common protein assays (Guttenberger et al., 1991). In this case the step-by-step protocol given by Guttenberger (1991) is followed.

Micro-dissectioning and enzymic cycling

Where single cells or cell-groups of specific function or stage of differentiation are the target of the investigation, methods are needed which permit the generation and quantitative analyses of these 'fractions'. Preparation of such small tissue fractions can be achieved by a method called micro-dissectioning of plant tissues. For the quantitative analyses of e.g. metabolite pools or enzyme activities of such small samples a method called enzymic cycling is employed (Lowry and Passonneau, 1972). Here we present data on the investigation of cambial tissue of Robinia pseudoacacia which differentiate into bark

and wood tissues, respectively. The method outlined has also been used for investigating the role of IAA in the differentiation of wood cells in Pinus sylvestris during earlywood and latewood formation (see chapter by Uggla and Sundberg in this volume).

Sample preparation

Equipment needed:	Chemicals needed:
Chisels	Dry ice
Dissectioning room (constant conditions 20 °C, 40% RH)	Liquid nitrogen
Freeze-drier (composed of a vacuum pump (Edwards E2M5), freezing trap (KF -2-110, Saur), and a freezer (–35 °C) with the sample compartment	Vacuum grease
Freezers (–80 °C, –25 °C)	
Glass evacuation tubes	
Glass vials with perforated caps (e.g. Whatman)	
Increment borer (5 mm diameter)	
Micro-knives made from broken edges of razor blades	
Nylon mesh (size, e.g. 50 μm, Züricher Beuteltuchfabrik)	
Saw	
Stereomicroscope (e.g. Leitz M6)	
Ultra-micro balance or glass fibre balance	

As for biochemical analyses, non-destructive plant (tree) collections and destructive sampling is performed. In order to preserve the metabolic state of the intact tissue, the samples are frozen in liquid nitrogen immediately after withdrawal from the trunk, transported on dry-ice and stored at –80 °C until further handling. For micro-dissectioning, freeze-dried sections (approx. 50 μm in thickness) containing the desired tissues are cut off directly from freeze-dried tissue blocks (Magel et al., 1996) or are obtained by cutting off cryo-sections from the deep-frozen tissue blocks by using a freezing microtome (e.g. HM 505 E microtome, Microm Laborgeräte) equipped with a steel knife and cooled to –25 °C and subsequent freeze-drying. The sections are collected in pre-cooled trays, freeze-dried for a week (see above) and stored in evacuation tubes at –20 °C. The sampling and anatomical characterisation of the tissues was performed by light microscopic investigations of the sections as shown in Uggla et al. (1996, 1998). For further details of this procedure, readers are referred to the chapter by Uggla and Sundberg in this volume.

Before sample sections are removed and used, the storage assembly is warmed to ambient RT in a constant atmosphere (20 °C, 40% RH) and admission of air should be at moderate rate.

Note – it is essential to ensure that the drying assembly has reached ambient temperature before it is opened. All further handling of the freeze-dried sections must be in a constant atmosphere (20 °C, 40% RH).

For dissectioning, the samples are transferred to a piece of plexiglas, which rests on the stage of a stereomicroscope. In order to minimise the generation of static electricity – exacerbated by the required low humidity and the need to work on hydrophobic surfaces – exposure of the surfaces and tools to a radiation source and/or spraying the equipment with ionised air are recommended (for detail see Lowry and Passonneau, 1972; Hampp et al., 1990). The desired tissue areas e.g. phloem and xylem oriented cells of the cambial layers (Magel et al., 1996)

are dissected by using micro-knives. For manufacturing these knives refer to Lowry and Passonneau (1972) and Hampp et al. (1990). In short, a razor blade is cut or broken into small splinters. For different types of microknives, splinters of different sizes are used. For achieving microknives of different flexibility (depending on the stiffness of the tissue), the splinters are glued to copper or steel wires of different diameters or to nylon bristles which are affixed to a handle (e.g. Pasteur pipette or dissecting needle). Dissected samples are transferred with 'hair points' and collected in a tray.

Estimation of sample mass depends on the sample size. For larger samples, it can be performed using commercially available balances which have a sensitivity of about 10 to 100 ng (e.g. Sartorius balance MP 8-1, or Sartorius ultramicro). For smaller sections, down to 10 pg, custom fabricated quartz fibre balances, so-called 'fishpole balances', are suitable. In principal, the weighing procedure of these balances is based on the deflection toward gravity caused by the mass of the sample. Calibration of these balances is with p-nitrophenol crystals. For detailed guidelines on their construction and applications, refer to Lowry and Passonneau (1972) and Hampp et al. (1990).

Extraction and quantification of pyridine nucleotides from *Robinia* cambial tissue by enzymic cycling

Note – all steps in which freeze-dried material is handled must be performed in constant conditions (see above).

Equipment needed:	Chemicals needed:
Adjustable microlitre pipettes and disposable tips	2-amino-2-methylpropanol 0.14 M pH 9.9
Block heater	2-oxoglutarate 400 mM
Microplate reader with heating device (SLT)	5'-ADP 5 mM
Microplates (flat bottom, Costar)	6-P-gluconate dehydrogenase (230 U/ml)
Teflon racks with reaction wells	Alcohol dehydrogenase (1440 U/ml)
Water bath or block heater for microtubes	BSA 0.18%
	Dry ice
	EDTA = ethylenediaminetetraacetic acid 0.28 mM
	Ethanol 3.5 M
	Glucose-6-phosphate 50 mM
	Glucose-6-phosphate dehydrogenase (224 U/ml)
	Glutamate dehydrogenase (480 U/ml)
	Glutamate-oxaloacetate-transaminase (480 U/ml)
	HCl 0.1 M

Liquid nitrogen Malate dehydrogenase (480 U/ml) $MgCl_2$ 28 mM NAD, NADH, NADP, NADPH Na-glutamate NH_4Cl 500 mM Oxaloacetate 200 mM ß-mercaptoethanol 0.1 M pH 10.8 ß-mercaptoethanol 80 mM Tris 0.25 M pH 8.1 Tris 0.5 M pH 8.0

Note – unless otherwise stated, all solutions are aqueous – enzymes are dialysed and resuspended in the respective sample buffer.

Extraction of pyridine nucleotides

a) Extraction of oxidised pyridine nucleotides (nicotinamide adenine dinucleotide (NAD), nicotinamide adenine dinucleotide phosphate (NADP))

Duration of procedure: approx. 30 min.

1   Place cell sections comprising 10 to 20 μg of dry weight into wells of teflon racks.
    Note – as sample size is small it is recommended to work under a stereo microscope.
2   Add 20 μl of 0.1 M HCl.
3   Extract in a block heater for 10 min at 95 ℃.
4   Cool on ice.

b) Extraction of reduced pyridine nucleotides (nicotinamide adenine dinucleotide (reduced) (NADH) and nicotinamide adenine dinucleotide phosphate (reduced) (NADPH))

Duration of procedure: approx. 30 min.

1   Place cell sections comprising 10 to 20 μg of dry weight into wells of teflon racks.
    Note – as sample size is small it is recommended to work under a stereo microscope.
2   Add 20 μl of 0.1 M β-mercaptoethanol pH 10.8.
3   Extract in a block heater for 10 min at 80 °C.
4   Cool on ice.

Measurement of pyridine nucleotides from *Robinia* cambial tissue by enzymic cycling

The aliquots taken for biochemical analyses contain picomole amount of pyridine nucleotides.

If these small amounts are to be measured with conventional equipment, signal amplification is needed. Enzymic cycling is theoretically capable of limitless amplification of any chemical that is cycled between two 'forms' (e.g. $NAD^+ \Leftrightarrow NADH$). Pyridine nucleotides are the most versatile because they can potentially be coupled to any biochemical reaction, and because the oxidised and reduced forms can be selectively destroyed.

In principle, a cycling reagent contains substrate in excess (Figure 1, A and C). The respective pyridine nucleotide is added, and being alternatively oxidised and reduced, it causes the accumulation of products (Figure 1, B and D). The cycling reagent does not distinguish between the oxidised and reduced forms. The rate of cycling depends on the total amount of catalyst (e.g. NAD^+ + NADH), which does not change during the cycling steps. As the concentration of pyridine nucleotides added is below the Michaelis constant, the rate of accumulation of B and D is proportional to the amount of pyridine nucleotides. We routinely run assays with 3000 to 5000 cycles/h which is less than 10% of the maximum rate. After terminating the cycling step by heating, the products formed are quantified

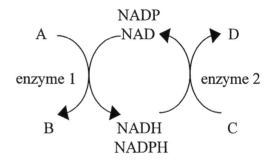

Figure 1   Reaction scheme for the enzymic cycling of pyridine nucleotides. Substrates (A, C) and enzymes (1, 2) depend on the type of cycling (NAD or NADP).

enzymically by coupling to the formation of pyridine nucleotides. (It is important to note that this last step is conducted in 200 μl volumes at μmol concentrations. Thus there is no interference from substances, including the cycled pyridine nucleotides, carried over from the cycling step.) The precise number of cycles in the cycling step as well as the precise amount of pyridine nucleotides in the samples can be determined by pyridine nucleotide standards. For detailed description and extensive background of this method, see Lowry and Passonneau (1972) and Hampp *et al.* (1984, 1990).

In the following, the quantification of NAD(P) and NAD(P)H in cambial tissue of *Robinia* is outlined. In order to eliminate artefacts by sample-specific quenching of the reaction, quantification of the sample endogenous NAD(P)(H) content is by internal standardisation. Four replicates were done for each of the sample blank, the sample, and the sample supplemented with the internal standard (2 pmol of NAD(P) or NAD(P)H). In addition, medium blanks were run.

a) Quantification of NAD and NADH

In the specific step, ethanol and oxaloacetete are consumed by the action of alcohol dehydrogenase and malate dehydrogenase, and malate and acetaldehyde accumulate stochiometrically to the amount of NAD(H) added. In the indicator step, the amount of malate formed is quantified enzymatically by coupling to the generation of NADH.
Duration of procedure: approx. 4 h.

1. Prepare a test mix of all reagents common to all the programmed reactions. Prepare only the quantity required for the total number of sample wells including controls; make sure that at least one extra aliquot is prepared to ensure adequate volume is available for the last well.
2. Combine the following amounts per well:
   8 μl of 0.5 M Tris buffer, pH 8.0
   3 μl of 3.5 M ethanol
   1 μl of 80 mM β-mercaptoethanol
   1 μl of 0.8% BSA
   1 μl of 80 mM oxaloacetate.
3. Dispense 14 μl of the test mix to the wells of the microplate.
4. Add 12 μl of ddH$_2$O to the sample blanks.
   Add 10 μl of ddH$_2$O to the samples.
5. Add 10 μl of 0.2 μM NAD or NADH to the samples supplemented with the internal standard.

6. Add 1 μl of malate dehydrogenase (480 U/ml) and 1 μl of alcohol dehydrogenase (1440 U/ml) to the samples and the samples supplemented with the internal standard.
7. Dispense 3 μl of the extract (see Extraction of NAD and NADH) to the sample blanks, the samples and the samples supplemented with the internal standard.
8. Mix on a shaker.
9. Incubate for 60 min at RT.
10. Stop reaction by heating for 20 min at 100 °C.
11. Cool on ice.
12. Add to each well 100 μl of the indicator reagent (0.14 M 2-amino-2-methylpropanol pH 9.9, 31 mM Na-glutamate, 1.2 mM NAD, 8 U/ml malate dehydrogenase, 1.4 U/ml glutamate-oxaloacetate-transaminase).
13. Incubate for 45 min at RT.
14. Measure absorbance spectrophotometrically at 340 nm in a microplate reader.

b) Quantification of NADP and NADPH

In the specific step, glucose-6-phosphate and α-ketoglutarate are consumed by the action of glucose-6-phosphate dehydrogenase and glutamate dehydrogenase, and glutamate and 6-P-gluconolactone accumulate stochiometrically to the amount of NADP(H) added. In the indicator step, the amount of 6-P-gluconolactone formed is quantified enzymatically by coupling to the generation of NADPH.
Duration of procedure: approx. 4.5 h.

1. Prepare a test mix of all reagents common to all the programmed reactions. Prepare only the quantity required for the total number of sample wells including controls; make sure that at least one extra aliquot is prepared to ensure adequate volume is available for the last well.
2. Combine the following amounts per well.
   8 μl of 0.25 M Tris pH 8.1
   1 μl of 50 mM glucose-6-phosphate
   1 μl of 5 mM 5'-ADP
   1 μl of 500 mM NH$_4$Cl
   1 μl of 400 mM 2-oxoglutarate.
3. Dispense 14 μl of the test mix to the wells of the microplate.
4. Add 12 μl of ddH$_2$O to the sample blanks.
   Add 10 μl of ddH$_2$O to the samples.
5. Add 10 μl of 0.2 μM NADP(H) to the samples supplemented with the internal standard.

6   Add 1 μl of glucose-6-phosphate dehydrogenase (224 U/ml) and 1 μl of glutamate dehydrogenase (480 U/ml) to the samples and the samples supplemented with the internal standard.

7   Dispense 3 μl of the extract (see Extraction of NADP and NADPH) to the sample blanks, the samples and the samples supplemented with the internal standard.

8   Mix on a shaker.

9   Incubate for 90 min at 37 °C.

10 Stop reaction by heating for 20 min at 100 °C.

11 Cool on ice.

12 Add to each well 100 μl of the indicator reagent (0.22 M Tris pH 8.1, 0.28 mM EDTA (ethylenediaminetetraacetic acid), 170 mM NH_4Cl, 28 mM $MgCl_2$, 1.2 mM NADP, 0.23 U/ml 6-P-gluconate dehydrogenase).

13 Incubate for 45 min at RT.

14 Measure absorbance of generated NADPH spectrophotometrically at 340 nm in a microplate reader.

Extraction and quantification of the catalytic activity of enzymes (glucose-6-phosphate, G6P-DH and 6-phosphogluconate dehydrogenase, 6PG-DH) from *Robinia* cambial tissue

Equipment needed:	*Chemicals needed:*
Adjustable microlitre pipettes	6-phosphogluconate 250 mM
Block heater	Glucose-6-phosphate 250 mM
Centrifuge/microfuge equipped with adapter for glass microtubes and cooling, at least 12000 g	Ice
Dental drill	Liquid nitrogen
Disposable tips for pipettes	$MgCl_2$ 200 mM
Glass microtubes, flat-bottomed	NADP
Microplate reader with heating device (SLT)	PVPP (MW 360000)
Microplates (flat-bottomed, Costar)	β-mercaptoethanol
Whirlimixer	Tris/borate 0.1 M / 0.3 M, pH 7.6
	Tris/MES 0.3 M pH 7.6

Note – unless otherwise stated, all solutions are aqueous.

a) Extraction of G6P-DH and 6PG-DH from Robinia *cambial tissue*

Duration of procedure: approx. 45 min.

1   Place cell sections comprising 200–500 μg of dry weight into flat-bottomed glass microtubes.

2   Add an equal amount of PVPP 360000.

3   Homogenise under liquid nitrogen-cooling with a dental drill (10000 rpm) to fine tissue powder.

4   Add 100 μl of ice-cold extraction medium (Tris/borate 0.1 M / 0.3 M, pH 7.6, 1 mM β-mercaptoethanol, 0.5 mM NADP).

5   Extract on ice for 20 min with repeated, gentle vortexing.

6   Centrifuge at 12000 x g for 10 min at 4 °C.

7   Aliquots of the clear supernatants are used as crude extract in the enzyme assays for the quantification of the enzyme activities.

b) Measurement of the catalytic activities of G6P-DH and 6PG-DH from cambial tissue

Assay

The measurement of the enzyme activity is based on published assays (e.g. Guttenberger *et al.*, 1993) and was adapted to the specific requirements of black locust tissues. Both enzyme activities can be recorded and calculated by the increase in absorbance at 340 nm due to the generation of NADPH. At least three replicates per sample are performed. Run blanks with ddH_2O instead of the specific substrate or with sample-free extraction buffer. The kinetics of the enzyme reactions are followed in a total volume of 100 μl in microtitre plates (important: the plates must be flat-bottomed).

Note – to avoid loss of enzyme activities, all handling must be done on ice.

Note – before routinely measuring enzyme activities, check the tissue-specific enzyme characteristics (e.g. temperature- and pH-optimum of the reaction, K_m-, K_i- and / or K_a-values of the enzyme for the substrates, inhibitors, and / or activator molecules, linearity of the assay, etc.).

Duration of procedure: approx. 90 min.

1   Prepare a test mix of all reagents common to all the programmed reactions. Prepare only the quantity required for the total number of wells; make sure that at least one extra aliquot is prepared to ensure adequate volume is available for the last well.

2   Combine the following amounts per well:
    20 μl of 0.3 M Tris/MES, pH 7.6
    5 μl of 200 mM NADP
    5 μl of 200 mM MgCl$_2$.

3   Dispense 30 μl of the test mix to each well of the microplate.

4   Add 10 μl of the clear supernatant (see Extraction).

5   Add 10 μl of ddH$_2$O to the blanks.

6   Start the reaction with 10 μl 250 mM glucose-6-phosphate or 6-phosphogluconate for the measurement of the catalytic activity of glucose-6-phosphate dehydrogenase or 6-phosphogluconate dehydrogenase, respectively.

7   Gently mix, and follow the enzyme activity by the formation of NADPH (recorded as an increase of the absorbance at 340 nm) for 60 min at 37 °C using a microplate reader system (see above).

8   Calculate rates from the linear part of the recorded kinetics, usually between 5 and 60 min of incubation, by using the software package supplied with the instrument.

APPLICATIONS AND RESULTS

Biochemical studies

For wood tissues, the successful application of biochemical and molecular methods is hampered by the very low percentage of living cells together with the rigidity and high phenolic content of the material. Using the methods outlined above has enabled us to investigate the distribution of e.g. the sucrose metabolising and flavonoid synthesising enzymes in stem tissues of deciduous trees. Here we present data on the activities of sucrose synthase (SuSy) and chalcone synthase (CHS) two enzymes which turned out to play a key role during heartwood formation in tissues of *Robinia pseudoacacia* L. (black locust).

Sucrose synthase

In herbaceous plants it is assumed that SuSy is involved in a futile cycling of sucrose (Pontis, 1978; Su, 1982). The predominant role of SuSy *in vivo*, however, is the cleavage of sucrose in net-importing tissues and thus the presence of SuSy is correlated with the need to provide precursors for structural (cell wall components; Su, 1965; Fry, 1988), as well as storage oligosaccharides and polysaccharides (Heim *et al.*, 1993).

Using the outlined methods, kinetic properties of SuSy from *Robinia* correspond well with published data for this enzyme from other dicots and monocots. Using the on-line detection method, possible interference by accumulation of released fructose which is a potent inhibitor of SuSy catalytic activity (K_i 4.65 mM) could be avoided.

In stems of black locust, SuSy provides tissue- and seasonally-specific energy and substrates for different differentiation processes. The dominance of SuSy in the youngest increment (outer sapwood, Figure 2) in spring is related to the high demand for cell wall material (cellulose and lignin) of this rapidly growing sink tissue (Su, 1965; Fry, 1988; Hauch and Magel, 1998). During summer months SuSy is involved, as for starch-storing tissues of annual plants (Heim *et al.*, 1993), in sucrose-starch conversion and thus in the accumulation of starch in the middle and inner parts of the stem. However, the most prominent role SuSy plays is in the innermost living tissues of the trunk during the transition of living sapwood tissue into dead heartwood. Here, the enhanced degradation of sucrose coincides in time and place with the increased activities of key enzymes of the pathways involved in the formation of phenolic compounds (see below and Magel *et al.*, 1991, 1994; Magel and Hübner, 1997) and the accumulation of phenolic heartwood extractives (Magel *et al.*, 1994). Thus, our data imply that SuSy, and thus sucrose metabolism via the sucrose synthase pathway, is not only characteristic of metabolic sinks associated with polysaccharide biosynthesis (ap Rees, 1974), but also with the biosynthesis of phenolic compounds.

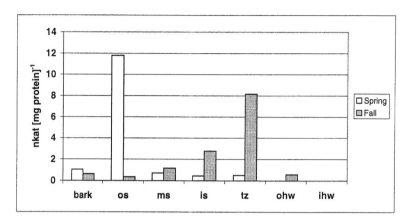

Figure 2   Catalytic activity of sucrose synthase in stem tissues of a single *Robinia pseudoacacia* L. tree cut during springtime (open bars) or fall (shaded bars). The variation of parallels differed less than 5% of the mean, thus no standard deviation is given. Key: os = outer, ms = middle, and is = inner sapwood, tz = sapwood-heartwood transition zone, ohw = outer and ihw = inner heartwood.

Chalcone synthase

The biosynthesis of flavonoids is characterised by the condensation of one molecule of 4-coumaroyl-CoA and three molecules of malonyl-CoA to naringenin chalcone (Hahlbrock and Scheel, 1989; Heller and Forkmann, 1988). This stepwise condensation is catalysed by CHS, the key enzyme of the various flavonoid pathways. Since flavonoids accumulate in the heartwood of black locust, it was hypothesised that their synthesis should be closely related to heartwood formation. However, as measuring of these enzyme activities failed, it was assumed that 'pathways others than the accepted ones should lead to the accumulation of phenolics'. Applying the methods outlined above, we succeeded in quantifying both catalytic activity and

enzyme specific protein (data not shown here) of the key enzyme of flavonoid biosynthesis. Our results showed that CHS was active almost exclusively at the sapwood-heartwood boundary (Figure 3). Highest activities of CHS were measured in trunks felled in September and November, just at the time when heartwood extractives accumulate.

Micro-dissectioning + enzymic cycling = quantitative histochemistry

Growth processes of wood axes result from the activity of vascular cambium which produces, dependent on the season and under genetic control, elements of the phloem in centrifugal and elements of the xylem in centripetal direction. Before the

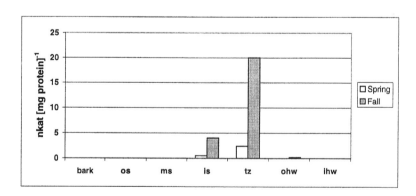

Figure 3   Catalytic activities of chalcone synthase in stem tissues of a single *Robinia pseudoacacia* L. tree cut during springtime (open bars) or fall (shaded bars). The variation of parallels differed less than 5% of the mean, thus no standard deviation is given. Key: os = outer, ms = middle, and is = inner sapwood, tz = sapwood-heartwood transition zone, ohw = outer and ihw = inner heartwood.

introduction of micro-dissectioning techniques, analyses of these important and very specific tissue zones were restricted to cytological and histochemical methods. Approaches to evaluate the differentiation of the individual tissues on a quantitative biochemical basis were hampered by the problem of obtaining samples of the differentiating cell layers without contamination by differentiated cells. Applying techniques of quantitative histochemistry (Lowry and Passonneau, 1972; Hampp et al., 1990), we succeeded in monitoring differences in metabolic pathways between phloem and xylem directed cambial derivatives

Phloem and xylem exhibit the most pronounced differences in the chemical nature of their cell walls. Hemicelluloses and cellulosic material dominate in the walls of the phloem cells, and lignin is additionally incrusted into the cell walls of xylem cells. As this should have consequences for the activity of related pathways, a first approach towards a biochemical characterisation of the cambial derivatives of *Robinia pseudoacacia* was to determine both the pool sizes of pyridine nucleotides and the levels of extractable activity of the two dehydrogenases.

Pyridine nucleotides in Robinia *cambial tissue*

Due to their multiple functioning (hydrogen-transferring co-factors of enzymes, energy production, and production or utilisation of reducing power), pyridine nucleotides constitute important regulatory factors in green and non-green tissues. Pools of pyridine nucleotides therefore play an

important role in various developmental processes, but their participation in differentiation and growth processes of the cambial zone of woody plants has been largely unexplored.

By applying quantitative histochemistry we were able to show that the total pool size of pyridine nucleotides is similar in the phloem (PD) and xylem (XD) oriented derivatives of the cambial zone of trees of *Robinia pseudoacacia* L. (Figure 4). However, the amount of NAD + NADH exceeded that of NADP + NADPH within the PD. A ratio of NADH : NAD of about 1 indicates the preponderance of catabolic pathways. In contrast, the NADP(H) system dominates in the XD zone and a ratio of NADPH : NADP of approx. 1 indicates increased rates of reductive biosyntheses (Magel et al., 1996).

G6P-DH and 6PG-DH in Robinia *cambial tissue*

As both reduction equivalents and precursors needed for the synthesis of lignin are derived from the oxidative pentose phosphate pathway, more information can be obtained from the quantification of the catalytic activities of the two key enzymes of this pathway, G6P-DH and 6PG-DH. Within the XD zone of the cambium, the extractable activities of both enzymes G6P-DH and 6PG-DH were greatly increased (Figure 5) compared with the outward-directed cells which differentiate into phloem tissue. Concomitantly with the preponderance of the NADP(H) system (Figure 4) this indicates that, as for herbaceous plants (Pyrke and ap Rees, 1976, 1977), the activity

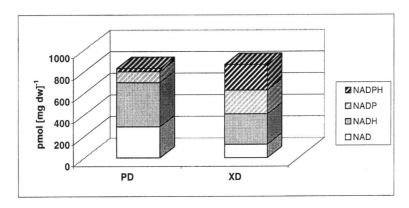

Figure 4 Content of di-phospho-pyridine nucleotides (NAD open bars, NADH closed bars) and tri-phospho-pyridine nucleotides (NADP light hatched bars, and NADPH dark hatched bars) in the cambial differentiation zone of one *Robinia pseudoacacia* L. tree. PD identifies the phloem-oriented, and XD the xylem-oriented daughter cells of the cambial mother cells. The variation of parallels differed less than 5% of the mean, thus no standard deviation is given.

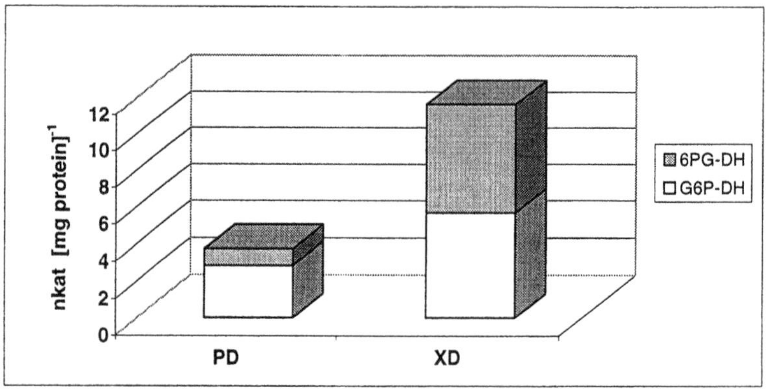

Figure 5 Catalytic activities of glucose-6-phosphate dehydrogenase (open bars) and 6-phosphogluconate dehydrogenase (closed bars) in the cambial differentiation zones of one individual *Robinia pseudoacacia* L. tree. PD identifies the phloem-oriented, and XD the xylem-oriented daughter cells of the cambial mother cells. The variation of parallels differed less than 5% of the mean, thus no standard deviation is given.

of the oxidative pentose phosphate pathway can be taken as a measure of lignin formation during xylogenesis.

Taken together, this first approach clearly shows that phloem-oriented and xylem-oriented cambial descendants exhibit distinct differences in their biochemical patterns even in the early stages of differentiation.

FUTURE PROSPECTS

Biochemical investigation of wood is hampered by the very specific properties of this tissue. Employing modern biochemical techniques and using purification systems, these difficulties can be overcome. However, techniques or methods which work with one of the tissues may not necessarily be usable for others, without modifications. With the help of quantitative histochemistry, tissues which serve different tasks and which show a highly co-ordinated arrangement in trees trunks can be investigated separately. Modifying these techniques down to the single cell level, the fate and contribution of single cells as part of the whole complexity of wood physiology will be highlighted.

ACKNOWLEDGEMENTS

The work was financed by grants from the Deutsche Forschungsgemeinschaft and the BMBF. I thank H. Bleuel and E. Dongus for technical assistance and S. Hauch for contributing to the methodological approaches.

REFERENCES

ap Rees, T. (1974) Pathways of carbohydrate breakdown in higher plants. In: *International review of biochemistry*, edited by D.H. Northcote, pp. 90–127. London: Buttersworth.

Bradford, M.M. (1976) A rapid and sensitive method for the quantitation of microgram quantities of protein utilising the principle of protein-dye binding. *Anal. Biochem.*, **72**, 248–254.

Britsch, L. and Grisebach, H. (1985) Improved preparation and assay of chalcone synthase. *Phytochem.*, **24**, 1975–1976.

Fry, S. (1988) *The growing plant cell wall: chemical and metabolic analysis*, pp. 333. UK: Harlow: Longman.

Guttenberger, M. (1991) Protein determination. In: *Cell biology. A laboratory handbook*, edited by J.E. Celis. Vol. 3, pp. 169–178. San Diego: Academic Press.

Guttenberger, M., Neuhoff, V. and Hampp, R. (1991) A dot-blot assay for the quantification of nanogram amounts of protein in presence of carrier ampholytes and other possibly interfering substances. *Anal Biochem.*, **196**, 99–103.

Guttenberger, M., Schaeffer, C. and Hampp, R. (1993) Kinetic and electrophoretic characterisation of NADP dependent dehydrogenases from root tissues of Norway spruce (*Picea abies* (L.) [Karst.]) employing a rapid one-step extraction procedure. *Trees*, **8**, 191–197.

Hahlbrock, K. and Scheel, D. (1989) Physiology and molecular biology of phenylpropanoid metabolism. *Ann. Rev. Plant Physiol.*, **40**, 347–369.

Hampp, R., Goller, M. and Füllgraf, H. (1984) Determination of compartmented metabolite pools by a combination of rapid fractionation of oat mesophyll protoplasts and enzymatic cycling. *Plant Physiol.*, **75**, 1017–1021.

Hampp, R., Rieger, A. and Outlaw, W.H. Jr. (1990) Microdissection and biochemical analysis of plant tissues. In: *Modern methods of plant analysis. New Series*, Volume 11, *Physical methods in plant sciences*, edited by

H.F. Linskens and J.F. Jackson., pp. 124–147. Heidelberg: Springer.

Hauch, S. and Magel, E. (1998) Extractable activities and protein content of sucrose phosphate synthase, sucrose synthase and neutral invertase in trunk tissues of *Robinia pseudoacacia* L. are related to cambial wood production and heartwood formation. *Planta*, **207**, 266–274.

Heim, U., Weber, H., Baumlein, H. and Wobus, U. (1993) A sucrose synthase gene in *Vicia faba* L.: Expression pattern in developing seeds in relation to starch synthesis and metabolic regulation. *Planta*, **191**, 394–401.

Heller, W. and Forkmann, G. (1988) Biosynthesis. In: *The flavonoids*, edited by J.B. Harborne. London: Chapman and Hall.

Hillis, W.E. (1987) *Heartwood and tree exudates*. Heidelberg: Springer.

Hinderer, W., Noe, W. and Seitz, H.U. (1983) Differentiation of metabolic pathways in the umbel of *Daucus carota. Phytochem.*, **22**, 2417–2420.

Lowry, O.H. and Passonneau, J.V. (1972) *A flexible system of enzymatic analyses*. London: Academic Press.

Magel, E.A. (2000) Biochemistry and physiology of heartwood formation. In: *Cell and molecular biology of wood formation*.edited by R. Savidge, J. Barnett and R. Napier, pp. 363–376. Oxford: Bios.

Magel, E., Bleuel, H., Hampp, R. and Ziegler, H. (1996) Pyridine nucleotide levels and activities of dehydrogenases in cambial derivatives of *Robinia pseudoacacia* L.. *Trees*, **10**, 325–330.

Magel, E.A., Claudot, A.C., Drouet, A. and Ziegler, H. (1991) Formation of heartwood substances in the stem of *Robinia pseudoacacia* L. I. Distribution of phenylalanine ammonium-lyase and chalcone synthase across the trunk. *Trees*, **5**, 203–207.

Magel, E. and Hübner, B. (1997) Distribution of phenylalanine ammonia-lyase (PAL) and chalcone synthase (CHS) within trunks of *Robinia pseudoacacia* L.. *Botanica Acta*, **110**, 314–322.

Magel, E.A., Jay-Allemand, C. and Ziegler, H. (1994) Formation of heartwood substances in the stem of *Robinia pseudoacacia* L. II. Distribution of non-structural carbohydrates and wood extractives across the trunk. *Trees*, **8**, 165–171.

Pontis, H.G. (1978) On the scent of the riddle of sucrose. *Trends Biochem Sci*, **3**, 137–139.

Pozueta-Romero, J., Yamaguchi, J. and Akazawa, T. (1991) ADPG formation by the ADP-specific cleavage of sucrose-reassessment of sucrose synthase. *FEBS Lett.*, **291**, 233–237.

Pryke, J.A. and ap Rees, T. (1976) The pentose phosphate pathway as a source of NADPH for lignin synthesis. *Phytochem.*, **16**, 557–560.

Pryke, J.A. and ap Rees, T. (1977) Activity of pentose phosphate pathway during lignification. *Planta*, **132**, 279–284.

Su, J.C. (1965) Carbohydrate metabolism in the shoots of bamboo *Leleba oldhami*. I. Preliminary survey of soluble saccharides and sucrose degrading enzymes. *Bot. Bull. Acad. Sin.*, **6**, 153–159.

Su, J.C. (1982) Some aspects of sucrose metabolism in plants. *Proc. Natl. Sci. Council*. Part B, **6**, 17–21.

Uggla, C., Mellerowicz, E.J. and Sundberg, B. (1998) Indole-3-acetic acid controls cambial growth in *Pinus sylvestris* (L.) by positional signalling. *Plant Physiol.*, **118**, 113–121.

Uggla, C., Moritz, T., Sandberg, G. and Sundberg, B. (1996) Auxin as a positional signal in pattern formation in plants. *Proc. Natl. Acad. Sci. USA*, **93**, 9282–9286.

5 Protein Analysis in Perennial Tissues

Wood Formation in Trees ed. N. Chaffey
© 2002 Taylor & Francis
Taylor & Francis is an imprint of the
Taylor & Francis Group
Printed in Singapore.

Tannis Beardmore[1]*, Suzanne Wetzel[2] and Carrie-Ann Whittle[3]

[1] NRCan Canadian Forest Service-Atlantic Region, 1350 Regent St. S., Fredericton, New Brunswick E3B 5P7, Canada;
[2] NRCan Canadian Forest Service-Ontario Region, 1219 Queen St. E., Sault Ste. Marie, Ontario P6A 5MZ, Canada;
[3] Department of Biology, University of Dalhousie, Halifax, Nova Scotia B3H 3J5 Canada.
* Author for correspondence

ABSTRACT

Methodologies for analysis of proteins isolated from perennial tissues are described, with particular reference to overcoming problems related to protein degradation, and presence of phenolic compounds. Techniques described cover the basic steps of protein isolation, quantification and characterisation using one-dimensional sodium dodecyl sulphate-polyacrylamide gel electrophoresis (SDS-PAGE). Also described is a technique which permits high-resolution protein characterisation, two-dimensional iso-electric focusing (IEF) electrophoresis, where protein is separated in the first dimension by IEF according to charge (iso-electric point, pI), then in the second dimension by SDS-PAGE, where protein is separated according to size (molecular weight, MW). In addition, two immuno-detection techniques, Western blotting and immuno-localisation are described in detail.

Key words: immunolocalisation, iso-electric focusing, one- and two-dimensional electrophoresis, protein, Western blotting

INTRODUCTION

The purpose of this chapter is to introduce the reader to basic techniques in protein analysis, starting from the collection of plant materials and successful extraction of proteins, to protein quantification and, finally, initial characterisation based on MW, iso-electric point, similarities in antigenicity and cellular localisation. The electrophoretic characterisation of proteins can provide information that is required for synthesising peptides, as well as give quick insights into potential similarities and differences between proteins. The two-dimensional iso-electric focusing-polyacrylamide gel electrophoresis (IEF-PAGE) technique, while technically challenging, permits the high-resolution separation of a large number of proteins. Techniques such as this allow one to compare changes in patterns of protein turnover (Chrispeels and Bollini, 1982), synthesis (Vierling and Key, 1985), accumulation (Hurkman and Tanaka, 1986), and protein modification (Langheinrich and Tischner, 1991). The use of antibodies is diverse and the focus herein is on Western blotting, which allows for identification of specific protein(s) and immunolocalisation of the protein, which provides further information related to potential protein function. While these techniques may not give definitive answers as to the physiological function of the protein, careful analysis of the proteins of interest by a variety of these techniques greatly helps in providing the background for deducing a potential role, which can then be tested experimentally.

Of the numerous protein isolation and analysis procedures available, the ones presented in this chapter have proven to be fast, reliable, and reproducible in perennial tissues, such as wood, buds, roots, and stems. Many detailed texts exist, such as the *Current Protocols in Protein Science* edited by Coligan *et al.* (1998), *The Protein Protocols Handbook* edited by Walker (1996), *Protein Analysis and Purification, Benchtop Techniques*, edited by Rosenberg (1996), should the reader want to broaden his or her background in this area.

TECHNIQUES

> It is recommended that gloves be worn through-out all the procedures described herein and that one becomes familiar with the safety procedures required for the handling and disposal of all chemicals.

Note – chemicals printed in bold type are considered in more detail in the relevant 'SOLUTIONS FOR ...' section.

Problems encountered in isolating protein from perennial tissues

A variety of problems can be encountered during the isolation and analysis of proteins from perennial tissues such as wood. However, the two main problems, discussed below, result from the presence of proteolytic enzymes (eg. proteases, peptidases or proteinases) and phenolic compounds.

Protein degradation

Proteolytic enzymes hydrolyse the peptide bond in proteins and polypeptides. Many of these enzymes are compartmentalised within vacuoles. When the tissue is homogenised, these enzymes are released into the extract, resulting in protein degradation. This damage can be minimised by:

1)   isolating the protein immediately after collecting the plant samples;
2)   the addition of protease inhibitors to extraction buffer (see below);
3)   reducing the number of freeze-thaw cycles experienced by the protein extract to a minimum (freeze sufficiently small working aliquots); and,
4)   keeping the protein extract on ice at all times.

There are many classes of proteolytic enzymes in plants; not all of them can be inhibited. Therefore, it is recommended that a mixture of the following protease inhibitors be added to the protein extracts. They can be purchased individually [e.g.: leupeptin (serine and cysteine protease inhibitor), phenylmethylsulphonyl fluoride (PMSF) (serine protease inhibitor) and pepstain A (aspartic (acid) protease inhibitors)], or as a mixture of several protease inhibitors (e.g. protease inhibitor cocktails, Roche Molecular Biochemical). Metalloproteinases

can be inhibited by the addition of chelators such as ethylenediaminetetraacetic acid (EDTA).
Note – PMSF is an inhibitor of acetylcholinesterase (neurotransmitter) and thus is very toxic: weigh this compound in a fume hood and wear gloves and a mask.

Phenolic compounds

Unfortunately, phenolic compounds are one of the most ubiquitous groups of plant metabolites that can reduce the quality and quantity of protein isolated from perennial tissues. Generally, plant extracts, which are brown, are said to contain phenolics. This colouration is due to the oxidation of phenolics to quinonic compounds (Bravo, 1998). These quinonic compounds are highly reactive and will polymerise with each other, oxidising other phenolics, and will bind to proteins resulting in protein aggregation, cross-linking and precipitation (reviewed by Loomis, 1974). This damage can be minimised by the addition of antioxidants such as L-ascorbic acid (*Populus* xylem, Christensen *et al.* 1998), (*Hevea brasiliensis* stem, Tian *et al.* 1998) and polymers which absorb phenolics such as soluble and insoluble polyvinylpyrrolidone (PVP) (roots of perennial weeds, Cyr and Bewley, 1990) and uncharged polystyrene Amberlite resins (e.g. Amberlite XAD-2, Sigma Co.) (Gray, 1978) to the protein isolation buffer.

Collection and storage of plant materials

The quality and yield of protein is influenced by the collection procedure and the time of collection (harvesting season). Plant materials should be placed on ice immediately after collection. Whenever possible, protein should be extracted from fresh, not frozen, tissues, to minimise protein degradation (discussed above). However, if necessary, plant tissues can be stored at –20 °C until extracted.

The time at which the plant materials are collected is an important consideration as there are seasonal changes in the amounts and types of protein present in perennial tissues (reviewed by Rowland and Arora, 1997). In the fall, many woody perennial plants in temperate regions induce the synthesis of numerous proteins, such as vegetative storage proteins (VSPs) (reviewed by Stepien *et al.* 1994), proteins related to endodormancy (reviewed by Rowland and Arora, 1997) in stem and root tissues (Wetzel and Greenwood, 1989; Langheinrich and Tischner, 1991), and terminal and lateral buds (Lang and Tao, 1990; Roberts *et al.* 1990; Binnie *et al.* 1994; Nir *et al.* 1986; Golan-Goldhirch, 1998; Lawrence *et al.* 1997). Also, the presence of undesirable compounds, such as proteases and phenolics (Burtin *et al.* 1998; Honkanen *et al.* 1999) can vary during the seasons.

Isolation of total protein from perennial tissues

Traditionally, proteins were classified according to their solubility (Osborne, 1924). Based on sequential extraction of the proteins, this classification system still provides the primary framework for current protein studies. This system is composed of four protein classes:

1) albumins (soluble in water and dilute buffers with a neutral pH);
2) globulins (soluble in salt solutions but insoluble in water);
3) glutelins (soluble in dilute acids or alkali solutions); and
4) prolamins (soluble in aqueous alcohols).

The buffer used in the isolation procedure described below will isolate all 4 classes of protein. However, for the most part, membrane-bound proteins will not be isolated using this procedure. Refer to Hurkman and Tanaka (1986) for the isolation of protein from plant membranes.

The isolation procedure discussed herein is for total protein, which can be used for subsequent electrophoretic analyses. The isolation of specific proteins can be done, but this requires information concerning the physical properties (e.g., iso-electric point (pI), solubility), chemical structure and biochemical properties of the protein(s) of interest. For information in this area, the reader is referred to Coligan *et al.* (1998).

Equipment needed:	Chemicals needed:
Centrifuge	Potassium hydrogen phosphate [K_2HPO_4]
Microfuge tubes (1.5–2.0 ml capacity)	Sodium chloride [NaCl]
Mortar and pestle	Liquid nitrogen
Spatula	Ice
Syringe	**Basic protein extraction buffer**
Hand-held homogeniser	
Freezer at –20 °C	

Duration of procedure: approx. 2 h for 10 samples.

1 Quickly grind 1 mg of fresh or frozen tissue in liquid nitrogen using a mortar and pestle until a fine powder develops.
2 Transfer powder into a chilled hand-held homogeniser using a spatula, and add 100–150 µl of basic protein extraction buffer.
3 Grind vigorously until a consistent thick liquid forms.
4 Transfer liquid to a chilled microfuge tube.
5 Centrifuge at 14,000 rpm for 30 minutes at 4 °C.
6 Place samples on ice, remove the clear suspension using a fine-tipped syringe, and place in a new microfuge tube [only a very small quantity of clear extract (20–50 µl) may be available for removal]. These samples constitute the primary extract of total protein and should be either used immediately, or frozen at –20 °C.

Organic solvent precipitation of proteins

There are a number of methods for concentrating total protein. Precipitation of proteins with acetone is recommended since it is simple and fast, and can be used to recover the majority of protein in a crude extract. Organic solvents, such as ethanol or acetone, can displace the water around the hydrophobic areas on the surface of protein. As a result, the water solubility of these proteins decreases resulting in aggregation and precipitation of the protein. However, hydrophobic proteins, such as those found in membranes, can only be completely solubilised in solvents. Methods for precipitation of hydrophobic proteins are discussed in Coligan *et al.* (1998).

Equipment needed:	Chemicals needed:
Microfuge tubes (1.5 or 2.0 ml capacity)	Acetone, ice-cold
Centrifuge	Ice

Duration of procedure: approx. 2 h.

1 Thaw protein extract on ice, add an equal volume of ice-cold acetone, mix by inverting, and let samples incubate on ice for approx. 1 h.

2   Centrifuge samples at 14,000 rpm for 5 min, discard supernatant, and let the test tubes air-dry.

3   Resuspend protein in the appropriate buffer for further one-dimensional or two-dimensional electrophoretic analysis (see below).

Protein quantification

The choice of method for quantifying proteins should be based on the presence of known, or potential, interfering substances in the protein extract, sensitivity required, and availability of equipment. The bicinchoninic acid (BCA) method for quantifying protein is recommended. The advantages that this procedure offers over the commonly used Bradford (Bradford, 1976) and Lowry (Lowry et al. 1951) methods are that it can be performed in the presence of detergents (i.e. SDS) and denaturing reagents (i.e. urea) (Smith et al. 1985).

BCA forms a complex with cuprous ion (Cu^{1+}) in an alkaline solution resulting in the formation of a stable coloured (intense purple) chromophore which has a maximum absorbance at 562 nm (Smith et al. 1985). The BCA reagents can be purchased in a kit from Bio-Rad, Sigma or Pierce, and it is recommended that the manufacture's instructions be followed. The assay can be scaled down and performed in microtitre plates (0.5–10 μg protein/ml). The procedure below is for the standard assay (0.1 μg–1.0 μg protein/ml).

Equipment needed:	Chemicals needed:
Test tubes (16 × 100 mm borosilicate) Spectrophotomer capable of measuring absorbance at 562 nm	BCA reagent A and B (contained in kit) Bovine serum albumen (BSA) (Cohn V Fractionate, Sigma, No. A2153): 1 mg/ml in H_2O stored in aliquots at –20 °C
Cuvettes (plastic disposable, quartz or glass) Water bath (37–60 °C) Pipettes (0.1, 1.0, and 2.0 ml)	

Note – BCA protein assay is sensitive to interference by compounds such as 100 mM EDTA, 1 mM DTT and 20% (w/v) ammonium sulphate, 2 M sodium acetate, 1 M sodium phosphate and 1.0 M glycine (see manufacturer's instruction for full details).

Duration of procedure: approx. 3 h for 20 samples.

1   Prepare a calibration curve ranging from 0.1 to 1.0 mg/ml BSA, diluted with the same buffer used to extract the protein.

2   Prepare samples in separate tubes by mixing 100 μl of protein extract, or buffer used to prepare the samples (reference standard) or BSA standard dilutions with 2 ml of BCA reagent A/B mix.

3   Incubate samples at 37 °C for 30 min, then cool samples to room temperature (RT) prior to measurement of absorbance.
      Note–cooling the samples can give rise to temperature gradients in the cuvettes that cause incorrect absorbance measurements, invert to mix samples.

4   Zero the spectrophotometer at 562 nm using the reference standard. Measure sample and standard absorbances at 562 nm.

5   Use a standard curve (net absorbance at 562 nm versus mg protein) to determine the protein concentration in the samples.

Comments

The assay is sensitive to incubation temperature and duration (step 3). Since the BCA assay is dependent on the amino acid composition of the protein, it determines *relative* protein concentration of similar protein samples, not the absolute concentration of the protein samples.

One-dimensional and two-dimensional iso-electric focusing gel electrophoresis

One- and two-dimensional gel electrophoretic techniques [sodium dodecyl sulphate polyacrylamide gel electrophoresis (SDS-PAGE), and iso-electric focusing (IEF), respectively] have been critical tools in the understanding of protein expression in woody plant tissues. Specifically, these electrophoretic techniques have been successfully utilised to:

i)   identify proteins controlling bud dormancy in poplar (*Populus* spp.) (Jeknic and Chen, 1999);

ii) characterise proteins related to drought resistance in maritime pine (*Pinus pinaster* Ait.) (Costa *et al.*, 1998);

iii) localise flower inhibitor and promoter proteins in citrus trees (*Citrus* spp.) (He *et al.*, 1998);

iv) describe proteins inherent to flower buds within peach trees (*Prunus persica* L. Batsch cv. Redhaven) (Faye and LeFloch, 1997);

v) determine changes in protein expression across seasons in foliage of western (*Pinus monticola* D. Don) and eastern white (*Pinus strobus* L.) pines (Ekramoddoullah *et al.*, 1995; Ekramoddoullah and Taylor, 1996); and,

vi) characterise seed storage proteins and compare to VSPs in poplar (*Populus* spp.) (Beardmore *et al.*, 1996).

From such studies, it can be appreciated that both one- and two-dimensional electrophoretic techniques can be used effectively to address fundamental developmental and physiological questions in woody plant taxa. Iso-electric focusing combined with SDS-PAGE can resolve hundreds of proteins with iso-electric points ranging between 4–8. It often is difficult to resolve basic proteins using this technique and a method referred to as nonequilibrium pH gradient electrophoresis (NEPHGE) can be used to resolve basic proteins (e.g. proteins with iso-electric points up to pH 10). This method is described by O'Farrell *et al.* (1977).

The objective of the following section is to describe in detail effective procedures for conducting SDS-PAGE and IEF in woody plant tissues. These protocols are based on techniques originally developed by Laemmli (1970) using electrophoretic equipment available from Bio-Rad (and the associated instruction manuals), with specific modifications for woody plants.

One-dimensional gel electrophoresis

One-dimensional gel electrophoresis under denaturing conditions (e.g. in the presence of 0.1% (w/v) SDS) separates proteins based on their MW as they migrate through an SDS-PAGE gel. This procedure is relatively simple to conduct and has minimal time requirements thereby making it a good mechanism for preliminary analysis of the protein complement in woody plants.

Equipment needed:

Glass syringe, 50 ml total volume
Orbital shaker
Mini-Protean II Electrophoresis Cell
(Bio-Rad-including accessories: lower buffer chamber and lid, gaskets, inner cooling core, casting stand, combs, spacers, glass plates, sandwich clamp assemblies)
Pasteur pipette
Power supply (175 V capacity)
Petri dishes (60 × 15 mm)
Plastic bags (small, sealable)
Freezer at –20 °C
Vortex mixer
Microfuge tubes (1.5 or 2.0 ml capacity)

Chemicals needed:

Protein MW standards (Bio-Rad, Sigma etc.)
Ammonium persulphate (AMPS)
N,N,N',N'-tetramethylethylenediamine (TEMED)

Acrylamide stock solution I
15% separation gel solution
Destain solution
Running buffer
Separation gel buffer
SDS-reducing buffer
Ethanol
Deionised water
Spacer gel solution
Spacer gel buffer
Stain solution
Ice

Comments/notes:

Numerous hazardous chemicals are involved in both SDS-PAGE and IEF, it is recommended that proper safety guidelines for specific reagents and equipment be determined prior to conducting these procedures and strictly adhered to.

Often the acrylamide gel solutions are degased just prior to their being poured into the glass plate sandwiches. This is done to remove oxygen within the solution, which can inhibit polymerisation. However, if the gel solution is *gently* mixed there is no need for degasing.

I. Steps for preparation of polyacrylamide gels

Duration of one-dimensional SDS electrophoresis procedure: 1 day.

1   Clean the glass plates of the Mini-Protean II apparatus with 70% (v/v) ethanol, rinse with deionised water, and dry thoroughly.

2   Assemble glass plate sandwiches using 0.75 mm spacers in the clamp assembly and place securely in the casting stand.

3   Prepare the 15% separation gel solution, mix gently, and transfer to the middle of the glass plate sandwich using a Pasteur pipette.

4   Once the separation gel has been transferred to glass plate sandwich, gently add 1–2 ml of deionised water along its surface (taking care not to disturb the interface), and allow gel to polymerise for 1 h at RT.

5   Remove water from the surface of the separation gel by inverting casting stand containing glass plate sandwiches.

6   Prepare spacer gel solution, mix gently and transfer using a Pasteur pipette to surface of the polymerised separation gel. Immediately, place a sample 0.75 mm comb between the glass plates (in the spacer gel) ensuring the bottom of the comb rests slightly above the separation gel. Allow polymerisation for at least 1 h at RT.

7   Remove comb slowly from the spacer gel, and fill the sample wells with running buffer.

8   Remove the glass plate sandwiches containing the gels from the casting stand and transfer them to the cooling core (as described in Bio-Rad's Mini-Protean II Instruction Manual).

II. Preparation of protein samples for electrophoresis

1   Combine 20 μl of total protein extract (20 μl containing approx. 80 μg total protein) with 60 μl of SDS-reducing buffer in a microfuge tube on ice and vortex briefly (greater dilution by reducing buffer may be applied).

2   Transfer mixture to boiling water for 5 min (generally 5 min is optimal, however, occasionally a woody plant sample may require up to 15 min in boiling water and/or an increase of SDS concentration in the reducing buffer to 4%).

3   After boiling, immediately transfer polypeptide mixture onto ice.

4   These samples may now be utilised immediately, or stored frozen at –20 °C for several months.

III. Sample loading and electrophoresis

1   Using a fine-tipped syringe, load 10–30 μg of total protein sample into each lane of the spacer gel (10–30 μl of sample in SDS-buffer solution) and approx. 10 μg of protein standard in a separate lane.
    Note – loading of samples may be most effectively accomplished by placing the tip of the syringe at the base of the well and slowly releasing its contents entirely from this position.

2   After all samples are loaded, transfer cooling core containing the gels and protein samples into the buffer tank. Add running buffer to the tank until the unit is half full (also ensure inner core is filled with enough running buffer to cover the top of the gels). Place the lid on the buffer tank, ensuring accurate connections of anode and cathode, and electrophorese at 175 V for 45–50 min. The electrophoresis unit should be turned off when the samples have migrated through the gel. This is determined by monitoring the progression of the dye, bromophenol blue, as it migrates down through the gel.

IV. Gel disassembly

1   After electrophoresis, dispose of running buffer from unit, and remove gel sandwich from the cooling core. Subsequently, recover

the gels from the glass plates by gripping the spacers, and gently pull the two plates apart. At this point, the gel can be used in Western blotting, or stained.

V. Gel staining: visualisation of proteins

1   Transfer each gel to a disposable Petri dish containing enough stain solution to cover the entire gel. Gently agitate on an orbital shaker for 2–24 h.
2   Transfer gels to destain solution for 2–12 h. Repeated changes of destain solution may be required. Once adequate visualisation of protein bands becomes apparent, remove from solution and photograph.
3   Gels may be stored in destain solution for up to one week, or stored in sealed plastic bags with sterile water at 4 °C for several months.

Comments/notes

The location of the protein in the gel can be determined using a number of stains, such as Coomassie blue (as described herein), or silver stain. The detection limit for Coomassie blue is approx. 0.1–1 μg/band; for silver staining, sensitivity is 1–10 ng/band, and is therefore used to detect smaller amounts of protein. There are a number of commercial silver stain kits available (Bio-Rad, Sigma etc.)

Two-dimensional IEF gel electrophoresis

For two-dimensional gel electrophoresis, polypeptides are separated based on two independent properties, first on their iso-electric points and second on their mobility (i.e. MW) on an SDS-polyacrylamide gel. The following protocol has proven effective for separating polypeptides in numerous tree species including maple (*Acer saccharinum*), poplar (*Populus balsamifera* L., *P. tremuloides* Michx.), and larch (*Larix* sp.).

Note – due to the sensitivity of this method to variation in the protocol, it is critical that the same brand and quantity of reagents be utilised for all samples involved within one comparative analysis.

Equipment needed:

Acid-resistant gloves
Absorbant tissues
Microfuge tubes (1.5 or 2.0 ml capacity)
Mini-Protean II Tube Cell (for first dimension IEF) (Bio-Rad-including accessories: tube cell module, sample reservoirs, capillary tube connectors, capillary tubes, casting tube, tube gel ejector, lower buffer chamber and lid)
Oven at 60 °C
Power supply (750V capacity)
Parafilm
Syringe
Tape (or clamp apparatus)

Chemicals needed:

AMPS
Chromerge Cleaning Solution (Fisher, No. 57712)
TEMED
Acrylamide stock solution II

First dimension gel solution
First dimension sample buffer
First dimension sample overlay buffer
Lower chamber buffer
Upper chamber buffer
Phosphoric acid

I. Procedure for IEF polyacrylamide gel preparation (using Mini-Protean II Tube Cell)

Duration of procedure for one-dimensional and then two-dimensional electrophoresis: one long day

1   One day prior to electrophoresis, soak all capillary tubes with Chromerge Cleaning Solution for 1 h.

2   Remove capillary tubes from Chromerge using acid-resistant rubber gloves, rinse thoroughly in deionised water, and dry overnight in an oven at 60 °C.
3   On the day of electrophoresis, prepare the casting stand for the first dimension IEF gel solution by sealing the bottom securely with Parafilm. Subsequently, fill the casting stand with the dry capillary tubes and mount the

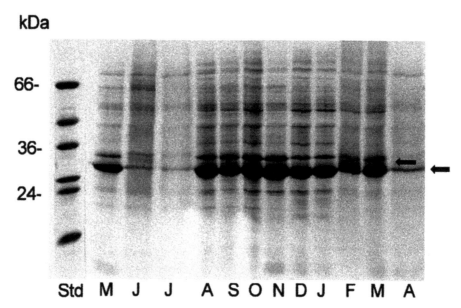

kDa

66-

36-

24-

Std M J J A S O N D J F M A

Figure 1 One-dimensional electrophoresis (SDS-PAGE) separation of total protein isolated from *Larix decidua* Mill.
bark on a monthly basis.
Months are labelled from the left starting with May through to April. Twenty micrograms of protein were loaded in
each lane. Gel was stained with Coomassie blue. Molecular weight standards (Std) were run on the far-left lane and are
in kilodaltons (kDa). Arrows on the right hand side mark vegetative storage proteins (VSPs). The major protein bands
are the 32- and 36-kDa VSPs, which are evident throughout the year. These proteins increased in prominence in August
and remained present in high concentrations throughout the winter. This VSP accumulation in the fall is likely due to
the degradation of leaf proteins and the subsequent transport of these N compounds to the bark, while in the spring,
catabolism of the storage proteins supports new growth.

unit upright, either by taping firmly to a wall
or by using a clamp apparatus.

5 Prepare first dimension gel solution. Mix
gently.

6 Immediately, and quickly, cast the gels into
capillary tubes by using either a syringe, or
the gel ejector (i.e. add a capillary tube to the
gel ejector). The solution will polymerise
within minutes.

7 Allow to polymerise for 1 h, remove capillary
tubes from casting unit, and clean with
absorbant tissues.

8 Attach a sample reservoir to the end of each
capillary tube using the tubing connectors.

9 Load a gel tube containing a sample reservoir
into each of the positions of the tube cell
module ensuring no leakage to the lower
chamber region.

10 Fill each capillary tube and its associated
gasket with upper chamber buffer using a
syringe. Place the syringe at the top of each
gel (within the capillary tube) and slowly

release enough buffer to fill the reservoir. The
gels are now ready to be loaded with protein
samples.

II. Sample preparation for IEF

1 Combine 30 μl of protein extract (containing
20–30 μg total protein, preferably in protein
extraction buffer) with at least 90 μl of first
dimension sample buffer in a fumehood.

2 Mix solutions gently by inverting the tubes
several times and let stand at RT for 5–10 min
(up to 30 min).

3 Samples are now ready to be loaded onto
gels, or may be stored frozen at –20 °C for
several months.
Note – it is imperative that the ratio of first
dimension sample buffer to protein extract be
at least 3:1 in order to maintain a sample
density that will ensure that the protein
sample remains below the overlay buffer.

III. Procedure for IEF sample loading and electrophoresis

1   Take up 10–30 μl of prepared protein sample into a fine-tipped syringe, and place the syringe tip at the base of the sample reservoir containing upper chamber buffer. Gently expel the entire sample from this position and sample should sink down to the gel surface.

2   After all the gels have been loaded, transfer 30 μl of first dimension sample overlay buffer into a clean syringe. Place the syringe tip at the uppermost point of the sample reservoir (within the upper chamber buffer located within the reservoir) and release the solution slowly.

3   Prior to transferring loaded gels to the buffer tank, fill a syringe with lower chamber buffer, and expel this solution at the bottom end of each capillary tube to minimise the potential for formation of air pockets at the bottom of the gels.

4   Transfer the tube cell module containing the loaded gels into the buffer tank. Fill the lower chamber with lower chamber buffer. Subsequently, using a pipette, fill the upper chamber with upper chamber buffer until the tops of all reservoirs are covered with solution.

5   Fill the corners of the upper chamber (i.e. the corners of the plastic bar across the middle of the upper chamber) with vacuum grease to prevent short-circuiting of the unit.

6   Place the lid on the electrophoretic unit and electrophorese at 500 V (no pre-focusing) for 30 min. Subsequently, increase the voltage to 750 V for 3 h (see Trouble-shooting for more detail pertaining to establishing the correct duration of electrophoresis).

7   Turn off the power source and remove the tube cell module from the lower tank. Remove the capillary tubes from the tube cell module.

8   Fill the gel ejector with running buffer and attach a capillary tube to its end. Gently eject gels from the capillary tubes and place in microfuge tubes (as described in the Mini-Protean II Tube Cell Instruction Manual). These gels may now be transferred to the second dimension immediately, or frozen at –20 °C.

Trouble-shooting

If gels slip out of capillary tubes during electrophoresis, this may indicate that the tubes have not been cleaned adequately. Prepare new tubes and soak longer in Chromerge Cleaning Solution. If the problem persists, it may be necessary to attach nylon mesh (e.g. a 1 × 1 cm square of pantyhose or stockings works very well) to the bottom end of each capillary tube using water proof tape to prevent gels from slipping out into the buffer tank during electrophoresis.

The electrophoresis time required for each protein to reach its iso-electric point varies. It may be necessary to first test a number of different electrophoresis times to ensure that all the proteins migrate to their iso-electric points. This is done by setting up an IEF run with approximately 10 IEF tubes, containing identical protein samples. Electrophorese the samples as described in the above method, but remove two tubes at the following times; 2, 3, 4, 5, 6 hr. Run these IEF gels in the second dimension as described below. By comparing the results of these gels one should be able to determine the optimum electrophoresis time.

Second-dimension electrophoresis of IEF gels

For the second dimensional separation, IEF gels undergo separation on a SDS-PAGE gel similar to the procedure described for one-dimensional gel electrophoresis (with some modifications).

Equipment needed:

Glass beaker (1 l, one per gel)
Mini-Protean II Electrophoresis Cell
(Bio-Rad, including accessories)
Power supply (175 V capacity)
Plastic bags (small, sealable)

Chemicals needed:

Acetic acid
Chromerge Cleaning Solution (Fisher,
No. C57712)
AMPS
Methanol

Weighing dish (large) Spatula	TEMED **Acrylamide stock solution I** **15% Separation gel solution** **Running buffer** **Separation gel buffer** **SDS-reducing buffer** Silver Stain Plus Kit (Bio-Rad, No. 161-0449)

IV. Procedure for second dimensional separation of IEF gels

1   Prepare a 15% polyacrylamide separation gel in a glass plate sandwich as described for one-dimensional gel electrophoresis above (but do not include a spacer gel).

2   Once the separation gel has polymerised, transfer the sandwich in its clamping assembly to the cooling core.

3   Fill the buffer chamber with running buffer until the surfaces of the gels are covered.

4   Thaw an IEF gel from a microfuge tube and place onto a large plastic disposable weighing dish containing enough SDS -reducing buffer to cover the gel. Incubate the gel in the buffer for approximately 10 min and then using a thin spatula straighten the gel onto the edge of the dish, and place directly onto the surface of the separation gel.

5   Electrophorese the unit at 175 V for 45 min in running buffer.

V. Visualisation of polypeptides

1   Remove gels from sandwich, and transfer to fixative solution (50% (v/v) methanol, 10% (v/v) acetic acid in high performance liquid chromatography (HPLC) grade water in a 1 l glass beaker) for a minimum of 20 min to prepare for silver staining.
    Note – ensure all glassware for staining has been soaked in Chromerge Cleaning Solution and rinsed thoroughly in HPLC grade water prior to use.

2   After fixation of polypeptides to the gels, rinse the gels for 3 × 10 min in HPLC-purified water. Rinse times longer than 10 min may be needed if discrete polypeptides are not detected on the gels.

3   While rinsing the gels, prepare reagents for silver staining as indicated in Bio-Rad's Silver Stain Plus Kit.

Note – it is essential that HPLC grade water be utilised in all silver stain solutions, including the fixative, developer, and stop reaction solutions.

4   Add the appropriate quantity of silver stain and development solution to the rinsed gel. Once the desired intensity is observed, remove all silver stain solution from the gel and add 10% acetic acid to stop the reaction.

5   Photograph gels. These gels may be stored in stop solution for several weeks, or transferred to clear plastic seal-tight bags with 1 ml of sterile water and stored at 4 °C for several months.

Purification of proteins for antibody production and methods for producing antibodies to the purified proteins

Antibody preparation from proteins of woody tissues is no different from that from herbaceous plants. For information pertaining to protein (antigen) purification for antibody production, the reader is referred to the books *Short Protocols in Molecular Biology* edited by Ausubel *et al.* (1992), *Current Protocols in Molecular Biology* edited by Ausubel *et al.* (1997) and *Using Antibodies* by Harlow and Lane (1999). These texts are also an excellent source of information concerning preparation, purification and production of monoclonal and polyclonal antibodies.

Western blotting and immunolocalisation of proteins

Western blotting

Western blotting or immunoblotting combines electrophoresis with immunodetection, where a specific antigen(s) (i.e. protein of interest) separated on a protein gel, is recognised by polyclonal or monoclonal antibodies. However, similarity of

MW, as determined by 1-D SDS-PAGE, does not confirm whether a protein is identical, or related, to another. But, Western blotting does allow one to determine whether one protein is related to another based on similar epitopes/antigenicity to the protein against which the antibody was raised. The use of Western blotting is advantageous when screening a large number of samples for presence of specific protein(s). With woody tissues this has been used to examine seasonal variation in VSP accumulation within the bark of poplar (*Populus* ssp.) (Coleman *et al.*, 1991; Langheinrich and Tischner, 1991; van Cleve and Apel, 1993), clonal variation in VSP accumulation in poplar (*Populus* spp.) bark (Langheinrich, 1993), variation within a genus or family (Wetzel 1990; Harms and Sauter, 1994) and tissue specificity of VSP accumulation in poplar (*Populus grandidentata*) (Beardmore *et al.*, 1996).

Western blotting of protein is divided into sections:

I)   One- or two-dimensional SDS gel electrophoresis of antigens (as described above).

II) Electroblotting of antigens (protein) onto nitro-cellulose. Proteins from the gel are transferred onto nitro-cellulose membrane using the submerged, or 'wet' method of transfer.

III) Immunoblotting, in which the location of the antigens is determined by probing the blot with the primary antibody (either a polyclonal or monoclonal antibody). To detect the primary antibody, a secondary antibody is attached to the primary antibody. The secondary antibody (anti-IgG antibody (e.g. rabbit anti-goat antibody)) is conjugated to an enzyme such as alkaline phosphatase (AP) or horseradish peroxidase. This procedure uses the AP-anti-Ig as the secondary antibody. For detection of the antigen, the AP catalyses a reaction which forms a coloured precipitate using a 5-bromo-4-chloro-3-indolyl phosphate (BCIP)/nitroblue tetrazolium (NBT) reagent, at the antigen-antibody complex site on the blot.

The procedure described is modified from that of Towbin *et al.* (1979). For a full description of primary, secondary antibodies and detection systems, the book *Using Antibodies*, edited by Harlow and Lane (1999), is recommended.

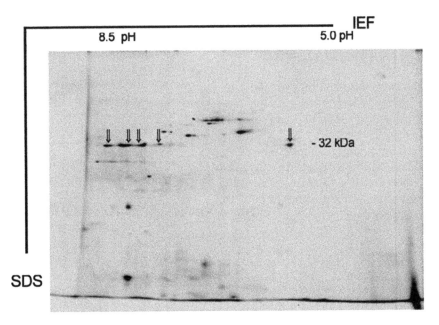

Figure 2 Two-dimensional electrophoretic (IEF and SDS-PAGE) separation of total polypeptides isolated from roots of *Populus grandidentata*.
Twenty micrograms of protein were separated in the first dimension by IEF using a pH range of 8.5–5.0, followed by SDS-PAGE in the second dimension. Gel was silver-stained. The arrows show the 32-kDa vegetative storage protein (VSP). There appear to be five basic isoforms of the 32-kDa VSP.

Equipment needed:	*Chemicals needed:*
Mini Protean II Cell (Bio-Rad, for gel size 7.5 × 10 cm)	Alkaline Phosphatase Conjugate Substrate Kit (Bio-Rad, No. 170-6432)
Mini Trans-Blot Cell (Bio-Rad) (also referred to as a submerged Western Transfer apparatus) for gel size 7.5 × 10 cm (NB. GIBCO-BRL/ Life Technologies also produced vertical gel electrophoresis systems which can be used in these procedures)	**Antibodies**
Orbital shaker	MgCl₂
Nitro-cellulose transfer membrane (0.45 mm pore size) (Bio-Rad, No. 162-0115)	Tris-base
Absorbant filter paper (Whatman 3 mm chromatography paper)	Na₂CO₃
Petri dishes (60 × 15 mm)	**Antibody buffer**
Forceps	**Blocking solution**
	Western transfer buffer
	Tris-buffered saline (TBS)
	Tris-buffered saline with Tween 20 (TTBS)
	Stain solution

Duration of procedure: over 2 days or 1 long day.

I. SDS-PAGE

1   Run samples on a mini SDS-polyacrylamide gel (as described above). The quantity of protein loaded per well should range from 10–30 μg.

2   After electrophoresis of the protein, prepare the gel by removing the stacking gel and cut off one corner of the gel to aid in identifying the lanes of protein.

Note – reducing agents may be used (i.e. 2-mecaptoethanol or dithiothreitol) in the protein extract. However, it should be noted that the epitopes of some antibodies might be destroyed by the use of reductants.

II. Electroblotting

1   Equilibrate the gels in Western transfer buffer for 30 min in a Petri dish. While the gel is equilibrating, cut the nitro-cellulose and 2 pieces of filter paper to a size that is just larger than the size of the gel (approx. 8 × 11 cm). Wet the nitro-cellulose membrane and filter papers by slowly sliding them in at a 45 ° angle into the Western transfer buffer and allow to soak for 15–30 min.

Note – throughout these procedures, the nitro-cellulose membrane should only be handled with forceps to avoid contamination that can be detected by staining of the blot.

2   Fill the Mini Trans-Blot tank half way with Western transfer buffer.

3   Assemble a sandwich in the gel cassette clamping system for the Trans-Blot Cell starting with the saturated fibre pad, followed by the filter paper, the gel, the nitro-cellulose membrane, filter paper again and the fibre pad (see Figure 3). Try to ensure that there are no bubbles between the nitro-cellulose and the gel. However, if bubbles are present, keep the gel and nitro-cellulose submerged in the buffer while rolling a glass test tube over the nitro-cellulose that is on top of the gel. This will gently push out any bubbles that are present. Then place the filter paper and fibre pad on top and close the gel cassette clamping system.

4   Insert the gel cassette clamping system vertically in the Trans-Blot tank containing the Western transfer buffer. The black plate of the gel cassette must be placed towards the negative cathode side of the Trans-Blot Cell to achieve transfer of proteins onto the nitro-

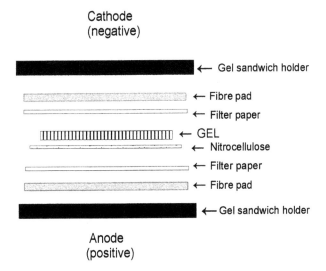

Figure 3   Sandwich assembly for electroblotting using the Trans Blot Cell.

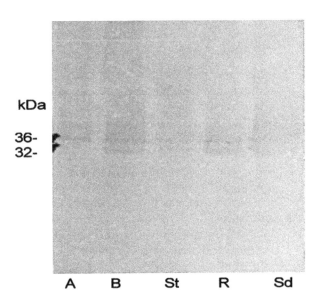

Figure 4   Western blot of one-dimensional electrophoresis (SDS-PAGE) separation of total proteins isolated from *Populus alba* plantlets and *P. grandidentata* seeds. A, apical leaves; B, basal leaves; St, stem; R, root; Sd, seed. Twenty micrograms of protein were loaded in each lane. The protein was transferred to nitro-cellulose and probed with the primary antibody specific to the 32-kDa *Salix* vegetative storage protein (VSP) (diluted 1:1000 (v/v) in the blocking solution). This antibody was then conjugated to the secondary antibody, a goat anti-rabbit IgG-alkaline phosphatase conjugate (diluted 1:3333 in TBS with 2% BSA). The VSP antibody was visualised using Alkaline Phosphatase Conjugate Substrate Kit (Bio-Rad) according to the manufacturer's instructions. This blot shows that the VSPs are present in all plantlet tissue examined and that the seed storage proteins also cross-reacted to the antibody, suggesting that these proteins exhibit some amino acid sequence similarity to the VSP.

cellulose membrane. Fill the Trans-Blot tank with Western transfer buffer to just above the level of the top row of circles on the gel holder cassette. In all transfers, the temperature will rise during the run. Therefore, it is recommended that the Trans-Blot Cell is placed on top of a container filled with ice, to keep the unit cool during the run. Alternatively, run the Cell in a cold room.

5   Connect the power supply and perform the transfer at constant 100 V for 60–90 min.

6   Stain the gel with Coomassie blue (as described above) to verify that the transfer of proteins to nitro-cellulose has taken place. Note – if there are a large number of protein bands remaining in the gel, and they are absent from the nitro-cellulose membrane, the transfer time may be increased. Pre-stained protein standards (i.e. Rainbow coloured protein MW markers, Amersham Science) may also be run on the gel; if protein transfer is successful, the standards should be present on the nitro-cellulose.

Trouble-shooting

The transfer of proteins can be slow if:

1)   the % acrylamide and the cross-linker (N, N'-methylene-bis-acrylamide) is high in the gel (e.g. >15% acrylamide gel is used);

2)   a gel thicker than 75 mm is used; and,

3)   proteins of interest have a high MW (i.e. greater than 60 kDa). If this is the case, we recommend that the transfer time be increased.

Also, it is recommended that a positive control be included on the gel (i.e. sample that is known to contain the antigen), and that a gel containing the protein samples is run and blotted to pre-immune primary antibody (for polyclonal antibodies serum (when the antibody is raised in mammals)) or egg yolks (for antibody raised in chickens) is taken from the animal prior to immunisation). A comparison of the blots of the pre-immune antibody with the primary antibody will identify background bands.

III. Immunoblotting: detection of antigen proteins on nitro-cellulose

1   After transfer of the proteins, carefully remove the nitro-cellulose from the sandwich and place it in a Petri dish containing either 50 ml of 3% (w/v) milk powder or 3% (w/v) gelatin blocking solution in TBS. Incubate either overnight at 4 °C, or for 60 min. at RT. This step blocks all unoccupied protein-binding sites on the nitro-cellulose membrane with a protein which will not react with the antibodies.

2   Remove nitro-cellulose from the blocking solution and wash with TTBS for 2 × 5 min. with gentle agitation at RT.

3   Transfer nitro-cellulose to a clean Petri dish containing the primary antibody in 50 ml of 3% (w/v) milk in TBS solution and incubate at RT for 60–90 min. with gentle agitation. The primary antibody in 3% (w/v) milk in TBS can be reused if stored at 4 °C.
    Note – the final primary antibody concentration in the TBS solution should be between 1–50 μg/ml. If the concentration of the antibody is unknown, try several dilutions in order to determine the correct range. For polyclonal antibodies, 1:100–1:1000 dilutions are recommended; for monoclonal antibodies, 1:10–1:1000 dilutions are suggested for the AP detection system.
    Note – denaturing electrophoresis is used in this procedure to separate the proteins. Therefore, the antibodies must recognise the denatured antigen. When polyclonal antibodies are used, this is usually not a problem, but some monoclonal antibodies do not recognise denatured antigens.

4   Wash the membrane with TTBS for 2 × 5 min.

5   Transfer washed membrane to second antibody in 50 μl 3% (w/v) milk TBS solution, incubate for 60 min at RT. Commercially available secondary antibodies (i.e. enzyme labelled secondary antibodies such as horseradish peroxidase-anti-Ig conjugate) are used in the TBS solution at concentrations of 0.5–5 μg/ml, which corresponds to dilutions ranging between 1:200 and 1:2000 of the commercial antibody.

6   Wash with TTBS for 2 × 5 min. and rinse with TBS to remove Tween 20 from the membrane.

7   Perform alkaline phosphatase detection. This step is conducted using a commercially available kit, such as Bio-Rad's Alkaline Phosphatase Conjugate Substrate Kit. Prepare the colour development solution immediately prior to use (according to the manufacturer's instructions), immerse the membrane in the development solution. Develop to the intensity required. Stop colour development by immersing membrane in distilled water.

Trouble-shooting

One of the most common problems is high background staining on the nitro-cellulose. This can be overcome in a variety of ways:

1)   the blot was insufficiently blocked with milk or gelatin, resulting in non-specific binding of the antibody to various antigens. Either increase the blotting time with milk or gelatin, or use another blocking agent (i.e. if you used milk, try gelatin);

2)   decrease the quantity of primary antibody used; and,

3)   decrease the time that the primary and secondary antibodies are incubated with the blot.

Immunolocalisation of proteins

Immunolocalisation commonly involves the use of a specific, unlabeled primary antibody to locate the antigen in the cell, followed by a secondary antibody, coupled to an easily identifiable reagent, which attaches to the primary antibody, thereby revealing its location. This is termed indirect detection. Direct detection is also possible and involves the use of antibodies specific to the antigens that have been purified and labeled with an easily detectable label. Indirect detection methods are more versatile and cost-effective since the secondary antibodies are commercially available and work for a variety of primary antibodies. This saves time and expense that is needed in the purification and labeling of every primary antibody used in direct detection. However, in some cases direct labeling may be preferred, as it can produce clearer signals and may be very useful when trying to compare location of several antigens in the same cell (Harlow and Lane, 1999).

Some recent studies involving immunolocalisation in woody tissues include the localisation of:

vegetative storage proteins (Sauter and van Cleve, 1990; Stepien and Martin, 1992; Wetzel and Greenwood, 1991; Lawrence et el. 1997); a dehydrin in peach (Wisniewski *et al.*, 1999); pectins in active cambium (Chaffey *et al.*, 1997b), dormant cambium (Chaffey *et al.*, 1998), differentiating xylem (Chaffey *et al.*, 1997a) and phloem (Chaffey *et al.*, 2000) of *Aesculus hippocastanum,* and poplar cambium (Guglielmino *et al.*, 1997; see also the chapter by Micheli *et al.*, in this volume); cytoskeletal proteins in angiosperm (see chapter by Chaffey in this volume) and gymnosperm (see chapter by Funada in this volume) trees; enzymes of lignification (see chapter by Samaj and Boudet in this volume); lignins (Ruel *et al.*, 1999), and xyloglucans (see chapter by Itoh in this volume).

Immunolocalisation studies are often the culmination of much work which has gone into extraction, purification and immunoblotting of the protein of interest. Once an antibody has been shown to satisfactorily recognise the protein in an immunoblot, it is a good candidate for immunolocalisation studies. However, on the immunoblot, the antibody reacts to a fully denatured protein, whereas in immunolocalisation, the antigens are much closer to their native state. Therefore, one cannot guarantee that an antibody which works well on an immunoblot will also do well in immunolocalisation.

The degree of success of immunolocalisation depends on the local antigen concentration, the accessibility of epitopes, and the type of detection method being employed. Careful, systematic controls are required for proper interpretations of results, and generally include incubation of the sections in pre-immune serum or solutions, as well as treatment with the second antibody alone (see also chapters by Chaffey, Funada, Micheli *et al.*, and Samaj and Boudet in this volume).

Overall, immunolabelling involves a series of washes and incubations over several days. Although specific recipes are usually required for immunolabelling particular cell components (see e.g. chapters by Chaffey, Funada, Micheli *et al.*, and Samaj and Boudet in this volume), the general procedure remains the same. Sections are first treated with a blocking solution, containing bovine serum albumen, gelatin, Carnation (or any non-fat) milk or other general protein, which will block epitopes which might otherwise cross-react with the primary antibody. Sections are then incubated with the primary antibody solution (or pre-

immune solution for controls), washed, and incubated with the secondary antibody solution. Secondary antibodies are usually conjugated to a fluorochrome, an easily detectable enzyme or, with increasing frequency over the past years, to colloidal gold which, for light microscopy studies is silver-enhanced. For a detailed discussion of the various options in immunolocalisation, the reader is referred to Chapter 6 in Harlow and Lane (1999).

Most of the work on immunolocalisation of non-structural proteins in perennial tissues has been conducted on angiosperms, in particular, poplar (e.g. Sauter *et al.*, 1988; Wetzel *et al.*, 1989) and peach (Wisniewski *et al.*, 1999). Generally such studies have been directed at finding out how the proteins are potentially related to nitrogen storage, dormancy and/or freezing tolerance. Tissues considered have been xylem and phloem parenchyma cells of stems and twigs; little work has been conducted on root tissue or bud tissues, presumably because of technical difficulties in sampling and extracting proteins. Despite the advent of many new analytical techniques, interference from phenolics and the inherent difficulties in sampling underground portions of the tree leave much room for future studies on the role of various proteins in perennial tissues.

I. Tissue sampling

When studying the role of proteins in perennial tissues, for many studies, year-round sampling is preferred, or at least, sampling during the growing season and dormant season to show fluctuation of the protein. Unfortunately, although this in itself is time-consuming, alternative methods such as growth room experiments and tissue-culture-grown plantlets can easily result in artefacts being introduced.

Generally, the same trees sampled for protein work should be sampled for microscopy work. For statistical purposes, several individual trees should be sampled and, from each tree, several samples taken (locations to be specified in advance). Because of the effort, time and expense involved, it is impractical to have large sampling sizes. However, a minimum of n = 3 individuals from any experiment, and hopefully at least one repetition of the experiment can avoid artefacts and misinterpretation of results. For our work on storage proteins in *Salix* (Wetzel *et al.*, 1991), we chose three indi-

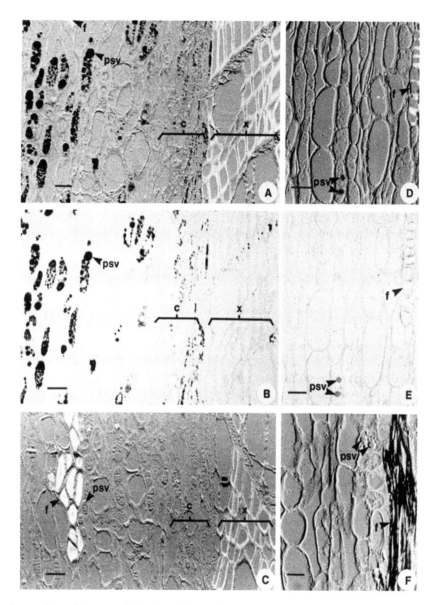

Figure 5 Silver-enhanced, gold immunolabelling of the *Salix* vegetative storage protein (VSP).
A–C   Radial sections of overwintering *Salix microstachya* shoots, illustrating from left to right the phloem parenchyma, cambial region, ray parenchyma and xylem. Protein storage vacuoles (psv) are found in phloem parenchyma, ray parenchyma and cambial cells. A and B, Section treated with anti-*Salix* VSP IgG seen under Nomarski (A) and bright-field (B) optics. Protein-storage vacuoles (psv) in the phloem parenchyma cells are densely labeled, indicating the presence of the VSP. Note in B that small protein-storage vacuoles in the cambial cells and xylem ray parenchyma are also labeled. C, As in A and B but treated with pre-immune IgG as control and viewed under Nomarski optics. No labeling is seen. D–F, Radial sections of *Salix microstachya* shoots harvested during the summer, concentrating on the phloem parenchyma. D and E, Section treated with anti-*Salix* VSP IgG seen under Nomarski (D) and bright-field (E) optics. A weak positive signal, indicating the presence of the VSP, is seen over occasional protein-storage vacuoles; f, fibre. F, As in D and E but treated with pre-immune IgG as control and viewed under Nomarski optics. All sections were silver-enhanced after IgG treatment. Key: c, cambial zone; f, fibre; x, xylem tissue. Bar, 20 μm.
Photograph courtesy of Dr S. Wetzel, and *Canadian Journal of Botany* (for actual paper see Wetzel *et al.*, 1989).

vidual trees at each sampling time. From these trees, three different branches (previous year's growth) were excised with clipping shears from the lower third of the crown in the south facing direction (this location was chosen for convenience in sampling).

Branches were cut into 1 cm lengths and quartered immediately (in the field) using a razor blade. Quartered samples were placed into pre-labeled vials containing fixative and then transferred to the laboratory for processing. In this way, degradation of proteins was minimised. To show seasonal variation in protein accumulation and mobilisation, sampling was done over a one-year period. In this way, it could be demonstrated that protein mobilisation occurred in the early summer months, followed by few or no storage proteins in the branches during the summer and, finally, accumulation in the winter months (Wetzel *et al.*, 1989) .

II. Tissue fixation, embedding and sectioning

The major differences between working with woody tissues as compared to non-woody tissues for immunolocalisation studies come into play during the first two steps, namely tissue preparation (fixation and embedding), and sectioning for either light microscopy studies or electron microscopy studies. Detailed procedures for this are considered in the chapters by Chaffey, Funada, Micheli *et al.*, and Samaj and Boudet in this volume. The fixation procedure must come as close as possible to retaining the antigen in its proper place, allowing access of the antibody to the antigen, maintaining the antigen in a form that is recognisable by the antibody, and preserving overall cell structure. A wide range of fixatives and embedding procedures are in common use, a decision on which one depends on a number of factors (see above-mentioned chapters).

Briefly, for the procedure described here, material was fixed in 3% (v/v) glutaraldehyde, 2% (v/v) acrolein and 1% caffeine (w/v) (Mueller and Greenwood, 1978) in 0.05 sodium cacodylate buffer, pH 7.4 for a minimum of 24 hours at 4 °C. Samples were rinsed three times for 2 h in cacodylate buffer, then twice in distilled water. They were then dehydrated through a graded water-ethanol series (50, 70, 80, 90, 95 and 100%, 2h each step), then infused with 25, 50, 75% (v/v) glycol-methacrylate (glycol-methacrylate made up as: 95% 2-hydroxyethyl methacrylate, 5% (v/v) poly-ethylene glycol 400, 0.3% (w/v) 2,2'-azo-bis-isobu-tyronitrile) (after Feder and O'Brien, 1968) and left in glycol-methacrylate overnight. Tissue segments were placed in small plastic capsules with fresh resin and polymerised at 65 °C overnight. Sections (radial and longitudinal), 0.75–2.0 μm were cut with a Sorvall JB-4 microtome using dry glass knives and mounted on glass slides by passing the slide over the flame of a bunsen burner, thereby securing the section to the slide.

III. Antibody preparation

As mentioned above, preparation of antibodies against the protein of interest is the same for trees as it is for any other plant. In work with *Salix* vegetative storage protein, our antibodies were produced in chickens, according to the procedure of Jensius *et al.* (1981). This technique has the advantage that the antibodies can be extracted from the eggs rather than serum. Production of chicken antibodies can also be carried out commercially, by sending purified proteins to companies such as Gallus Immunotech in Canada.

IV. Antibody binding

Once sections have been prepared, they are incubated in the antibody solution. Since more antibody leads to better labelling, but also stronger background labelling, the challenge in this step is to find the compromise levels of antibody concentrations that give strong signals, but which do not significantly promote non-specific staining. This can be determined by doing serial dilutions of each antibody preparation. In general, a good starting point for the primary antibody concentration is between 0.1 and 10 μg/ml. Dilutions of 1:10, 1:100, 1:1000, etc. can then be used to find the optimal concentrations.

In work on *Salix*, after considerable trial and error, we used fairly dilute antibody solutions (0.1 μg/ml). The vegetative protein is abundant in large amounts during the winter and labels well.

V. Detection

Detection systems include fluorescence, enzyme and gold. Each system has advantages and disadvantages. The steps below detail the procedure when using colloidal gold; other chapters in this volume consider fluorescence-labelling (e.g. Chaffey, Funada, Micheli *et al.*). Colloidal gold particles conjugated with a wide range of anti-immunoglobulin antibodies are available commercially (e.g. Bio-Rad, Amersham). Colloidal-gold can also be conjugated to secondary antibodies in-house (for detailed procedures, see Slot and Geuze,

1985). Colloidal-gold should be washed before use to remove free protein. Silver-enhancement of the gold label makes this technique very useful for light microscopy. Fortunately, silver enhancement kits are available from a variety of manufacturers (Bio-Rad, Amersham).

Immunolocalisation for light microscopy

Equipment needed:	Chemicals needed:
Sections of material prepared for light microscopy and dried onto gelatin-coated slides	Colloidal Gold Detection Kit (Bio-Rad, No. 170-6517)
Glass Petri dishes	**Rinse solution**
Filter paper	**Blocking solution**
25 μl automatic pipettor	**Antibody and control solutions**
Pasteur pipettes	**DPX** mounting medium
Timer	
Coplin jars	
Light microscope with differential interference contrast (DIC, Nomarski) optics	
Coverslips	

Duration of procedure: approx. 1 day.
Note – for incubation steps, place small drops of the relevant solution onto the sections on the slide, and incubate in a moist chamber (sealed Petri dish with damp filter paper at the bottom). The complete procedure can be done in the laboratory at RT.

1   Stand slide in rinse solution in Coplin jar for 10 min. This step pre-wets the sections; if necessary, slides can also be stored overnight in this solution.
2   Shake excess rinse solution from slide.
3   Place a drop of the blocking solution on each section, incubate for 10 min.
4   Shake off excess solution and replace slide in rinse solution in Coplin jar (from step 1 above), leave for 10 min.
5   Remove excess rinse solution, place 1 drop (25 μl) of appropriately diluted primary anti-body – or control – solution on each section, incubate for 1 h.
6   Wash off antibody solution with rinse solution using a Pasteur pipette.
7   Then stand slide in Coplin jar containing rinse solution, changing solution three times in 15 min.
8   Shake off rinse solution, add one drop (25 μl) of appropriately diluted secondary antibody to each section, and incubate for 30 min.
9   Wash off secondary antibody with rinse solution using a Pasteur pipette.
10 Place slide in Coplin jar with clean rinse solution. Let stand for 10 min, changing solution 2 times during this time.
11 Wash slide three times with distilled, deionised water, 1 min. per wash.
12 Air-dry slides.
13 Perform silver-enhancement according to kit manfacturer's instructions.
14 Mount coverslips in DPX.
15 Observe sections using differential interference contrast (DIC, or Nomarski) and bright field optics.

SOLUTIONS FOR ELECTROPHORESIS AND IMMUNOBLOTTING

Note   [*], can store at RT
        [**], store at 4 °C

Acrylamide stock solution I [**]
| 29.2% (w/v) acrylamide | 29.2 g |
| 0.8% (w/v) bis-acrylamide | 0.8 g |

Fill to 100 ml with HPLC grade water. Store in the dark for a maximum of 30 days.

Acrylamide stock solution II [**]
| 28.4% (w/v) acrylamide | 28.4 g |
| 1.6% (w/v) bis-acrylamide | 1.6 g |

Fill to 100 ml with HPLC grade water. Store in the dark for a maximum of 30 days.
Note – a higher concentration of bis-acrylamide may be applied to yield slightly different resolution for individual sample types (up to 5.5% (w/v) bis-acrylamide has been effective for some samples).

Antibodies: [**]
1.   Primary antibody: antibody directed against the protein of interest.

2. Secondary antibody: anti-immunoglobin G (IgG) conjugate. The type of IgG depends on what animal the primary antibody was raised in (i.e., if chickens were used, then the secondary antibody conjugate would be chicken IgG) and the detection system used (i.e. alkaline phosphatase (AP) or horse-radish peroxidase). In this procedure, the secondary antibody anti-IgG is conjugated to AP (Bio-Rad, or Promega), abbreviated as AP-anti-IgG.

Antibody buffer
1% (w/v) gelatin in **TTBS**, warm to dissolve gelatin, cool before use.

Basic Protein Extraction buffer[1]**
25 mM KH_2PO_4, 1.0 M NaCl, deionised water, pH 7.0
[1], If the presence of phenolics is a concern, the following may be added to the extraction buffer:
0.05 M ascorbic acid, and 2% (w/v) soluble PVP or Amberlites according to the manufacturer's instructions (e.g. quantity can range from 1–2 gm hydrated weight of Amberlite/gm of fresh tissue).
[1], If problems with protease degradation are of concern add 1 mM PMSF and 10 μM leupeptin or use a protease inhibitor cocktail (i.e. Roche Molecular Biochemical) according to the manufacturer's instructions and add 1 mM EDTA.

Blocking solution
3% (w/v) gelatin or 3% (w/v) Carnation non-fat powdered milk in TBS.

Destain solution*
40% (v/v) methanol	400 ml
10% (v/v) acetic acid	100 ml

Make up to 1 l with deionised water.

First dimension gel solution
9.2 M urea	5.5 g

1.33 ml of **Acrylamide stock II**
2% (v/v) Triton X-100	2 ml of 10% Triton-X 100
ampholyte 3/10.5	100 μl
ampholyte 4/6	400 μl
ampholyte 6/8	500 μl
ampholyte 8/10.5	400 μl
0.01% (w/v) AMPS	10 μl 10% of freshly made AMPS
0.25% (v/v) TEMED	25 μl TEMED

Dissolve urea in approximately 4 ml HPLC grade water, add other ingredients, except AMPS and TEMED. It can be difficult to dissolve the urea and this solution can be warmed in a water bath (≤ 45 °C) with gentle swirling. Once the urea is dissolved bring the final volume of the solution up to 10 ml with HPLC grade water. Add AMPS and TEMED, gently swirl and cast immediately.
Note–due to the high concentration of TEMED, the gel will polymerise very quickly, and casting should occur as soon as possible.

First dimension sample buffer
9.5 M urea	5.7 g
2.0% (v/v) Triton-X 100	2.0 ml 10% Triton X-100
5% (v/v) 2-mercaptoethanol	0.5 ml
ampholyte 3/10.5	100 μl
ampholyte 4/6	400 μl
ampholyte 6/8	500 μl
ampholyte 8/10.5	400 μl

Dissolve urea with approximately 4 ml HPLC grade water. It can be difficult to dissolve the urea and this solution can be warmed in a water bath (≤ 45 °C) with gentle swirling. Once the urea is dissolved, add remaining ingredients and bring the final volume of the solution up to 10 ml with HPLC grade water. Store 0.5 ml aliquots at –20 °C.

First dimension sample overlay buffer
4.5 M urea	5.41 g
ampholyte 3/10.5	100 μl
ampholyte 4/6	400 μl
ampholyte 6/8	500 μl
ampholyte 8/10.5	400 μl
bromophenol blue	500 μl of 0.05% (w/v) bromophenol blue solution

Dissolve urea in approximately 8 ml HPLC grade water. It can be difficult to dissolve the urea and this solution can be warmed in a water bath (≤ 45 °C) with gentle swirling. Once the urea is dissolved add remaining ingredients, and dilute to 20 ml with HPLC grade water. Aliquot solution into 0.5 ml units and freeze at –20 °C.

Lower chamber buffer
1.36 ml H_3PO_4 (85% solution in water) per 2 l deionised water.

Running buffer*

1.5% (w/v) Tris base, pH 8.3	15 g
7.2% (w/v) Glycine	72 g
0.5% (w/v) SDS	5 g

Fill to 1 l with deionised water.

15% Separation gel solution (two mini-gels)
7.5 ml **Acrylamide stock I**
5 ml **Separation gel buffer**
2.5 ml HPLC grade water
2.5 ml of freshly made 0.24% (w/v) AMPS
2.5 ml 0.3% (v/v) TEMED

Combine acrylamide stock, separation gel buffer, and HPLC grade water. Swirl gently. Add AMPS and TEMED immediately prior to use.

Separation gel buffer**

| 1.5M Tris-HCl, pH 8.8 | 23.6 g |
| 0.4% (w/v) SDS | 0.4 g |

Fill to 100 ml with HPLC grade water

SDS-reducing buffer**

0.0625 M Tris-HCl, pH 6.8	0.99 g
10% (v/v) glycerol	10 ml
2% (w/v) SDS	2 g
5% (v/v) 2-mercaptoethanol	5 ml
0.05% (w/v) bromophenol blue	0.05 g

Fill to 100 ml with HPLC grade water. Store in fume hood due to 2-mercaptoethanol.

Spacer gel buffer*

| 0.5M Tris-HCl pH 6.8 | 7.9 g |
| 0.4% (w/v) SDS | 0.4 g |

Fill to 100 ml with HPLC grade water

Spacer gel solution (two mini-gels)
1 ml **Acrylamide stock I**
2.5 ml **Spacer gel buffer**
1.25 ml of freshly made 0.24% (w/v) AMPS
4 ml HPLC grade water
1.25 ml of 2.0% (v/v) TEMED

Combine acrylamide stock, spacer gel buffer and HPLC grade water. Swirl gently. Add AMPS and TEMED immediately prior to use.

Stain solution*

0.1% (w/v) Coomassie Brilliant Blue R-250	1 g
40% (v/v) methanol	400 ml
10% (v/v) acetic acid	100 ml

Fill to 1 l with deionised water

Tris buffered saline (TBS)**
20 mM Tris-HCl, 500 mM NaCl, pH 7.5

Tris buffered saline with Tween 20 (TTBS)**
0.5 ml Tween 20 to 1 l of TBS

Upper chamber buffer
1.6 g of NaOH per 1 l of deionised water.

Western transfer buffer
20% (v/v) methanol, 25 mM Tris-HCl, 192 mM glycine, pH 8.3

SOLUTIONS FOR IMMUNOLOCALISATION

Antibodies

Primary antibody solution (keep on ice)
Concentration of 0.01 μg/ml in **rinse solution** (e.g. chicken-anti-VSP, made in-house). For your own material, it will be necessary to experiment to find the optimal antibody concentration. We recommend the range 0.1–10 μg/ml as a good starting point.

Control solution (keep on ice)
Made as for primary antibody, except containing pre-immune serum instead of antibody.

Gold-conjugated secondary antibody solution (keep on ice)
Concentration of 1 μg/ml in **rinse solution**. A good starting point for testing secondary antibodies is 1–50 μg/ml (e.g. 12 nm -colloidal gold (Amersham, No. RPN400) tagged to rabbit-anti-chicken IgG, Bio Rad, No. 1706460). The procedure for conjugation of colloidal gold to antibody is described in Slot and Geuze (1985). A range of commercially available colloidal gold secondary antibodies can be obtained (e.g. Sigma, Amersham), thereby eliminating one step in the lab.

Blocking solution
3XPBS-Tween-glycine-BSA:

PBS (phosphate-buffered saline, Sigma, No. P4417) 3x concentrated from manufacturer's instructions, pH 7.4
0.2% (v/v) polyoxyethylenesorbitan monolaureate (Tween 20)
0.2% (w/v) glycine
2% (w/v) bovine serum albumen (BSA, Sigma, Fraction V, No. A9647)

Rinse solution
3XPBS-Tween-glycine:

PBS (3x concentrated)
0.2% (v/v) Tween 20
0.2% (w/v) glycine

DPX (Electron Microscopy Sciences, No. 13510) Commercial permanent mounting medium containing distyrene (10 g), dibutyl phthalate (5 ml) and xylene (35 ml).

APPLICATIONS

Significant contributions to woody and perennial plant physiology have been made with the use of protein characterisation, analytical techniques and procedures. These contributions have ranged from identification and characterisation of, vegetative storage proteins, which function as temporary stores of carbon and nitrogen (Sauter *et al.*, 1988; Wetzel *et al.*, 1989); proteins related to endodormancy (e.g. dehydrin-like proteins, Arora and Wisniewski, 1994; Artlip *et al.*, 1997; Golan-Goldhirsh, 1998), lignification (Christensen *et al.*, 1998), disease resistance; and heavy metal-induced proteins (Przymusinski and Gwozodz, 1999). In much of this work, polyacrylamide gel electrophoresis has resulted in the identification of the protein(s) of interest. Antibodies are then often raised to the protein(s) allowing for the screening of a larger number of samples and for the localisation of the protein(s).

Work of this type necessitated the improvement or development of techniques associated with protein analysis, to overcome problems associated with the isolation of protein from perennial tissues (e.g. phenolics, proteases). Increasingly, in an attempt to see how trees relate to biotic and abiotic stresses such as climate change, studies in the area of protein and gene expression will continue to make a significant contribution to plant physiology.

Foremost amongst applications for immunolocalisation studies lies the importance of proteins in perennial tissues. Recent studies looking at expression of dehydrin in peach trees promise to lead to new understanding of tree response to environmental stresses resulting in cellular dehydration (Wisniewski *et al.*, 1999). Visualising the location of the protein of interest greatly aids in the formulation of hypotheses as to its role. Similarly, a knowledge of the localisation of vegetative storage proteins led to the confirmation that they may be similar in function to seed storage proteins (Beardmore *et al.*, 1996). The old phrase, 'a picture is worth a thousand words', summarises the value of immunolocalisation of proteins in perennial tissues since this technique can elegantly corroborate that which is surmised from biochemical studies and, thereby, help to determine the course of further research studies.

More commercial applications of immunolocalisation to trees will likely be hampered by the fact that it is a difficult technique, requiring skill in both microscopy, as well as in biochemistry to produce the antibody and to determine proper concentrations for labelling. However, the detailed protocols provided here, and in other chapters in this volume, show that the technique can be performed even in these demanding tissues and, considering the value of understanding gained by visualisation and continuing research into the area, commercial applications will no doubt follow. With the recent trend to more responsibly and intensively manage forest resources, commercial tree improvement and bio-engineering programmes are two examples where immunolocalisation techniques no doubt will prove invaluable.

FUTURE PROSPECTS

One area of advancement that has been made by the companies which supply electrophoresis equipment is the availability of pre-mixed gel solutions (from e.g. Amersham, Pharmacia, Biotech, Bio-Rad, Life Technologies) and pre-cast gels (from Amersham, Pharmacia, Biotech, Bio-Rad). These products greatly reduce preparation time and aid in providing consistent reproducibility of the final results. However, ultimately, the quality of the results is dependent on the quality of the protein extract. The area which is still the most cumbersome when working with woody tissue is that of phenolic interference. Hopefully, advancements will continue within this area as more research is conducted on woody perennials.

Immunolocalisation will undoubtedly be used with increasing frequency in future years. Concern over the impacts of global climate change and environmental stresses on forests will likely promote research at the individual tree level. Increased understanding of the physiological responses of trees to environmental stresses will be required before one can predict with accuracy

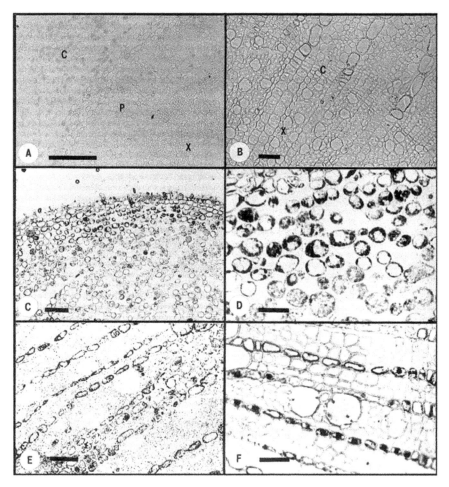

Figure 6 Current use of how immunolocalisation contributes to the understanding of physiological events in perennial tissues: immunolocalisation of peach dehydrin (PCA60) in shoot tissues of peach collected in January. Labelling studies utilised a dehydrin antibody directed against a synthetic 20-mer peptide representing the carboxy-terminus consensus sequence of dehydrins. Transverse sections of twigs were prepared for light microscopy as described by Wisnieswski *et al.* (1999): sections were blocked for 1 h (0.01% (v:v) SDS in PBS), rinsed with PBS and incubated for 2 h with anti-dehydrin IgG (1:50 (v:v) dilution), rinsed and incubated for 2 h in goat-anti-rabbit IgG conjugated to 10 nm gold particles (1:100 (v:v) dilution). For determination of non-specific labelling, the pre-immune serum was substituted for the dehydrin antibody (A). Sections were silver-enhanced for 3–5 min. A. Control section probed with pre-immune serum. B. Control section probed with a mixture of dehydrin antibody and the synthetic peptide used to generate the antibody. C. Positive labelling of dehydrin in cortical tissues. D. General distribution of dehydrin in cortical cells. Note absence of label in vacuoles. E. Positive labelling of dehydrin in xylem ray parenchyma cells of peach. F. General distribution of dehydrin in xylem ray parenchyma cells. Note absence of label from vacuoles. Key: C, cortical tissue; P, phloem tissue; X, xylem tissue. Bars: 50 μm (A); 10 μm (B, D, E, F); 20 μm (C). Photograph courtesy of Dr M. Wisniewski and *Physiologia Plantarum* (for actual paper see Wisniewski *et al.*, 1999).

forest responses at the landscape or global level. Activation and/or deactivation of genes in response to various stresses will foster protein and immunohistochemical work. In addition, molecular genetics is continually gaining momentum in

forest research. Here, too, immunohistochemistry will have an important role. Being able to visualise any modified gene products and confirm cellular localisation will be vital to studies on expression of foreign gene products.

ACKNOWLEDGEMENTS

The authors would like to acknowledge Ms K. Forbes, and Drs M. Wisniewski and R. Smith for their review of this chapter.

REFERENCES

Arora, R. and Wisniewski, M.E. 1994. Cold acclimation in genetically related (sibling) deciduous and ever-green peach (*Prunus persica* [L] Batsch). 2. A 60-kilo-dalton bark protein in cold-acclimated tissues of peach is heat stable and related to the dehydrin family of proteins. *Plant Physiol.*, **105**, 95–101.

Artlip, T.S., Callahan, A.M., Bassett, C.L. and Wisniewski, M.E. 1997. Seasonal expression of a dehydrin gene in sibling deciduous and evergreen genotypes of peach (*Prunus persica* [L] Batsch). *Plant Mol. Biol.*, **33**, 61–70.

Ausubel, F.M., Brent, R., Kingston, R.E., Moore, DD., Seidman, J.G., Smith, J.A. and Struhl, K. ((eds.))(1992) *Short Protocols in Molecular Biology, A Compendium of Methods form Current Protocols in Molecular Biology*, second edition. New York, Greene Publishing Associates and John Wiley and Sons.

Ausubel, F.M., Brent, R., Kingston, R.E., Moore, D.D., Seidman, J.G., Smith, J.A. and Struhl, K. (eds.)(1997) *Current Protocols in Molecular Biology*, Volumes 1–3. New York: John Wiley and Sons, Inc.

Beardmore, T., Wetzel, S., Burgess, D. and Charest, P. (1996). Characterisation of seed storage proteins in *Populus* and their homology with *Populus* vegetative storage proteins. *Tree Physiol.*, **16**, 833–840.

Binnie, S.C., Grossnickle, D.R. and Roberts, D.R. (1994) Fall acclimation patterns of interior spruce seedlings and their relationship to changes in vegetative storage proteins. *Tree Physiol.*, **14**, 1107–1120.

Bradford, M.M. (1976) A rapid and sensitive method for quantitation of microgram quantities of protein utiliz-ing the principle of protein-dye binding. *Anal. Biochem.*, **72**, 248–254.

Bravo, L.M. (1998) Polyphenols, chemistry, dietary sources, metabolism and nutritional significance. *Nutrition Reviews*, **56**, 317–333.

Burtin, P., Jay-Allemand, C., Charpentier, J-P. and Janin, G. (1998) Natural wood colouring processes in *Juglans* sp. (*J. nigra*, *J. regia* and hybrid *J. nigra* 23 × *J. regia*) depends on native phenolic compounds accumulated in the transition zone between sapwood and heartwood. *Trees*, **12**, 258–264.

Chaffey, N.J., Barlow, P.W. and Barnett J.R. (1997a) Formation of bordered pits in secondary xylem vessel elements of *Aesculus hippocastanum* L., an electron and immunofluorescent microscope study. *Protoplasma*, **197**, 64–75.

Chaffey, N.J., Barlow, P.W. and Barnett, J.R. (1998) A seasonal cycle of cell wall structure is accompanied by a cyclical rearrangement of cortical microtubules in fusiform cambial cells within taproots of *Aesculus hippocastanum* L. (Hippocastanaceae). *New Phytol.*, **139**, 623–635.

Chaffey, N.J., Barlow, P.W. and Barnett, J.R. (2000) Structure-function relationships during secondary phloem development in *Aesculus hippocastanum*, micro-tubules and cell walls. *Tree Physiol*, **20**, 777–786.

Chaffey, N.J., Barnett, J.R. and Barlow, P.W. (1997b) Endomembranes, cell walls and cytoskeleton, aspects of the biology of the vascular cambium of *Aesculus hippocastanum* L. *Int. J. Plant Sci.*, **158**, 97–109.

Chrispeels, M.J. and Bollini, R. (1982) Characteristics of membrane-bound lectin in developing *Phaseolus vulgaris* cotyledons. *Plant Physiol.*, **70**, 1425–1428.

Christensen, J.H., Bauw, G., Welinder, K.G., Montagu, M.V. and Boerjan, W. (1998) Purification and charac-terization of peroxidases correlated with lignification in poplar xylem. *Plant Physiol.*, **118**, 125–135.

Coleman, G., Chen, T.H.H., Ernst, S.G. and Fuchigami, L., 1991. Photoperiod control of poplar bark storage protein accumulation. *Plant Physiol.*, **96**, 130–136.

Coligan, J.E., Dunn, B.M., Ploegh, H.L., Speicher, D.W. and Wingfield, P.T. (eds.)(1998) *Current Protocols in Protein Science*, Volume I, Chapter 4, Extraction, Stabilization and Concentration, pp. 4.1.1–4.5. John Wiley, Inc., NY.

Costa, P., Bahrman, N., Frigerio, J.M., Kremer, A. and Plomion, C. (1998) Water-deficit-responsive proteins in maritime pine. *Plant Mol. Bio.*, **38**, 587–596.

Craig, S. and Goodchild, D.J. (1984) Periodate-Acid treatment of sections permits on-grid immunogold localisation of pea seed vicilin in ER and golgi. *Protoplasma*, **122**, 35–44.

Cyr. D.R., and Bewley, J.D. (1990) Proteins in the roots of perennial weeds of chicory (*Cichorium intybus* L.) and dandelion (*Taxacum officinale* Weber) are associated with overwintering. *Planta*, **182**, 370–374.

Ekramoddoullah, A.K.M., and Taylor, D.W. 1996. Seasonal variation of western white pine (*Pinus monti-cola* D. Don) foliage proteins. *Plant Cell Physiol.*, **37**, 189–199.

Ekramoddoullah, A..K.M., Taylor, D.W. and Hawkins, B.J. (1995) Characterisation of a fall protein of sugar pine and detection of its homologue associated with frost hardiness of western white pine needles. *Can. J. For. Res.*, **25**, 1137–1147.

Faye, F. and LeFloch, F. (1997) Adenosine kinase of peach tree flower buds, Purification and properties. *Plant Physiol.*, Biochem. **35**, 15–22.

Feder, N. and O'Brien, T.P. (1968) Plant microtechnique, some principles and new methods. *Am. J. Bot.*, **55**, 123–142.

Golan-Goldhirch, A. (1998) Developmental proteins of *Pistacia vera* L. bark and bud and their biotechno-logical properties, A Review. *J. Food Biochem.* **22**, 375–382.

Gray, J.C. (1978) Absorption of polyphenols by polyvinylpyrrolidone and polystyrene resins. *Phytochem.*, **17**, 495–497.

Guglielmino, N., Liberman, M., Jauneau, A., Vian, B., Catesson, A.M. and Goldberg, R. (1997) Pectin immunolocalisation and calcium visualization in dif-ferentiating derivatives from poplar cambium. *Protoplasma*, **199**, 151–160.

Harlow, E. and Lane, D. (eds)(1999) *Using Antibodies.* Cold Spring Harbor, NY: Harbor Laboratory Press.

Harms, U. and Sauter, J.J. (1994) Biochemical and immunological investigations on vegetative storage proteins of Taxodiaceae species. *J. Plant Physiol.,* **143**, 601–908.

He, S., Deng, L., Li, Y. and Liao, R. (1998) Effects of floral promotion or inhibition treatments on flowering of citrus trees and protein fractions in buds. *J. Tropical and Subtropical Bot.,* **6**, 125–130.

Honkanen, T., Haukijo, E. and Kitunen, V. (1999) Responses of *Pinus sylvestris* branches to simulated herbivory are modified by tree sink/source dynamics and by external resources. *Functional Ecology,* **13**, 126–140.

Hurkman, W.J. and Tanaka, C.K. (1986) Solubilization of plant membrane proteins for analysis by two-dimensional gel electrophoresis. *Plant Physiol.,* **81**, 802–806.

Jeknic, Z. and Chen, T. (1999) Changes in protein profiles of poplar tissues during the induction of bud dormancy by short-day photoperiods. *Plant Cell Physiol.,* **40**, 25–35.

Jensenius, J.C., Andersen, I., Hau, J., Crone, M. and Koch, C. (1981) Eggs: conveniently packaged antibodies. Methods for purification of yolk IgG. *J. of Immunol. Methods,* **46**, 63–68.

Laemmli, E.K. (1970) Cleavage of structural proteins during the assembly of the head bacteriophage T4. *Nature,* **227**, 680–685.

Lang, G.A. and Tao, J. (1990) Analysis of fruit bud proteins associated with plant dormancy. *HortSci.,* **25**, 1068.

Langheinrich, U. (1993) Clonal variation in apical growth and content in vegetative storage proteins in *Populus. Trees,* **7**, 242–249.

Langheinrich, U. and Tischner, R. (1991) Vegetative storage proteins in poplar. *Plant Physiol.,* **97**, 1017–1025.

Lawrence, S.D., Greenwood, J.S., Korhnak, T.E. and Davis, J.M. (1997) A vegetative storage protein homolog is expressed in the growing shoot apex of hybrid poplar. *Planta,* **203**, 237–244.

Loomis, M.D. (1974) Overcoming problems of phenolics and quinones in the isolation of plant enzymes and organelles. *Methods Enzymol.* **31**, 528–544.

Lowry, O.H., Rosenbrough, N.J., Farr, A.L. and Randall, R.J. (1951) Protein measurement with the Folin phenol reagent. *J. Biol. Chem,* **193**, 265–275.

Mueller, W.C. and Greenwood, A.D. (1978) The ultra-structure of phenolic-storing cells fixed with caffeine. *J. Exp. Bot,* **29**, 757–764.

Nir, G., Shulman, Y., Fanberstein, S. and Lavee, S. (1986) Changes in the activity of catalase (EC 1.11.1.6) in relation to the dormancy of grapevine (*Vitis vinifera* L.) Buds. *Plant Physiol.,* **81**, 1140–1142.

O'Farrell, P.Z., Goodman, H.M., and O'Farrell, P.H. (1977) High resolution two-dimensional electrophoresis of basic as well as acidic proteins. *Cell,* **12**, 1133–1142.

Osborne, T.B. (1924) *The Vegetable Proteins.* London: Longmans, Green.

Przymusinski, R. and Gwozdz, E.A. (1999) Heavy metal-induced polypeptides in lupin roots are similar to pathogenesis-related proteins. *J. Plant Physiol.,* **154**, 703–708.

Roberts, D.R., Toivonen, P. and McInnis, S.M. (1990) Discrete proteins associated with overwintering of interior spruce and Douglas-fir seedlings. *Can. J. Bot.,* **69**, 437–440.

Rosenberg, I.M. (1996) *Analysis and Purification,* Benchtop Techniques. Boston: Birkhauser, 434 pp.

Rowland, L.J. and Arora, R. (1997) Proteins related to endodormancy (rest) in woody perennials. *Plant Sci.,* **126**, 119–144.

Ruel, K., Burlat, V. and Joseleau, J-P, (1999) Relationship between ultrastructural topochemistry of lignin and wood properties. *IAWA,* J., **21**, 203–211.

Sauter, J.J. and van Cleve, B. (1990) Biochemical, immunochemical, and ultrastructural studies of protein storage in poplar (*Populus × canadensis* 'robusta') wood. *Planta,* **183**, 92–100.

Sauter, J.J., Van Cleve, B. and Apel, K. (1988) Protein bodies in ray cells of *Populus × canadensis* Moench 'robusta'. *Planta,* **173**, 31–34.

Sauter, J.J., van Cleve, B., and Wellenkamp, S. (1989) Ultrastructure and biochemical results on the localization and distribution of storage proteins in a poplar tree and in twigs of other tree species. *Holzforschung,* **43**, 1–6.

Slot, J.W. and Geuze, H.J. (1985) A new method of preparing gold probes for multiple-labelling cytochemistry. *Eur. J. Cell Bio.,* **38**, 87–93.

Smith, P.K., Krohn, R.I., Hermanson, G.T., Mallia, A.K., Gartner, F.H., Provenzano, M.D., Fujimoto, E.K., Goeke, N.M., Olson, B.J. and Klenk, D. (1985) Measurement of protein using bicinchonininic acid. *Anal. Biochem.,* **150**, 76–85.

Stepien, V., Martin, F. (1992) Purification, characterization and localization of the bark proteins of poplar. *Plant Physiol Biochem.,* **30**, 399–407.

Stepien, V., Sauter, J.J. and and Martin, F. (1994) Vegetative storage proteins in woody plants. *Plant Physiology Biochemistry,* **32**, 185–192.

Towbin, H., Staehelin, T. and Gordon, J. (1979) Electrophoretic transfer of proteins from polyacrylamide gels to nitro-cellulose sheets, procedure and some applications. *Proc. Natl. Acad. Sci. USA,* **76**, 4350–4354.

van Cleve, B. and Apel, K. (1993) Induction by nitrogen and low temperatures of storage-proteins synthesis in poplar trees exposed to long days. *Planta,* **189**, 157–160.

van Cleve, B., Clausen, S. and Sauter, J.J. (1988) Immunolocalisation of a storage protein in poplar wood. *J. Plant Physiol.,* **133**, 371–374.

Vierling, E., Key, J.L. (1985) Ribulose 1, 5–bisphosphate carboxylase synthesis during heat shock. *Plant Physiol.,* **78**, 155–162.

Walker, J.M. (ed.)(1996) *The Protein Protocols Handbook.* Totowa, NJ: Humana Press, 809 pp.

Wetzel, S. (1990) Proteins as an overwintering storage form of N in temperate forest tree species. Ph.D. Thesis. University of Guelph, Guelph ON.

Wetzel, S., Demmers, C. and Greenwood, J.S. (1989) Seasonally fluctuating bark protein are a potential form of nitrogen storage in 3 temperate hardwoods. *Planta*, **178**, 275–281.

Wetzel, S. and Greenwood, J.S. (1989) Proteins as a potential nitrogen compound in bark and leaves of several softwoods. *Trees*, **3**, 149–153.

Wetzel, S. and Greenwood, J.S. (1991) The 32-kilodalton vegetative storage protein of *Salix microstachya* Turz. *Plant Physiol.*, **97**, 771–777.

Wisniewski, M., Webb, R., Balsamo, R., Close, T.J., Yu, X-M. and Griffith, M. (1999) Purification, immunolocalisation, cryoprotective, and antifreeze activity of PCA60, A dehydrin from peach (*Prunus persica*). *Physiol. Plant.*, **105**, 600–608.

6 The Use of GUS Histochemistry to Visualise Lignification Gene Expression *In Situ* during Wood Formation

Wood Formation in Trees ed. N. Chaffey
© 2002 Taylor & Francis
Taylor & Francis is an imprint of the
Taylor & Francis Group
Printed in Singapore.

Simon Hawkins[1]*, Gilles Pilate[2], Eric Duverger[1], Alain Michel Boudet[3] and Jacqueline Grima-Pettenati[3]

[1] *Antenne Scientifique Universitaire de Chartres, Université d'Orléans, 21 rue de Loigny la Bataille, F-28000 Chartres, France;*
[2] *Amélioration, Génétique et Physiologie Forestière, INRA, Avenue de la Pomme de Pin BP20619, F-45160 Ardon, France;*
[3] *UMR 5546 CNRS-UPS, Pôle de Biotechnologie Végétales, 24 chemin de Borde Rouge, BP17, Auzeville, F-31326 Castanet Tolosan, France*
* Author for correspondence

ABSTRACT

A number of different techniques can be used to characterise the spatial (and temporal) expression patterns of genes in plants. In this chapter, various protocols, associated techniques and tips are described so as to enable the reader to use GUS-histochemistry in order to analyse promoter expression patterns *in situ* in tree species. As an example we describe the use of this technique to characterise the *in situ* expression pattern of the cinnamyl alcohol dehydrogenase (CAD) promoter from *Eucalyptus gunnii* Hook. in transgenic poplar. The corresponding enzyme catalyses the final step in the formation of monolignols – the monomers of the three-dimensional phenolic polymer lignin, which is a major component of the cell wall in wood and other lignified tissues. The chapter describes how the use of GUS histochemistry has contributed (and is contributing) to our understanding of the molecular and cellular basis of wood formation, and presents potential future areas of research and developments of new techniques.

Key words: cinnamyl alcohol dehydrogenase, genetic transformation, GUS, lignin, poplar, wood

INTRODUCTION

Characterisation of gene expression patterns *in situ*

A number of techniques (Figure 1) can be used to characterise the spatial expression patterns of genes (i.e. in what tissues and cells the genes are expressed), depending upon whether the analyses are performed at the transcriptional (mRNA) level or translational (protein) level. 'Low resolution' techniques include microdissection followed by either Northern blotting or reverse transcription-polymerase chain reaction (RT-PCR) (transcriptional level) or, at the translational level, by Western blotting or enzymological studies. Such approaches are capable of localising gene expression to a particular organ or tissue (including complex tissues such as xylem and phloem) but they do not have sufficient resolution to localise gene expression at the cellular level. While the 'medium resolution' technique of 'tissue blotting' (Ye and Varner, 1991; Ye, 1996; see also the chapter by Micheli *et al.* in this volume) can greatly

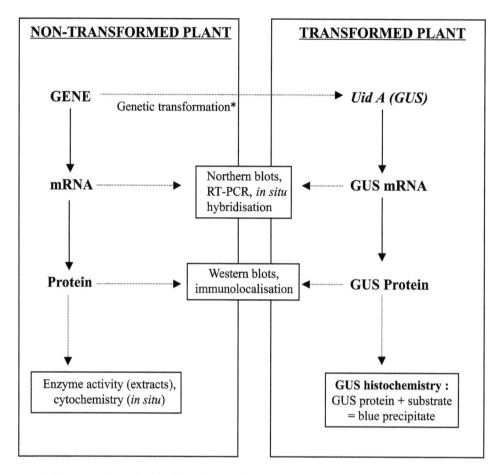

NON-TRANSFORMED PLANT	TRANSFORMED PLANT

GENE .... Genetic transformation* ....→ *Uid A (GUS)*

mRNA ....→ Northern blots, RT-PCR, *in situ* hybridisation ←............ **GUS mRNA**

Protein ....→ Western blots, immunolocalisation ←............ **GUS Protein**

Enzyme activity (extracts), cytochemistry (*in situ*)

GUS histochemistry : GUS protein + substrate = blue precipitate

* Plant transformed with chimeric gene (promoter from gene under investigation + reporter gene coding sequence)

Figure 1   Different techniques for localising gene expression.

improve the quality of the spatial information obtained, information at the cellular level can only be obtained through the use of 'high resolution' techniques.

As with low resolution techniques, high resolution techniques can be employed at the transcriptional level (for example *in situ* hybridisation (ISH) – see the chapter by Regan and Sundberg in this volume – or *in situ* PCR) or at the translational level (for example immunolocalisation – see the chapter by Samaj and Boudet in this volume – or enzyme cyto-chemistry – see the chapter by Micheli *et al.* in this volume). Another extremely powerful high resolution technique is that of reporter-gene histo-chemistry (Figure 1). In this technique the promoter of the gene under investigation is fused to a reporter gene such as *uid a* (*GUS*) and the chimeric gene construct introduced into a model plant by genetic transformation. In the transformed plant the promoter is activated by cell-/tissue-specific transcription factors (themselves produced/activated in response to developmental/environmental signals), leading to the synthesis of the β-glucuronidase (GUS) protein in those cells where the promoter is active. The cellular location of the GUS protein (and hence activity of the promoter) is revealed by the addition of a colourless substrate (5-bromo-4-chloro-3-indolyl ß-D-glucuronide or 'X-gluc') which is converted by the GUS protein to a blue-coloured, insoluble product (di-X indigo) which is then precipitated in the cell.

In general, plants have little or no endogenous GUS (glucuronidase) activity, but in the rare cases where this is so, modifications of the incubation conditions can usually be made so as to eliminate such activity (Hänsch *et al.*, 1995).

Genetic transformation of trees

As indicated above, one pre-requisite for the utilisation of reporter-gene histochemistry is that the species under investigation is genetically transformable. The investigation of wood development at the molecular level can be carried out in species which show little secondary xylem formation (and indeed there are some very good arguments as to why species such as *Arabidopsis thaliana*, for example, should be used – see e.g. Chaffey, 1999b, 2001). Nevertheless, the analysis of wood formation *in situ*, in tree species is necessary in order to verify the models and hypotheses established from the study of non-tree species. The first genetic transformation of a tree species (hybrid poplar) was reported in 1987 (Fillatti *et al.*, 1987). Since then, although different economically-important tree species (for example, eucalyptus, apple, oak) have become targets for transformation, poplar has become the biological model of choice for molecular studies in angiosperm tree species (see the Introduction to this volume). Poplar presents a number of advantages for such studies: it is readily transformable, easy to manipulate *in vitro* and possesses a small genome (see also Chaffey, 1999a, 2001, and the chapter entitled. 'An introduction to the problems of working with trees' in this volume for further details). In addition, an expressed sequence tag (EST) library has also been generated (Sterky *et al.*, 1998). Poplar can be easily transformed by *Agrobacterium*-mediated transformation (Fillatti *et al.*, 1987; Leplé *et al.*, 1992) or, less easily, by direct gene-transfer techniques (McGown *et al.*, 1991; Chupeau *et al.*, 1994).

Genetic transformation in tree species has been used to directly modify/improve certain economically important traits, for example, tolerance to insect predators (Leplé *et al.*, 1995), resistance to herbicides (Brasileiro *et al.*, 1992) and lignin metabolism (Baucher *et al.*, 1996, Grima-Pettenati and Goffner, 1999). In addition, such a technique has proved to be extremely useful in the analysis of complex phenomena at the molecular level, either by studying the phenotypic effects of up-/down-regulation of a given enzyme (strategy sense/antisense) or by using reporter genes to characterise the spatio-temporal expression patterns of gene promoters.

The GUS reporter gene system

Of the various reporter genes (for example, lacZ, luciferase, etc.) utilised for investigating gene expression in plants, the GUS system has become the most widespread – mainly because of its relative simplicity and the fact that it can provide important information about the biological role of the gene and gene product under investigation.

The *uid a* gene codes for a bacterial β-glucuronidase (GUS). In the classical histochemical test (Jefferson *et al.*, 1987), the substrate 5-bromo-4-chloro-3-indolyl–β–D-glucuronide (X-gluc) is hydrolysed to yield a soluble, colourless intermediate (5-bromo-4-chloro-indoxyl or XH) which then immediately undergoes oxidative dimerisation to form the coloured product 5,5'-dibromo-4, 4'-dichloro-indigo (diX-indigo). The end product, diX-indigo, forms stable, blue crystals which are not affected by fixation, dehydration or embedding and which are clearly visible under the light microscope.

We have used GUS histochemistry and other *in situ* techniques to characterise the expression patterns of 'lignification genes' (see Applications section below) *in situ* in woody plants.

TECHNIQUES

Genetic transformation and regeneration

While the characterisation of gene expression profiles by GUS histochemistry is relatively non-labour intensive in comparison with other *in situ* techniques (e.g. *in situ* hybridisation, immunolocalisation), the preceding steps (isolation of genomic clones, fusion of promoter to reporter gene coding sequence, transformation and regeneration of the target plant) which permit the utilisation of this powerful technique are certainly very labour intensive. Figure 2 gives an overview of the transformation and regeneration procedure that we have used to regenerate transgenic poplar. Explants from INRA hybrid poplar clone 717 1B4 (*Populus tremula* L. × *P. alba* L.) are easily transformed by co-cultivation with *Agrobacterium tumefaciens* strain C58 pMP90 containing the desired binary vector, as described by Leplé *et al.* (1992).

Molecular characterisation

Although perhaps not strictly necessary for the utilisation of the GUS histochemical test itself, the molecular characterisation of potential transformants is nevertheless essential for a correct interpretation of the results. Transformants are characterised by the use of two main methods – the

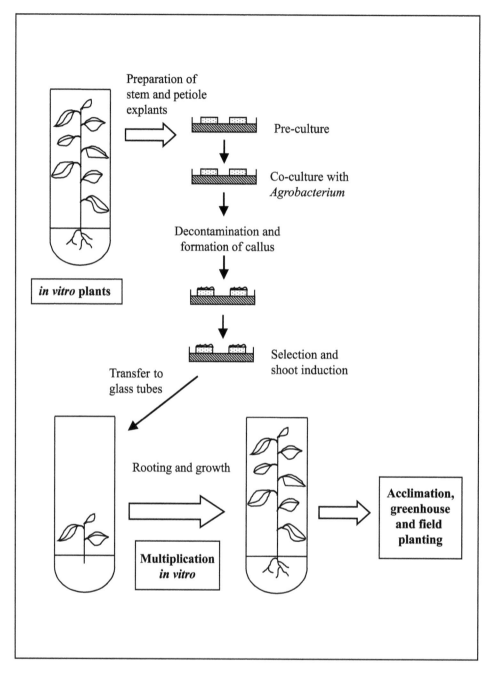

Figure 2   Overview of the transformation protocol used for generating transgenic poplar plants (for further details see: Leplé *et al.*, 1992).

polymerase chain reaction (PCR) and Southern analysis.

PCR

PCR (together with Southern analysis) is used to confirm that the potential transformant contains the transgene and also to verify (initially) that the left and right borders of the T-DNA have functioned correctly. Box 1 gives the sequences of the forward and reverse primers, together with the cycle conditions, that we have used to successfully amplify a 550 bp fragment from the coding sequence of the *uid a* gene.

Box 1: PCR conditions and primers

PCR reactions are performed in a volume of 50 μl containing reaction buffer (50 mM KCl, 100 mM Tris-HCl pH 9.0, 0.1% Triton X-100), 1.5 mM MgCl₂, 0.2 mM dNTP mix, 2 mM primers, 1U Taq DNA polymerase (Promega) and genomic DNA (0.25 μg). DNA is amplified using the following cycles: initial denaturing at 94 °C for 4 min, then 30 cycles of denaturation at 94 °C for 45 sec, annealing at 55 °C for 1 min, and extension at 72 °C for 45 sec. Amplified products are viewed on 1% agarose tris-borate-EDTA (TBE) gels by ethidium bromide staining.

Primers for *uid a* coding sequence (amplification product = 550 bp)
5′ primer: TAT ACG CCA TTT GAA GCC G (3727 – 3746)
3′ primer: AAG CCA GTA AAG TAG AAC GGT (4236 – 4215)

Southern analysis

Southern analysis and a positive PCR result constitute proof that the target plant has been transformed. Southern analysis is also used to estimate the number of copies of the transgene incorporated. This is important since variations in the level of transgene expression can be related to the number of copies incorporated (Hobbs, 1990) and it has been shown that gene-silencing events are very often related to high transgene copy numbers (Matzke and Matzke, 1995).

In addition, since the position of transgene integration in the genome has also been shown to affect the level and pattern of expression (Hobbs, 1990), it is important to analyse the expression patterns of a number of individual transformants. Box 2 shows a protocol that we have used to successfully characterise transgenic poplar.

Box 2: Southern blotting conditions

Genomic DNA is isolated from leaves of *in vitro*-grown plants using the Dneasy plant kit (Qiagen) according to the manufacturer's instructions. 2.5 μg DNA is digested separately with appropriate restriction enzymes and separated on a 0.8% agarose tris acetate-EDTA(TAE) gel. DNA is then transferred

to a positively-charged nylon membrane (Boehringer) using a vacuum transfer apparatus (Appligene). Membranes are then hybridised overnight at 42 °C with 15 ng.ml⁻¹ digoxigenin (DIG)-labelled DNA. Following washing, bound probe is revealed by chemiluminescence (CDP-Star) according to manufacturer's instructions (Boehringer). DIG-labelled GUS probes are synthesised by PCR using the primers detailed in Box 1.

Further details on these two techniques can be found in standard molecular biology laboratory books (for example, Sambrook *et al.*, 1989).

Sample preparation, histochemistry and microscopy

Sample preparation

The conducting and supporting cells making up the axial system of the xylem in trees are extremely elongated in the vertical dimension (vessel elements of 300–400 μm and phloem fibres with a length greater than 1 mm in young hybrid poplar stems) and this aspect of their anatomy *must* be taken into account when preparing samples for GUS histochemistry. Standard sample preparation techniques commonly used for herbaceous plants, which entail the cutting of thin (50–100 μm) discs *prior* to incubation with GUS reaction medium, are unsuitable for woody plants. The utilisation of such protocols leads to the cutting of the elongated cells in several places resulting in the formation of 'open tubes' (Figure 3 a) with the subsequent loss of cytoplasmic contents (including the GUS protein!). In such conditions an incomplete, or, worse, a misleading, picture (Figures 7 C, D) of gene expression patterns is obtained (Hawkins *et al.*, 1997a).

In consequence, it is recommended (at least for stem/root samples) that the samples taken for subsequent incubation with GUS reaction medium (GRM) are between 1.5 and 2 mm long in the vertical axis (Figure 3 b). In order to facilitate substrate penetration, it is also recommended that 'pie' sections (Figure 3 b) be utilised rather than complete discs – this permits the entry of reaction medium

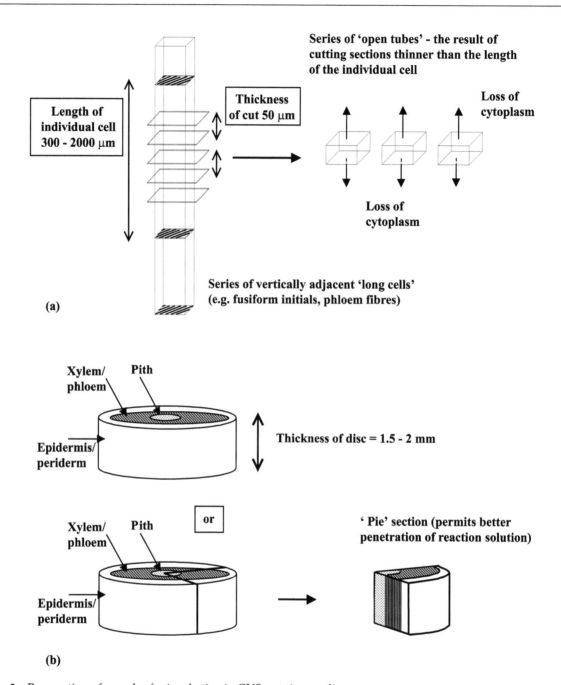

Figure 3 Preparation of samples for incubation in GUS reaction medium
a) diagram illustrating the problems associated with the use of a typical technique utilised for herbaceous plants;
b) recommended approach for preparing samples from stems/roots of poplar plants (*in vitro*, greenhouse and field-grown).

into the sample via several surfaces (the epidermis/periderm being relatively impermeable) and reduces the length of the diffusion pathway. Sections are best prepared from the stems/roots of young plants and branches of older plants by using fresh razor blades. In the case of older plants, 'pre-samples' (e.g. short lengths of branches) can be removed using a combination of secateurs and hacksaws prior to the preparation of samples for incubation in GRM.

GUS histochemistry and pre-treatments

Once prepared, samples are placed directly into suitable containers (e.g. multi-well plates, Eppendorfs, glass vials) containing GRM (see Box 3), and the blue colouration is then left to develop at 37 °C in the dark (e.g. multi-well plates can be simply covered by a sheet of aluminium foil). Since the X-gluc reagent is expensive, the reaction is generally developed in a small volume, for example 1 ml in a multi-well plate can be used for up to 5 samples.

Box 3: GUS histochemistry

Transfer samples to GUS reaction medium (GRM)[1] in an appropriate vessel (e.g. multi-well plaque). Cover samples with aluminium foil and incubate at 37 °C until sufficient blue colour has developed (2 h – overnight).

[1]GUS reaction medium:
Although GRM can be made up and stored overnight at 4 °C for use the next day, we generally prepare GRM fresh by mixing together the following stock solutions. If necessary, 0.25% Triton X-100 can be added to the GRM to facilitate substrate penetration.

Stock solution (concentration)		vol. (μl) stock per ml GRM	Final concentration
0.5 M NaPO$_4$ buffer, pH 7	(5x)	200	0.1 M
100 mM Na$_2$ EDTA, pH7[3]	(10x)	100	10 mM
5 mM K$^+$ Ferricyanide[4]	(10x)	100	0.5 mM
5 mM K$^+$ Ferrocyanide[4]	(10x)	100	0.5 mM
10 mM X-gluc[5]	(20x)	50	0.05 mM
sterile dH$_2$O[2]		450	

[2] To avoid precipitation, always add the water first.
[3] It is necessary to add solid NaOH to increase the pH thereby allowing the Na$_2$EDTA to go into solution.
[4] The ferric salts are necessary to limit intercellular diffusion (De Block and Debrouwer, 1992). These solutions can be stored for up to one month, if kept away from light, and at 4 °C.
[5] Dissolve 52 mg of 5-bromo-4-chloro-3-indolyl-β-D-glucuronic acid (X-gluc) in 10 ml DMF (dimethylformamide – Attention, DMF dissolves polystyrene!), and store in 0.5 ml aliquots in Eppendorfs at –20 °C.

In some cases, pre-treatments (see Box 4), for example with cold 90% acetone (to prevent wound-induced gene induction) or with a 0.5% paraformaldehyde solution (pre-fixation), can be used. However, in the authors' experience, care should be taken with such treatments since they can interfere with the activity of the GUS enzyme (a two-minute acetone treatment of *in vitro* transgenic poplar plants proved sufficient to eliminate all GUS activity). If such pre-treatments are used, the samples should be rinsed (2 × 15 min) in phosphate buffer before transfer to GRM. Vacuum infiltration, or the inclusion of a non-ionic detergent (Triton X-100) in the GRM can also be used to improve substrate penetration, especially in 'light' tissues (e.g. leaf samples), which tend to float on the surface of the solution.

Box 4: Pre-treatment

Pre-fixation
Pre-treat samples in 0.5% *p*-formaldehyde prepared in 100 mM phosphate buffer, pH 7 (0–30 min, depending upon the organ/tissue type), then rinse twice (2 × 15 min) in 100 mM phosphate buffer pH 7 before transfer to GRM.

Acetone pre-treatment[1]
Pre-treat the sample in cold (stored at 4 °C) 90% acetone (0–30 min, depending upon the organ/tissue type), then rinse twice (2 × 15 min) in 100 mM phosphate buffer pH 7 before transfer to GRM.

[1]Acetone pre-treatment
The acetone pre-treatment was used in our experiments to prevent wound-induced induction of the CAD promoter (Hemerly *et al.*, 1993). This treatment can also facilitate the penetration of the substrate (by dissolving cutin in the cuticle) – as well as fixing the GUS (and other proteins) by precipitation. The necessity of such a treatment is determined by the nature of the material. In our case, for example, the use of such a treatment with *in vitro* plants resulted in the complete inhibition of the GUS reaction (presumably due to 'over-fixation' of the GUS protein). With such material, we eliminated this step and included Triton X-100 (0.25%) in the GRM to facilitate substrate penetration.

Preparation of sections for observation

After development of the blue colouration, one has the choice of either not embedding the material or else embedding the material (e.g. in paraffin or glycol methacrylate resin). In the first case samples can be either directly observed (after cutting free-hand sections if desired) or else fixed and stored for later observation. In the authors' experience, the classical botanist's fixative – formalin/acetic acid/(ethyl) alcohol (FAA) – has many advantages (easy to prepare and stable). Following fixation (2–48 h, although samples can be left much longer) samples can be transferred to 70% ethanol and stored almost indefinitely at room temperature (RT) in screw-top bottles, Eppendorfs etc. (take care that the alcohol does not efface the labels!). The transfer to ethanol renders subsequent manipulations less disagreeable (from an olfactory point of view!) than with FAA (utilise a chemical hood as much as possible).

If samples are not fixed, it is necessary to stop the reaction and development of further colouration – this is simply done by transferring the sample to GRM without X-gluc and then to distilled water. Details of the possibilities following GUS histochemistry for non-embedded material are given in Figure 4.

If samples are to be embedded, then the material must be fixed. Details of the possibilities following GUS histochemistry for embedded material are given in Figure 5.

The choice of whether to make hand sections, vibrating microtome sections, or to embed plant material and make microtome sections, depends largely upon the degree of resolution required and the time that one is prepared to dedicate to this part of the experiment, and also upon the strength of the signal (which in turn depends upon the penetration of the substrate and the incubation time and temperature). For example, hand sections (approx. 50 μm), cut with a fresh safety razor blade, are relatively quick to prepare and usually provide sufficient resolution to determine the expression pattern in different *tissues* (Figure 7E). However, it is usually more difficult to determine the *cellular* expression pattern in such sections. If one wishes to characterise the cellular expression pattern, it is probably better to embed the material and to make microtome sections (Figures 7 F, G, H). In practice, the best strategy is probably to use a combination of different types of sections since each type of section provides different and complementary information – for example, while thin or semi-thin sections can provide the best resolution at the cellular level, the very thinness of the section may result in a weak signal being overlooked.

One problem often encountered in the observation of samples is the masking of the blue colouration by plant pigments. Pigments such as chlorophyll are generally easily removed by passage through an ethanol series (in the authors' experience, a passage to 95% ethanol and incubation overnight removes the chlorophyll and allows clear visualisation of the GUS blue colouration in poplar. However, in other tree species rich in phenolics (e.g. walnut – C. Breton, personal communication), oxidative polymerisation (Craig, 1992) can cause excessive browning and mask the blue colour. In this case, treatment with a mixture of chloral hydrate, lactic acid and phenol known as chlorallactophenol CLP) (Beeckman and Engler, 1994) can be used.

Hand sections are best cut with a fresh razor blade (scalpel blades or other knives are not appropriate) which should be changed often (trees are full of

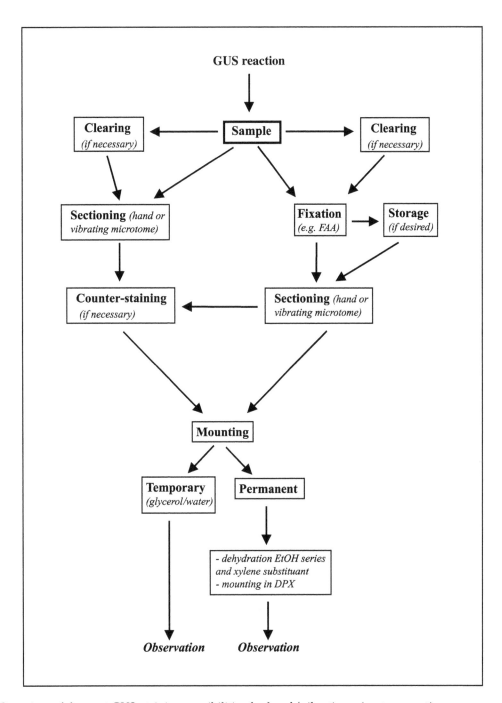

Figure 4 Overview of the post-GUS-staining possibilities for hand/vibrating microtome sections.

lignified and crystal-containing cells which will quickly blunt the edge of a blade!). In the case of relatively woody stems and roots, do not try and cut complete 'discs' – they will be too thick to provide much information, and, since the organs are (relatively!) symmetric, the same information can be gained from a semi-circle/quarter with the added advantage that, at the edges, the section will be sufficiently thin to provide resolution at the cellular level. In order to avoid crushing fragile organs such as young roots or petioles, they can be quickly embedded in 3% agarose (to avoid creating a soggy mess (!) samples stored in ethanol/FAA should first be rehydrated) in a flat latex mould.

Standard sections (50–100 μm) can be cut from young, not-too-woody stems using a vibrating microtome – although in the authors' experience, it is just as efficient, and probably less time-consuming, to prepare hand sections.

Tissue samples can also be fixed and embedded in either paraffin (Figure 5 and Box 5) or glycol methacrylate resin (Figure 5 and Box 6) using standard histological protocols. Provided that the appropriate precautions are taken against RNases, the paraffin protocol can also be used to embed samples for *in situ* hybridisation (Hawkins *et al.*, 1997b) and immunolocalisation (Samaj *et al.*, 1998).

Box 5: Fixation and embedding (paraffin)

Fixation
Remove samples from GRM and transfer to fixative[1] for the appropriate time[2].

Dehydration
Remove samples from fixative and dehydrate. The dehydration protocol depends upon the fixative and the embedding medium used, the protocols given below are for paraffin embedding.

Dehydration protocol for samples fixed in p-formaldehyde/glutaraldehyde[1]
The samples are dehydrated through a series of ethanol/tertiary butyl alcohol (TBA)[3] over a period of three days in baths as follows:

Dehydration – day 1
Rinse 1: PBS[4] (2 × 45 min)
Rinse 2: 100 mM NaCl$_2$ (45 min)
15% alcohol bath (85 ml[5] 100 mM NaCl + 15 ml[5] 95% EtOH) – 90 min[6]
30% alcohol bath (70 ml 100 mM NaCl + 30 ml 95% EtOH) – 90 min
50% alcohol bath (50 ml 100 mM NaCl + 40 ml 95% EtOH + 10 ml TBA) – 90 min
70% alcohol bath (30 ml 100 mM NaCl + 50 ml 95% EtOH + 20 ml TBA – overnight*

Dehydration – day 2
85% alcohol bath (15 ml 100 mM NaCl$_2$ + 50 ml 95% EtOH + 35 ml TBA) – 90 min
95% alcohol bath (45 ml 95% EtOH + 55 ml TBA) – 90 min
100% alcohol bath 1 (25 ml 100% EtOH + 75 ml TBA) – 90 min
100% alcohol bath 2 (100 ml TBA) – 90 min
100% alcohol bath 3 (100 ml TBA) – overnight*

Dehydration – day 3
100% alcohol bath 4 (100 ml TBA) – 90 min

Dehydration protocol for samples fixed in FAA[1].

Dehydration – day 1
50% EtOH bath[7] (50 ml 95% EtOH + 50 ml dH$_2$O) – 2 × 90 min
50% alcohol bath (40 ml 95% EtOH + 10 ml TBA + 50 ml dH$_2$O) – 90 min
70% alcohol bath (50 ml 95% EtOH + 20 ml TBA + 30 ml dH$_2$O) – overnight*

Dehydration – day 2
85% alcohol bath (50 ml 95% EtOH + 35 ml TBA + 15 ml dH$_2$O) – 90 min
95% alcohol bath (45 ml 95% EtOH + 55 ml TBA) – 90 min
100% alcohol bath 1 (25 ml 100% EtOH + 75 ml TBA) – 90 min
100% alcohol bath 2 (100 ml TBA) – 90 min
100% alcohol bath 3 (100 ml TBA) – overnight*

Dehydration – day 3
100% alcohol bath 4 (100 ml TBA) – 90 min

Embedding (paraffin)

The samples (fixed in either *p*-formaldehyde/glutaraldehyde or FAA) are prepared for paraffin infiltration by transferring them to TBA/paraffin oil (1:1, v/v, 90 min). Fill plastic moulds with liquid Paraplast Plus (Sigma), allow it to solidify (but not cool completely), then transfer the sample to the top of the Paraplast together with a few drops of TBA, and place in the oven at 58 °C. Leave the sample for 1 h and then pour off the liquid Paraplast and replace it with fresh liquid Paraplast. Renew with fresh liquid Paraplast, leave for 24–48 h before removing the moulds and allowing the blocks to solidify at room temperature.

As in all histological manipulations, care should be taken so as not to efface the markings on the samples/moulds – otherwise you could find yourself with a large number of anonymous samples! Once solidified, the paraffin blocks can be stored indefinitely at room temperature before cutting sections.

Notes

[1] Fixative

We have used two main fixatives – details of which are given below – although most other fixatives are probably also compatible with the GUS histochemistry technique

- *p-formaldehyde/glutaraldehyde:* 4% *p*-formaldehyde, 0.5% glutaraldehyde in PBS[4]. Prepare *fresh* fixative from 10% *p*-formaldehyde stock (add 10 g *p*-formaldehyde (Attention: attacks mucous membranes!) to 70 ml H_2O plus 1 drop 10 M NaOH (to help dissolve the *p*-formaldehyde) and heat in a water bath at 70 °C for 20 min. This stock can be stored for up to 1 month at 4 °C).
- *Formalin:acetic acid:(ethyl) alcohol (FAA):*10% formalin (solution of formaldehyde at 37%, stabilised with methanol):5% glacial acetic acid:60% EtOH. FAA can be stored indefinitely in a screw-top bottle.

[2] Fixation time

The fixation time depends upon the type of organ/tissue being fixed as well as the size of the sample. In the authors' experience non-woody organs (young leaves, flowers etc.) are adequately fixed after a few hours, whereas lignified organs (stems, roots etc.) generally require longer times (24–48 h).

[3] TBA (tertiary butyl alcohol = 2-methyl-2-propanol)

TBA has a high fusion point (approx. 25 °C) and, when pure, freezes at relatively high temperatures (around 10 °C). As a result, its utilisation in winter (or cold laboratories!) can be a problem as there is a real risk that it will freeze around (and within) the sample, resulting in tissue and cell damage. In order to avoid this problem, we generally keep the bottle of TBA firmly closed in an oven at 37 °C, and use it directly from the oven. If the bottle of TBA freezes, it can be thawed-out in a water bath or oven at 60 °C. When mixed with ethanol, the TBA does not freeze.

[4] PBS

Phosphate-buffered saline: 100 mM NaCl, 10 mM phosphate buffer, pH 7.

[5] Volumes

The volumes shown are for a stock solution of 100 ml, the actual volumes used for the baths depend upon the number of samples; we generally use around 5 ml for 10 samples in a glass vial.

[6] Bath times

The relatively long times quoted are necessary to ensure the penetration of the dehydrating solutions into poplar stem sections, but can be varied depending upon the nature of the tissue.

[7] 50% EtOH bath

This apparently 'backwards' fixation step is necessary to eliminate the formalin before continuing with the dehydration.

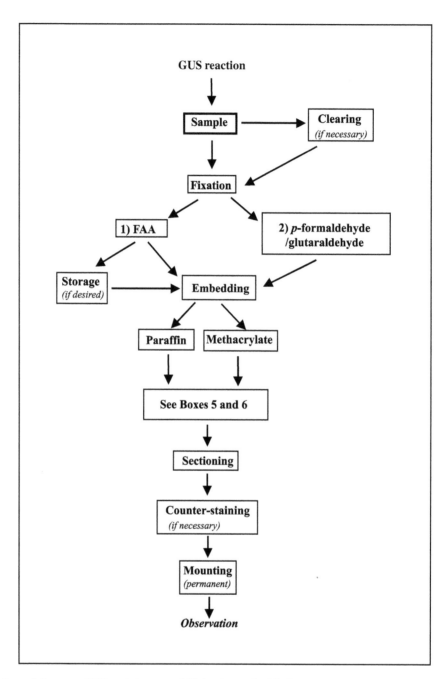

Figure 5   Overview of the post-GUS-staining possibilities for embedded sections.

Samples can be equally embedded in methacrylate, which offers very good preservation of cells and cell structures (see also the chapter on 'Wood microscopical techniques' in this volume). Box 6 gives details of the protocol that we have used to embed samples obtained from transgenic poplar growing in the field.

Initial experiments with Unicryl™ (British BioCell International) have also shown its compatibility with GUS histochemistry in poplar and the fact that, according to the manufacturer, it can also be used for *in situ* hybridisation and immunolocalisation suggests that this resin could prove very useful for the characterisation of gene expression patterns in poplar.

Box 6: Fixation and embedding (glycol methacrylate resin – Technovit 7100 (Kulzer))

Fixation
Following overnight GUS histochemistry, samples were fixed in FAA for a minimum of 48 h.

Dehydration
Samples were dehydrated in a graded water:ethanol series as follows.

Dehydration – day 1
70% EtOH bath (70 ml 95% EtOH + 30 ml dH$_2$O) – 2 × 90 min
90% EtOH bath (90 ml 95% EtOH + 10 ml dH$_2$O) – 90 min
90% EtOH bath (90 ml 95% EtOH + 10 ml dH$_2$O) – overnight

Dehydration – day 2
100% EtOH bath (100 ml 100% EtOH) – 3 × 90 min
100% EtOH bath (100 ml 100% EtOH) – overnight

Infiltration
Transfer samples to freshly prepared infiltration solution (1 g hardener per 100 ml Technovit 7100 solution) and leave overnight

Embedding
Remove samples from the infiltration solution and transfer to freshly prepared embedding solution (1 ml hardener II per 15 ml infiltration solution) in the mould – work quickly since the 'pot life' of the embedding solution is approximately 4 min. Leave the samples to polymerise overnight at RT.

Sectioning and observation
Samples are best cut on a sledge microtome at a thickness of 2 μm. Following fixation to the microscope slides (see Box 7), samples can be counter-stained with Safranin 0 (see Box 8), as for paraffin sections, before permanent mounting and microscopic observation.

Paraffin sections can be cut on rotary microtomes; methacrylate sections are best cut on sledge microtomes or ultramicrotomes. Standard steel or disposable steel blades can be used for paraffin sections while tungsten carbide, glass or diamond knives should be used with methacrylate (or Unicryl) sections. For stem and root samples we generally cut paraffin sections at a thickness of 10 μm and for methacrylate sections at 2–10 μm (depending upon the resolution required/signal strength).

As with all microscopic observations, complementary information and confirmation of cellular expression patterns can be obtained with different types of sections and longitudinal (radial/tangential) sections are especially useful for studies of gene expression in wood (Figure 7 H). Such sections can also be prepared free-hand.

Sections are then fixed to slides before further processing and observation – details are given in Box 7.

Box 7: Cutting and mounting of sections

Free-hand sections
Temporary mounting. Position section on a (clean) microscope slide in a few drops of glycerol/dH$_2$O 1:1 (v/v) and cover with a (clean) cover slip. Such preparations can be kept for several hours or overnight.

Permanent mounting (if sections are not too thick). Dehydrate section rapidly (1 minute per bath) in a series of ethanol baths (30%, 50%, 70%, 90%, 95%, 100% × 2,), transfer to 'xylene replacement' (e.g. Histochoice clearing agent – Sigma; Histolemon – Carlo-Erba) (2 × 1 min), and then place section (carefully, with forceps) in a drop(s) of DPX mountant (e.g. Lemonvitrix – Carlo-Erba). Cover with a clean cover slip and allow to dry (2 h – overnight) at room temperature in a fume cupboard.

Paraffin sections
Transfer (carefully) paraffin 'ribbons' to a 'pool' of dH_2O on a pre-coated slide[1] on a hot-plate (approx. 40 °C). Allow the water to evaporate and then transfer slides to an oven overnight at 42 °C. The ribbons will then be stuck to the slide, which can then be stored more or less indefinitely as long as high temperatures are avoided.

Methacrylate sections
Transfer each individual section with forceps to a drop of dH_2O on a pre-coated slide[1] and then continue as for the paraffin sections.

[1] **Pre-coated slides**
Slides are pre-coated by dipping so as to improve the adherence of the sections to the microscope slide. We have used both poly-L-lysine-coated slides (0.5 mg ml^{-1} in dH_2O) and Biobond™ (British BioCell)-coated slides successfully.

After attachment of the embedded sections to the slides, a number of different possibilities exist prior to observation.

- Deparaffination and DPX mounting (paraffin sections)
- Deparaffination, counter-staining and DPX mounting (paraffin sections)
- Direct mounting in DPX (methacrylate sections)
- Counter-staining and mounting in DPX (methacrylate sections)

Light microscopic observations

Although some workers have tried to use the GUS system to investigate the localisation of proteins (GUS fusion proteins) at the ultrastructural level, the observation by Caissard *et al.* (1994) that di-X microcrystals were generally associated with various cytomembranes and lipid inclusions, even when a cytosolic location was expected, and that non-negligible intracellular diffusion took place, has limited the use of the GUS system for electron microscopy.

Under the light microscope, with bright-field optics, the product of the β-glucuronidase reaction is blue and can therefore be readily detected. In semi-thin and thin sections, it is often difficult to clearly discern the thin primary walls of certain cells (e.g. initials in the vascular cambium) and in this case a light counter-stain with Safranin O (which colours the cell walls pink) is useful (Figures 7 F, H, K). Box 8 gives details of the protocol that we use.

Box 8: Counter-staining cell walls with Safranin O

Counter-staining of methacrylate sections
Transfer slides to Safranin O solution[1] diluted 1:10, v/v with dH_2O (10 sec[2])
Wash slides in running tap water (3 min)
Rinse slides in dH_2O (5 sec)
Dry slides on a hot plate at 80 °C
Mount sections in DPX

Counter-staining of paraffin sections
Transfer slides to Safranin O solution[3] and leave overnight
Rinse slides in 95% EtOH (containing 3 drops 11 M HCl) until no more colour leaches out (generally approx. 30 sec)
Rinse slides in 95% EtOH (containing 3 drops of ammonia solution) for 15 sec
Rinse slides in 100% EtOH
Rinse slides in clove oil rinse[4]
Rinse slides in xylene substitute
Mount slides in DPX

Notes

1   Safranin O solution

For the methacrylate sections, we used the Safranin O solution provided in the Unicryl Staining kit (British BioCell International).

2   Staining times

The staining time has to be evaluated empirically for each sample-type

3   Safranin O solution

For the paraffin sections we used a classical (Johansen, 1940) stock solution of Safranin O (1 g Safranin O dissolved in 50 ml ethylene glycol monomethyl ether (EMMEG – in glassware, Attention – EMMEG dissolves plastic!), add 25 ml 95% EtOH and 25 ml dH$_2$O, then add 1 g sodium acetate and 2 ml formalin). Although this solution works extremely well, it would probably be worthwhile experimenting with the standard solution used for the methacrylate sections.

4   Clove oil rinse

Mix 1 part clove oil:2 parts 100% ethanol:2 parts xylene substitute.

The reaction product can also be viewed (in semi-thin and thin sections) using dark-field optics in which case it appears red against a black background (Figure 7 G). The utilisation of such optics also offers the advantage that cell walls (including those of fusiform and ray initials) are coloured silver-white and so the use of a counter stain is not necessary.

TROUBLE-SHOOTING, CAVEATS AND INTERPRETATIONS

Trouble-shooting and caveats

Table 1 details some of the problems that can be encountered, together with their (possible!) causes and solutions.

Table 1   Possible problems and causes in visualising the GUS colouration, and potential solutions

Problem	*Possible cause* – and solution
Little/no blue colouration	*Plant not transformed* – repeat experiment with positive control (definite GUS-expressing transformant); verify transformation by PCR
Little/no blue colouration	*Pre-treatment (fixation) too-long/strong* – repeat experiment without pre-treatment
Little/no blue colouration	*Weak promoter/gene not active* – verify GUS expression level by fluorimetry, measure corresponding 'enzyme' activity
Little/no blue colouration	*No/slight substrate penetration* – pre-dissect sample, pre-treat sample (e.g. acetone), add Triton X-100, DMSO (10%) to reaction medium, vacuum-infiltrate sample
Little/no blue colouration	*Section too thin* – cut thicker sections
Little/no blue colouration	*Insufficient incubation time/temperature* – increase incubation time and temperature (but not above 37 °C)
Too much colouration	*Strong promoter activity* – reduce incubation time/temperature
Colouration in unexpected tissues/cells	*Artefact or discovery* (!) – repeat experiment, verify by other techniques
Colouration in unexpected tissues/cells	*Wound-induced expression* – pre-treat sample with acetone
Blue colouration difficult to detect	*Blue masked by pigments/phenols* – clear sample (e.g. ethanol, CLP)
Blue colouration difficult to detect	*Blue masked by counter-stain* – reduce strength and counter-stain incubation time
Red colouration not visible with dark-field	*Sections too thick* – use thinner sections

The optimal conditions for the GUS histo-chemistry test have to be determined empirically for each species, organ/tissue and physiological state. Factors affecting the intensity of the blue colouration can be divided into basically 2 different categories:

- 'physical' (substrate penetration, over-fixation of GUS protein)
- 'molecular' (copy number, integration position, promoter strength and activity).

Many of the problems associated with substrate penetration have been discussed earlier (see the section entitled Sample preparation, histochem-istry and microscopy above).

At the molecular level the intensity of the blue colouration is fundamentally related to the quan-tity of the GUS protein produced via transcription and subsequent translation. The level of transcrip-tion is itself affected by the process of transgene integration. For example, variations between individual transformants have been related to the position of integration on the chromosome (Dean et al., 1988; An, 1989; Hobbs et al., 1990, Spiker and Thompson, 1996), as well as to the number of copies of the transgene incorporated (Hobbs et al., 1990; Bhattacharyya et al., 1994). In the first case, it is hypothesised that the transgene can either be incorporated into a transcriptionally-active, non-condensed region of the chromatin (euchromatin), in which case it is expressed, or else into a tran-scriptionally-inactive, condensed region (hete-rochromatin), in which case it is less likely to be actively expressed.

In the second case, multiple copies of the same sequence can result in the inactivation of the trans-gene – a mechanism known as 'gene-silencing' or 'homology-dependent gene silencing' (HdGS) since the mechanisms involved depend upon sequence homology (Matzke and Matzke, 1995).

It is also necessary to remember whether one is working with a homologous system (e.g. study of a promoter from a tobacco gene, in tobacco), or with a heterologous system (e.g. study of a promoter from an Arabidopsis gene in tobacco).This is import-ant since it has been shown (Shufflebottom et al., 1993; Cappellades et al., 1996) that the expression pattern of a given promoter in a heterologous system is not necessarily identical to that in a homologous system.

Finally, some investigations (Douglas et al., 1991; Sieburth and Meyerowitz, 1997) have shown that not all the regulatory regions of a gene – necessary for the control of its expression – are contained in the 5′ promoter region, and may be found, for example, either in the 3′ non-coding region of the gene or else in exons. In the case where not all the regulatory regions are contained in the promoter, the expression pattern – as determined by GUS his-tochemistry – may well be incomplete.

Verification and interpretation of results

Probably the easiest way of making sure that the blue colouration is due to GUS activity and is not artefactual is to determine the latter by fluorimetry in protein extracts of the organ/tissue under investigation using 4-methyl-umbelliferyl-D-glucuronide (MUG) as the substrate (Jefferson et al., 1987).

Another biologically valuable way of verifying that the blue colouration is not an artefact is to measure the activity/amount of the corresponding enzyme/protein. Since a blue colouration suppos-edly indicates that the promoter of the gene of interest is active, then presumably, the 'real' gene is also active and producing its protein which can then be measured. Such activity can be measured either in extracts from the whole organ or, if possi-ble, in individual tissues separated by dissection. For example, we (Hawkins and Boudet, 1994; Hawkins et al., 1997b) were able to show that the blue colouration identified in cells of the phloem and pith of poplar transformed with a chimeric CAD-GUS construct was probably related to the enzyme activity in the corresponding tissues, sug-gesting that the colouration was not an artefact. Western blots (as well as Northern blots or RT-PCR for the corresponding mRNA) can equally be used to verify the presence of the product of the gene of interest.

If the 'real' gene of interest is active then the corresponding mRNA and protein will be pro-duced. In this case, the spatial expression patterns of the gene – as indicated by reporter gene histo-chemistry – can be verified by in situ hybridisation (see the chapter by Regan and Sundberg in this volume) and/or immunolocalisation (see the chapter by Samaj and Boudet in this volume). Such an approach is also important since, as indicated earlier, not all the regulatory regions of a gene are necessarily contained in the promoter. Caution should, however, be used in interpreting these results since the different techniques have different sensitivities and are subject to different experimental constraints.

In interpreting the results of GUS histochemistry experiments it is important to remember that the GUS protein is not the same as the 'real' protein produced by the gene under study. The GUS protein is stable (De Block and Debrouwer, 1992) and its turnover rate is very probably lower than that of the 'real' protein thereby permitting its accumulation in the cell. In consequence, the gene-expression pattern obtained is *not* a 'snap-shot' image of promoter activity – but is rather a composite picture composed of the activity of the promoter prior to (and during) the preparation of the sample and its incubation in the reaction medium. In this context, techniques such as *in situ* hybridisation or immunolocalisation probably give a more precise idea of 'instantaneous' expression. However, it is probably this 'accumulation' of GUS protein (together with the less extensive manipulation) that is responsible for the greater 'sensitivity' of the GUS test in comparison with other *in situ* techniques. Table 2 illustrates this point in relation to the CAD 2 gene of eucalyptus.

APPLICATIONS – CHARACTERISATION OF LIGNIFICATION GENE EXPRESSION PATTERNS *IN SITU*

We have used the techniques described above to obtain detailed, high resolution information concerning the spatio-temporal expression patterns of genes coding for enzymes of the lignin biosynthetic pathway (Figure 6). Lignin is the phenolic polymer that reinforces and waterproofs the walls of the conducting cells (vessel elements, tracheids) and supporting cells (fibres) of the xylem tissue in higher plants. Indeed, it is the lignin that is responsible for the 'woodiness' of woody plants! The deposition of lignin in the cell walls of developing fusiform (and ray) initials is an integral and major event in wood formation. If we wish to have a thorough understanding of the molecular and cellular basis of wood formation in trees, then an understanding of the cellular and molecular basis of lignification is also necessary.

Such information is also important within a biotechnological context since the amount and type of lignin present affects the suitability of the raw material for industrial use. For example, during the production of wood pulp for the fabrication of paper, lignin has to be removed from the cellulose

Table 2 Patterns of CAD 2 expression in poplar stems revealed by different *in situ* localisation techniques

Tissue	Localisation Cell/tissue	GP 1	GP 2	ISH	Immuno.
Pith	Parenchyma/future sclereids?	+	+	–/+	–/+
Primary xylem	All cell types	+	+	–/+	–/+
Secondary xylem	Cambium (fusiform initials)/differentiating xylem (fibres, axial parenchyma, vessel elements)	–	+	+	+
Secondary xylem	Cambium (ray initials) and ray parenchyma	+	+	+	+
Secondary phloem	Fibres	–	+	+	+
Secondary phloem	CCC	+	+	+	+
Secondary phloem	Companion cells	–	+	–	–/+
Cortex	Druses/sclereids	+/–	+/–	–/+	–
Periderm	Phellogen	–	+	–	–/+
Periderm	Phelloderm	–	+	–	–/+

Notes:
+ = gene expression detected; – = gene expression not detected; +/– = probable gene expression (weak signal); –/+ = potential signal (weak signal or artefact ?)
GP 1 = GUS protocol 1 (sections made *prior* to incubation in GUS reaction medium – see the section entitled 'Sample preparation, histochemistry and microscopy');
GP 2 = GUS protocol 2 (sections made *after* incubation in GUS reaction medium – see the section entitled 'Sample preparation, histochemistry and microscopy').
ISH = *in situ* hybridisation (Hawkins *et al.*, 1997; Lacombe *et al.*, 1997)
Immuno. = immunolocalisation (Samaj *et al.*, 1998)
CCC = chambered crystalliferous cells (Trockenbrodt, 1995)

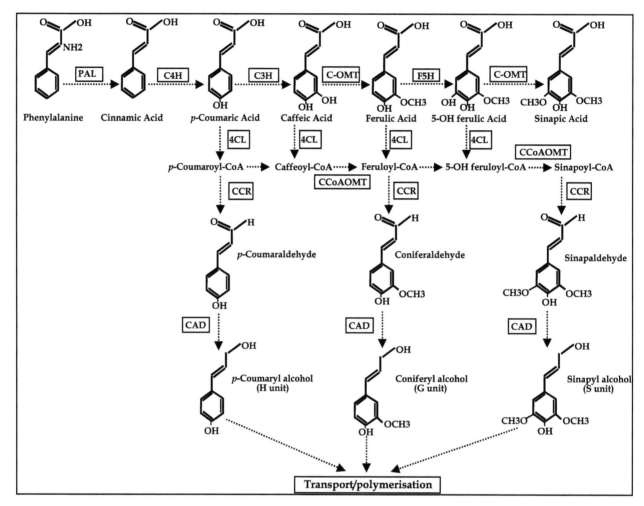

Figure 6 The 'classical' lignin biosynthetic pathway. Monolignols (p-coumaryl alcohol, coniferyl alcohol and sinapyl alcohol) are polymerised to form the 3-dimensional lignin polymer.
Key: PAL: phenylalanine ammonia-lyase; C4H: cinnamate 4-hydroxylase; C3H: coumarate 3-hydroxylase; C-OMT: caffeate/5-hydroxyferulate O-methyltransferase; F5H: ferulate 5-hydroxylase; 4CL: 4-coumarate co-enzymeA ligase; CCoAOMT: caffeoyl-co-enzymeA O-methyltransferase; CCR: cinnamoyl co-enzymeA reductase; CAD: cinnamyl alcohol dehydrogenase.

fibres in order to produce high quality paper that does not discolour with age. Presently, this separation is achieved by treatment with chemicals and is relatively expensive in both economic and environmental terms. In consequence, a number of biotechnological research programmes have been set up with the aim of modifying the amount/composition of the lignin molecule through a genetic engineering approach so as to make the extraction of lignin easier (Grima-Pettenati and Goffner, 1999).

One potential drawback with such an approach however, is that lignin also plays an important role in defence. Lignin is constitutively present in the outer protective bark layer (periderm) of many tree species and it is synthesised in response to pathogen attack and mechanical wounding (Vance *et al.*, 1980; Walters, 1992). In consequence, it is possible that plants down-regulated in their ability to synthesise lignin, or in which the composition of the lignin molecule has been altered, could be more susceptible to pathogen attack and abiotic stress. One solution to this potential problem would be to use tissue-specific, instead of constitutive, promoters to direct transgene expression so as to restrict the modifications in targeted enzyme activity to defined tissues (e.g. xylem, but not periderm). However, such a strategy supposes that we have available a promoter (or are able to construct a synthetic one) whose expression pattern is perfectly characterised at the tissue and cellular level and which is capable of directing transgene expression in trees in the required manner.

As part of our investigation into the process of lignification at the cellular level, we were trying to answer the following questions – where (in which cells and tissues) and when are the genes coding for enzymes of the lignification pathway expressed? The responses to such questions are important for a number of reasons.

First, it provides strong circumstantial evidence for the involvement of a gene in a given physiological process, for example, the expression of a 'lignification' gene would be expected to be spatially associated with lignifying tissues.

Secondly, it is known that the monomeric composition of the lignin molecule varies not only at the tissue and cellular level, but also between what can be described as 'developmental' lignin (deposited in cell walls as a 'normal' part of growth and development) and 'stress' lignin (deposited in response to abiotic and biotic factors) (Lange *et al.*, 1995). In such a context, a legitimate question is whether such variations are brought about by the differential expression of different genes in different tissues/physiological situations – the characterisation of the expression pattern of a given gene (s) is obviously one way of trying to answer such a question. Such information is important since, in conjunction with detailed *in situ* analysis (Müsel *et al.*, 1997; Stewart *et al.*, 1997) of cell lignin composition, it could enable us to correlate gene expression and lignin type paving the way to subsequent directed modifications of lignin at the tissue/cellular level.

Thirdly, a complete understanding of the spatio-temporal expression pattern is important for the identification of *cis*-acting promoter sequences responsible for the binding of proteins necessary for transcription.

Finally, a detailed knowledge of the patterns of tissue-specific and/or inducible gene expression is important in the larger biotechnological context of plant transformation. It is becoming clear that an optimal exploitation of such technology will depend upon the targeted and or inducible expression – via fully characterised promoters – to particular tissues/cells (Gatz and Lenk, 1998). This is particularly important in the case of strategies aimed at independently modifying lignin in different cells/tissues.

The use of high resolution, *in situ* techniques (Boudet *et al.*, 1995; Hawkins *et al.*, 1997 a, b; Lacombe *et al.*, 1997; Samaj *et al.*, 1998) has enabled us to extend our understanding of the expression of lignification genes to the cellular (and subcellular level). For example, Feuillet *et al.* (1995), using GUS histochemistry, had shown that cinnamyl alcohol dehydrogenase (CAD) expression in poplar was associated with lignified tissues and, more precisely, that in xylem cells, the CAD gene was expressed in ray cells but not in the differentiating cells (vessels, fibres) undergoing lignification. This observation led to the development of a theory of 'cell-cooperation' in which monolignols (lignin monomers) were synthesised in ray cells and transported to the sites of assembly in other cells (Feuillet *et al.*, 1995). However, the development of techniques taking into account the anatomy of woody angiosperms (Hawkins *et al.*, 1997b) showed that the CAD gene *is* expressed in those cells (vessels, fibres) that are undergoing lignification. This observation suggests that lignifying cells produce, at least initially, the monolignols necessary for their own lignification. Localisation (Lacombe *et al.*, 1997; Hawkins *et al.*, 1997a) by *in situ* hybridisation of the expression of another two key lignin enzymes (cinnamyl co-enzymeA reductase – CCR and caffeic acid *O*-methyltransferase – C-OMT) to developing xylem vessels and fibres would appear to support this hypothesis, as does the immunolocalisation of the phenylalanine ammonia lyase (PAL) protein (Osakabe *et al.*, 1996) and the C-OMT protein (Douglas *et al.*, 1997) to vessels and fibre cells in xylem poplar.

GUS histochemistry is also proving extremely useful in determining the biological roles of different members of 'lignin multigene families'. For example, Hu *et al.* (1998) used this technique (as well as Northern analysis and enzymological studies of the recombinant proteins) to investigate the role of two different 4-coumarate: CoA ligase (4CL) genes (Pt4CL1, Pt4CL2) cloned from aspen (*Populus tremuloides*).

The observation that Pt4CL1 promoter activity is associated with a blue colouration in lignifying xylem tissue in transgenic tobacco, whereas Pt4CL2 promoter activity is associated with the epidermal cells of stems and leaves led the authors to suggest that Pt4CL1 was involved in the formation of the cinnamoyl-CoA thioesters leading to monolignol formation, whereas Pt4CL2 was

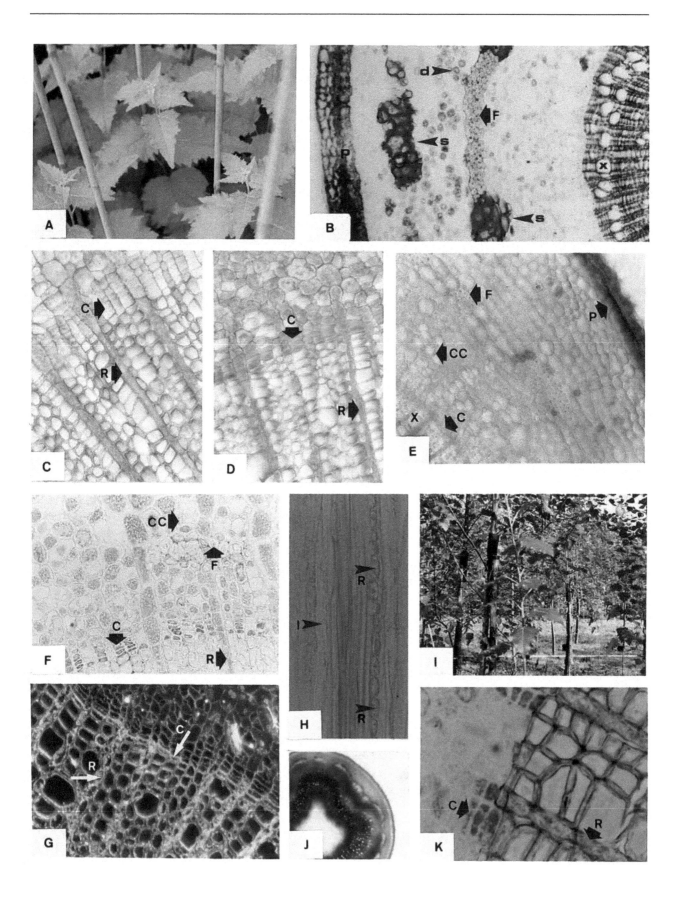

Figure 7   The use of GUS histochemistry to visualise lignification gene expression *in situ* during wood formation.
A) Transgenic poplar (INRA clone 717 1B4, *Populus tremula* L. x *P. alba* L.) transformed with the EuCAD-GUS construct growing under greenhouse conditions.
B) Phloroglucinol-stained hand-cut cross-section of a young twig from a transgenic poplar tree transformed with the EuCAD-GUS construct and growing under field conditions. Lignified cell walls are stained red; periderm (P), sclereids (S), fibres (F), druses (D), xylem (X); (x 100).
C) Promoter GUS histochemistry in a hand-cut cross-section of a stem from a greenhouse-grown transgenic poplar plant transformed with the EuCAD-GUS construct. Thin sections (50–100 μm) were prepared prior to incubation in GUS reaction medium (see text -– sample preparation – for details). GUS activity (blue colouration) is *only* detected in the xylem rays (R) but *not* in the vascular cambium (C); (x 200).
D) Promoter GUS histochemistry in a hand-cut cross-section of a stem from a greenhouse-grown transgenic poplar plant transformed with the EuCAD-GUS construct. Sections were prepared *after* incubation in GUS reaction medium (see text -– sample preparation – for details). GUS activity is detected in both xylem rays (R) and vascular cambium (C); (x 200).
E) Promoter GUS histochemistry in a hand-cut cross-section of a stem from a greenhouse-grown transgenic poplar plant transformed with the EuCAD-GUS construct illustrating the different areas of EuCAD promoter activity. GUS activity is detected in the xylem (X), phloem companion cells (CC), fibres (F) and periderm (P); (x 100).
F) Promoter GUS histochemistry in a paraffin-embedded semi-thin (10 μm) cross-section counterstained with Safranin O. GUS activity is detected in the vascular cambium (C), xylem rays (R), phloem companion cells (CC) and fibres (F); (× 200).
G) Promoter GUS histochemistry in a paraffin-embedded semi-thin (10 μm) cross-section (not counter-stained) viewed under dark-field optics. GUS activity (red colouration under dark field optics) is detected in the vascular cambium (C) and xylem rays (R); (x 200).
H) Promoter GUS histochemistry in a paraffin-embedded semi-thin (10 μm) tangential-longitudinal-section counterstained with Safranin O. GUS activity is detected in individual cells (R) of xylem rays as well as in a fusiform initial (I); (× 200).
I) Transgenic poplar (INRA clone 717 1B4, *Populus tremula* L. × *P. alba* L.) transformed with the EuCAD-GUS construct growing under field conditions (INRA Orléans, France).
J) Promoter GUS histochemistry in a hand-cut cross-section of a young twig harvested in July from a field-grown transgenic poplar tree transformed with the EuCAD-GUS construct; (× 10).
K) Promoter GUS histochemistry in a methacrylate-embedded semi-thin (10 μm) cross-section of a young twig harvested in March from a field-grown transgenic poplar tree transformed with the EuCAD-GUS construct. GUS activity is detected in the vascular cambium (C) and xylem rays (R); (× 400).

involved in the formation of non-lignin-related biosynthesis (e.g. flavonoids). This hypothesis was further supported by the observation that Pt4CL2, but not Pt4CL1, promoter activity was associated with a blue colouration in the floral organs. Subsequent analysis of the promoter sequences revealed that while 3 *cis*-acting elements (boxes P, A, L), which are characteristic of all known PAL and 4CL promoters, could be found in Pt4CL1, such elements were absent from the Pt4CL2 promoter sequence.

Similar approaches have also been adopted for investigating the biological roles of individual members of other multigene families in the phenylpropanoid pathway (Figure 6). For example, PAL is the most extensively studied member of this pathway. The corresponding gene(s) show a complex expression pattern and are induced by UV light, fungal elicitors and wounding, as well as being differentially-regulated in different organs, tissues and cells (Dixon and Paiva, 1995). For

example, in transgenic tobacco, the bean PAL2 gene is expressed in a number of different tissues including differentiating xylem while the PAL3 gene is not expressed in the xylem (Shufflebottom *et al.*, 1993). Similarly, in parsley plants and suspension cultures, PAL genes 1, 2 and 3 are induced by UV, elicitor treatment and wounding in both stems and leaves while the PAL4 gene is only induced by elicitor treatment and wounding in roots (Logemann *et al.*, 1995).

Recently, other genes of the phenylpropanoid pathway have also been cloned including cinnamate 4-hydroxylase (C4H) (Bell-Lelong *et al.*, 1997; Mizutani *et al.*, 1997; Ye, 1997), 4CL (Lee and Douglas, 1996), ferulate 5-hydroxylase (F5H) (Meyer *et al.*, 1996) and a cytochrome P450 monooxygenase (Osakabe *et al.*, 1999). In some cases, detailed cell-specific expression patterns of the cloned genes have been investigated by GUS histochemistry (Bell-Lelong *et al.*, 1997; Ye, 1997). However, as is the case with many other expression studies of phenylpropanoid genes, tree species were not used and caution should perhaps be taken in extrapolating the observed expression patterns in the secondary xylem of plants such as *Arabidopsis* to the more complex situation observed in trees. The increasing use of poplar as an easily-transformable model species should permit the detailed study of the expression patterns of these newly-cloned genes in a tree species.

Recent work (Meyer *et al.*, 1998; Zhong *et al.*, 1998; Osakabe *et al.*, 1999), looking at how the coordinated expression of different genes might be involved in controlling the monomeric composition of the lignin molecule is proving to be particularly exciting. Such approaches, together with detailed promoter expression studies, should help us to see whether and how the controlled expression of these genes can control the structure of lignin. As stated earlier, such information is important in a biotechnological context since the monomeric composition of the lignin molecule influences the suitability of the raw material for subsequent industrial use.

As part of a larger project at the Orléans INRA research station (France) on the long-term stability of transgene expression in trees, some of us (Hawkins *et al.*, 1997c) are using GUS histochemistry to investigate the seasonal expression patterns of the eucalyptus CAD promoter in field-grown transgenic poplar (Figure 7 I). Since CAD enzyme activity can be used as a biochemical marker of lignification and is known to show seasonal-dependent variations in activity, the CAD-GUS construct can be used as a visual marker of lignification and should provide important information about the seasonal expression of this key gene at the tissue and cellular level (Figures 7 J, K). Such information should enable us to correlate the observed cytological changes in cambial activity (Catesson, 1994; Farrar and Evert, 1997) with changes in gene expression thereby allowing us to better understand the molecular aspects of wood formation – in a tree species – and, in field conditions!

CONCLUSIONS AND FUTURE PROSPECTS

The last five years have seen much progress in our understanding of lignification. The cloning of many of the genes which code for enzymes involved in key steps has provided us with powerful new tools with which to investigate molecular aspects of this important biological process. These tools have been exploited in a number of complementary research strategies including: a) the investigation of spatial, temporal and inducible patterns of phenylpropanoid gene expression, b) down-regulation of given enzyme activity together with subsequent chemical analysis of the metabolites likely to be affected (e.g. lignin monomeric composition), and c) research for new genes playing a role in xylogenesis and lignification through the screening of *Arabidopsis* mutants (see also Chaffey, 2001).

The growing importance of the first experimental approach is demonstrated by the increasing number of research groups that are adopting high resolution *in situ* techniques as part of their strategy to investigate the molecular biology of cambial development and wood formation (as well as other physiological processes) in trees. For example at the recent IUFRO meetings in Quebec (1997) and Oxford (1999), a number of research teams presented work which included the use of such techniques.

Until recently, the potential of *Arabidopsis* mutants for the study of cell wall formation has not been fully exploited, despite the fact that the characterisation (Chapple *et al.*, 1992) of a mutant affecting F5H activity and the subsequent cloning of the corresponding gene (Meyer *et al.*, 1996) have clearly demonstrated the efficiency of such an approach. Recently however, projects have been initiated for the isolation of genes implicated in cell wall formation and lignification and several mutants showing abnormal xylem cell phenotypes have been characterised (Turner and Somerville, 1997; Jouanin *et al.*, 1999). It can be expected that such an approach should result in the cloning of new genes involved in the process of lignification whose expression patterns can then be studied by the techniques detailed in this and other chapters of this book.

The growing use of confocal microscopy techniques (see also the chapter by Funada, and the 'Wood microscopical techniques' chapter in this volume) to make 'optical sections' and to reconstruct the vascular system in 3-D (Dharmawardhana *et al.*, 1992; Travis *et al.*, 1997) should reduce the time necessary for the preparation of samples and prove invaluable in the establishment of a rapid and efficient screening process necessary for such an approach.

One disadvantage of the GUS histochemistry technique is that gene expression patterns cannot be followed directly in living material – samples must be isolated, the substrate introduced into the tissues and cells of the sample, and the reaction left to develop before the colouration can be detected. The development of a new reporter gene – GFP (Green Fluorescent Protein) (Chalfie *et al.*, 1994; Niedz *et al.*, 1995; Chiu *et al.*, 1996), which permits real-time observations of dynamic changes in living cells promises to be extremely valuable. This protein, isolated from the jellyfish *Aequorea victoria*, is capable of emitting green fluorescent light (509 nm) when excited by either blue (475 nm) or UV (395 nm) light and can be used to study gene expression, subcellular localisation, and trafficking of proteins and organelles *in vivo* (Haseloff *et al.*, 1997; Köhler, 1998).

ACKNOWLEDGEMENTS

SH would like to thank the teams of Jacqueline Grima-Pettenati and Alain Boudet (Toulouse), and Gilles Pilate (Orléans) for the warm welcome he received during the course of the work described in this chapter. Drs G. Engler and J. de Almeida Engler are thanked for their introduction to various aspects of 'molecular histology' and Nigel Chaffey is gratefully acknowledged for having had the original idea of the 'wood cook book'.

REFERENCES

An, G., Mitra, A., Choi, H.K., Costa, M.A., An, K., Thornburg, R.W. and Ryan, C.A. (1989) Functional analysis of the 3′ control region of the potato-inducible proteinase inhibitor II gene. *Plant Cell*, **1**, 115–122.

Baucher, M., Chabbert, B., Pilate, G., Van Doorsselaere, J., Tollier, M-T., Petit-Conil, M., Cornu, D., Monties, B., Van Montagu, M., Inzé, D., Jouanin, L. and Boerjan, W. (1996) Red xylem and higher lignin extractability by down-regulating a cinnamyl alcohol dehydrogenase in poplar (*Populus tremula* × P. *alba*). *Plant Physiology*, **112**, 1479–1490.

Beeckman, T. and Engler, G. (1994) An easy technique for the clearing of histochemically stained plant tissue. *Plant Molecular Biology Reporter*, **12**, 37–42.

Bell-Lelong, D.A., Cusumano, J.C., Meyer, K. and Chapple, C. (1997) Cinnamate-4-hydroxylase expression in *Arabidopsis* – regulation in response to development and the environment. *Plant Physiology*, **113**, 729–738.

Bevan, M., Shufflebottom, D., Edwards, K., Jefferson, R. and Schuch, W. (1989) Tissue- and cell-specific activity of a phenylalanine ammonia-lyase promoter in transgenic plants. *European Molecular Biology Organisation Journal*, **8**, 1899–1906.

Bhattacharyya, M.K., Stermer, B.A. and Dixon, R.A. (1994) Reduced variation in transgene expression

from a binary vector with selectable markers at the right and left T-DNA borders. *Plant Journal*, **6**, 957–968.

Boerjan, W, Christensen, J.H., Meyermans, H., Chen, C., Baucher, M., Van Doorsselaere, J., Petit-Conil, M., Lapierre, C., Leplé, J.C., Pilate, G., Jouanin, L. and Van Montagu, M. (1997) Understanding lignin biosynthesis in poplar through genetic engineering. *Joint meeting of the IUFRO Working Parties 2 04–07 and 2 04–06. Somatic Cell Genetics and Molecular Genetics of Trees*, Quebec, Canada, 12–16 August.

Boudet, A.M., Lapierre, C. and Grima-Pettenati, J. (1995) Biochemistry and molecular biology of lignification. *New Phytologist*, **129**, 203–236.

Brasileiro, A.C.M., Tourneur, C., Leplé, J.C., Combes, V. and Jouanin, L. (1992) Expression of the mutant *Arabidopsis thaliana* acetolacetate synthase gene confers chlorsulfuron resistance to transgenic poplar plants. *Transgene Research*, **1**, 133–141.

Caissard, J.C., Guivarc'h, A., Rembur, J., Azmi, A. and Chriqui, D. (1994) Spurious localizations of diX-indigo microcrystals generated by the histochemical GUS assay. *Transgenic Research*, **3**, 176–181.

Cappellades, M., Torres, M.A., Bastisch, I., Stiefel, V., Vignols, F., Bruce, W.B., Peterson, D., Puigdomenech, P. and Rigau, J. (1996) The maize caffeic acid O-methyltransferase gene promoter is active in transgenic tobacco and maize plant tissues. *Plant Molecular Biology*, **31**, 307–322.

Catesson, A.M. (1994) Cambial ultrastructure and biochemistry: Changes in relation to vascular tissue differentiation and the seasonal cycle. *International Journal of Plant Sciences*, **155**, 251–261.

Chaffey, N.J. (1999a) Cambium: old challenges – new opportunities. *Trees*, **13**, 138–151.

Chaffey, N.J. (1999b) Wood formation in forest trees: from *Arabidopsis* to *Zinnia*. *Trends in Plant Science*, **4**, 203–204.

Chaffey, N.J. (2001) Cambial cell biology comes of age. In *Trends in European Forest Tree Physiology Research*, edited by S. Huttunen, J. Bucher, P. Jarvis, R. Matyssek and B. Sundberg. (in press). Kluwer.

Chalfie, M., Tu, Y., Euskirchen, G., Ward, W.W. and Prasher, D.C. (1994) Green fluorescent protein as a marker for gene expression. *Science*, **263**, 802–805.

Chapple, C.C.S., Vogt, T., Ellis, B.E. and Somerville, C.R. (1992) An *Arabidopsis* mutant defective in the phenylpropanoid pathway. *Plant Cell*, **4**, 1413–1424.

Chiu, W-L., Niwa, Y., Zeng, W., Hirano, T., Kobayashi, H. and Sheen, J. (1996) Engineered GFP as a vital reporter in plants. *Current Biology*, **6**, 325–330.

Chupeau, M.C., Pautot, V. and Chupeau, Y. (1994) Recovery of transgenic trees after electroporation of poplar protoplasts. *Transgenic Research*, **3**, 13–19.

Craig, S. 1992. The GUS reporter gene. Application to light and transmission electron microscopy. In *GUS Protocols: Using the Gus Gene as a Reporter of Gene Expression*, edited by S.R. Gallagher. San Diego: Academic Press, Inc.

Dean, C., Jones, J., Favreau, M., Dunmuir, P. and Bedrook, J. (1988) Influence of flanking sequences on variability in expression levels of an introduced gene in transgenic tobacco plants. *Nucleic Acids Research*, **16**, 9267–9283.

De Block, M. and Debrouwer, D. (1992) *In situ* enzyme histochemistry on plastic-embedded plant material. The development of an artefact-free-β-glucuronidase assay. *Plant Journal*, **2**, 261–266.

Dharmawardhana, D.P., Ellis, B.E. and Carlson, J.E. (1992) Characterisation of vascular lignification in *Arabidopsis thaliana*. *Canadian Journal of Botany*, **70**, 2238–2244.

Dixon, R.A. and Paiva, N.L. (1995) Stress-induced phenylpropanoid metabolism. *Plant Cell*, **7**, 1085–1087.

Douglas, C.J., Ites-Morales, M-E., Ellard, M., Paszkowski, U., Hahlbrock, K. and Dangl, J.L. (1991) Exonic sequences are required for elicitor and light activation of a plant defense gene, but promoter sequences are sufficient for tissue specific expression. *European Molecular Biology Organisation Journal*, **10**, 1767–1775.

Douglas, C.J., Lee, D., Allina, S., Meyer, K., Chapple, C., Dharmawardhana, P., MacDonald, K., Grey-Mitsumuni, M., Carlson; J. and Ellis B. (1997) Coniferin-specific beta glucosidase and 4-coumarate: CoA ligase genes: targets for lignin modification in gymnosperm and angiosperm trees. *Joint meeting of the IUFRO Working Parties 2 04–07 and 2 04–06. Somatic Cell Genetics and Molecular Genetics of Trees*, Quebec, Canada, 12–16 August.

Farrar, J.J. and Evert, R.F. (1997) Seasonal changes in the ultrastructure of the vascular cambium of *Robinia pseudoacacia*. *Trees*, **11**, 191–202.

Feuillet, C., Lauvergeat, V., Deswarte, C., Pilate, G., Boudet, A. and Grima-Pettenati J. (1995) Tissue- and cell-specific expression of a cinnamyl alcohol dehydrogenase promoter in transgenic poplar plants. *Plant Molecular Biology*, **27**, 651–667.

Fillatti, J.J., Sellmer, J., McCown, B., Haissig, B. and Comai, L. (1987) *Agrobacterium* mediated transformation and regeneration of *Populus*. *Molecular Gene and Genetics*, **206**, 192–196.

Gatz, C. and Lenk, I. (1998) Promoters that respond to chemical inducers. *Trends in Plant Science*, **3**, 352–358.

Guivarc'h, A., Caissard, J.C., Azmi, A., Elmayan, T., Chriqui, D. and Tepfer, M. (1996) *In situ* detection of expression of the *gus* reporter gene in transgenic plants: ten years of blue genes. *Transgenic Research*, **5**, 281–288.

Grima-Pettenati, J. and Goffner D. (1999) Lignin genetic engineering revisited. *Plant Science*, **145**, 51–65.

Hänsch, R., Koprek, T., Mendel, R.R. and Schultze, J. (1995) An improved protocol for eliminating endogenous β-glucuronidase in barley. *Plant Science*, **105**, 63–69.

Haseloff, J., Siemering, K.R., Prasher, D.C. and Hodge, S. (1997) Removal of a cryptic intron and subcellular localization of green fluorescent protein are required to mark transgenic Arabidopsis plants brightly. *Proceedings of the National Academy of Sciences, USA*, **94**, 2122–2127.

Hawkins, S. and Boudet, A. (1994) Purification and characterisation of cinnamoyl alcohol dehydrogenase iso-

forms from the periderm of *Eucalyptus gunnii* Hook. *Plant Physiology*, **104**, 75–84.

Hawkins, S.W., Boudet, A. and Grima-Pettenati, J. (1997a). Tissue-specific expression of caffeic acid-O-methyltransferase (C-OMT) and lignin monomeric composition in eucalyptus. *Compte-rendus du Troisième Colloque de la Société Française de Physiologie Végétale*, Toulouse, France. Edited by J.C. Pech, A. Latache and M. Bouzagen, pp. 221–222. Paris: SFPV.

Hawkins, S.W., Samaj, J., Lauvergeat, V., Boudet, A. and Grima-Pettenati, J. (1997b) Cinnamyl alcohol dehydrogenase (CAD): Identification of new sites of promoter activity in transgenic poplar. *Plant Physiology*, **113**, 321–325.

Hawkins, S.W., Leplé, J.C., Tuffereau, B., Pugieux, C. and Pilate, G. (1997c) Stability of transgene expression in trees. *Compte-rendus du Troisième Colloque de la Société Française de Physiologie Végétale*, Toulouse, France. Edited by J.C. Pech, A. Latache and M. Bouzagen, pp. 397–398. Paris: SFPV.

Hemerly, A.S., Ferreira, P., de Almeida Engler, J., van Montagu, M., Engler, G. and Inzé, D. (1993) *Cdc2a* expression in Arabidopsis is linked with competence for cell division. *Plant Cell*, **5**, 1711–1723.

Hobbs, S.L.A., Kpodar, P. and DeLong, C.M.O. (1990) The effect of T-DNA copy number, position and methylation on reporter gene expression in tobacco transformants. *Plant Molecular Biology*, **15**, 851–864.

Jefferson, R.A. (1987) Assaying chimeric genes in plants: the GUS gene fusion system. *Plant Molecular Biology Reporter*, **5**, 387–405.

Johannsen, D.A. (1940) *Plant Microtechnique*. McGraw-Hill: New York.

Jouanin, L., Goujon, T., Ferret, V., Pollet, B. and Lapierre, C. (1999) *Arabidopsis thaliana*: a model plant for lignification studies. *Forest Biotechnology '99. Joint meeting of: The International Wood Biotechnology Symposium and The IUFRO Working Party 2. 04–06 Molecular Genetics of Trees*, Oxford, UK, 11–16 July.

Kohler, R.H. (1998) GFP for *in vivo* imaging of subcellular structures in plant cells. *Trends in Plant Science*, **3**, 317–320.

Lacombe, E., Hawkins, S., Van Doorsselaere, J., Piquemal, J., Goffner, D., Poeydomenge, O., Boudet, A.M. and Grima-Pettenati, J. (1997) Cinnamoyl CoA reductase, the first committed enzyme of the lignin branch biosynthetic pathway: cloning, expression and phylogenetic relationships. *Plant Journal*, **11**, 429–441.

Lange, B.M., Lapierre, C. and Sandermann, H. (1995) Elicitor-induced spruce stress lignin. Structural similarity to early developmental lignins. *Plant Physiology*, **108**, 1277–1287.

Lee, D. and Douglas, C.J. (1996) Two divergent members of a tobacco 4-coumarate:coenzyme A ligase (4CL) gene family – cDNA structure, gene inheritance and expression, and properties of recombinant proteins. *Plant Physiology*, **112**, 193–205.

Leplé, J.C., Brasileiro, A.C.M., Michel, M.F., Delmotte, F. and Jouanin, L. (1992) Transgenic poplars: expression of chimeric genes using four different constructs. *Plant Cell Reports*, **11**, 137–141.

Leplé, J.C., Bonadé-Bottino, M., Augustin, S., Pilate, G., Dumanois Lê Tan, V., Delplanque, A., Cornu, D. and Jouanin, L. (1995) Toxicity to *Chrysommmela tremulae* (Coleoptera: Chrysomelidae) of transgenic poplars expressing a cysteine proteinase inhibitor. *Molecular Breeding*, **1**, 319–328.

Logemann, E., Parniske, M. and Hahlbrock, K. (1995) Modes of expression and common structural features of the complete phenylalanine ammonia-lyase gene family in parsley. *Proceedings of the National Academy of Sciences, USA*, **92**, 5905–5909.

Matzke, M.A. and Matzke, A.J.M. (1995) How and why do plants inactivate homologous (trans)genes? *Plant Physiology*, **107**, 679–685.

McCown, B.H., McCabe, D.E., Russel, D.R., Robison, D.J., Barton, K.A. and Raffa, K.F. (1991) Stable transformation of *Populus* and incorporation of pest resistance by electric discharge particle acceleration. *Plant Cell Reports*, **9**, 590–594.

Meyer, K., Shirley, A.M., Cusumano, J.C., Bell-Lelong. D. and Chapple, C. (1996) Ferulate-5-hydroxylase from Arabidopsis thaliana defines a new family of cytochrome P450-dependant monooxygenases. *Proceedings of the National Academy of Sciences, USA*, **93**, 6869–6874.

Mizutani, M., Ohta, D. and Sato, R. (1997) Isolation of a cDNA and a genomic clone encoding cinnamate 4-hydroxylase from Arabidopsis and its expression manner *in planta*. *Plant Physiology*, **113**, 755–763.

Müsel, G., Schindler, T., Bergfeld, R., Ruel, K., Jacquet, G., Lapierre, C., Speth, V. and Schopfer, P. (1997) Structure and distribution of lignin in primary and secondary cell walls of maize coleoptiles analyzed by chemical and immunological probes. *Planta*, **201**, 146–159.

Niedz, R.P., Sussman, M.R. and Satterlee, J.S. (1995) Green fluorescent protein: an *in vivo* reporter of plant gene expression. *Plant Cell Reports*, **14**, 403–406.

Osakabe, Y., Nanto, K., Kitamura, H., Kawai, S., Kondo, Y., Fujii, T., Takabe, K., Katayama, Y. and Morohoshi, N. (1996) Immunocytochemical localization of phenylalanine ammonia lyase in tissues of *Populus kitakamiensis*. *Planta*, **200**, 13–19.

Samaj, J., Hawkins, S., Lauvergeat, V., Boudet, A. and Grima-Pettenati, J. (1998) Immunolocalisation of cinnamyl alcohol dehydrogenase 2 correlates well with cell-specific expression of GUS fusions in transgenic poplar shoots. *Planta*, **204**, 437–443.

Sambrook, J., Fritsch, E.F. and Maniatis, T. (1989) *Molecular Cloning: A Laboratory Manual*, Vols. 1, 2 and 3, 2nd edn. Cold Spring Harbor, NY: Cold Spring Harbor Laboratory.

Shufflebottom, D., Edwards, K., Schuch, W. and Bevan, M. (1993) Transcription of two members of a gene family encoding phenylalanine ammonia-lyase leads to remarkably different cell specificities and induction patterns. *Plant Journal*, **3**, 835–845.

Sieburth, L.E. and Meyerowitz, E.M. (1997) Molecular dissection of the AGAMOUS control region shows that cis elements for spatial regulation are located intragenically. *Plant Cell*, **9**, 355–365.

Smith, C.G., Rodgers, M.W., Zimmerlin, A., Ferdinando, D. and Bolwell, G.P. (1994) Tissue and subcellular immunolocalisation of enzymes of lignin synthesis in differentiating and wounded hypocotyl tissue of French bean (*Phaseolus vulgaris* L). *Planta*, **192**, 155–164.

Spiker, S. and Thompson, W.F. (1996) Nuclear matrix attachment regions and transgene expression in plants. *Plant Physiology*, **110**, 15–21.

Sterky, F., Regan, S., Karlsson, J., Hertzberg, M., Rohde, A., Holmberg, A., Amini, B., Bhalerao R., Larsson, M., Villarroel, R., Van Montagu, M., Sandberg, G., Olsson, O., Teeri, T.T., Boerjan, W., Gustafsson, P., Uhlén, M., Sundberg, B. and Lundeberg, J. (1998) Gene discovery in the wood-forming tissues of poplar: Analysis of 5,692 expressed sequence tags. *Proceedings of the National Academy of Sciences, USA*, **95**, 13330–13335.

Stewart, D., Yahiaoui, N., McDougall, G.J., Myton, K., Marque, C., Boudet, A.M. and Haigh, J. (1997) Fourier-transform infrared and Raman spectroscopic evidence for the incorporation of cinnamaldehydes into the lignin of transgenic tobacco (*Nicotiana tabacum* L) plants with reduced expression of cinnamyl alcohol dehydrogenase. *Planta*, **201**, 311–318.

Subramaniam, R., Reinold, S., Molitor, E.K. and Douglas, C.J. (1993) Structure, inheritance and expression of hybrid poplar (*Populus trichocarpa* × *Populus deltoides*) phenylalanine ammonia-lyase genes. *Plant Physiology*, **102**, 71–83.

Travis, A.J., Murison, S.D., Perry, P. and Chesson, A. (1997) Measurement of cell wall volume using confocal microscopy and its application to studies of forage degradation. *Annals of Botany*, **80**, 1–11.

Trockenbrodt, M. (1995) Calcium oxalate crystals in the bark of *Quercus robur, Ulmus glabra, Populus tremula* and *Betula pendula*. *Annals of Botany*, **75**, 281–284.

Turner, S.R. and Somerville C.R. (1997) Collapsed xylem phenotype of Arabidopsis identifies mutants deficient in cellulose deposition in the secondary cell wall. *Plant Cell*, **9**, 689–701.

Vance, C.P., Kirk, T.K. and Sherwood, R.T. (1980) Lignification as a mechanism of disease resistance. *Annual Review of Phytopathology*, **18**, 259–288.

Walters, M.H. (1992) Regulation of lignification in defense. In *Genes Involved in Plant Defense*, edited by F.M.T. Boller, pp. 327–352. Vienna: Springer.

Ye, Z.H. (1996) Expression patterns of the cinnamic acid 4-hydroxylase gene during lignification in *Zinnia elegans*. *Plant Science*, **121**, 133–141.

Ye, Z.H. and Varner, J.E. (1991) Tissue-specific expression of cell wall proteins in developing soybean tissues. *Plant Cell*, **3**, 23–37.

7 High Resolution *In Situ* Hybridisation in Woody Tissues

Wood Formation in Trees ed. N. Chaffey
© 2002 Taylor & Francis
Taylor & Francis is an imprint of the Taylor & Francis Group
Printed in Singapore.

Sharon Regan[1*] and Björn Sundberg[2]

[1]*Department of Biology, Carleton University, 1125 Colonel By Drive, Ottawa, Ontario K1S 5B6, Canada*
[2]*Department of Forest Genetics and Plant Physiology, Swedish University of Agricultural Sciences, SE-901 83 Umeå, Sweden*
*Author for correspondence

ABSTRACT

In situ hybridisation is an extremely powerful method for demonstrating the products of gene expression within cells. However, most of the published protocols for this technique are designed specifically for herbaceous species, and are largely unsuitable for trees. Woody plants, and wood-forming tissues in particular, have a number of peculiarities – such as non-specific binding of probes to xylem cell walls, and problems of adequately fixing the cytoplasm. This chapter describes a recently developed fast-freezing, freeze-substitution method for *in situ* hybridisation which has been specifically designed for trees.

Key words: cryo-immobilisation, freeze-substitution, gene expression, wood-forming tissues

INTRODUCTION

The analysis of gene expression patterns is an important step in the understanding of gene function and *in situ* hybridisation (ISH) permits their analysis at a cellular level. The technique involves a combination of histological preservation of tissue with the molecular detection of RNA. In general, tissue is fixed, embedded and sectioned to reveal the (ultra)structural features of the tissues. The location of specific RNAs is detected by the hybridisation of a specific labelled ribonucleotide probe.

Most protocols for ISH involve formaldehyde-fixation and paraffin-embedding. However, stem tissues of poplar, and probably also other trees, are not suitable for this type of sample preparation due to the excessive non-specific binding of the probes to the secondary cell walls, particularly those of the xylem cells. The fixation and embedding protocol described here uses glutaraldehyde-fixation and embedding in LR White acrylic resin. Preparation of samples by this regime results in high resolution ISH and reduces the problem of non-specific binding to the cell walls.

Another feature of the poplar stem which makes it more difficult to study by ISH is the tendency of cell contents, including RNA, to be displaced during conventional fixation protocols. We have found that the RNA in the vascular cambium is easily lost, and the RNA in the phloem, especially within the sieve-tube elements, can be displaced within the cell due to phloem-surging. For these reasons, a plunge-freezing, freeze-substitution method has been developed for ISH in wood stems, which cryo-immobilises the RNA prior to fixation. This results in superior preservation of cell contents and improved retention of RNA in tissues of the cambial region.

The protocol described here is based on the method of Leitch *et al.* (1994), with specific modifications for the analysis of gene expression in woody tissues as recently outlined by Regan *et al.* (1999).

TECHNIQUES

General precautions for working with RNA

One of the most difficult tasks involved with ISH is preventing RNase contamination, and the general precautions for controlling RNases should be carried out as described by Sambrook *et al.* (1989).

RNases are very stable enzymes that effectively degrade RNA. It should be presumed that RNase is ubiquitiously present as it is found in plant tissues, on hands, and objects which have come into contact with plants and hands. For this reason, all solutions and materials that will be used for ISH are treated to eliminate potential RNase contamination. As RNase is resistant to autoclaving, more effective measures are needed to remove the enzyme from solutions and materials.

The most effective method of removing RNase from solutions is treatment with diethylpyrocarbonate (DEPC). The details for each solution are presented within the protocol, but the general rule is that all solutions, except those which contain Tris, should be treated with 0.1% DEPC. The solution should be incubated at 37 °C for at least 2 h and then autoclaved for 15 min at 15 lb/sq. in., on the liquid cycle.

For solutions containing Tris, 1% DEPC needs to be added. Alternatively, reserve a separate bottle of Tris for the preparation of solutions for ISH, and add Tris after all other components have been treated with DEPC, and the solution has been autoclaved.

RNases can be removed from glassware and metalware by baking at 180 °C for at least 2 h. Sterile disposable plasticware can be used directly, while reusable plasticware must either be treated with chloroform or filled/submerged with water containing 0.1% DEPC and incubated and autoclaved as described for solutions. For plasticware which cannot be autoclaved, soak in water with DEPC as above, then rinse with DEPC-treated (and autoclaved) water to remove DEPC. To prevent RNase contamination from hands, wear gloves at all times and change often. Finally, reserve a separate set of micropipetters, which have not previously been used with RNase, for this work.

Recording the details of the experiment

Because there are several steps involved in any single ISH, it is impossible to perform the protocol identically each time. Therefore, it is advisable to record the exact details of the steps, since small changes in the protocol could have profound effects on the results. At the end of this chapter are examples of suggested itinerary tables for recording the details of all steps in the protocol. We recommend filling out these itinerary tables with the exact times of incubations for each experiment, and

keeping them for future reference (and trouble-shooting!).

PREPARATION OF SENSE AND ANTISENSE PROBES

For the accurate analysis of gene expression patterns, two probes are needed for ISH. Unlike the analysis of gene expression by northern blot analysis, where the specificity of the probe can be assessed based on the size of the hybridised transcript, the accuracy of an *in situ* probe can only be assessed if a control probe is also used. For this reason, it is common to make RNA probes, and to make two versions from the same template, an antisense probe and a sense probe. The antisense RNA probe is able to bind to the endogenous RNA due to complementary base pairing and is therefore the experimental probe. The sense RNA probe is unable to bind to the endogenous RNA and therefore serves as a control probe, revealing potential sites within the tissue able to attract the RNA probes non-specifically.

Formation of Templates

Equipment needed:	Chemicals needed:
Microcentrifuge	Appropriate restriction enzymes/buffers
Incubator at 37 °C	Plasmid purification kit or reagents
Agarose gel electrophoresis kit	
Eppendorf tubes (1.5 ml)	

Duration of procedure: approx. 5 h.

1   Your gene of interest must be cloned into an appropriate vector, e.g. Bluescript SK (Stratagene), such that one RNA polymerase promoter is located at the 5′ end of the gene (T3 in Bluescript SK) and the other RNA polymerase promoter is located at the 3′ end of the gene (T7 in Bluescript SK). After cloning of your gene into the vector, use routine protocols to isolate the plasmid. Either a miniprep protocol or a larger-scale protocol works fine, the only limitation is that the plasmid preparation procedure cannot include an RNase step, as this will interfere with probe preparation. For some templates

it is necessary to fill in the 5′ or 3′ overhangs left from the restriction digest. Use standard protocols such as Klenow or T4 DNA polymerase to fill in (Sambrook *et al.*, 1989)
Note – some non-specific binding of the probe during ISH can arise when transcripts are produced from Bluescript because the sequence in the multicloning site between *Sac*1 and *Xba*1 is homologous to a ribosomal RNA.

2 From the plasmid preparation, set up two restriction digests in Eppendorf tubes. In one, linearise the plasmid with any restriction enzyme that recognises a site at the 5′ end of the gene, this will produce the antisense (experimental) template. In the other, linearise the plasmid at the 3′ end of the gene to produce the sense (control) template.

3 Your templates must be perfectly linearised, as small amounts of undigested plasmid in your template preparation will interfere with the transcription. After digestion of the plasmid, purify the linearised plasmid by electrophoretic separation in an agarose gel (0.9%). If using a mechanism such as Gene Clean (Bio 101, Inc.) to purify the DNA from the gel piece, it is critical to include an extra centrifugation to ensure there is no carry-over of the DNA binding medium (Glass Milk) from the Gene Clean step, since it may interfere with the formation of labelled RNA probes.

4 Insert a photo of your linearised templates in the probe preparation itinerary.

Digoxigenin labelling of RNA transcripts

Equipment needed:	Chemicals needed:
Freezer at –20 °C	T3/SP6 DIG-RNA labelling kit (Boehringer Mannheim, No. 1175 025)
Microcentrifuge	RNasin
Agarose gel electrophoresis kit	tRNA
Eppendorf tubes (1.5 ml)	Ethanol Ammonium acetate

Duration of procedure: 3 h plus overnight precipitation.

Before starting, warm all components except the enzymes (T3, T7, SP6, and RNasin) to room temperature (RT) because the spermidine in the reaction buffer may precipitate the template if the components are cold. Digoxigenin (DIG)-labelling of RNA transcripts is carried out according to manufacturer's instructions.

For a typical reaction, 0.5 to 1.0 μg of linearised DNA is required. Since it is often difficult to measure accurately the DNA concentration by spectrophotometer, we generally compare the concentration of 1 μl of template with a known concentration of molecular weight marker after gel electrophoresis.

Transcription reaction

1 Set up the following reaction mixture in a 1.5 ml Eppendorf tube:

DNA	0.5–1.0 μg in 1–14 μl
Transcription buffer (10X)	2 μl
NTPs (DIG-labelled)	2 μl
DEPC-water	to 18 μl

2 Mix by flicking the tube with a finger and spin briefly in a microcentrifuge to collect contents in the bottom of tube.

3 Add the following to the reaction mixture:

RNA polymerase (T3, T7 or SP6)	1 μl
RNase inhibitor (RNasin)	1 μl

4 Mix by pipetting up and down and swirling with tip of pipette.

5 Incubate at 37 °C for 30 min–2 h.

6 Remove 1 μl, check the quality and concentration of the labelled probe on an 1% agarose gel by standard procedures (Sambrook *et al.*, 1989), and insert the photo into the probe preparation itinerary.
Note – a good transcription reaction generally results in 10-fold more labelled RNA than template. For an unknown reason, we find the T3 polymerase is less effective than the T7 polymerase in producing RNA transcripts.
Note – for the analysis of RNA concentration by gel electrophoresis, we find that the typical loading dyes (especially bromophenol blue) run at about the same position as the transcripts and often mask the ethidium bromide staining. To avoid this, we use either a very dilute concentration of bromophenol blue

(0.1–1% of standard recipe, Sambrook *et al.*, 1989), or use another dye, such as orange G, which has a much faster mobility in agarose gels and will not co-locate with the transcripts.

7   Stop the reaction and precipitate RNA by adding:

1X MS buffer	75 μl
tRNA (100 μg/μl)	2 μl
3.8 M ammonium acetate	100 μl
100% cold (–20 °C) ethanol	600 μl

1X MS buffer:
10 mM Tris, pH 7.5
10 mM MgCl$_2$
50 mM NaCl

8   Incubate at –20 °C overnight, or at –80 °C for at least 1 h

9   Centrifuge at maximum speed for 15 min at 4 °C (depending on the make of the centrifuge, maximum speed is between 10–13,000 rpm), and discard the supernatant. You should have a clearly visible pellet in the bottom of the tube at this point.

10   Wash the pellet by adding 500 μl of a cold (–20°C) 70% ethanol: 30% 0.15 M NaCl mixture, and centrifuge again for 15 min at 4 °C.

11   Air-dry the pellet.

12   Resuspend pellet in 50 μl DEPC-water.
     Note – it is better to resuspend the pellet in water at this time since Tris buffer could interfere with the carbonate-hydrolysis.

13   To remove the template after transcription, 40 units of RNase-free DNase (Boehringer Mannheim) can be added and incubated for 15 min at 37 °C.
     Note – we generally omit this step if we have a good transcription reaction (at least 10X more RNA than template) since the relatively small amounts of DNA present do not appear to interfere with the ISH.

Carbonate-hydrolysis of probe

Equipment needed:	Chemicals needed:
Freezer at –80 °C	Ethanol
Freezer at –20 °C	Acetic acid
Eppendorf tubes (1.5 ml)	DEPC-treated Tris, ethylene diamine tetraacetic acid (EDTA) (DEPC-treated TE)
Microcentrifuge	NaCl
	Sodium acetate
	Sodium carbonate
	Sodium bicarbonate

Duration of procedure: 1 h plus overnight or 2 h precipitation.

When initiating a new ISH experiment, we typically hydrolyse only 1/2 of the probe, so that the final probe mixture for ISH contains 1/2 full length probe and 1/2 hydrolysed probe. Once we have achieved a hybridisation signal, the choice of probe (hydrolysed vs unhydrolysed) is determined empirically.

1   Remove 25 μl and place in a clean 1.5 ml Eppendorf tube.

2   Add 25 μl of 200 mM carbonate buffer (pH 10.2), mix, and incubate at 60 °C for the appropriate length of time (see below).

2X carbonate buffer:
80 mM NaHCO$_3$
120 mM Na$_2$CO$_3$
Should be pH 10.2, if made correctly

3   Add: 5 μl   10% acetic acid
             6 μl   3 M sodium acetate, pH 5.2
             156 μl   100% ethanol (–20 °C)

4   Incubate at –20 °C overnight, or at –80 °C for 2 h.

5   Centrifuge for 15 min in a microcentrifuge at 4 °C at top speed.

6   Wash pellet with 500 μl of ice-cold 70% ethanol:0.15 M NaCl mixture.

7   Air-dry the pellet.

8   Resuspend pellet in 25 μl DEPC-treated TE, and store at –80 °C until needed.
     Note – although RNA probes should be stable indefinitely if stored at –80 °C, we generally use the probes within 1 month.

Calculating the hydrolysis time for RNA probes
(from Leitch *et al.*, 1994)

$$t = \frac{Li - Lf}{K \times Li \times Lf}$$

Where: t = time
          K = rate constant (0.11 Kb/min)
          Li = initial length (Kb)
          Lf = final length (Kb)

The optimum length for *in situ* probes is between 150–500 bp. Thus, if your insert is 1.5 Kb, the hydrolysis time can be determined as follows:

For 150 bp:

$$\frac{1.5 - 0.15}{0.11 \times 1.5 \times 0.15} = 54.5 \text{ min}$$

For 500 bp:

$$\frac{1.5 - 0.5}{0.11 \times 1.5 \times 0.5} = 12.1 \text{ min}$$

Therefore, a hydrolysis time between 12 min and 55 min will give a probe of the correct size. We would use about 30 min as the hydrolysis time, but will need to adjust this time depending on the amount of background signal.

TISSUE PREPARATION

Growth of hybrid aspen

The protocol described here has been developed for hybrid aspen (*Populus tremula* × *P. tremuloides*). Hybrid aspen are grown in a greenhouse with an 18 h photoperiod, a temperature of 22/17 °C (day/night), and a relative humidity of at least 70%. Natural daylight was supplemented with light from HQi-TS 400W/DH metal halogen lamps (Osram). Plants were watered daily and fertilised once a week with a 1:100 dilution of SUPERBAS (Hydro Supra AB).

Fixation of tissue

Equipment needed:	Chemicals needed:
Liquid nitrogen dewar	Liquid nitrogen
Freezer at –80 °C	Methanol
Freezer at –20 °C	DEPC-water
Polypropylene embedding capsules (8 mm, flat-bottomed, TAAB, No. C095)	LR White acrylic resin (TAAB, Medium grade, No. L012)
50 ml polypropylene tubes	Glutaraldehyde 25% EM grade
Large beaker (2–4 l capacity)	
Rotator	
Razor blades	

Duration of procedure: 1–4 weeks for freeze substitution and 1 week for embedding.

1 Place the fixative (1.25% glutaraldehyde in 100% methanol) in a 50 ml sterile disposable polypropylene tube and store at –70 to –80 °C for several hours before adding the samples.

Note – the fixation method presented here works well for poplar stems, but there are several other fixative variations that could, and should, be tested in your own tissue. The type of fixative and the concentration needed to accurately immobilise RNA could vary in other tissues and other trees. It is therefore worthwhile to test a couple of other fixation conditions with your tissue to assess the effect of fixation on ISH. The following fixatives would be useful to test:

> 2% formaldehyde instead of 1.25% glutaraldehyde
> 2% formaldehyde in addition to 1.25% glutaraldehyde

2 Cut the stem from the tree, place the cut end immediately in a large beaker of water, and recut the stem beneath the water level. In general we use a stem that is 1.5–2.0 cm in diameter at its base.

3 Starting at the bottom of the stem piece, cut an 8–10 cm length (which will be used for samples), and recut the bottom of the stem under water.

4 Working quickly, using a single-edged knife, cut pieces of stem as described in Figure 1 and plunge immediately in liquid nitrogen. Repeat on another portion of the stem piece, or cut a fresh 8–10 cm length from the stem. We generally sample throughout the stem so that expression analysis can be performed on young and older stem pieces.

Note – the integrity of the samples is checked both after cutting and after freezing in liquid nitrogen. The most common problem is separation of the bark from the xylem due to breakage in the cambial region (see also the chapter 'An introduction to the problems of working with trees' in this volume). Discard any broken samples.

5 Place the frozen samples into the cold fixative, gently mix the fixative/samples by inverting the tube several times, and return to the –80 °C freezer. The amount of tissue should not exceed 1/3 of the volume of the fixative. Since very few rotating devices

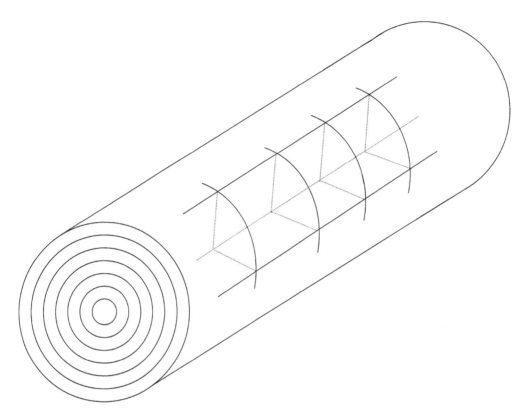

Figure 1 Sampling strategy for the isolation of intact poplar wood. Begin by making 8–10 radial slices in the stem, (only 4 slices are shown in the figure) about 0.3–0.5 cm apart, but only half-way through the stem. Then make two longitudinal cuts, intersecting all the radial cuts, and separated by about 0.5 cm. The two longitudinal cuts should be made such that the cuts meet within the stem, usually in the pith, and gently remove the pie-shaped pieces.

function at –80 °C, the samples are mixed by hand several times per day by gently inverting the tube.

6   Freeze-substitution should continue for several days; the time must be determined empirically for each tissue. We recommend performing test fixation times of 1, 2, 3 and 4 weeks and assessing the anatomy and ISH signal at each time.

7   Following freeze-substitution, the samples are gradually warmed to 4 °C by transferring to –20 °C for 1 day, on ice for 1 day, and then kept at 4 °C for the remainder of the embedding procedure.

Note – If you are using green tissue, you can get an indication of whether the samples have been kept at –80 °C for the duration of the freeze-substitution by looking at the colour of the fixative. If the tissue has remained sufficiently cold during the fixation, the solution will be colourless. If the fixation solution is green (due to leakage of chlorophyll), this indicates that the samples have warmed during the freeze-substitution, and the tissue may not be adequately fixed. To prevent warming of the tissue during fixation, mix the samples as quickly (and gently) as possible and ensure that the freezer itself does not warm up due to repeated opening by multiple users. Once the samples are moved to –20 °C, the fixative should become green.

8   Remove the fixative and replace with 100% methanol pre-cooled to 4 °C and continually mix on a rotator for 1 h at 4 °C. Gradually infiltrate with LR White acrylic resin by incubating in the following resin: 100% methanol mixes, 25%:75%, 50%:50% and 25%:75% for at least 6 h each at 4 °C.

Note – it may be necessary to degas the LR White: methanol solutions to improve infiltration. In these cases, a gentle vacuum (water aspirator, for example) is applied for 5–10 min after samples are placed in the new mixture, and then the samples are incubated at 4 °C for the remaining time.

9   Five changes in pure LR White for a least 6 h each at 4 °C are needed to ensure the complete removal of the methanol and infiltration of the tissue.

10 Orient the samples in the bottom of an embedding capsule so that transverse or tangential longitudinal sections can be produced. Avoid introducing air into the LR White as it will interfere with polymerisation. Fill with fresh LR White to the top of the capsule, insert a label (markings with pencil on high quality writing paper work best), close the lid, and polymerise at 65 °C for 15 h.

Sectioning of samples for *in situ* hybridisation

Equipment needed:	Chemicals needed:
Microtome	DEPC-treated water
Glass or diamond knives	
Slide warmer at 60 °C	
Silanised slides (DAKO A/S, No. S3003)	
Knife boats	
Sterile Pasteur pipettes	
Paint brush	

Duration of procedure: depends on sectioning skills.

1   Trim the resin block faces to an area no greater than 0.5 cm², and cut 2 μm sections. Float the sections onto DEPC-treated water in the knife boat.
    Note – we use a Microm HM350 microtome, but any other type that can accurately section at 2 μm is suitable.

2   Transfer the sections to a pool of DEPC-treated water on silanised slides (prewarmed to approx. 60 °C on a slide warmer) using a DEPC-treated paintbrush. Allow sections to stretch on the warm water for several min before removing any excess water with a sterile Pasteur pipette.

3   Heat-fix sections to the slides at 60 °C for at least 12 h.
    Note – we generally prepare fresh sections immediately before each ISH to reduce the risk of RNase contamination of the slides and

cover the slide warmer to prevent dust from landing on the slide.
Note – we have tried several types of slides but only silanised slides retain the sections throughout the ISH protocol. We find that the Dako slides are superior to silanised slides prepared in the lab.

IN SITU HYBRIDISATION

Pre-treatment and pre-hybridisation

Equipment needed:	Chemicals needed:
Slide-staining dishes	DEPC-treated water
Stir plate	Pronase
Glass beaker (400 ml)	Glycine
Microwave oven	*p*-formaldehyde
Pipettes	Triethanolamine
pH indicator papers	Acetic anhydride
Polypropylene slide boxes	Polyvinylpyrrolidone (PVP)
Incubator at 55 °C	NaOH
	H₂SO₄
	NaCl
	NaH₂PO₄.H₂O
	Na₂HPO₄

Pre-treatment

Duration of procedure: approx. 3 h.

For the pre-treatment, we use the Tissue-Tek II celcon slide staining dishes (Histolab AB, No. 4456) which can be repeatedly autoclaved and are safe for solvents. A solvent-resistant slide rack is supplied with the trays.

To prepare the staining dishes for the pre-treatment, 100 ml of water with DEPC is added, shaken, and stored at 37 °C for at least 12 h, and autoclaved on liquid cycle. The slide rack is treated in one of the slide dishes, containing enough water plus DEPC to cover the portion of the rack that comes in contact with the slides (approx. 200 ml).

When performing the pre-treatment, start by making the 4% *p*-formaldehyde, and also be sure to remove the pronase from the freezer so that it is thawed when needed. All steps are carried out at RT and the 4% *p*-formaldehyde and the acetic anhydride steps must be used in the fume hood.

For each step, it is important that you *gently* dip the rack in each solution 20 times, and then leave for the specified time. Vigorous dipping could result in the loss of sections.

> Acetic anhydride is used in the following procedure.
>
> It is very dangerous, since it can explode when it comes into contact with water.
>
> Therefore, the following precautions must be followed:
>
> 1 Always use in the fume hood, and keep the sash down as low as possible.
> 2 Use a clean *dry* pipette to remove the required amount from the bottle, check that the autoclave has not left residual water in the tip of the pipette.
> 3 Only add acetic anhydride to the triethanolamine solution, never add it to pure water. Be certain that the triethanolamine is fully mixed before adding the acetic anhydride.

1 Place slides in the DEPC-treated slide rack and incubate in **phosphate-buffered saline (PBS)** for 2 min.
2 Transfer to 0.125 mg/ml pronase for 10 min.

3 Stop the pronase reaction by transferring to 0.2% glycine for 2 min.
4 Fix samples in 4% *p*-formaldehyde for 10 min.
5 Incubate twice in PBS for 2 min each.
6 Incubate slides in two changes of 0.5% acetic anhydride in 0.1 M triethanolamine to completely block the positive charges on the sections.

Note – take note of the specific instructions for preparing the acetic anhydride as outlined above and in the table for stock solutions for pre-treatment.
7 Wash the sections in PBS for 2 min.

Pre-hybridisation

Duration of procedure: approx. 5 min plus at least 1 h incubation.

1 Transfer slides to a DEPC-treated polypropylene slide box (Histolab AB, No. 0530), 5 slides/box, making sure that the section-side of the slides does not face the wall of the slide box.
2 Add sufficient **pre-hybridisation solution** to each slide box to cover the sections on the slides.
3 Incubate slides at 55 °C for at least 1 h.

Stock solutions for pre-treatment and pre-hybridisation

Stock Solutions	Dilution for pre-treatment
10X Phosphate-buffered saline (PBS) for 1l 1.3M NaCl 74 g 0.07M Na$_2$HPO$_4$ 9.94 g 0.03M NaH$_2$PO$_4$.H$_2$O 4.14 g DEPC-treat as usual	20 ml of 10X PBS, make up to 200 ml with DEPC-ddH$_2$O
10X Pronase Buffer 0.5M Tris-HCl, pH 7.5 0.05M EDTA in DEPC-H$_2$O	20 ml 10X pronase buffer to 200 ml with DEPC-H$_2$O + 0.6 ml pronase
Pronase (Sigma type XIV, No. P6911) Make up to 40 mg/ml in DEPC-H$_2$O Pre-digest by incubating for 4 h at 37 °C Store at –20 °C in 1 ml aliquots	

10% Glycine
10 g in 100 ml H_2O, DEPC-treat as usual

3.4 ml 10% glycine + 20 ml 10X PBS to 200 ml with DEPC-ddH_2O

4% Formaldehyde
See details in 'Preparation of *p*-formaldehyde'

2M Triethanolamine
Cannot be autoclaved. So pour 74.5 g (in a 50 ml sterile tube) triethanolamine into an autoclaved bottle, then make up to 250 ml with DEPC-dd H_2O adjust pH to 8 using HCl. (Note, this takes a lot of HCl.) The solution should become clear

8.75 ml 2M triethanolamine + 165.5 ml DEPC-ddH_2O + 0.85 ml acetic anhydride
Note – the 0.1 M triethanolamine solution can be prepared in advance but the acetic anhydride should be added just before use (in a fume hood). With the 0.1 M triethanolamine stirring, add the acetic anhydride quickly, continue to stir for a couple seconds and quickly transfer the solution to the appropriate tray

10X SALTS
3M NaCl
0.1 M Tris-HCl, pH 6.8
0.1 M $NaPO_4$ buffer
50 mM EDTA
Store at RT

For 1 litre
175.3 g NaCl
7.8 g $NaH_2PO_4.H_2O$
7.1 g Na_2HPO_4
100 ml of 0.5M EDTA solution
Add DEPC, autoclave then add 100 ml of 1M solution Tris-HCl, pH 6.8

Pre-hybridisation solution
0.1% polyvinylpyrrolidone (PVP) in 1X salt

Preparation of *p*-formaldehyde

Duration of procedure: approx. 1 h.
This fixative must be made fresh from powder or it will not be effective. It is used in the fixation of tissue prior to embedding and in the pre-treatment of tissue for ISH.
To make:

1 Make 200 ml of 1X PBS in a 400 ml beaker.
2 Add 2 pellets of NaOH.
3 Heat in the microwave until mixture is approx. 60 °C.
4 Add 8 g of *p*-formaldehyde (in the fume hood), and swirl mixture to dissolve.
5 Cool the solution to RT.
6 Adjust to pH 7 with H_2SO_4 (not HCl, since this will release a carcinogen).

We generally check the solution with pH indicator paper to decrease the risk of introducing RNases.

HYBRIDISATION

Equipment needed:	Chemicals needed:
Incubator at 55 °C	Formamide
Slide mailing boxes	tRNA
Sections on silanised slides	Deionised formamide
Water bath at 80 °C	Denhardt's solution
RNase-free cover slips	Dextran sulphate (DS)
Hybridisation box	Ice
Sterile forceps	Reservoir solution

Duration of procedure: approx. 3 h.

Note – the hybridisation temperature needs to be determined empirically, but 55 °C is a good starting temperature. Increase the temperature if there is signal with the sense probe.

Riboprobes are most commonly used at a concentration of 0.1–0.3 ng/μl/Kb probe length. If you have hydrolysed 1/2 of your probe, make sure you have mixed it with the unhydrolysed probe.

The amount of probe added to the slide depends on the concentration of the probe. You need to estimate the concentration from the gel you ran after the transcription reaction. Generally, if the transcription is good, only 1 μl of probe is needed per slide.

Since it is difficult to measure accurately the concentration of the probe, it is very important that the concentration of the sense and antisense probes are the same in each experiment. Again, you must estimate the concentration of the probes from the gel after the transcription reaction. If, for example, one of the probes is about 1/2 as concentrated as the other, then you would need to add 2 μl per slide of the less concentrated probe, and 1 μl per slide of the more concentrated probe. Therefore a good picture of your probe following the transcription reaction is essential to the success of the following steps.

Preparation of probe mix

For 1 slide (example)

   1 μl probe
   7 μl ddH$_2$O
   8 μl deionised formamide

If more than one slide is treated with the same probe, we recommend making a master mix of the probe for n + 1 slides (where n is the number of slides). Therefore, if you intend to hybridise 4 slides, you should make a mixture for 5 slides.

Before adding the probe to the hybridisation buffer, heat the probe mix to 80 °C for 2 min (to denature the probe and thereby enable it to access the RNA in the tissue), cool on ice and spin down, then add to the hybridisation buffer and mix thoroughly.

HYBRIDISATION BUFFER

Components		1	8 (9)	12 (13)	16 (17)
10X salts	(μl)	10	90	130	170
deionised formamide	(μl)	40	360	520	680
50% dextran sulphate	(μl)	20	180	260	340
100 mg/ml tRNA	(μl)	1	9	13	17
50X Denhardt's solution	(μl)	2	18	26	34
ddH$_2$O, sterile and filtered	(μl)	7	63	91	119
TOTAL VOLUME	(μl)	80	720	1040	1360

When we make hybridisation mixes that will be applied to several slides, we make extra solution because there is always some pipetting error. So, if you have 8 slides, then you would be advised to make sufficient hybridisation buffer for 9 slides.

Dextran sulphate (DS) is very viscous and very difficult to pipette. Follow these tips to ensure that it is measured correctly:

1   Heat the DS to 37 °C to make the solution more liquid and easier to pipette.
2   When making the hybridisation buffer, add the DS last.
3   When pipetting, place pipette tip into DS and wait about 30 sec so that the DS has plenty of time to enter the pipette tip completely.
4   When adding the DS to the rest of the hybridisation buffer solutions, pipette up and down several times to ensure all the DS has been added to the hybridisation buffer. Vortex vigorously, invert the tube, and tap it to ensure that all the DS is suspended in the rest of the buffer.

1   Once the hybridisation buffer is ready, add the freshly denatured probe to make the final hybridisation mix (for each slide, the hybridisation mix includes 80 μl hybridisation buffer plus 16 μl probe mix).

2   Remove a slide from the pre-hybridisation solution using sterile forceps. Shake the pre-hybridisation liquid off and remove any that remains with a clean tissue so that the slide is almost dry, and place on a clean surface.

3   Add 90–95 μl of hybridisation mix to the centre of the slide, and spread it with the tip of your pipette, but without touching the sections.

4   Using sterile forceps, remove the plastic film from the RNase-free coverslips (Hybrislip, Surgipath Medical Industries Inc., No. 002HS22) and gently place the coverslip on top of the slide.

5   Hybridise at 55 °C overnight in a hybridisation chamber containing reservoir solution.

Note – there are several ways to create a hybridisation chamber; we use plastic sandwich boxes and glue the top of a microtitre plate to the bottom of the box (Figure 2). Six slides rest on the microtitre plate in each box. To maintain a humid environment in the chamber, 20–30 ml of reservoir solution is carefully poured into the reservoir around the microtitre plate. To avoid dilution of the hybridisation mixture, it is important to make sure the reservoir solution does not come in contact with the slides.

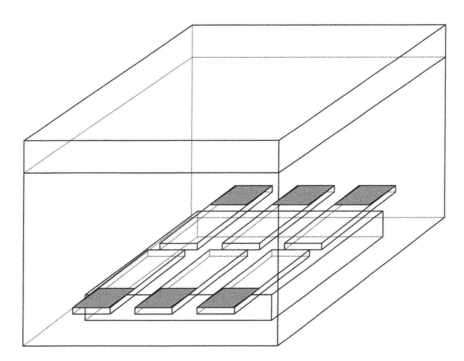

Figure 2 Hybridisation chamber. Glue the top of a microtitre plate to the bottom of a plastic box, such as a clear plastic sandwich box. Six slides fit on to the microtitre plate top as shown. The reservoir solution is carefully poured into the reservoir, and the box is tightly closed.

STOCK SOLUTIONS FOR HYBRIDISATION

Stock solutions	Preparation of stock solutions
10X SALTS 3M NaCl 0.1M Tris-HCl, pH 6.8 0.1M NaPO$_4$ buffer 50 mM EDTA Store at RT	*For 1 litre* 175.3 g NaCl 7.8 g NaH$_2$PO$_4$.H$_2$O 7.1 g Na$_2$HPO$_4$ 100 ml of 0.5M EDTA solution Add DEPC, autoclave then add 100 ml of 1M Tris-HCl, pH 6.8
tRNA (Sigma type XXI, No. R4251) 100 mg/ml	100 mg/ml in DEPC-H$_2$O aliquot and store at –20 °C
Formamide Deionise for hybridisation buffer Use straight from bottle in washes	*Deionised formamide* 15 ml formamide 2–3 ml mixed bed resin (Bio Rad) mix for several hours, then filter aliquot and store at –80 °C
50X Denhardt's solution	*For 100 ml* 1 g Ficoll (Type 400) 1 g polyvinylpyrrolidone 1 g bovine serum albumen (Sigma, Fraction V) make up to 100 ml with DEPC-H$_2$O aliquot and store at –20 °C
50% Dextran sulphate (DS)	*For 10 ml* 5 gm and DEPC-H$_2$O to 10 ml start with about 4 ml of DEPC-water stirring with a sterile stir bar, add the DS slowly until completely dissolved, then add the remaining water to a total volume of 10 ml aliquot and store at –20°C
20X SSC 3.0 M NaCl 0.3 M Na$_3$ citrate	*Reservoir solution* 2X SSC 50% formamide

POST-HYBRIDISATION WASHING OF SLIDES

Equipment needed:	Chemicals needed:
Rotary shaker Incubator with shaker at 55 °C Incubator with shaker at 37 °C	Formamide RNase Wash buffer NTE 1 X PBS

Duration of procedure: approx. 2.5 h.

The following is the protocol for the post-hybridisation washing of slides which includes an RNase step to reduce the signal from non-specifically bound probe. In general, we perform ISH without RNase, to decrease the potential for contaminating the lab. However, when a clear signal cannot be achieved without RNase, this protocol is followed. To perform the post-hybridisation washes without RNase, omit steps 5–7.

Note – solutions used after the RNase step are not treated with DEPC.

1   Dip slides one at a time in wash buffer until the coverslips come off.
    Note – it is important that the cover slips are not prised off as this will also cause the sections to fall off.
2   Wash for 4 × 15 min in wash buffer at RT.

3   Wash for 5 min at 37 °C in NTE.
4   Wash for 5 min at 37 °C in NTE.
5   Incubate for 30 min at 37 °C in NTE with RNase. Just before use add 0.7 ml of 10 mg/ml RNase to 350 ml of NTE.
6   Wash for 5 min at 37 °C in NTE.
7   Wash for 5 min at 37 °C in NTE.
8   Wash for 5–30 min at 55 °C in wash buffer.
    Note – the timing for this step needs to be determined empirically for each probe or tissue. We generally start with a short washing step and proceed to longer washing if there is too much non-specific binding.
9   Wash for 5 min at RT in 1X SSC.
    Note – higher stringency can be achieved by decreasing the salt concentration in this step to 0.5X or 0.1X SSC.
10 Wash for 5 min in 1X PBS at RT.
11 Replace with fresh 1X PBS and store at 4 °C overnight.

SOLUTIONS FOR POST-HYBRIDISATION WASHES

Stock Solutions	Components
Wash buffer 2X SSC 50% formamide Make fresh De-gas before use.	35 ml of 20X SSC 175 ml formamide to 350 ml with sterile water
20X SSC 3.0M NaCl 0.3M Na$_3$ citrate Make in advance, autoclave. Store at RT	175.3 g NaCl 88.2 g sodium citrate to 1 litre with ddH$_2$O
10X NTE 5M NaCl 100 mM Tris-HCl, pH 7.5 10 mM EDTA Make in advance, autoclave Store at RT	292.2 g NaCl 15.76 g Tris-HCl 3.7 g EDTA (disodium) to 1 litre with dd H$_2$O Mix everything, then adjust to pH 7.5
RNase A (Sigma type I-A, No. R4875) 10 mg/ml Store at –20 °C	Make stock solution in water. Aliquot

DIG-LABELLED PROBE DETECTION

Equipment needed:	*Chemicals needed:*
Slide boxes Slow shaker Aluminium foil Glass cover slip	DIG buffers 1–6 Blocking reagent DIG antibody Bovine serum albumen NBT and BCIP tablets *In situ* mounting media

Duration of procedure: approx. 5 hr plus 1–2 days for detection.

Note – all solutions and chemicals are stored at RT, except for the blocking reagent, NBT and BCIP tablets, DIG antibody and BSA which are stored at 4 °C. There is no need to DEPC-treat the solutions in these steps.

All incubations are performed at RT on a rotary shaker set at 20–30 rev/min.

Place slides in slide-mailing boxes for 5 slides, making sure that the sections do not face the sides of the box. Alternatively, place the slides on the bottom of a large square Petri plate (25 cm × 25 cm). For each change of solution (except for the antibody solution, DIG buffer 4), add 10 ml of the next solution, swill to make sure the slides are covered, decant, replace with fresh solution and leave for the appropriate time.

1   Incubate for 5 min in 10 ml of Buffer 1.
2   Incubate for 1 h in 10 ml of Buffer 2.
3   Incubate for 1 h in 10 ml of Buffer 3.
4   Incubate for 1 h in 10 ml of Buffer 4.
5   Wash 4 times, 20 min each in 10 ml of Buffer 3.
6   Incubate for 5 min in 10 ml of Buffer 1.
7   Incubate for 5 min in 10 ml of Buffer 5.
8   Incubate in 10 ml of buffer 6, cover with aluminium foil to keep dark and on a very slow shaker for as long as necessary. The hybridisation signal appears dark blue to black (Figure 3) and the control slide should have very little hybridisation signal.
Note – we generally incubate for 1–2 days, but check the slides often for evidence of hybridisation signal. When there is a clear signal with the antisense probe, stop the

reaction, as well as the corresponding sense-probe-hybridised slides.
9   To stop the reaction, wash each slide separately with a gentle stream of ddH$_2$O.
10   Add a drop of *in situ* mounting media and gently place a glass cover slip on top.
Note – we get the best coverage of the sections if the mounting media spreads out slowly on its own rather than applying pressure to the cover slip. Therefore, add a drop of the mounting media, apply a cover slip and wait for the media to spread.
Note – this mounting media safely preserves the product of the alkaline phosphatase reaction, and the slides can be stored indefinitely. We generally find the sections are quite wrinkled at this point and are therefore very difficult to photograph. However, the sections flatten out over time and the results can be better photographed several weeks after adding the mounting media.

SOLUTIONS FOR DIG-LABELLED PROBE DETECTION

DIG Buffer 1	100 mM Tris-HCl 150 mM NaCl, pH 7.5 (20 °C) Can be stored at RT indefinitely
DIG Buffer 2	Buffer 1 containing 0.5% (w/v) blocking reagent (Boehringer Mannheim, No. 1096 176) Let dissolve for 1 h at 60–70 °C, the solution remains turbid. Make fresh daily
DIG Buffer 3	Buffer 1 containing: 1% (w/v) BSA (Sigma, No. A-7906), and 0.3% (v/v) Triton X-100 Make fresh daily
DIG Buffer 4	Buffer 3 containing Anti-digoxigenin-AP (Fab fragments, Boehringer Mannheim, No. 1093 274) at a concentration of 1:3000
DIG Buffer 5	This buffer is made of 2 solutions. A precipitate is formed if they are stored mixed together. To prevent this from happening, make up the 2 solutions separately at a 10X concentration and store them at RT. On the day of hybridisation, mix them together at 1X final concentration. 1X solution #1: 100 mM Tris-HCl, 100 mM NaCl, pH 9.5 (20 °C) 1X solution #2: 50 mM MgCl$_2$
DIG Buffer 6	Dissolve one NBT/BCIP tablet in 10 ml of water and use directly, or dilute with 40 ml of **Buffer 5**
NBT/BCIP	nitroblue tetrazolium chloride and 5-bromo-4-chloro-3-indolyl-phosphate, toluidine salt ready-to-use tablets (Boehringer Mannheim, No. 1697 471)
In situ **mounting media**	0.1M sodium phosphate buffer, pH 7.0, 1 mg/ml sodium azide, 75 g/l gelatin (stir until dissolved), and 50% glycerol

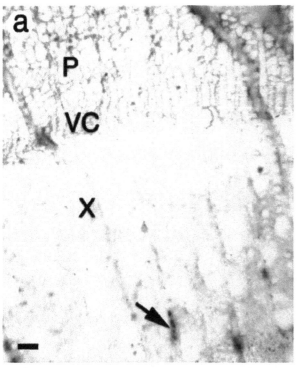

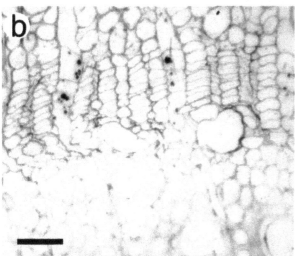

Figure 3 Typical *in situ* hybridisation results in the poplar stem. The hybridisation of the DIG-labelled probe appears dark blue. Probe is antisense cinnamyl alcohol dehydrogenase hybridised on hybrid aspen. The specific hybridisation signal is shown in a, and at a higher resolution in b. Because this signal is at high resolution, the DIG labelling can be detected specifically in the cytoplasm, which is pressed against the side of the cell by the very large vacuoles in these cells (arrow). Key: P, phloem; VC, vascular cambium; X, xylem. Bars are 40 µm.

Applications

Although this method is high resolution, it is not as sensitive as conventional ISH methods since only RNA exposed on the surface of the plastic sections is available for hybridisation. Conventional ISH protocols use paraffin-embedded tissue which are sectioned at 10 µm and the paraffin is removed before hybridisation, which exposes the entire thickness of the section to the probe. Since ISHs are inherently difficult procedures to perform, we recommend that you first perform an easier standard protocol on a paraffin-embedded soft tissue (such as a leaf), and ensure that you can obtain clear positive results, before trying this protocol.

FUTURE PROSPECTS

Incorporation of ISH results is becoming more common in publications, especially in light of submission requirements for journals such as *The Plant Cell* which will no longer accept gene expression analysis based solely on promoter-reporter gene fusions, unless there are supporting ISH data. However, since ISHs are laborious and require a fair amount of expertise in both histochemical and molecular biological methods, there have been some recent advances which may make them more 'user friendly'. Recently, whole-mount ISH techniques have been adapted to plant tissues (Engler *et al.*, 1994). Unlike the protocol described here, the whole-mount technique enables detection of transcripts in whole pieces of tissue, without any sectioning, thereby eliminating the histochemical portion of the protocol. Whole-mount ISHs do not provide as high a resolution as methods involving sectioning, but in many cases a lower resolution expression pattern is sufficient. Studies with whole-mount ISHs have only been reported in a limited number of organs in herbaceous plants, but it is likely only a matter of time before these techniques can be adapted to woody plants and wood-forming tissues!

ACKNOWLEDGEMENTS

We thank Philippe Label for Figure 2, and Vince Francheschi, Veronica Bourquin, Philippe Label, Christian Breton and Wout Boerjan for helpful discussions during the development of this protocol. This research was supported by the Swedish Council for Forestry and Agricultural Research, the Swedish

Natural Sciences Research Council, the Foundation for Strategic Research and a Visiting Fellowship from the Kempe Foundation to S.R.

REFERENCES

de Almeida Engler, J., Van Montagu, M. and Engler, G. (1994) Whole-mount messenger RNA *in situ* hybridization in plants. *Plant Mol. Biol. Rep.*, **12**, 319–329.

Leitch, A.R., Schwarzacher, T., Jackson, D. and Leitch, I.J. (1994) *Microscopy Handbook: In situ hybridisation.* Oxford: Royal Microscopical Society.

Regan, S., Bourquin, V., Tuominen, H. and Sundberg, B. (1999) Accurate and high resolution *in situ* hybridisation analysis of gene expression in secondary stem tissues. *Plant J.*, **19**, 363–369.

Sambrook, J., Fritsch, E.F. and Maniatis, T. (1989) *Molecular Cloning: A Laboratory Manual.* Cold Spring Harbour: Cold Spring Harbor Laboratory Press.

APPENDIX

Probe preparation itinerary

Date of template preparation:
Date of transcription reaction:

Digestion of Template   insert gel photo here

Name of gene:

Sense probe linearised with ____ and transcribed with ____
Antisense probe linearised with ____ and transcribed with ____

Synthesis of transcripts

Component	Expected volume	Actual volume
DNA	1–14 μl	
Transcription buffer (10X)	2 μl	
DIG labelled dNTPs	2 μl	
RNA polymerase T3 or T7	1 μl	
RNase inhibitor (RNasin)	1 μl	
DEPC-water	to balance (13–0 μl)	
Total volume	**20 μl**	**20 μl**

Incubate at 37 °C for 30 min to 2 h
Success of transcription   insert gel photo here

Itinerary for freeze-substitution and embedding of tissue in LR White

Date:
Tissue:

Solution	Time	Temp.	Actual condition
1.25% glutaraldehyde in 100% methanol	1–4 weeks	–80 °C	
100% methanol	1 h	4 °C	
25% LR White: 75% absolute methanol	6 h	4 °C	
50% LR White: 50% absolute methanol	6 h	4 °C	
75% LR White: 25% absolute methanol	6 h	4 °C	
100% LR White	6 h	4 °C	
100% LR White	6 h	4 °C	
100% LR White	6 h	4 °C	
100% LR White	6 h	4 °C	
100% LR White	6 h	4 °C	
Polymerise	15 h	65 °C	

Pre-treatment of tissues

Date of pre-treatment:
Type of tissue:
Type of fixation and embedding:

Treatment	Concentration	Temp.	Usual time	Changes, if any
PBS		room	2 min	
Pronase	0.125 mg/ml	room	10 min	
Glycine	0.2%	room	2 min	
PBS		room	2 min	
Formaldehyde	4% in PBS, pH 7	room	10 min	
PBS		room	2 min	
PBS		room	2 min	
Acetic anhydride	0.5% in 0.1 M triethanolamine	room	10 min	
Acetic anhydride	0.5% in 0.1 M triethanolamine	room	10 min	
PBS		room	2 min	
Transfer slides to mailing boxes, making sure that your sections do not face the walls of the box				
Pre-hybridisation	0.1% PVP in 1X salts	55 °C	at least 1 h	

Itinerary for post-hybridisation washes

Date:

Solution	Expected time	Actual time
2 X SSC	until coverslips come off	
2 X SSC	15 min, RT	
2 X SSC	15 min, RT	
2 X SSC	15 min, RT	
2 X SSC	15 min, RT	
NTE (pre-RNase)	5 min, 37 °C	
NTE (pre-RNase)	5 min, 37 °C	
NTE (plus RNase)	30 min, 37 °C	
NTE (post-RNase)	5 min, 37 °C	
NTE (post-RNase)	5 min, 37 °C	
Wash buffer	5 min, 55 °C	
1% SSC	5 min, RT	
1X PBS	5 min, RT	
1X PBS	store at this stage, 4 °C	

Itinerary for DIG detection

Date:

Solution	Expected time	Actual time
DIG Buffer 1	5 min	
DIG Buffer 2	1 h	
DIG Buffer 3	1 h	
DIG Buffer 4	1 h	
DIG Buffer 3	20 min	
DIG Buffer 3	20 min	
DIG Buffer 3	20 min	
DIG Buffer 3	20 min	
DIG Buffer 1	5 min	
DIG Buffer 5	5 min	
DIG Buffer 6	1–2 days	
ddH$_2$O	gentle stream	
In situ mounting media		

Wood Formation in Trees ed. N. Chaffey
© 2002 Taylor & Francis
Taylor & Francis is an imprint of the
Taylor & Francis Group
Printed in Singapore.

8 Random Amplification of Polymorphic DNA and Reverse Transcription Polymerase Chain Reaction of RNA in Studies of Sapwood and Heartwood

Elisabeth Magel[1*], Siegfried Hauch[1] and Luigi De Filippis[2]

[1] *Physiological Ecology of Plants, University of Tübingen, Auf der Morgenstelle 1, D-72076 Tübingen, Germany*
[2] *Department of Environmental Sciences, University of Technology, Sydney, P. O. Box 123 , Broadway NSW 2007, Australia*
* *Author for correspondence*

ABSTRACT

Determination of genetic differences and levels of gene expression in mature and old woody tissues based on morphological and anatomical characteristics, levels of metabolites or enzymatic activity is often difficult. However, the use of molecular markers allows the assessment of polymorphic (genetic and biochemical) variation amongst tissues in individuals, and between closely related species directly at the DNA and RNA level. Unfortunately, such techniques have not been generally applied to the bark and wood of mature trees. Here we describe the technique of random amplification of polymorphic DNA (RAPD) by the polymerase chain reaction (PCR) and its application to the analysis of the relationship between the bark and variously aged wood zones of *Robinia pseudoacacia*. Additionally, we describe the use of differential display reverse transcription (DDRT)-PCR to screen *Eucalyptus microcorys* for genes which have been down-regulated as a result of salinity. Moreover, we used reverse transcription (RT)–PCR in order to get information about the regulation of a specific enzyme (sucrose synthase) in *Robinia pseudoacacia* stems during heartwood formation. We demonstrate that the procedures and protocols developed are applicable to all tissue types tested from leaves and bark, right through to the inner heartwood zones. Our results show that RAPD-PCR, DDRT-PCR, and RT-PCR are versatile and sensitive methods, capable of detecting not only genomic changes (i.e. at the DNA level), but also post-transcriptional changes (i.e. at the RNA level) in trees.

Key words: arbitrary primers, bark, DDRT-PCR, DNA polymorphism, *Eucalyptus*, gene regulation, RAPD-PCR, RNA differences, *Robinia*, RT-PCR, wood

INTRODUCTION

Progress in the genetics, physiology, biochemistry and identification of different accessions of woody trees until recently has relied on phenotypic assays of genotype, most commonly based on morphological traits recorded in the field, or laboratory methods such as metabolite levels, protein electrophoresis and isozyme analysis (Crawford, 1990).

With the advent of molecular genetic techniques, DNA- and RNA- based procedures have been proposed for hybrid and species identification, and metabolic state and gene expression studies (Moran et al., 1990; Mou et al., 1994). DNA and RNA methodologies more directly reflect the relatedness, genetic diversity and metabolic expression of the sampled material, and analyses can be carried out on single cells through to complete individuals.

Arbitrarily primed polymerase chain reaction (PCR) techniques have now been used successfully to detect changes in the DNA or RNA in a number of plant species (Micheli and Bova, 1997). Welsh and McClelland (1990) and Williams *et al.* (1990) reported an identification technique based on the amplification of random genomic DNA sequences by PCR using single, short arbitrary primers, commonly called random amplified polymorphic DNA (RAPD)-PCR. Liang and Pardee (1992) and Welsh *et al.* (1992) have also used arbitrary primers to detect differentially expressed genes in a procedure called differential display reverse transcription by PCR (DDRT-PCR). In 1987, Veres and co-workers published a technique which used PCR to amplify mRNA sequences from cDNA. By utilising defined sets of arbitrary primers (usually ten nucleotides in length), RAPD-PCR and DDRT-PCR are able to generate amplified products which can be used as genetic and metabolic expression markers. On the other hand, by utilising defined appropriate longer oligonucleotide primers, RT-PCR is able to generate amplified products which can be used as expression markers for a specific enzyme or protein.

The use of RAPD-PCR to investigate genetic variation in forest taxa has been reported before in white spruce and Douglas fir (Carlson *et al.*, 1991), red maple (Krahl *et al.*, 1993), Norway spruce (Heinze and Schmidt, 1995), *Eucalyptus* (Chen and De Filippis, 1996) and *Robinia* (De Filippis and Magel, 1998). In fact, Tulsieram *et al.* (1992) and Grattapaglia *et al.* (1995) have constructed partial linkage (chromosome) maps of white spruce and *Eucalyptus* respectively, using RAPD markers. DDRT-PCR to our knowledge has not been used to detect differences in expression of genes in woody plants, but has been used to show differential expression of a

number of genes in herbaceous (Hannappel *et al.*, 1995; Torelli *et al.*, 1996) and crop (Oh *et al.*, 1995; Rymerson *et al.*, 1995) plants. Up to now, RT-PCR has been used predominantly in fruit and forest trees to detect pathogens of the aerial parts (i.e. leaves). As far as we are aware the first reported use of RT-PCR as a tool for evaluation of the regulation of fundamental basic pathways in wood tissues is that of Hauch and Magel (1998).

TECHNIQUES

Note – if a commercially available kit, such as the QIAGEN RNeasy or pGEM-T vector system were chosen, we refer to the detailed step-by-step instructions and protocols supplied with the kits by the manufacturer.

RAPD-PCR

One of the key features of RAPDs is that minimal DNA template is required. However, the purity of the DNA has been implicated in lack of success in RAPD-PCR. In particular, reproducibility of band patterns and inhibition of the PCR process itself by carry-over of water-soluble compounds like polysaccharides, phenolics and proteins. We describe here a rapid CTAB miniprep preparation for genomic DNA which has been modified from a bacterial miniprep method by Doyle (1991), and which has been used successfully to isolate DNA in a number of woody plant species.

Equipment needed:

Adjustable microlitre disposable tip micropipettes
Agarose gel electrophoresis (horizontal – mini submerged type GT, Bio-Rad)

Centrifuge / microfuge to at least 12000 g

Disposable, autoclaved tips for pipettes
Gel scanning system (e.g. Elscript 400 – Hirschmann; optional automatic computerised especially useful for mini gel system analysis)

Chemicals needed:

Acetic acid fixative solution 10%
Acrylamide (40% (w/v) stock: 38 g acrylamide + 2 g bis-acrylamide (BIS) / 100 ml distilled water)
Agarose gels (1.5–2.0%) made up in TBE buffer (see below)
Ammonium persulphate (AMPS; 10% w/v)
Arbitary random primers (10 base, kits are available commercially), e.g. University of British Columbia (UBC) or Operon Technology (OP) kits; base sequence of the decamers used:

UBC325	5' TCTAAGCTCG 3'
UBC327	5' ATACGGCGTC 3'
UBC328	5' ATGGCCTTAC 3'

Laminar flow hood or biohazard cabinet to 'PC2' (PC2 is not strictly required)

Pestle and mortar or microtubes with micro-tube pestle

Petri dishes (glass)

Photographic box with bottom illumination (slide viewer) for silver staining

Photographic camera system with UV filters for recording stained gels

Polyacrylamide gel electrophoresis kit (vertical type – mini-system)

Power supply (50–200 V)

Reaction (micro) tubes (sterilised)

Shaker (horizontal)

Spectrophotometer

Sterile gloves

Thermalcycler (preferably with a heated lid, e.g. Crocodile III, Appligene or Personal Cycler, Biometra)

UV trans-illuminator (for ethidium bromide stained DNA)

Water bath or block heater for microtubes

Chloroform (keep refrigerated)

Chloroform:isoamylalcohol (24:1), keep refrigerated

CTAB buffer (2% CTAB (cetyltrimethyl-ammonium-bromide), 1.4 M NaCl, 20 mM EDTA (ethylenediaminetetraacetic acid), 100 mM Tris-HCl pH 8.0, 1% v/v β-mercaptoethanol, 1% w/v polyvinyl pyrrolidone PVP-40)

Deionised or distilled water, sterilised by autoclaving

Developer solution (30 g Na_2CO_3 + 2 mg $Na_2S_2O_3$) in 1l distilled water, plus 1.5 ml of 37% formaldehyde

DNA size markers: λ *Hind* III; pBR322 *Mva* I

dNTPs (stock – 10 mM of each) in TE buffer (below) dATP, dCTP, dGTP, dTTP

Ethanol (100 and 70%) (keep refrigerated)

Ethidium bromide

Gel loading buffer (10 X), 50% (w/v) glycerol, 0.2% (w/v) bromophenol blue in TE buffer (below); store at 4 °C, longer at –20 °C

Gelatin 10 mg/l in bidistilled water

Genomic DNA (stock of 5–25 $\mu g/\mu l$) in TE buffer (below)

Isopropanol (keep refrigerated)

$MgCl_2$ stock (25 mM)

Polyacrylamide gels (5–8%) made up in TBE buffer (below)

Proteinase K (10 mg/ml in 50 mM Tris-HCl, pH 8.0, 1 mM $CaCl_2$)

RNase A (in 10 mg/ml in Tris-HCl, pH 8.0, 14 mM NaCl)

Sodium dodecyl sulphate (SDS); 10% in bidistilled water

Silver nitrate solution (1g $AgNO_3$) in 1l distilled water, plus add 1.5 ml of 37% formaldehyde

Taq DNA polymerase (stock of 1 Unit/μl, Biotech International)

Taq polymerase buffer (10 X; usually supplied with the Taq enzyme)

TBE buffer (10 X; 89 mM Tris, 89 mM boric acid, pH 8.0, 2 mM EDTA)

TE buffer (10 mM Tris-HCl, pH 8.0, 1 mM EDTA)

N, N, N′, N′ -tetramethylethylenediamide (TEMED)

Note – unless otherwise stated, all solutions are aqueous.

Duration of procedure (approx.):

DNA isolation half day
PCR preparation 2 hours
PCR amplification 5 hours
Gel electrophoresis 1–2 hours

DNA isolation

1   Freeze-dried wood and bark samples (100 mg) are placed in 2 ml tubes.
    Note – all subsequent steps should be performed in a fume hood.

2   Aliquot 500 μl of CTAB buffer into the micro-tubes and heat to 65 °C for 2–3 min.

3   Homogenise the tissue using the minitube pestle.
    Note – the pestle may be reused after decontamination with 10% free chlorine bleach or 10% SDS, followed by subsequent washes in distilled water (at least 3 washes).

4   Incubate tubes at 60 °C for 30 min.

5   After incubation, add 3 μl of proteinase K, and incubate for 90 min at 37 °C.

6   Add 500 μl of an equal volume of chloroform:isoamyl alcohol extraction solution to tubes and mix (8–10 inversions).
    Note – placing the woody tissue on a horizontal shaker (about 500 rpm) for 10 min may increase the DNA yield.

7   Centrifuge at 12000 × g for 5 min at room temperature (RT).

8   Pipette off the top aqueous layer (about 400 μl) into a fresh microtube.

9   Add 2/3 the volume of cold isopropanol and mix gently to precipitate DNA.

10 Centrifuge at 12000 × g for 2 min at RT.

11 Add 1000 μl of chilled 70% ethanol and invert tubes several times to precipitate DNA.
    Note – allowing the samples to settle at RT, or 4 °C for up to 1 h, may increase the DNA yield. Longer periods will not affect DNA quality, but only increases yields very slightly.

12 Centrifuge at 12000 × g for 5–10 min at RT.

13 Decant the supernatant, air-dry the pellets (but do not allow the pellets to dry completely), and resuspend the isolated DNA in 50–500 μl of TE buffer (depending on the concentration of DNA required).
    Note – to assist DNA dissolution in the TE buffer, tubes can be placed in a water bath at 37 °C for 60 min.

14 Once the DNA has fully dissolved, add RNase A to a final concentration of 10 μg/ml, and incubate at 37 °C for 2 min.

15 Check for DNA quality and purity by agarose gel electrophoresis (see below) and by spectrophotometry, i.e. ratio of absorption 260 nm/280 nm.

16 Quantify DNA by spectrophotometry (50 × OD of DNA solution at 260 nm × dilution of DNA solution = DNA content in μg/ml).

RAPD-PCR

The original RAPD assay used a single arbitrary primer (9–10 bases in length), which is usually sufficient to produce fingerprint patterns. Although the sequence of the primers is arbitrarily chosen, two basic criteria must be satisfied; they should have a minimum G+C content of 40% (generally 50–60% G+C used), and be devoid of palindromic sequences (Williams et al., 1990, 1993). If using combinations of two arbitrary primers, the two primers must be designed so as to avoid complementation between them. Success can be achieved using both single or pairwise arbitrary primers; but here only results from the use of single 10-base arbitrary primers are presented.

Optimisation of the PCR reaction would normally require each variable, i.e. $MgCl_2$, dNTP, primer(s), DNA amounts and polymerase enzyme, to be tested independently. In practice, RAPD-PCR reactions are now designed to add excess dNTPs and primer(s), so testing of these two reagents is not required. The amounts of template DNA and polymerase enzyme are set within narrow limits which are optimal for most situations. However, the $MgCl_2$ concentration used must be optimised for each primer/combination of primers (Park and Kohel, 1994).

In order to minimise risks of contamination (e.g. amplified DNA carry-over and DNA from airborne contamination), the following measures should be taken when preparing RAPD reaction tubes:

(a) Prepare all solutions and reactions at least in a laminar flow cabinet.
(b) Devote a set of adjustable micropipettes only to the laminar flow cabinet.
(c) Wear sterile gloves during the period of set-up of tubes.
(d) Each experiment should include duplicate tubes, and template-free controls.

1   Prepare a master mix of all reagents common to all the programmed reactions. Prepare only the quantity required for the total number of sample tubes including controls; make sure that at least one extra aliquot is prepared to ensure adequate volume is available for the last tubes.

2   In a sterile master mix tube, combine the following amounts per sample:

   4.0 μl of 25 mM $MgCl_2$ (volume to be ajusted for different $MgCl_2$ levels)
   2.5 μl of 10 × Taq polymerase buffer
   5.0 μl of 2 mM dNTPs
   0.25 μl of gelatine
   0.25 μl of RNase A

   Add distilled water to a final volume of 20.0 μl per sample (if a single arbitary primer is used) or 19.0 μl (if using two arbitary primers) and mix well.

3   Dispense 20 μl (or 19 μl) of the master mix to each reaction microtube.

4   Add 1.0 μl of 6 μM primer(s) and 1 μl template DNA (5–25 ng).

5   Gently mix the reaction mixtures by tipping each tube; briefly spin the tubes to remove air bubbles (this is normally an optional procedure).

6   Add 1 unit of Taq DNA polymerase enzyme per tube.

7   Overlay the mixtures with one drop of mineral oil (not necessary if the thermocycler has a heated lid).

8   Place all tubes in the thermocycler programmed for:
   1 cycle of 94 °C, 5 min; 35 °C, 1 min
   40 cycles of 94 °C, 1 min; 35 °C, 1 min; 72 °C, 2 min.

9   Immediately the thermocycler has finished, place all tubes on ice and store at 4 °C.

Gel electrophoresis and DNA staining

The amplified DNA fragments can be separated either on agarose gels stained with ethidium bromide (separates a wider range of molecular weight DNA, but ethidium bromide is less sensitive) or on polyacrylamide gels stained with silver (separates a narrower size range of molecular weight DNA, but the sensitivity of silver is considerably higher). Both these methods are simple, fast, low cost, and avoid radioactivity.

Agarose / ethidium bromide

1   Prepare a 1.5–2.0% agarose gel in 1 × TBE buffer (it is advisable to use a comb with teeth as thin as possible for sharp bands).

2   Add 1 μl of 10 × loading buffer to 9 μl aliquot of each amplified sample. Take care to ensure that the pipette tip goes to the bottom of the tube so as not to draw up oil if an oil-overlay has been used.

3   Load all of the 10 μl samples, and 10 μl of a molecular size DNA marker on the gel.

4   Run the gel in 1 × TBE buffer at 100 V (no higher than 100 V is recommended for better resolution), until the dye front has covered a distance of at least 6 cm from the wells; in practice the dye is run to the edge of the agarose gel.

5   Stain the gel with ethidium bromide for 30 min and then destain the gel in deionised water for a further 30 min, with at least two changes of the water.

6   View and photograph the gel on a UV transilluminator with a wavelength of at least 254 nm.
   Note – if you plan to recover bands from the gel, shorter wavelengths are most harmful to the DNA.

Polyacrylamide / silver nitrate or ethidium bromide

1   Prepare a 5–8% polyacrylamide gel in 1 × TBE buffer according to the manufacturer's instructions for preparing the polyacrylamide gels.

2   Add 1 μl of 10 × loading buffer to 9 μl aliquot of each amplified sample (diluted to a suitable factor by trial and error beforehand). If an oil overlay has been used, take care to ensure that the pipette tip goes to the bottom of the tube so as not to draw up oil.

3   Load all the samples (2–5 μl each), and 1 μl of a molecular size DNA marker to gels.

4   Run the gel in 1 × TBE buffer at 50 V (no higher than 50 V is recommended for better resolution), until the dye has run off the bottom edge of the gel.

5   Stain the gels with ethidium bromide as detailed above, or,

6   Otherwise stain the gels with silver nitrate as detailed below (steps 7–14).

7   Place the gels in glass Petri dishes on a light box (slide-viewing light box), where the light is from below the gels.

8   Fix the gels in 10% acetic acid for 20 min.

9   Siphon off the fixing solution from the gels, and wash carefully twice with deionised water at 20 sec intervals.

10 Stain gels with silver nitrate/formaldehyde solution for 30 min.

11 Siphon off the solution, and wash gels carefully twice with deionised water at 20 sec intervals.

12 Develop gels under light (on the light box) with the developer solution for 2–5 min, or until the bands are best detected by eye.

13 Stop the developing procedure quickly by placing in 10% acetic acid, leave in this solution for 5 min to fix fully.

14 Remove the 10% acetic acid and replace with fresh 30% acetic acid for long term storage at 4 °C.

15 Photograph gels on the light box using high speed black and white film (we recommend Kodak Tri × Pan TX4 400 ASA or Ilford HP5 400 ASA).

DDRT – PCR

Until recently, isolation of differentially expressed genes could only be achieved through approaches based on either differential screening of cDNA libraries or construction of subtracted cDNA libraries (Sterky *et al.*, 1998). The disadvantage presented by these approaches are that they are laborious and time consuming, and require large amounts of RNA. Additionally, both of the screening methods above detect only abundant mRNAs (0.1% or greater) and limited numbers of RNA samples can be compared. Most of these limitations have been overcome recently in an innovative approach by extending the random amplification and PCR strategies already established in DNA fingerprinting to RNA (Liang and Pardee, 1992; Welsh *et al.*, 1992).

Equipment needed:

Adjustable microlitre disposable tip micropipettes
Agarose gel electrophoresis kit (horizontal – mini submerged type GT, Bio-Rad)
Centrifuge / microfuge to at least 12000 g
Disposable, autoclaved tips for pipettes
Gene Vac CVP vacuum pump
Laminar flow hood or biohazard cabinet to 'PC2' (PC2 is strictly not required)
Pestle and mortar or microtubes with microtube pestle

Chemicals needed:

Acetic acid fixative solution 10% aqueous
Acrylamide (40% (w/v) stock; 38 g acrylamide + 2 g BIS/100 ml distilled water)
AMPS 10% (w/v)
Anchored oligonucleotide ($dT_{12}AG$)
Chloroform (keep refrigerated)
Chloroform:isoamyl alcohol (24:1), keep refrigerated
CTAB:Phenol buffer – (100 mM Tris-HCl, pH 8.0; 1% (w/v) CTAB; 0.1 M LiCl; 10 mM EDTA; 0.5% (w/v); PVP-40; 14.3 mM β-mercaptoethanol) **:** (phenol, pH 8.0); 1:1 (v/v)

Photographic box with bottom illumination (slide viewer) for silver staining

Photographic camera system with UV filters for recording stained gels

Polyacrylamide gel electrophoresis (vertical type – protean IIxi cell system, Bio-Rad)

Power supply (50–200 V)

Reaction (micro) tubes (sterilised)

Sequencer (e.g. automated 377 DNA sequencer, PE Applied Biosystem)

Shaker (horizontal)

Spectrophotometer

Sterile gloves

Thermalcycler (preferably with a heated lid, e.g. Crocodile III, Appligene or Personal Cycler, Biometra)

UV transilluminator (for ethidium bromide stained DNA)

Water bath or block heater for microtubes

Deionised or distilled water, sterilised by autoclaving

Denaturing agarose gels (2.0%) made up in TBE buffer (below) with 7 M urea

Denaturing polyacrylamide gels (5–6%), made up in TBE buffer (below) with 7 M urea (urea/polyacrylamide solution may require filtering)

Diethylpyrocarbamate (DEPC)-water:3 M sodium acetate (3:1 v/v) (pH 5.2)

Developer solution (30 g Na_2CO_3 + 2 mg $Na_2S_2O_3$) in 1l distilled water, plus 1.5 ml of 37% formaldehyde

DEPC- treated distilled water (0.1% v/v)

Plasmid DNA purification system (miniprep) – WIZARD (Promega)

DNAse (RNase free)

DNA size markers: ΦX174 *Hae* III

dNTPs (stock – 10 mM of each) in TE buffer (below) dATP, dCTP, dGTP, dTTP

Ethanol (100 and 70%) (keep refrigerated)

Ethidium bromide

Formamide (95%), 10 mM EDTA (pH 8.0)

Gel loading buffer (5 X), 50% (w/v) glycerol, 0.1% (w/v) xylene cyanol (below); store at 4 °C, longer at –20 °C

Gelatin 10 mg/l in bidistilled water

Isopropanol (keep refrigerated)

LiCl 4 M

$MgCl_2$ stock (25 mM)

pGEM-T vector system (Promega)

Phenol (pH 8.0)

Phenol : chloroform, 1:1 (v/v)

Poly (A^+) oligonucleotide $(dT)_{12–18}$

Polyacrylamide gels (5–8%) made up in TBE buffer (below)

QIAGEN RNeasy Plant Mini Kit (RNA isolation kit)

Reverse transcriptase (MMLV, Promega)

Reverse transcriptase first-strand cDNA synthesis kit (Gibco BRL Superscript)

SDS, 10% aqueous

Selected arbitary 10-mer primers from OPB kit, the base sequence used were:

OPB2　5′ TGATCCCTGG 3′
OPB6　5′ TGCTCTGCCC 3′
OPB10　5′ CTGCTGGGAC 3′

Silver nitrate solution (1g $AgNO_3$ in 1l distilled water, plus 1.5 ml of 37% formaldehyde)
Taq DNA polymerase (stock of 1 Unit/μl, Biotech International)
Taq polymerase buffer (10 X; usually supplied with the Taq enzyme)
TBE buffer (10 X; 89 mM Tris, 89 mM boric acid, pH 8.0, 2 mM EDTA)
TE buffer (10 mM Tris-HCl, pH 8.0, 1 mM EDTA)
TEMED
Urea (7 M)

Note – unless otherwise stated, all solutions are aqueous.
Duration of procedure (approx.):
  RNA isolation   – chemical method 1.5 days,
                   – Qiagen kit half day
  cDNA preparation 2 hours
  PCR amplification 5 hours
  Gel electrophoresis 4–5 hours
  Band cutting and elution half day
  Ligation, transformation, purification and sequencing about one week

Total RNA isolation

All methods using PCR amplification by arbitrary primers are sensitive to template quality. Thus, success requires undegraded RNA to be used as template, free of chromosomal DNA. DNA contamination must be removed by DNase (free of RNase) treatment. It should also be noted that the isolation of poly (A^+) RNA is neither required nor advisable if the differential display method is adopted.

CTAB-Phenol

1 Place fresh or freeze-dried powders of leaf material into 2 ml microtubes.
   Note – all subsequent steps should be performed in a fume hood.
2 Aliquot 500 μl of homogenisation CTAB-phenol buffer into the microtubes and heat to 80 °C for a maximum of 5 min.
3 Homogenise the tissue with a minitube pestle.
   Note – the pestle may be reused after decontamination with 10% free chlorine bleach or 10% SDS, followed by at least 3 washes in distilled water.
4 Add 250 μl of chloroform:isoamyl alcohol (24:1) extraction solution to tubes and vortex for 30 sec (be careful in releasing any pressure that might build up).
5 Incubate tubes at RT for 30 min.
6 Centrifuge at 12000 $\times$ g for 5 min at RT.
7 Pipette the top aqueous layer (about 400 μl) into a fresh microtube.
8 Gently mix 400 μl of 4 M LiCl into the microfuge and allow the RNA to precipitate overnight.
9 The following morning, centrifuge the tubes at 12000 $\times$ g for 2 min at RT, discard the supernatant.
10 Add 250 μl of 4 M LiCl, vortex the tubes.
11 Centrifuge at 12000 $\times$ g for 5 min at RT.
12 Discard the LiCl-washing, repeat the washing procedure.
13 Dissolve RNA pellets in DEPC-water: 3M sodium acetate (1:3 v/v), pH 5.2 and precipitate with 2 volumes of ice cold 100% ethanol.
14 Place the mixtures at -20 °C for at least 4 h.
15 Centrifuge at 12000 $\times$ g for 5 min at –20 °C.
16 Decant the supernatant and add 1 ml 70% ethanol.
17 Centrifuge at 12000 $\times$ g for 2 min at RT.
18 Discard the ethanol and repeat the 70% ethanol step and centrifugation.
19 Allow the RNA pellets to air-dry and resuspend in 20–30 μl of DEPC-treated water (volume required depends on the concentration of RNA required).
   Note – to assist in dissolution of the RNA in the water, tubes can be placed in a water bath at 65 °C for 15 min.

20 Once the RNA has fully dissolved, store it at –70 °C for later analysis.

21 Check for RNA quality and purity by agarose gel electrophoresis (see below) and by spectrophotometry, i.e. ratio of absorption 260 nm/280 nm.

22 Quantify RNA by spectrophotometry (40 × OD of DNA solution at 260 nm × dilution of RNA solution = content of RNA in μg/ml).

QIAGEN RNeasy plant RNA mini kit

Follow the manufacturer's detailed step-by-step protocol.

First strand cDNA synthesis

The first step after RNA isolation from tissues is to amplify subsets of mRNA into partial cDNA sequences. Reverse transcription is used with primers designed to select only a fraction of the RNA for synthesis of first strand cDNA. In the original differential display protocol, synthesis of first strand cDNA was initiated by an anchored oliogo-dT primer. This primer contained a stretch of 11–12 Ts plus 2 additional 3′ bases, and primes reverse transcription only from a sub-population of poly-adenylated tailed RNAs to the specificity produced by the two additional 3′ bases (theoretically assesses 1/12 of the total mRNA pool). The resulting cDNA is then amplified using the same oligo-dT primer plus an arbitrary 10-base oligonucleotide primer, thus generating a collection of products which are derived from the 3′ end of a subset of the polyadenylated RNAs. Recently, one- or two-base anchored oligo-dT primers, and two arbitrary 10-base primers have been successfully used after cDNA synthesis in differential display by ourselves, but data from the use of two 10-base primer combinations are not presented here.

The manufacturer's detailed step-by-step protocol was followed for the Superscript Pre-amplification system (Gibco BRL) and the oligo (dT)$_{12-18}$, and MMLV reverse transcriptase supplied. The PCR conditions were also as set out in the instruction manual.

DD-PCR

The low stringency conditions throughout the amplification step can also generate spurious products, although higher stringency conditions can be

used as described by Zhao *et al.* (1995). It is therefore crucial to verify that the observed differences in RNA samples are not due to differences in quality or concentration of template. This can be achieved by titration of each RNA template, where only products occurring at two or more concentrations are chosen, and not occurring at any concentrations of the other RNA samples. Multiple independent samples are also frequently required, and a control performed leaving out only reverse transcriptase in order to ensure that the RNA samples are not contaminated by genomic DNA must always be included.

In order to minimise risks of contamination (e.g. from carry-over of amplified DNA and from air-borne contamination), the following measures should be taken when preparing DD-PCR reaction tubes:

(a) Prepare all reactions at least in a laminar flow cabinet.
(b) Devote a set of adjustable micropipettes to the laminar flow cabinet.
(c) Wear sterile gloves during the set-up of tubes.
(d) Each experiment should include duplicate tubes and template-free controls.

---

1 Prepare a master mix (general method is detailed in the RAPD-PCR section, however with the omission in this case of RNase and gelatin) of all reagents common to all the programmed reactions.
Note – prepare only the quantity required for the total number of sample tubes, including controls, but make sure that at least one extra aliquot is prepared to ensure that sufficient volume is available for the last tube.

2 Dispense 19 μl of master mix to each reaction microtube.

3 Add 1–3 μl of first strand cDNA synthesised product to microtubes.

4 Add 1 μmol/l of the oligo (dT)$_{12}$AG anchored primer to each microtube.

5 Add 0.25 μmol/l of a single oligonucleotide 10-base arbitary primer(s).

6 Gently mix the reaction mixtures by tipping each tube; and briefly spin the tubes to remove air bubbles (this is normally an optional procedure).

7   Add 1.5 units of Taq DNA polymerase per tube.

8   Place all tubes in the thermocycler programmed for:

   1 cycle of 96°C, 3 min
   40 cycles of 94 °C, 1 min; 42 °C, 2 min; 72 °C, 1 min.

9   Immediately the thermocycler has finished, place all tubes on ice before vacuum-drying in a GenVac CVP pump.

10   Resuspend dry mixtures in 10 μl of 95% (v/v) formamide and 10 mM EDTA (pH 8) with 0.1% xylene cyanol dye.

Gel electrophoresis and DNA staining

Resolution of bands is achieved by denaturing polyacrylamide gel electrophoresis (PAGE) which allows high resolution in profiles containing a large number of products (Bauer *et al.*, 1993). Radioactivity is still the most commonly adopted approach for band visualisation using either [^{32}P] or [^{35}S] dNTPs, ethidium bromide does not allow enough sensitivity, and fluorescent labelling is still technically difficult and expensive. However, in our case we found silver staining a suitable alternative method to radioactivity or fluorescence.

Agarose/ethidium bromide (band elution)

1   Prepare 2.0% denaturing low melting point agarose gels in a similar manner to the RAPD-PCR section, except that 7 M urea is added to the agarose.

2   Run the gel in 1 × TBE buffer at 40 V (no higher than 40 V is recommended, not only for better resolution, but also because of the low melting point of the agarose), until the dye front has covered most of the length of the gel.

3   Stain gels with ethidium bromide and view on a UV trans-illuminator as stated before. Remember, if you plan to recover bands from the gel shorter wavelengths are most harmful to the DNA.

Polyacrylamide/silver nitrate (differential display)

1   Prepare a 6% denaturing polyacrylamide gel in 1 × TBE buffer according to the manufacturer's instructions, as for the RAPD-PCR section, except that 7 M urea is added to the stock acrylamide.

2   Run the gel in 1 × TBE buffer at 200 V (this higher voltage is only recommended for this larger gel system for better resolution, and also because it has a circulating water system for cooling), until the dye has run off the bottom of the gel.

3   Stain the gels with silver nitrate as detailed in the section on RAPD-PCR above.

4   Store the gels in fresh 30% acetic acid at 4 °C, and,

5   Photograph on the light box with high speed black and white film (we recommend Kodak Tri × Pan TX4 400 ASA or Ilford HP5 400 ASA).

Elution of differential bands

Once differentially expressed bands have been obtained, confirmation of the bands is essential. Although several alternative simple strategies have already been mentioned, the most common is still to use the re-amplified fragment as a probe in Northen blotting. Northen blots do not always give the desired results when mRNA is in low abundance, and can give no signal. RNase protection assay may be used, it requires small amounts of RNA, is simple, relatively fast and sensitive enough to detect low abundance mRNA (Hummel *et al.*, 1997). We use an alternative method of confirmation, by cloning and sequencing the differentially expressed products to avoid false positives (Jones *et al.*, 1997).

Note – to minimise the formation of pyrimidine dimers under prolonged UV exposure, a glass plate is placed between the UV source and gel.

1   Excise bands to be cloned and sequenced exactly from the agarose with a sterile fine scalpel and place in sterile microtubes for weighing.

2   Add 1 μl of TE buffer per mg of fresh weight of the gel in the microtubes.

3   Heat tubes to 65 °C for 10 min in a water bath to melt the agarose.

4   Extract the mixture with phenol (pH 8.0), then phenol:chloroform (1:1 v/v), and finally only chloroform to remove the agarose.

5   Precipitate aqueous DNA with 0.1 volume of 3 mM sodium acetate and 1 volume of cold isopropanol, and allow to stand overnight at −20 °C.

6   Centrifuge the DNA at 12000 × g for 10 min.
7   Add 1 ml of 70% ethanol to rinse the DNA, allow to air-dry, and resuspend in 20 μl TE buffer.

Cloning differentially displayed fragments

The manufacturer's detailed step-by-step protocol for the pGEM-T vector system (Promega) was followed using competent cells of *E. coli* JM109. Ligation and transformation were carried out according to instructions, and positive colonies were screened using the PCR conditions as also set out in the instruction manual.

Plasmid DNA purification

Plasmid DNA purification was carried out using Promega's Wizard Plus Miniprep DNA purification system. For each miniprep, one Wizard minicolumn was prepared, and plasmid DNA was purified according to the manufacturer's instructions.

DNA sequencing

Purified plasmid DNA was sequenced using dye terminator sequencing in the forward and reverse directions using an automated 377 DNA sequencer (PE Applied Biosystem). All sequences were analysed using the BLAST sequence similarity computer search features of GenBank (Benson *et al.*, 1999).

RT-PCR

Equipment needed:	*Chemicals needed:*
Adjustable microlitre disposable-tip micropipettes	Agarose gels (1.5–2.0%) made up in TBE buffer (see below)
Agarose gel electrophoresis (horizontal – mini submerged type)	Deionised or distilled water, sterilised by autoclaving
Centrifuge / microfuge to at least 12000 g	DEPC-treated distilled water (0.1% v/v)
Disposable, autoclaved tips for pipettes	DNA size markers: ΦX174 *Hae*; λ *Hind* III; pBR322 *Mva* I
Laminar flow hood or biohazard cabinet 'PC2' (PC2 is strictly not required)	dNTPs (stock – 10 mM of each) in TE buffer (below) dATP, dCTP, dGTP, dTTP
Pestle and mortar or microtubes with microtube pestle	Ethidium bromide
Photographic box with bottom illumination (slide viewer) for silver staining	Gel loading buffer (10 X), 50% (w/v) glycerol, 0.2% (w/v) bromophenol blue in TE buffer (below); store at 4 °C, longer at –20 °C
Photographic camera system with UV filters for recording stained gels	$MgCl_2$ stock (25 mM)
Power supply (50–200 V)	Miniprep DNA purification system – WIZARD (Promega)
Reaction (micro) tubes (sterilised)	pGEM-T vector system (Promega)
Sequencer (e.g. automated 377 DNA sequencer, PE Applied Biosystem)	Polyacrylamide gels (5–8%) made up in TBE buffer (below)
Shaker (horizontal)	Primers for amplification of sucrose synthase specific transcript were: 5′ primer CAYGGITAYTTYGCICA and as 3′ primer GGRTAYTTNGTYTTYTC (pos. 885–902 and 1347–1364 in MVf SUCS4/4, from Heim *et al.*, 1993)
Spectrophotometer	QIAGEN RNeasy Plant Mini Kit (RNA isolation kit)

Sterile gloves	Reaction buffer for RT-PCR reaction (0.75 M Tris-HCl, pH 9.0, 0.2 M $(NH_4)_2SO_4$, 0.1% Tween-20)
Thermalcycler (block base, preferably with a heated lid, e.g. Crocodile III, Appligene)	Reverse Transcriptase first-strand cDNA synthesis kit (Ready To Go T-Primed First-Strand Kit, Pharmacia)
UV trans-illuminator (for ethidium bromide stained DNA)	Taq DNA polymerase (stock of 1 Unit/μl; Eurogentec)
Water bath or block heater for microtubes	Taq polymerase buffer (10 X; usually supplied with the Taq enzyme)
Whirlimixer	TBE buffer (10 X; 89mM Tris, 89 mM boric acid, pH 8.0, 2 mM EDTA)
	TE buffer (10mM Tris-HCl, pH 8.0; 1 mM EDTA)
	Tissue total RNA in DEPC treated distilled water (stock of 0.5–2 μg/25 μl)

Note – unless otherwise stated, all solutions are aqueous.
Duration of procedure (approx.):
    RNA isolation half day
    cDNA preparation 2 hours
    PCR amplification 5 hours
    Gel electrophoresis 4–5 hours
    Ligation, transformation, purification and sequencing about one week

Total RNA isolation

QIAGEN RNeasy Plant RNA Mini Kit
The manufacturer's detailed step-by-step protocol was followed for extraction of total RNA, with the following changes:

1   Place 100 mg of freeze-dried wood material in a 2 ml cup.
2   Add 800 μl (instead of 450 μl as outlined in the protocol) of lysis buffer RLT (included in the kit) and vortex.
3   From this point onward, the manufacturer's protocol was followed.
    Note – in step 4 of the manufacturer's protocol, the volume of alcohol to add is half the volume of the lysis buffer you started with – in our assay 400 μl. Check for quality of RNA by agarose gel electrophoresis and spectroscopy (see Total RNA isolation section above).

First strand cDNA synthesis

First strand cDNA synthesis was performed employing the 'Ready To Go T-Primed First-Strand Kit' (Pharmacia, total volume 33 μl). Due to the very low RNA content of the wood, total RNA (between 500 ng and 2 μg) was used for first strand cDNA synthesis. Handling was as outlined in the manual with an additional final incubation step for 1 h at 41 °C.

RT-PCR reaction for sucrose synthase specific transcript from *Robinia pseudoacacia* L. wood

Note – as outlined in the RAPD-PCR and DDRT-PCR section, in order to minimise risks of contamination the following measures should be taken when preparing RT-PCR reaction tubes:

(a)   Prepare all reactions at least in a laminar flow cabinet.
(b)   Devote a set of adjustable micropipettes to the laminar flow cabinet.
(c)   Wear sterile gloves during the set-up of tubes.
(d)   Each experiment should include duplicate tubes and template-free controls.

1   Prepare a master mix of all reagents common to all the programmed reactions.
    Note – prepare only the quantity required for the total number of sample tubes including

controls; but make sure that at least one extra aliquot is prepared to ensure sufficient volume is available for the final tube.

2   In a sterile master mix tube, combine the following amounts per sample:

   10.0 μl of reaction buffer 10x (0.75 M Tris-HCl pH 9.0, 0.2 M $(NH_4)_2SO_4$, 0.1% Tween-20)
   6.0 μl 25 mM $MgCl_2$
   4.0 μl of 4 mM dNTPs
   74.0 μl of distilled water
   0.2 μl of primer-mix (100 pmol each)

3   Dispense 95 μl of the master mix to each reaction microtube.

4   Add 1–10 μl of first strand cDNA synthesised product as target template to microtubes, and gently mix the reaction mixtures by tipping each tube; briefly spin the tubes to remove bubbles (this is normally an optional procedure) place all tubes in the thermocycler programmed for:

   1 cycle: 2 min, 92 °C; 45 sec, 39 °C; 1 min, 72 °C
   10 cycles: 1 min, 91 °C; 45 sec, 40 °C; 1 min, 72 °C
   10 cycles: 1 min, 91 °C; 45 sec, 40 °C; 1.5 min, 72 °C
   10 cycles: 1 min, 91 °C; 45 sec, 40 °C; 2 min, 72 °C
   10 cycles: 1 min, 91 °C; 45 sec, 40 °C; 3 min, 72 °C
   1 cycle: 5 min, 72 °C
   1 cycle: 2 min, 25 °C.

5   Immediately the thermocycler has finished, place all tubes on ice.

Gel electrophoresis and DNA staining

Agarose / ethidium bromide

1   Prepare a 1% agarose gel in 1 × TBE Buffer.
2   Add 1 μl of 10 × loading buffer to 9 μl aliquots of each amplified sample.
3   Load all of the 10 μl samples and 10 μl of a molecular size DNA marker on the gel.
4   Run the gel in 1 × TBE buffer at 100 V until the dye front has covered most of the length of the gel.
5   Stain gels with ethidium bromide and view on a UV trans-illuminator as stated before.

Elution of differential bands

QIAGEN QIAEX-KIT
The manufacturer's detailed step-by-step protocol was followed for elution and purification of the specific amplification product.

Cloning of the SuSy specific fragment
As for DDRT-PCR step above.

Plasmid DNA purification
As for DDRT-PCR step above.

DNA Sequencing
As for DDRT-PCR step above.

APPLICATION

Use of RAPD-PCR to detect differences in genomic DNA extracted from different zones in *Robinia pseudoacacia* L. trees.

DNA extracted via the CTAB method ranged from 17–0.5 μg DNA/g dry weight; the yield from bark was significantly higher and about three times greater than that extracted from outer sapwood. The amount of DNA extracted from heartwood was significantly lower than that extracted from sapwood, with the transition zone DNA levels being in between these two zones. The extracted DNA was contaminated with RNA. Although this RNA was at times persistent and partially resistant to RNase treatment, it too was eventually removed after two or three separate treatments with RNase A. The $A_{260/280}$ ratios were between 1.88 and 1.48, which are towards the lower end of acceptable limits, and generally decreased with age of tissues; although the levels of contamination were not sufficiently high to halt the PCR reaction (De Filippis *et al.*, 1996). The $A_{260/230}$ ratios were generally low and below 2, which may indicate contamination from phenolics and polysaccharides, however, the compounds causing the higher non-specific absorbance at 230 nm did not inhibit the PCR reaction.

Initially, with some primers, inconsistencies between replicates were apparent during the RAPD-PCR amplification reactions. This problem was easily eliminated by diluting any PCR reaction inhibitors in the template DNA preparation with TE buffer. This is a common approach to diluting-out possible inhibitors in a sample, especially with the use of low quality DNA, and was essential for most of the DNA extracted from woody plants. A concentration of 2–4 mM MgCl₂ was found to be optimum for the primers used over 40 cycles, but each primer had its own optimum concentration of MgCl₂. Without this optimisation step, inconsistent banding of replicates was evident (Figure 1A, lanes d, e and f, g). The other chemical components of the RAPD-PCR reaction were present in sufficient concentration for polymorphic DNA bands to be adequately and consistently amplified with all of the DNA used. Primers which do not produce amplified DNA fragments must also be discarded, provided that the lack of bands was not due to a non-optimised MgCl₂ concentration.

The two accessions (i.e. trees A and B) of *Robinia pseudoacacia* were compared. Using primer UBC 325, differences in RAPD patterns with both agarose and polyacrylamide gel separation were seen (Figure 1A, lanes b, c, d and f, h, i; Figure 1B, lane b and lanes d, e). Band patterns from the polyacrylamide gel also demonstrate the susceptibility of this sensitive separation and staining method to under- or over-dilution of the amplified DNA applied to the micro-wells (Figure 1B, lanes c, d, e

and f, g). The fingerprinting/profiling patterns for tree B are very similar from lane c to lane g (Figure 1B), but the best separation patterns for outer and inner sapwood is demonstrated by a 1:2 DNA dilution (Figure 1B, lanes d, e and f). Using bark and sapwood DNA from trees A and B, primer UBC 328 also showed distinct polymorphic band differences between the two individual trees (indicated by arrows) (Figures 2A and 2B).

Although the overall fingerprinting patterns were similar (Figure 2A, lanes b, c and d, e; Figure 2B, lanes c, d and e, f), it was evident from Figure 2 that, within the one individual tree (i.e. tree B), there were minor band differences (arrowed) between wood zones. Use of slightly older wood zones confirmed the existence of differences between trees A and B (Figure 2C, lanes c, d, and e, f). Using primer UBC 327, there were band differences between the outer and inner transition zones (Figure 2D, lanes c and d), and also between the outer and inner heartwood (Figure 2D, lanes f and g) even in the one tree (especially look at bands between 1.06 and 0.38 Kbp DNA).

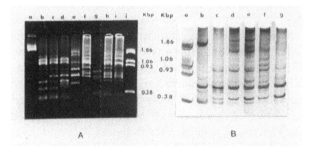

Figure 1 RAPD-PCR analysis of DNA from *Robinia pseudoacacia* bark and wood using arbitrary primer UBC 325; separated in 2% agarose and stained with ethidium bromide (A), or separated in 8% polyacrylamide and stained with silver (B). RAPD-PCR reactions were run over 40 cycles using 10–15 ng genomic DNA. **A:** Lane a = size marker (λ *Hind* III); lane b = wood ms, tree A; lane c = wood it, tree A; lane d = bark, tree A; lane e = bark, tree A (MgCl₂ not optimised); lane f = bark, tree B; lane g = bark, tree B (MgCl₂ not optimised); lane h = wood ot, tree B; lane i = wood is, tree B; lane j = size marker (pBR322 *Mva* I). **B:** Lane a = size marker (pBR322 *Mva* I); lane b = wood is, tree A (DNA diluted 1:2); lane c = wood is, tree B (DNA diluted 1:4); lane d = wood is, tree B (DNA diluted 1:2); lane e = wood os, tree B (DNA diluted 1:2); lane f = wood ot, tree B (DNA diluted 1:2); lane g = wood ot, tree B (DNA diluted 1:4). Key: Sapwood (s)-outer (o), middle (m), inner (i); Transition zone (t)-outer (o), inner (i); Heartwood (h)- outer(o), middle (m), inner (i).

Use of DDRT-PCR to screen *Eucalyptus microcorys* for genes which have been down-regulated as a result of salinity.

Although the RNA isolation procedure adapted from Verwoerd *et al.* (1989), which used SDS and phenol, was very effective in isolating RNA, the RNA was degraded, and the SDS/cell slurry became brown during incubation. Even after phenol/chloroform extraction there was no noticeable reduction in the brown colouration of the RNA pellet; which may have been due to high plant phenolics. The final RNA isolation procedure used was a modified CTAB-Phenol method adapted from Verwoerd *et al.* (1989) and Shellie *et al.* (1997). The total RNA extracted was relatively clean (by colour) and undegraded, and the method was reasonably efficient. The total RNA extracted from *Eucalyptus* shoots was visualised and quantified on a 0.8% agarose gel, and ranged between 10.5 and 36.9 μg RNA/g fresh weight (A₂₆₀/₂₈₀ of 1.83–1.92). Several commercially available kits were used, but difficulties were experienced; often the amount of RNA extracted was small, and the RNA from woody plants was mostly degraded. The only kit which produced the amount and quality of total RNA comparable to the CTAB-Phenol method was the QIAGEN

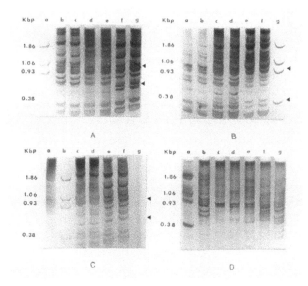

Figure 2 RAPD-PCR analysis of DNA from *Robinia pseudoacacia* bark and wood using arbitrary primers UBC 328 (A,B,C) or arbitrary primer UBC 327 (D), separated in 8% polyacrylamide and stained with silver. RAPD-PCR reactions were run over 40 cycles using 10–15 ng genomic DNA. **A:** Lane a = size marker (pBR322 *Mva* I); lanes b–c = wood os, tree B; lanes d–e = wood ms, tree B; lanes f–g = wood ms, tree A. **B:** Lane a–b = wood it, tree A; lanes c–d = wood is, tree A; lanes e–f = bark, tree B; lane g = size marker (pBR322 *Mva* I). **C:** Lane a = *Nicotiana tabacum*; lane b = size marker (pBR322 *Mva* I); lanes c–d = wood is, tree B; lanes e–f = wood is, tree A; lane g = negative control (all PCR components except genomic DNA). **D:** Lane a = size marker (pBR322 *Mva* I); lane b = wood os, tree A; lane c = wood ot, tree B; lane d = wood it, tree B; lane e = wood os, tree A; lane f = wood ih, tree B; lane g = wood oh, tree B. Key: Sapwood (s) — outer (o), middle (m), inner (i); Transition zone (t) – outer (o), inner (i); Heartwood (h) – outer (o), middle (m), inner (i). Arrowheads indicate distinct polymorphic band differences.

RNeasy Mini Plant Kit (used for RNA extraction in RT-PCR application, see below). We recommend this kit for woody plant tissues.

The primers that were used for differential display were an anchored oligo dT primer ($T_{12}AG$), and one of the random decamers from the OPB series kit. Figure 3A represents a typical DDRT-PCR gel stained with silver using $T_{12}AG$ and the decamer OPB 2, whereas Figure 3B is the DDRT-PCR profile using $T_{12}AG$ and OPB 6 and OPB 10. These primer combinations appeared to produce good profiles for comparison, although there was some band-smearing in some lanes with the

decamer OPB 2. The differential display profiles of Figure 3 show clearly that most genes (or cDNA fragments) were down-regulated in the *Eucalyptus* shoots adapted to elevated levels of salt, and that the profiles of the sensitive and adapted shoots were different (and could be considered a successful basis for RNA fingerprinting). Note that the amplified differential cDNA required a dilution factor of 1:16, and the intense down-regulated bands in salt adapted plants at 900 and 470 bp with OPB 2 (Figure 3A). On the other hand with primers OPB 6 and OPB 10 there were no apparent cDNA expression changes between salt sensitive and salt adapted plants (Figure 3B).

Due to the complexities of the silver-staining, acetic acid-fixation and the polyacrylamide, differential bands of interest were extremely difficult to elute and purify from these gels (and probably impossible to elute given the chemistry of silver ion and nucleic acids). Therefore, an alternative method of purifying the cDNA fragments had to be found before cloning. Since many of the bands

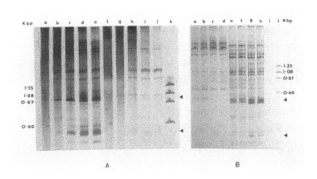

Figure 3 DDRT-PCR analysis of total RNA from salt-sensitive or salt-tolerant *Eucalyptus microcorys* leaves using an anchored oligo primer ($T_{12}AG$–for both gels A and B), plus arbitrary primers OPB 2 (for A); OPB 6 or OPB 10 (for B). Gels were separated in 8% polyacrylamide and stained with silver, and PCR reactions were run over 40 cycles using initially 1–5 μg total RNA. **A:** Lanes a–e = salt-sensitive; lanes f–j = salt-adapted; lanes a,f = no dilution; lanes b,g = 1: 2 dilution; lanes c,h = 1: 4 dilution; lanes d,i = 1: 8 dilution; lanes e,j = 1: 16 dilution; lane k = size marker (φX174 *Hae* III). **B:** Lanes a, b, e, f = salt-sensitive; lanes c, d, g, h = salt adapted; lanes a–d = primer OPB 6; lanes e–h = primer OPB 10; lane i = negative control (all PCR components except cDNA); lane j = size marker (φX174 *Hae* III). Arrowheads indicate down-regulated bands in salt adapted plants at 900 and 470 bp with OPB 2 (Figure 3A) and no apparent cDNA expression changes between salt sensitive and salt adapted plants with primers OPB 6 and OPB 10 (Figure 3B).

in the differential display gels were reasonably intense, the products were run on low melting point agarose gels (several lanes), the contents of each lane eluted by the method described above (modified from Sambrook et al., 1989), the DNA from multiple lanes combined, and finally the DNA was successfully cloned with the pGEM plasmid vector system.

Positive colonies containing the eluted insert were PCR-amplified with the M13 forward and reverse primers. From subsequent positive banding patterns, clonal colonies were selected for plasmid purification. In our investigations with *Eucalyptus,* a differentially expressed cDNA fragment of 900 bp was successfully sequenced (Figure 3A, arrowed); and was found to be 84% homologous in base sequence to α-tubulin of *Arabidopsis thaliana.*

Use of RT-PCR to get an insight into the regulation of one enzyme of sucrose cleavage, sucrose synthase (SuSy), which seems to play a key role in wood formation in *Robinia pseudoacacia* L.

Information about the function or presence of sucrose-metabolising enzymes in stem tissues of deciduous trees is largely lacking. For wood tissues, successful application of biochemical and molecular methods is hampered by the very low percentage of living cells together with the rigidity and high phenolic content of the material. Overcoming these difficulties (see techniques and applications sections above), we have investigated the distribution of the sucrose-metabolising enzymes in stem tissues of a deciduous tree, *Robinia pseudoacacia* L. As SuSy appeared to be a key enzyme (Hauch and Magel, 1998), its specific gene expression was studied in more detail by RT-PCR techniques. In order to find possibilities for semi-quantitation of the RT-PCR products, a reference template was constructed from a full-length SuSy cDNA clone from *Vicia faba* (MVfSUCS4/4, kindly provided by Dr. Hans Weber, IPK, Gatersleben, Germany; Heim et al., 1993) by incubation of 1 μg of the plasmid with *Eco32i* and *XbaI* (1U each, MBI) in a total volume of 10 μl. After ligation (Sambrook et al., 1989), the plasmid was transformed into competent *E. coli* cells (JM 109; Promega). Clones which revealed a 270 bp PCR-fragment (SUS270) after amplification using the SuSy specific primers, were selected. For semi-

quantitative RT-PCR, 130 ng of this reference fragment was co-amplified with 15 ng of total first-strand-cDNA from *Robinia* wood, containing unknown amounts of target template, by employing the conditions outlined above. The two cDNA fragments (SUS270 and SUS450) were separated on an agarose gel and stained with ethidium bromide.

The presence of SuSy transcript was analysed during cambial growth and heartwood formation. The deduced amino acid sequence of the 450 bp RT-PCR fragment (SUS450), which was amplified from RNA preparations from wood tissues by using the designed primers, revealed more than 90% homology with the corresponding sequence of a SuSy cDNA clone from *Vicia faba* (MVfSUSC4/4, Heim et al., 1993; data not shown). Thus, the primers used were specific for SuSy mRNA from *Robinia pseudoacacia* tissues. RT-PCR of total RNA preparations from individual trunk tissues with a reference fragment (SUS270) showed that SuSy-specific mRNA (SUS450) was highest in extracts from the youngest sapwood collected in July (Figure 4). With further maturation of the tissue, SuSy-specific mRNA became a negligible fraction of total mRNA. In the sapwood-heartwood transition zone, SuSy-specific mRNA was low in April, increased from July until the end of September and decreased again in the dying tissue until November (Figure 4). In both tissues, the cambial growth zone (data not shown) and the sapwood-heartwood transition zone, SuSy transcripts are correlated with the amounts of enzyme specific protein and its catalytic activity, both in time and type of tissue.

FUTURE PROSPECTS

RAPD-PCR future prospects lie in more-detailed analysis and sequencing of detected polymorphic RAPD bands, and this may provide more information about the growth and development of trees by comparing RAPD sequences with increasingly available genetic (gene) data in computer banks. Recent critical analysis of RAPD-PCR has focused on studies in which co-migrating bands have proven not to be homologous in nucleotide sequence, and divergence from homology can be as high as 30% (Rieseberg, 1996). This has considerably decreased the utility of RAPDs in comparing both closely related and distant species because the method and scoring of bands will tend to over-estimate differences.

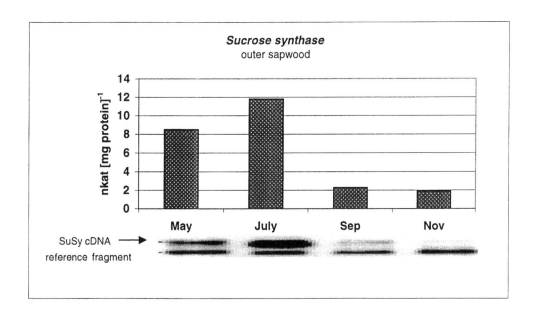

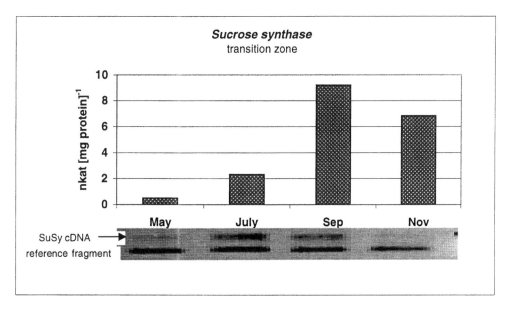

Figure 4   Semi-quantitation of SuSy mRNA from the outer sapwood (upper panel) and the sapwood-heartwood transition zone (lower panel) of *Robinia* stems collected in May, July, September (Sep), and November (Nov) by RT-PCR. Similar amounts of target cDNA were co-amplified with a reference fragment (SUS270) designed from a SuSy full-length clone (MVfSUCS4/4). The arrows indicate the amplified cDNA fragment SUS450.

A limitation of DDRT-PCR is that the mRNA species detected do not usually comprise the complete set of coding regions (Bertioli *et al.*, 1993), and this must be borne in mind when interpreting differentially expressed data. However, arbitrary primer design technology is advancing rapidly, and combinations of arbitary primers that are 50%

GC rich 15- to 21-mer in length, coupled to higher annealing temperatures, are proving useful in detecting low abundance long RNA sequences. The availability of a rapidly expanding sequence database and bioinformation render all information derived from DDRT-PCR increasingly useful. Automation of the differential display protocol has been contemplated, but it has not streamlined the process by which verification is effected. New and improved Northern hybridisation procedures may overcome this in future; Reverse Northern Hybridisation already shows good potential (Mou et al., 1994).

The use of RT-PCR in wood research will hopefully increase during the next few years. Much more information at the gene expression level is needed to understand the profound basic pathways characteristic for wood physiology and to create tools which could be targets for biotechnological approaches leading to the introduction of desired wood properties. However, using these methods, problems created by various species-specific wood components might arise and have to be overcome.

ACKNOWLEDGEMENTS

The authors would like to acknowledge financial support by the Deutsche Forschungsgemeinschaft (DFG) and Australian Research Council (ARC).

REFERENCES

Bauer, D., Müller, H., Reich, J., Riedel, H., Ahrenkiel, V., Warthoe, P. and Strauss, M. (1993) Identification of differentially expressed mRNA species by an improved display technique (DDRT-PCR). *Nucleic Acids Res.*, **21**, 4272–4280.

Benson, D.A., Boguski, M.S., Lipman, D.J., Ostell, J., Ouellett, B.F.F., Rapp, B.A. and Wheeler, D.L. (1999) GenBank. *Nucleic Acids Res.*, **27**, 12–17.

Bertioli, D.J., Schlichter, U.H.A., Adam, M.J., Burrows, P.R., Steinbiß, H.H. and Antoniw, J.F. (1995) An analysis of differential display shows a strong bias towards high copy number mRNAs. *Nucleic Acids Res.*, **23**, 4520–4523.

Carlson, J.E., Tulsieram, L.K., Glaubitz, J.C., Luk, V.W.K., Kauffeldt, C. and Rutledge, R. (1991) Segregation of random amplified DNA markers in F₁ progeny of conifers. *Theor. Appl. Genet.*, **83**, 194–200.

Chen, D.C. and De Filippis, L.F. (1996) Application of genomic DNA and RAPD-PCR in genetic analysis and fingerprinting of various species of woody trees. *Aust. For.*, **59**, 37–46.

Crawford, D.J. (1990) *Plant molecular systematics: Macromolecular approaches.* New York: John Wiley.

De Filippis, L.F., Hoffmann, E. and Hampp, R. (1996) Identification of somatic hybrids of tobacco generated by electrofusion and culture of protoplasts using RAPD-PCR. *Plant Sci.*, **121**, 39–46.

De Filippis, L.F. and Magel, E. (1998) Differences in genomic DNA extracted from bark and from wood of

different zones in *Robinia* trees using RAPD-PCR. *Trees*, **12**, 377–384.

Doyle, J.J. (1991) DNA protocols for plants. In *Molecular techniques in taxonomy*, edited by G.M. Hewitt, A.W.B. Johnston and J.P.W. Young. New York: Springer Verlag.

Grattapaglia, D., Bertolucci, F.L. and Sederoff, R.R. (1995) Genetic mapping of QTLs controlling vegetative propagation in *Eucalyptus grandis* and *E. urophylla* using a pseudo-testcross strategy and RAPD markers. *Theor Appl. Genet.*, **90**, 933–947.

Hannappel, U., Balzer, H.-J. and Ganal, M.W. (1995) Direct isolation of cDNA sequences from specific chromosomal regions of the tomato genome by the differential display technique. *Mol. Gen. Genet.*, **249**, 19–24.

Hauch, S. and Magel, E.A. (1998) Extractable activities and protein content of sucrose phosphate synthase, sucrose synthase and neutral invertase in trunk tissues of *Robinia pseudoacacia* L. are related to cambial wood production and heartwood formation. *Planta*, **207**, 266–274.

Heim, U., Weber, H., Baumlein, H. and Wobus, U. (1993) A sucrose synthase gene in *Vicia faba* L.: Expression pattern in developing seeds in relation to starch synthesis and metabolic regulation. *Planta*, **191**, 394–401.

Heinze, B. and Schmidt, J. (1995) Monitoring genetic fidelity *vs* somaclonal variation in Norway spruce (*Picea abies*) somatic embryogenesis by RAPD analysis. *Euphytica*, **85**, 341–345.

Hummel, R., Jorgensen, M., Bevort, M., Gronborg, M. and Lefers, H. (1997) Verification of differential display results by RNase protection, Chapter 32. In *Fingerprinting methods based on arbitrarily primed PCR*, edited by M.R. Micheli and R. Bova. Heidelberg: Springer Verlag.

Jones, S.W., Decheng, C., Weislou, O.S. and Esmaeli-Azad, B. (1997) Generation of multiple mRNA fingerprints using fluorescence-based differential display and an automated DNA sequencer. *Bio Techniques*, **22**, 536–543.

Krahl, K.H., Dirr, M.A., Halward, T.M., Kochert, G.D. and Randle, W.M. (1993) Use of single-primer DNA amplifications for the identification of red maple (*Acer rubrum* L.) cultivars. *J. Environ. Hort.*, **11**, 89–92.

Liang, P. and Pardee, A.B. (1992) Differential display of eukaryotic messenger RNA by means of the polymerase chain reaction. *Science*, **257**, 967–971.

Micheli, M.R. and Bova R. (1997) *Fingerprinting methods based on arbitrarily primed PCR.* Heidelberg: Springer.

Moran, G.F., Bell, J.C. and Prober, S. (1990) The utility of isozymes in the systematics of some Australian tree groups. *Aust. Sys. Bot.*, **3**, 47–57.

Mou, I., Miller, J., Wang, F. and Califour, L. (1994) Improvements to the differential display method for gene analysis. *Biochem. Biophys. Res. Comm.*, **199**, 564–569.

Oh, B.-J., Balint, D.E. and Giovannoni, J.J. (1995) A modified procedure for PCR-based differential display and demonstration of use in plants for isolation of genes related to fruit ripening. *Plant Mol. Biology Rep.*, **13**, 70–81.

Park, Y.H. and Kohel, R.J. (1994) Effect of concentration of MgCl₂ on random-amplified polymorphic DNA polymorphism. *BioTechniques*, **16**, 652–655.

Rymerson, R.T., Bodnaryk, R.P., Haber, S. and Procunier, J.D. (1995) Arbitrary primed RNA fingerprinting in plants. *Biotech. Techniques*, **9**, 563–566.

Sambrook, J., Fritsch, E.F. and Maniatis, T. (1989) *Molecular cloning: A laboratory manual* Volume 1. (2nd ed). Cold Spring Harbor: Laboratory Press, USA.

Shellie, K.C., Meyer, R.D. and Mirkov, T.E. (1997) Extraction of total RNA from melon mesocarp tissue. *Hort. Science*, **32**, 134.

Sterky, F., Regan, S., Karlsson, J., Hertzberg, M., Rohde, A., Holmberg, A., Amini, B., Bhalerao, R., Larsson, M., Villarroel, R., Van Montague, M., Sandberg, G., Olsson, O., Teeri, T.T., Boerjan, W., Gustafsson, P., Uhlén, M., Sundberg, B. and Lundeberg, J. (1998) Gene discovery in the wood-forming tissues of poplar: Analysis of 5692 expressed sequence tags. *Proc. Natl. Acad. Sci. USA*, **95**, 13330–13335.

Torelli, A., Soragni, E., Bolchi, A., Petrucco, S., Ottonello, S. and Branca, C. (1996) New potential markers of *in vitro* tomato morphogenesis identified by mRNA differential display. *Plant Mol. Biol.*, **32**, 891–900.

Tulsieram, L.K., Glaubitz, J.C., Kiss, G. and Carlson, J.E. (1992) Single tree genetic linkage mapping in conifers using haploid DNA from megagametophytes. *Bio Technology*, **10**, 686–690.

Veres, G., Gibbs, R.A., Scherer, S.E. and Caskey, C.T. (1987) The molecular basis of the sparse fur mouse mutation. *Science*, **237**, 415–417.

Verwoerd, T.C., Dekker, B.M.M. and Hoekema, A. (1989) A small-scale procedure for the rapid isolation of plant RNAs. *Nucleic Acids Res.*, **17**, 2362.

Welsh, J., Chada, K., Dalal, S.S., Cheng, R., Ralph, D. and McClelland, M. (1992) Arbitrary primed PCR fingerprinting of RNA. *Nucleic Acids Res.*, **20**, 4965–4970.

Welsh, J. and McClelland, M. (1990) Fingerprinting genomes using PCR with arbitrary primers. *Nucleic Acids Res.*, **18**, 7213–7218.

Williams, J.G.K., Hanafey, M.K., Rafalski, J.A. and Tingey, S.V. (1993) Genetic analysis using random amplified polymorphic DNA markers. *Methods in Enzymol.*, **218**, 704–740.

Williams, J.G.K., Kubelik, A.R., Livak, K.J., Rafalski, J.A. and Tingey, S.V. (1990) DNA polymorphisms amplified by arbitrary primers are useful as genetic markers. *Nucleic Acids Res.*, **18**, 6531–6535.

Zhao, S., Ooi, S.L. and Pardee, A.B. (1995) New primer strategy improves precision of differential display. *BioTechniques*, **18**, 845–850.

SUPPLIERS OF EQUIPMENT AND REAGENTS

Agar Scientific Ltd
66a Cambridge Road
Stansted
Essex
CM24 8DA
UK
Tel.:   +44-(0)1279-813519
Fax: +44-(0)1279-815106

Aldrich Chemical Co.
(Sigma-Aldrich Co. Ltd)
The Old Brickyard
New Road
Gillingham
Dorset
SP8 4XT
UK
Tel.:   0800-717181/+44-(0)1747-822211
Fax: 0800-378538/+44-(0)1747-823779
Internet: http://www.sial.com/aldrich/

American Can Co.
8770 W. Bryn Mawr Avenue
Chicago
Illinois 60631
USA
Tel.:   +1-773-399-3000
Fax: +1-773-399-3193
Internet: http://www.ancgi.com

Amersham Pharmacia Biotech Ltd
Amersham Place
Little Chalfont
Buckinghamshire
HP7 9NA
UK
Tel.:   0800-515313 (orders)/0800-616928
        (enquiries)/+44-(0870)-6061921
        (switchboard)
Fax: 0800-616927 (orders)
Email: uk_custserve@amersham.co.uk
        (enquiries)

Amersham Pharmacia Biotech Inc.
800 Centennial Avenue
PO Box 1327
Piscataway
New Jersey 08855
USA
Tel.:   +1-800-526-3593
Fax: +1-877-295-8102
Internet: http://www.apbiotech.com

Appligene Oncor
Parc d'innovation
rue Geiler de Keysersberg BP 72
67402 Illkirch Cedex
France
Tel.:   +33-(0)3-88-67-54-25
Fax: +33-(0)3-88-67-19-45
Email: serviceclients@appligene-oncor.fr
Internet: http://www.oncor.com/prod-app.htm

Balzers Instruments
Postfach 1000
Fl-9496 Balzers
Liechtenstein
Tel.:   +41-75-3884111
Fax: +41-75-3885414
Internet: http://www.balzers.com

Beckman (See Beckman Coulter GmbH)

Beckman Coulter GmbH
Siemenstrasse 1
D-85716 Unterschleissheim-Lohhof
Germany
Tel.:   +49-89-35-870-0
Fax: +49-89-35-870-490
Internet: http://www.beckman.com/

Bio 101 Inc.
1070 Joshua Way
Vista
California 92083
USA
Tel.:   +1-800-424-6101
Fax: +1-760-598-0116

Bioblock Scientific
Parc d'innovation BP 111
Bd Sébastien Brant
67403 Illkirch Cedex
France
Tel.: +33-(0)3-88-67-14-14
Fax: +33-(0)3-88-67-11-68
Internet: http://www.bioblock.com

BioCell (see British BioCell International)

Bio-Rad Laboratories
2000 Alfred Nobel Drive
Hercules
California 94547
USA
Tel.: +1-510-741-1000
Fax: +1-510-741-5800
Internet: http://www.discover.bio-rad.com

Biosis
Two Commerce Square
2001 Market Street
Suite 700
Philadelphia
Pennsylvania 19103-7095
USA
Tel.: +1-800-523-4806 (from USA and
Canada)/+1-215-587-4800 (worldwide)
Fax: +1-215-231-7401
Email: info@mail.biosis.org
Internet: http://wwww.biosis.org

Biosupplies Australia Pty Ltd
PO Box 835
Parkville
Victoria 3052
Australia
Fax: +61-3-9347-1071
Email: enquiries@biosupplies.com.au
Internet: http:www.biosupplies.com.au/

Biotech Australia
28 Barcoo Street
Roseville
Sydney
NSW 2069
Australia
Tel.: +61-(0)2-9928-8800
Fax: +61-(0)2-9928-8899
Email: enquiries@bioaust.com.au
Internet: http://www/bioaust.com.au

Biotech International (see Biotech Australia)

Bio-Tek Instruments
Werner ron Siemens Straße 1
D-85375 Newfahrn
Germany
Tel.: +49-(0)8165-905-302
Fax.: +49-(0)8165-905-240
Internet: http://www.biotek.de

Boehringer Mannheim
Roche Diagnostics GmbH
Sandhofer Straße 116
D-68305 Mannheim
Germany
Internet: http://biochem.roche.com

Boehringer Mannheim SA
2 avenue du Vercors BP 59
F-38242 Meylan Cedex
France
Tel.: +33-(0)4-76-76-30-87
Fax: +33-(0)4-76-76-46-90

B Braun Biotech International GmbH
Schwarzenberger Weg 73–79
D-34209 Melsungen
Germany
Tel.: +49-5661-71-3400
Fax: +49-5661-71-3702
Email: bbi.info@bbraun.com
Internet: http://www.bbraunbiotech.com

British BioCell International
Golden Gate
Ty Glas Avenue
Cardiff
CF4 5DX
Wales
Tel.: +44-(0)29-2074-7232
Fax: +44-(0)29-2074-7242

Cameca FRANCE
103 boulevard Saint Dennis BP6
92403 Courbevoie Cedex
France
Tel.: +33-(0)1-43-34-62-00
Fax: +33-(0)1-43-34-63-50
Internet: http://www.cameca.fr

Cappel Research Products
1263 South Chillicothe Road
Aurora
Ohio 44202
USA
Tel.: +1-800-279-5490
Fax: +1-216-562-2642
Internet: http://www.icnbiomed.com/immuno/
cappel.html

Carlo Erba Reactifs
Chaussée du Vexin
Parc d'Affaires des Portes
BP 616
F-27106 Val de Reuil Cedex
France
Tel.: +33-(0)2-32-09-20-00
Fax: +33-(0)2-32-09-20-20
Internet: http://www.gazettelabo.tm.fr/hybride/
nom/c/carlo.htm

Costar, Inc.
Corning
New York 14831
USA
Tel.: +1-978-635-2200
Fax: +1-978-635-2476

Dako A/S (See Dakopatts AB)

Dakopatts AB
Box 13
SE-125 21 Älvsjö
Sweden
Tel.: +46-8-5562-06-00
Fax: +46-8-5562-06-19
Internet: http://www.dakopatts.se

Dojindo Laboratories
2025-5 Taburu
Mashiki-machi
Kamimishaki-gun
Kumamoto 861-2202
Japan
Tel.: +81-96-286-1515
Fax: +81-96-286-1525
Internet: http://www.dojindo.com/

Eastman Kodak Company
Internet: http://www.kodak.com

Edwards High Vacuum Int.
Manor Road
Crawley
West Sussex
RH10 2LW
UK
Tel.: +44-(0)293-528844
Fax: +44-(0)293-533453

Electron Microscopy Sciences
PO Box 251
321 Morris Road
Fort Washington
Pennsylvania 19034
USA
Tel.: +1-215-646-1566
Fax: +1-215-646-8931
Email: sgkcck@aol.com
Internet: http://www.emsdiasum.com/

EMS (see Electron Microscopy Sciences)

Eurogentec Bel SA
Parc scientifique du Sart Tilman
B-4102 Seraing
Belgium
Tel.: +32-4-366-01-50
Fax: +32-4-365-51-03
Internet: http://www.eurogentec.com/

Fisher-Bioblock Scientific
Parc d'innovation BP 111
Bd Sébastien Brant
F-67403 Illkirch Cedex
France
Tel.: +33-(0)3-88-67-14-14
Fax: +33-(0)3-88-67-11-68
Internet: http://www.bioblock.com

Fisher Scientific
Im Heiligen Feld 17
D-58239 Schwerte
Germany
Tel.: +49-1805-258221
Fax: +49-1805-258223
Email: auftrag@de.fishersci.com

Fuji Film Co. Ltd
26–30, Nishiazabu 2-chome
Minato-ku
Tokyo 106-8620
Japan
Tel.:   +81-3-3406-2201
Fax: +81-3-3406-2575
Email: purchase@tokyo.fujifilm.co.jp
Internet: http://home.fujifilm.com/

Gallus Immunotech
Rr 3
CND-Fergus
Ontario
N1M 2W4
Canada
Tel.:   +1-519-843-2918
Fax: +1-519-843-2918
Internet: http://www.gallusimmunotech.com/

GenVac (See General Vacuum, Inc.)

General Vacuum, Inc.
190A Alpha Drive
Cleveland
Ohio 44143
USA
Tel.:   +1-888-646-9986
Fax: +1-440-646-9987
Internet: http://www.genvac.com/

Gibco BRL
Life Technologies SARL
BP 96
F-95613 Cergy Pontoise Cedex
France
Tel.:   +33-(0)1-34-64-54-40
Fax: +33-(0)1-30-37-50-07
Internet: http://www.lifetech.com/

Gibco BRL
Life Technologies GmbH
Tel.:   +49-721-61890
Fax: +49-721-6189500
Internet: http://www.lifetech.com, and
http://search.lifetech.com/

Harlan, Sera-lab Ltd
Hillcrest
Dodgeford Lane
Belton
Loughborough
LE12 9TE
UK
Tel.:   +44-1530-222123
Fax: +44-1530-224970

Heraeus Kulzer, Inc.
4315 S. Lafayette Blvd
South Bend
Indiana 46614
USA
Tel.:   +1-219-291-0661
Fax: +1-219-291-7248
Internet: http://kulzer.com/home.html

Hirschmann Instruments GmbH
Kramerstraße 17
D-82061 Neuried
Germany
Tel.:   +49-89-759-2206
Fax: +49-89-755-9304
Email: hirschmann.instruments@t-online.de

Histolab AB
Hulda Lindgrens gatan 6
SE-421 31 Göteborg
Sweden
Tel.:   +46-31-709-30-30
Fax: +46-31-709-30-40
Internet: http://www.histolab.se

Hydro Agr AB
Bos 516
SE-261 24 Landskrona
Sweden
Tel.:   +46-(0)4-187-61-00
Fax: +46-(0)4-185-83-46
Email: Lars.Johannesson@hydro.com

Supra AB (See Hydro Agr AB)

ICN Pharmaceuticals Ltd
1Elmwood
Chineham Business Park
Basingstoke
Hampshire
RG24 8WG
UK
Tel.:   0800-282474/+44-1-256-374-620
Fax: 0800-614735/+44-1-256-374-621
Email: sales@icnbiomed.com
Internet: http://www.icnbiomed.com

IKA-Werke GmbH & Co. KG
Postfach 1263
D-79217 Staufen
Germany
Tel.:   +49-7633-831-0
Fax: +49-7633-831-98
Email: sales@ika.de
Internet: http:www.ika.de

JEOL Ltd
1–2 Musashino 3-chome
Akashima
Tokyo 196-85558
Japan
Tel.:   +81-(0)-42-542-21-87
Fax: +81-(0)-42-546-57-57
Email: jeoltky@jeol.co.jp
Internet: http://www.jeol.co.jp

Kodak (see Eastman Kodak Company)

Komatsu Electronics Inc.
2597 Shinomiya
Hiratsuka-shi
Kanagawa-ken 254-8543
Japan
Tel.:   +81-463-22-8724
Fax: +81-463-23-3679
Internet: http://www.komatsu-electronics.co.jp

Konica Corporation
Shinjuku Nomura Building, 1-26-2
Nishi-Shinjuku
Shinjuku-ku
Tokyo 163-0512
Japan
Tel.:   +81-3-3349-5251
Fax: +81-3-3349-8998
http://www.konica.co.jp/english/

Kontron (see Bio-Tek Instruments)

Kosaka Laboratory Ltd
3-49-2 Kanamachi
Katsushika-ka
Tokyo 125-0042
Japan
Tel.:   +81-3-3607-1186
Email: tokyom@kosakalab.co.jp
Internet: http://www.kosakalab.co.jp
[Note: Japanese only]

Kubota Inc.
3-29-9 Hongo
Bunkyo-ku
Tokyo 113-0033
Japan
Tel.:   +81-3-3815-1331
Fax: +81-3-3814-2574
Internet: http://www.kubotacorp.co.jp/

Leica Microsystems Heidelberg GmbH
Im Neuenheimer Feld 518
D-69120 Heidelberg
Germany
Tel.:   +49-6221-41480
Fax: +49-6221-414833
Email: support@llt.de
Internet: http://www.llt.de/

Marivac Ltd
5821 Russell Street
Halifax
Nova Scotia
B3K 1X5
Canada
Tel.:   +1-800-565-5821
Fax: +1-902-455-4007
Email: marivac@ns.sympatico.ca
Internet: http://www3.ns.sympatico.ca/marivac/

MBI Fermentas GmbH
Franz Antoni Straße 22
D-68789 St. Leon-Rot
Germany
Tel.:   +49-6227-55853
Fax.:   +49-6227-53694
Internet: http://www.fermentas.com

MBI Fermentas Ltd
Graiciuno
Vilnius 2028
Lithuania
Tel.:   +370-2-64-1279
Fax: +370-2-64-3436
Internet: http://fermentas.lt

Meiwa Shoji Co., Ltd
2-4-25 Sentai
Sumiyoshi-ku
Osaka, 558-0047
Japan
Tel.: +816-6674-222
Fax: +816-6674-2323 or 816-6673-1211
Internet: http://www/meiwanet.co.jp

Merck Ltd
Merck House, Poole
Dorset
BH15 1TD
UK
Tel.: +44-(0)1202-669700
Fax: +44-(0)1202-665599

Microm Laborgeräte GmbH
Robert Bosch Straße 49
D-69190 Walldorf
Germany
Tel.: +49-6227-83-60
Fax: +49-6227-83-61-11
Internet: http://www.zeiss.de

Millipore
25 rue de Madrid
Parc d'affaires SLIC – BP 7406,
F-38074 St Quentin Fallavier Cedex
France
Tel.: +33-(0)4-74-94-45-83
Fax: +33-(0)4-74-95-57-50
Email: tech_service:Millipore.com

Molecular Probes, Inc.
4849 Pitchford Avenue
Eugene
Oregon 97402-9144
USA
Tel.: +1-541-465-8300
Fax: +1-541-344-6504
Internet: http://www.probes.com/

Okenshoji Co. Ltd
12–17 Ginza 6-chone Cho-ku
Tokyo 104-0061
Japan
Tel.: +81-3-3571-2879
Fax: +81-3-3572-6075
Internet: http://www.okenshoji.co.jp/

Olympus Optical Co. Ltd
Shinjuku 2-3-1
Shinjuku
Tokyo 163-0914
Japan
Tel.: +81-3-3340-2211
Fax: +81-3-3340-2130
Internet: http://www.olympus.co.jp

Osram AB
Rudanvägen 1
Box 504
SE-136 25 Haninge
Sweden
Tel.: +46-(0)8-707-44-01
Fax: +46-(0)8-707-44-40

PE Applied Biosystems (see PE Biosystems)

PE Biosystems
Biosystems GmbH
Brunnenweg 13
D-61331 Weiterstadt
Germany
Tel.: +49-06150-101-0
Fax: +49-06150-101-101
Internet: http://www.pebio.com/

Pelco Ltd (see Ted Pella, Inc.)

Ted Pella, Inc.
PO Box 492477
Redding
California 96049-2477
USA
Tel.: +1-530-243-2200
Fax: +1-530-243-3761
Email: sales@tedpella.com
Internet: http://tedpella.com/

Pharmacia Biotech Europe GmbH
Parc Technologique
rue René Razel
Saclay
F-91898 Orsay Cedex
France
Tel.: +33-(0)1-69-35-67-00
Fax: +33-(0)1-69-41-96-77

Pharmacia
Munzingerstraße 9
D-79111 Freiburg
Germany
Tel.: +49-(07)61-4903-0
Fax: +49-(07)61-4903-246
Internet: http://www.apbiotech.com/

Pierce Chemical Company
3747 North Meridian Road
PO 117
Rockford
Illinois 61105
USA
Tel.: +1-800-874-3723
Fax: +1-815-968-7316
Internet: http://www.piercenet.com/

Polylabo
10 rue de la Durance BP 36
F-67023 Strasbourg Cedex 1
France
Tel.: +33-(0)3-88-65-80-20
Fax: +33-(0)3-88-39-74-41
Internet: http://www.polylabo.com

Polysciences, Inc.
400 Valley Road
Warrington
Pennsylvania 18976
USA
Tel.: +1-800-523-2575
Fax: +1-800-343-3291
Internet: http://www.polysciences.com/

Prolabo
54 rue Roger Salengro
F-94126 Fontenay-sous-Bois Cedex
France
Tel.: +33-(0)1-45-14-85-00
Fax: +33-(0)1-45-14-86-16
Internet: http://www.prolabo.fr

Promega
Parc de activité des Verriéres
24 chemin des Verriéres
F-69260 Charbonniéres
France
Tel.: +33-(0)4-37-22-50-50
Fax: +33-(0)4-37-22-50-10
Internet: http://www.euro.promega.com

Qiagen
3 avenue du Canada
LP 809
F-91974 Courtaboeuf Cedex
France
Tel.: +33-(0)1-60-92-09-20
Fax: +33-(0)1-60-92-09-25
Internet: http://www.qiagen.com

Radiometer Analytical SA
75 rue d'Alsace
F-69627 Villeurbanne Cedex
Lyon
France
Tel.: +33-(0)4-78-03-38-38
Fax: +33-(0)4-78-68-88-12
Email: radiometer@nalytical.fr
Internet: http://www.radiometer.tm.fr/

Roche Molecular Biochemicals
201 Boulevard Armand Frappier
Laval
Québec H7V 4A2
Canada
Tel.: +1-450-686-7171
Fax: +1-450-686-7010
Internet: http://biochem.roche.com

Sakura Finetek USA, Inc.
1750 West 21 4th Street
Torrance
California 90501
USA
Tel.: +1-310-972-7800
Fax: +1-310-972-7888
Email: gilles.lefebvre@sakuraus.com

Sakura (trade name, see Konica Corporation)

Sartorius AG
D-37075 Göttingen
Germany
Tel.: +49-551-308-0
Fax: +49-551-308-289
Internet: http://www.sartorius.de/

Saur Laborbedarf
Carl-Zeiss-Straße 58
D-72770 Reutlingen
Germany
Tel.: +49-7121-54008
Fax: +49-7121-54000

Sds
ZI de Valdonne
F-13124 Peypin
France
Tel.: +33-(0)4-42-32-41-41
Fax: +33-(0)4-42-72-41-62

Shimadzu Corporation
International marketing division
3 Kanda-Nishikicho 1-chome
Chiyoda-ku
Tokyo 101
Japan
Tel.: +81-(0)3-219-5641
Fax: +81-(0)3-219-5710

also supplied by:
Roucaire Paris BP 65
20 avenue de l'Europe
F-78143 Velizy-Villacoublay Cedex
France
Tel.: +33-(0)1-39-46-96-33

Sigma-Aldrich, P.O. Box 14508
St. Louis
Missouri 63178
USA
Tel.: +1-800-262-9141
Fax: +1-800-325-5052
Internet: http://www.sigma-aldrich.com

Sigma-Aldrich Co. Ltd.
Fancy Road
Poole
Dorset
BH12 4QH
UK
Tel.: +44-1202-733114
Fax: +44-1202-715460
Email: ukcustsv@eurnotes.sial.com

SLT
TECAN Deutschland GmbH
Theodor Storm Straße 17
D-74564 Crailsheim
Germany
Tel.: +49-07951-94170
Fax: +49-07951-5038
Internet: http://www.tecan.de

Sony Corporation
6-7-35 Kitashinagawa
Shinagawa-ku
Tokyo 141-0001
Japan
Tel.: +81-3-5448-3311
Fax: +81-466-31-2595
Internet: http://www.world.sony.com

SPI Supplies/CANADA
Box 187
Postal Station "T"
Toronto
Ontario, M5B 4A1
Canada
Tel.: +1-416-787-9193
Fax: +1-416-781-0249
Email: spi@titan.tcn.ca
Internet: http://www.2spi.com/

Stratagene Europe
Gebouw California
Hogehilweg 15
1101 CB Amsterdam Zuidoost
The Netherlands
Tel.: 0800-0230448
Fax: 0800-0230447
Internet: http://www.stratagene.com

Stratagene GmbH
Postfach 105466
D-69044 Heidelberg
Germany
Tel.: +49-01308-40911
Fax: +49-01307-62088
Internet: http://www.stratagene.com

Surgipath Medical Industries Inc.
Europe Ltd
Venture Park
Stirling Way
Bretton
Peterborough
PE3 8YD
UK
Tel.: +44-(0)1733-333100
Fax: +44-(0)1733-331111
Internet: http://www.surgipath.com

TAAB Laboratories Equipment Ltd
3 Minerva House
Calleva Park
Aldermaston
Berkshire
RG7 8NA
UK
Tel.: +44-(0)118-981-7775
Fax: +44-(0)118-981-7881
Email: sales@taab.co.uk
Internet: http://www.microscopy-
uk.org/prodir/taab.htm

Vector Laboratories
3 Accent Park
Bakewell Road
Orton Southgate
Peterborough
PE2 6XS
UK
Tel.: +44-(0)1733-237999
Fax: +44-(0)1733-237119
Email: vector@vectorlabs.co.uk

Wako Pure Chemical Industries Ltd
3-1-2, Doshumachi
Chuo-ku
Osaka 540-8605
Japan
Tel.: +81-6-6203-3741
Internet: http://www.wako-chem.co.jp/
index.htm

Warner-Lambert Co. Ltd
Shick Division of Warner-Lambert Company
201 Tabor Road Morris Plains
New Jersey 07950-2693
USA
Tel: +1973-385-2000
Internet: http://www.wlkk.co.jp/index.html

Yamato Kohki Co. Ltd
2-14-43 Hizaori-cho
Asaka-shi
Saitama 351-0014
Japan
Tel.: +81-48-465-2314
Fax: +81-48-465-0029

Yamato Scientific Co. Ltd
2-1-6, Nihonbashi Honcho
Chuo-ku
Tokyo 103-0023
Japan
Tel.: +81-3-3231-1112
Internet: http://www.yamato-
net.co.jp/index.html

Carl Zeiss
Jena 07740
Germany
Tel.: +49-3641-64-1616
Fax: +49-3641-64-3144
Internet: http://www.zeiss.de

Züricher Beuteltuchfabrik
CH-8803 Rueschlikon
Switzerland
Tel.: +41-(0)1-724-6511
Fax: +41-(0)1-724-1525

TAXA REFERRED TO IN THIS BOOK

INDEX

phloem
  axial elements   131
  companion cells   208
  cytoskeleton of   131
  elements   208
  fibres   208
  maceration of   24
  parenchyma   208
  ray cells   131
  sieve elements   131
phloroglucinol/HCl (Wiesner reaction)   18–19
  recipe   18
phosphate buffer   54, 67, 69, 190
  recipe   69, 190
  recommended for electron microscopy   69
phosphate-buffered saline (PBS), recipe   57, 129, 148, 205,
    281, 304
phosphate-buffered saline plus bovine serum albumen
    (PBSA), recipe   57, 128
phosphate-buffered saline plus bovine serum albumen
    (PBSB), recipe   149
6-phosphogluconate dehydrogenase (6PG-DH)   229,
    238–9, 241–2
  assay   238–9
phosphotungstic acid (PTA)   51
  recipe   51
  staining of plasmalemma   51
'Pie' sections   275, 276, 302
pinhole   144, 151
piperazine-N,N'-bis-[2-ethylsulphonic acid] (PIPES)
    buffer see buffers
pits
  bordered   130, 131, 154
  contact   130, 131
  cross field   154
  role of cytoskeleton in development of   130, 131, 154
  simple   130, 131
pixel size   103, 150
plasmalemma (synonym plasma membrane)   58, 77, 78
plasmalemmal detachment   106
plasmalemmal invaginations   58, 78
plasmatubules (PTs)   58–9
plasmid, DNA purification   329
plasmodesmata
  distribution   43
  high voltage electron microscopy of   48
  staining with zinc-iodide-osmium-tetroxide (ZIO)   48
  staining with osmium-tetroxide-potassium-ferricyanide
    48
plasmolysis   77
plunge-freezing   297
polarised light see microscopical techniques
polyacrylamide gel electrophoresis (PAGE)   248, 253
polyacrylamide gels, preparation of   250
polyclonal antibodies see antibodies
polyethyleneglycol (PEG) (phenol scavenger)   231
poly-L-lysine adhesive see adhesives
polymerase-chain reaction (PCR)   274, 275, 319, 322
polymerisation of butyl-methylmethacrylate resin,
    problems of   129–30
polyphenol-containing material, interference   182
polyvinylpyrrolidone (PVP)   231, 303

insoluble (phenol scavenger)   231
soluble (phenol scavenger)   231
polysaccharides, test for   49, 183
ponceau 2R-periodic acid, stain for proteins   31
poplar
  model angiosperm/hardwood tree   3, 10, 212
  reasons why a good model tree   10–11
'pore system' (in walls)   92
positional signal   224
post-hybridisation   308–9
  solutions   309
post-translational modifications of tubulin see tubulin
potassium pyro-antimonate precipitation method   102,
    108
pre-fixation   121, 148, 203, 277
pre-hybridisation   303–5
pre-immune serum   186, 188, 189
pre-mixed gel solutions   265
pre-mix Spurr resin kits see Spurr epoxy resin
pre-prophase band (PPB)   114
pre-treatment, with acetone   278
primary antibody/ies
  anti-actin   125, 148
  anti-arabinan (LM6)   185, 190
  anti-callose   21, 55
  anti-cinnamyl alcohol dehydrogenase (CAD)   205
  anti-dehydrin   266
  anti-galactan (LM5)   185, 190
  anti-homogalacturonan (2F4)   185
  anti-pectin (JIM5)   53, 55, 185, 190
  anti-pectin (JIM7)   53, 55, 185, 190
  anti-tubulin   125, 148
  anti-vegetative storage proteins (VSPs)   261
  anti-xyloglucan   95–6
  UBIM22   125
primary (P) cell wall
  cambial cells   91
  cellulose microfibril orientation in   152
  enlarging cells   91
  intact tissue   86
  tissue culture   88
  visualisation   86
probe, for in situ hybridisation
  antisense (= control)   298
  control   298
  experimental   298
  sense (= experimental)   298
promoters
  constitutive   288
  tissue-specific   288
pronase   303, 304
pronase buffer   304
propidium iodide   32
  staining of nucleic acids   32, 35
  staining of organelles   35
propylene oxide   45, 46, 68, 70, 101, 102, 162, 168
protein(s)
  analysis   245–66
  cell wall-bound   192, 194
  classes of   247
  degradation, and minimisation thereof   246
  determination of   234

vegetative storage proteins (VSPs)   246, 249, 252, 257,
      260
vessel elements   130, 131
vibrating microtome *see* microtome
video-imaging   35
visible light (VIS)
      absorption spectra   2
      -microspectrometry   171–2
visualisation, of primary wall   86

wax
      paraffin *see* paraffin wax
      Paraplast (plus)   281
      qualities   25
      Steedman's *see* Steedman's wax
Western blotting   204, 205, 245, 254–7
Western transfer buffer   256, 264
whole-mount *in situ* hybridisation   311
Wiesner reaction for lignin *see* phloroglucinol/HCl
'wire-loop method'   166–7
wood
      biochemistry of   229–42
      compression   153

delignification of   94
formation, changes during   5
fibres   21, 23, 34, 47, 51, 130, 131, 132, 133, 208, 287
heartwood   229, 239, 240
juvenile   2
maceration of   21, 23
quantitative histochemistry of   229–42
sapwood   239, 240
tension   153
ultrastructure   41–60, 65–79
uses of   1

xylem ray cells
      contact   131
      isolation   131
xyloglucan, immunolocalisation of   95

Z-helix   152, 153
Z-series   150
zinc-iodide-osmium-tetroxide (ZIO)   48, 58
      recipe   58
      staining of endomembranes   48
Zinnia mesophyll system   3, 135

Milton Keynes UK
Ingram Content Group UK Ltd.
UKHW050441111024
449327UK00050B/439